Flood and Coastal Erosion Risk Management

I0894577

This book is a successor to and replacement for the highly respected manual and handbook on the benefits of flood and coastal risk management, produced by the Flood Hazard Research Centre at Middlesex University, UK, with support from Defra and the Environment Agency. It builds upon a previous book known as the 'Multi-Coloured Manual' (2005), which itself was a synthesis of the Blue (1977), Red (1987) and Yellow (1992) manuals. As such, it expands and updates this work, to provide a manual of assessment techniques of flood risk management benefits, indirect benefits and coastal erosion risk management benefits.

It has three key aims. First, it provides methods and data which can be used for the practical assessment of schemes and policies. Second, it describes new research to update the data and improve techniques. Third, it explains the limitations and complications of benefit–cost analysis, to guide decision-making on investment in river and coastal risk management schemes.

All the authors work at or in conjunction with the Flood Hazard Research Centre (FHRC) at Middlesex University, London, UK. The FHRC has a distinguished 40-year history of interdisciplinary research in this field. It has been commended for this by two Chief UK Government Scientific Advisors, Sir David King and Sir John Beddington, and by the award of a prestigious Queen's Anniversary Prize. This book is an output of new research projects and activities carried out under the joint sponsorship of the UK's Environment Agency and the Department of Environment, Food and Rural Affairs following the severe flooding in the UK in 2007.

Flood and Coastal Erosion Risk Management

A Manual for Economic Appraisal

Edmund Penning-Rowsell, Sally Priest, Dennis Parker, Joe Morris, Sylvia Tunstall, Christophe Viavattene, John Chatterton and Damon Owen

in partnership with Environment Agency

mcm

Department for Environment Food & Rural Affairs

Routledge
Taylor & Francis Group
LONDON AND NEW YORK

fhrc
Flood Hazard Research Centre

First published in paperback 2024

First published 2013
by Routledge
4 Park Square, Milton Park, Abingdon, Oxon OX14 4RN

and by Routledge
605 Third Avenue, New York, NY 10158

Routledge is an imprint of the Taylor & Francis Group, an informa business

British Library Cataloguing-in-Publication Data
A catalogue record for this book is available from the British Library

Library of Congress Cataloging-in-Publication Data
Flood and coastal erosion risk management : a manual for economic appraisal / Edmund Penning-Rowsell, Sally Priest, Dennis Parker.
pages cm
Includes bibliographical references and index.
1. Flood forecasting. 2. Flood control—Evaluation. 3. Flood damage—Evaluation. 4. Beach erosion—Forecasting. 5. Coast changes--Forecasting. 6. Risk management. I. Penning-Rowsell, Edmund.
GB1399.2.F53 2014
363.34'932—dc23 2013020121

ISBN: 978-0-415-81515-4 (hbk)
ISBN: 978-1-03-291976-8 (pbk)
ISBN: 978-0-203-06639-3 (ebk)

DOI: 10.4324/9780203066393

Typeset in Perpetua
by Keystroke, Station Road, Codsall, Wolverhampton

Contents

List of illustrations

FIGURES

TABLES

List of boxes

Key contributors and acknowledgements

KEY CONTRIBUTORS

The authors wish to recognise the particular contributions of David Cotterell and Jacqui Cotton in steering the research reported here. In addition, Colin Green contributed again to Chapter 2, and Sue Tapsell, Head of the Flood Hazard Research Centre, undertook much of the research reported here on the health effects of flooding. Joanna Pardoe gave invaluable assistance at the start of our research in 2010, and Lucy Caple and Josie Joyce helped to manage the task of manuscript production. These contributions were all vital to the publication of this new 'Multi-Coloured Manual'.

ACKNOWLEDGEMENTS

Over the last forty years, many people have contributed to our series of manuals, published in 1977 (the 'Blue Manual'), in 1987 (the 'Red Manual'), in 1992 (the 'Yellow Manual') and in 2005 (the 'Multi-Coloured Manual'). We record here our acknowledgement of all their contributions.

Particularly important for their contribution to this new volume have been members of the 2011–13 Environment Agency Steering Group: David Cotterell (Chair), Jacqui Cotton (EA Project Manager), Paul Cobbing, Steve Goring, Nick Haigh, Julian James, Stefan Laeger, Vania Paccagnan, Mike Steel, David Thomas and Harry Walton. In addition, we are grateful to the following individuals for their input and guidance in shaping the new MCM website: Luke Ballantyne, Roy Cotgrove, Mike Dobson, Andy Hardstaff, Steve King, Fola Ogugyoye, Nigel Sanders, Adam Schofield and Carolann Simmonds. Owen Tarrant and Doug Whitfield have helped in the assessment of individual chapters and the review process generally.

The updating of the methodology and datasets for residential properties, and developing the new methodology for damage assessment in non-residential properties, would not have been possible without the valuable contributions from Chris Netherton of the National Flood School and quantity surveyors Neil Boothby of CBL and Trevor Masterson of Bell and Masterson. We would also like to express our thanks to those involved in the flood victim discussion groups in Bewdley, Ruthin and Banbury, with special reference to the co-ordinating assistance of Gillian Holland, Dave Humphries and Carol Mawle.

In producing the previous version of the 'Multi-Coloured Manual' the members of the 2001–05 Defra/EA Advisory Committee provided support and guidance: David Richardson (Chair), Nick Haigh, Peter Borrows, David Calderbank, Mike Child, Beth Greenaway, Paul Leonard, Paula Orr, Lucy Toman, Bill Watts, Mike Brewer, Clare Twigger-Ross and Joanne Riley. Those who helped with the parallel Handbook were Kevin Andrews, Karl Hardy and Paul Murby of Defra, and David Cotterell and Bill Watts of the Environment Agency, together with Keith Cole of the Local Government Association.

We are grateful to the following individuals and organisations for taking part in discussion meetings over many years to review the usefulness of the

'Yellow', 'Red' and 'Blue' manuals and providing assistance since then. Their comments and suggestions were invaluable in guiding our work on this Manual. The titles and organisations for which they work are given as at the time when we contacted them; they are not necessarily correct at the time of publication.

- Adam Schofield – Project Engineer, Halcrow, Peterborough
- Alan Clark – Director, Geotechnics, High-Point Rendel, London
- Alan Williams – Coastal Engineering Consultant, Wirral
- Alistair Templeton – Principal Engineer, Babtie, Newcastle
- Andrew Dannat – Associate, Lewin Fryer and Partners, Hampton
- Ben Hamer – Senior Maritime Engineer, Halcrow, Swindon
- Brian Harvey – Professional Engineer, Halcrow, Swindon
- C.T. Marshall – Consultant Engineer, Arup, Leeds
- Colin A. Read – Managing Consultant, High-Point Rendel, London
- D. Stirling – Senior River Engineer, Arup, Leeds
- David Harlow – Principal Engineer (Coast Protection), Bournemouth Borough Council
- David Joy – Principal Engineer, West Dorset District Council
- Dick Thomas – Director, Coastal and Rivers, Posford Duvivier, Peterborough
- Ged Herby – Group Leader, Coast Protection, Wirral District Council
- Graham Lymbery – Sefton Metropolitan Borough Council
- Greg Guthrie – Principal Engineer, Posford Duvivier, Peterborough
- Ian Hutchison – Engineer, Posford Duvivier, Haywards Heath
- J. O'Connor – Chief Engineer/Associate Director, Head of Environment, W.S. Atkins, Leeds
- John Andrews – Divisional Director, Coastal and Rivers, Posford Duvivier, Haywards Heath
- John Riby – Scarborough Borough Council
- Mark Fletcher – Associate Director, Arup, Leeds
- Matthew Walker – Senior Coastal Engineer, Scott Wilson, Basingstoke
- Neal Turner – Group Manager, Client Services, Bournemouth Borough Council
- Neil Allsop – Assistant Engineer, West Dorset District Council
- Nicholas Pettit – Engineer, Posford Duvivier, Peterborough
- Nick Bean – Graduate Engineer, Posford Duvivier, Haywards Heath
- Nick Browning – Principal Drainage and Environmental Engineer, West Dorset District Council
- Peter Brooks – Senior Engineer/Sea Defences, Canterbury City Council
- Peter Frew – North Norfolk DC
- Peter Myers – Divisional Director, Bullens, Bradford
- Phil Simcox – Director, Ian Howick and Partners, Exmouth
- Phillip McLoughlin – Senior Graduate Engineer, Bullens, Bradford
- Richard Mocke – Project Manager, Halcrow, Swindon
- Richard Samphir – Group Engineer, Technology Leader, Project Appraisal, W.S. Atkins, Swansea
- Robert Beck – Engineering Manager, Shepway District Council
- Stephen Fort – Managing Consultant, High-Point Rendel, London
- Steve Webster – Principal Engineer, Halcrow, Exeter
- Toby Wilson – Civil Engineer, Halcrow, Exeter
- Will McBain – Senior Engineer, Arup, Leeds

We would like to thank Paul Patterson, Senior Assistant Engineer, and Waveney District Council who sponsored the Corton study, and the following for their other strategic and specialist advice:

- Adam Spencer – Economic Adviser, Department for Transport
- Alan Savory – Senior Consultant, the Independent Park Home Advisory Service
- Clive E. Munnings – Personal Claims Manager, Corporate Claims, AXA Insurance
- David Noble – Association of Drainage Authorities
- David Whensley – Technical Director, Energy Network Association
- Jackie Bennet – Association of British Insurers
- Jan Brooke – Environmental Consultant, Market Harborough
- John Dora – John Dora Consulting Limited
- Julia Humphrey – Senior Analyst, Climate Change and Water Quality, Ofwat

- Kate Avery – Sustainability Specialist (Climate Change Adaptation), Network Rail
- Keith Smith – Librarian, Ofgem
- Matt Cullen – Policy Adviser (Flooding), Association of British Insurers
- Nicholas Adjei – Head of Climate Change and Water Quality, Ofwat
- Nigel Holmes – Alconbury Consultants
- Paul Deakin – Halcrow, Swindon
- Peter Von Lany – Halcrow, Swindon
- Richard Gould – Analyst, Office of Rail Regulation
- Robbie Craig – Policy Officer, Defra
- Steve A. Sampson – Chief Building Surveyor, Building Repair Network, Royal and Sun Alliance

Within the Environment Agency, Gary Lane and Patricia Farrell are thanked for sponsoring the autumn 2000 research which contributed to Chapter 6, as is Tony Finnegan of the (then) DTLR, who gave invaluable assistance with regard to payments to local authorities under the Bellwin scheme, also referred to in Chapter 6.

We are grateful for the assistance given by Network Rail for data on rail disruption and to staff at North Yorkshire County Council for providing us with disruption costs (Chapter 6).

Numerous FHRC colleagues over the years have contributed to the research, data collection and production of the various manuals. These include Dr Clare Johnson, Dr Maureen Fordham, Dr Simon Oakes and Dr Peter Winchester.

At Middlesex University, we would like to give special thanks to Marion Locke for her administrative support, and Emily Potter and Peter Goodman for their data-gathering contributions, and the School of Health and Social Science's Technical Unit for their assistance with the generation of floor plans and other diagrams.

Acronyms and abbreviations

AA	Automobile Association
AADs	Annual Average Damages
ABI	Association of British Insurers
ACC/ISGWR	UN Administrative Coordination Committee Intersecretariat Group for Water Resources
ADS	Above Design Standard
AEP	Annual Exceedance Probability
AES	Agri-Environment Scheme
ALC	Agricultural Land Classification
ARV	Average Remaining Value
AST	Appraisal Summary Table
AWB/HMWB	Artificial or Heavily Modified Water Bodies
B&Bs	Bed and Breakfast accommodation
BAP	Biodiversity Action Plan
BCA	Benefit–cost analysis
BCP	Business Continuity Planning
BCR	Benefit–cost ratio
'Blue Manual'	Penning-Rowsell, E.C. and Chatterton, J.B. (1977) *The Benefits of Flood Alleviation: A Manual of Assessment Techniques*, Gower Publishing, Aldershot
BPEO	Best Practical Environmental Option
BWL	Barrier With Locks
CBA	Cost Benefit Analysis
CCW	Countryside Council for Wales
CEA	Cost-Effectiveness Analysis
CEH	Centre for Ecology and Hydrology
CFMPs	Catchment Flood Management Plans
CHaMPs	Coastal Habitat and Management Plans
CPI	Consumer Price Index
CV	Contingent Valuation
CVM	Contingent Valuation Method
DCLG	Department for Communities and Local Government
Defra	Department for Environment, Food and Rural Affairs
DEM	Digital Elevation Model
DETR	Department of the Environment, Transport and the Regions

DfT	Department for Transport
DM	Do Minimum
DMAs	District Meter Areas
DNO	Distribution Network Operator
DoE	Department of Education
DQS	Data Quality Score
DTLR	Department of Transport, Local Government and the Regions
EA	Environment Agency
ECLAC	Economic Commission for Latin America and the Caribbean
ENA	Energy Networks Association
ESTDAM	Non-GIS-based model developed by FHRC in the 1970s to calculate damages (now obsolete)
ETR	Engineering Technical Report
EWATS	Extreme Weather Action Teams
FCERM	Flood and Coastal Erosion Risk Management
FCERM-AG	Flood and Coastal Erosion Risk Management Appraisal Guidance
FDGiA	Flood Defence Grant in Aid
FLOODsite	Integrated Flood Risk Analysis and Management Methodologies (EU funded project, Contract Number: GOCE-CT-2004-505420)
FOCs	Freight Operating Companies
FRM	Flood Risk Management
FWAGs	Farming and Wildlife Advisory Groups
FWR	Foundation for Water Research
FWRBP	Flood Warnings Response Benefit Pathways
GDP	Gross Domestic Product
GEP	Good Ecological Potential
GES	Good Ecological Status
GIS	Geographic Information System
HLF	Heritage Lottery Fund
HLS	Higher Level Stewardship
IBCR	Incremental Benefit–Cost Ratio
IEH	Institute for Environment and Health
IFP	Indicative Floodplain
LIDAR	Light Detection and Ranging
MAFF	Ministry of Agriculture, Fisheries and Food
MCA	Multi-Criteria Analysis
MCH	'Multi-Coloured Handbook': *The Benefits of Flood and Coastal Risk Management: Handbook of Assessment Techniques*
MCM	'Multi-Coloured Manual': *The Benefits of Flood and Coastal Risk Management: A Manual of Assessment Techniques*
MCM-Online	The companion website to this publication: http://www.mcm-online.co.uk
MDSF	Modelling and Decision Support Framework
NAAR	National Appraisal of Assets at Risk
NaFRA	National Flood Risk Assessment
NAO	National Audit Office
NBO	Next Best Option
NEA	National Ecosystem Assessment
NHS	National Health Service
NPD	National Property Dataset
NPV	Net Present Value

NRA	National Rivers Authority
NRD	National Receptor Database
NRPs	Non-Residential Properties
ODPM	Office for the Deputy Prime Minister
Ofgem	The Office of Gas and Electricity Markets
ONS	Office for National Statistics
PAG	Project Appraisal Guidance
PARs	Project Appraisal Reports
PF	Partnership Funding
PLP	Property-Level Protection
PVd	Present Value of damages
RAC	Royal Automobile Club
RBMPs	River Basin Management Plans
'Red Manual'	Parker, D.J., Green, C.H. and Thompson, P.M. (1987) *Urban Flood Protection Benefits: A Project Appraisal Guide*, Gower Technical Press, Aldershot
RPA	Risk and Policy Analysts Ltd
SAC	Special Area of Conservation
SDS	Strategic Direction Statement
SEA	Strategic Environmental Assessment
SEMD	Security and Emergency Measures Direction
SFVI	Social Flood Vulnerability Index
SMART objectives	Specific, Measurable, Achievable, Realistic and Timely
SMEs	Small and Medium-sized Enterprises
SMMT	Society of Motor Manufacturers and Traders
SMPs	Shoreline Management Plans
SPA	Special Protection Area
SSSI	Site of Special Scientific Interest
STW	Sewage Treatment Works
SUDS	Sustainable Urban Drainage Systems
TE2100	Thames Estuary 2100 (http://www.environment-agency.gov.uk/static/documents/Leisure/SE_TE2100_briefing.pdf)
The 'Green Book'	HM Treasury (2003) *The Green Book: Appraisal and Evaluation in Central Government*, HM Treasury, London
TOCs	Train Operating Companies
VAT	Value Added Tax
VOA	Valuation Office Agency
VOC	Vehicle Operating Costs
VOE	Value of Enjoyment
WAAD	Weighted Annual Average Damages
WDRM	Warning-Dependent Resistance Measures
WFD	Water Framework Directive
WIRM	Warning-Independent Resistance Measures
WIRMB	Warning-Independent Resistance Measures Benefits
WLC	Whole-Life Cost
WLMPs	Water Level Management Plans
WTP	Willingness To Pay
'Yellow Manual'	Penning-Rowsell, E.C., Green, C.H., Thompson, P.M., Coker, A.M., Tunstall, S.M., Richards, C. and Parker, D.J. (1992) *The Economics of Coastal Management: A Manual of Benefit Assessment Techniques*, Belhaven/Wileys, London

Foreword

This Manual has been produced by the Flood Hazard Research Centre (FHRC) at Middlesex University, under the Modelling and Risk theme of the joint Defra and Environment Agency research and development programme.

It draws on collaboration between the FHRC, Defra, the Environment Agency and other stakeholders, and we recommend its use and the accompanying MCM-Online for benefit assessment as part of flood and coastal erosion risk management appraisal. Application of the principles, methods and supporting data provides clarity and robust evidence to underpin investment decisions. This updated version of the Manual provides users with:

- A completely new set of data on the potential flood damage to non-residential properties (such as businesses, industry and public buildings) vulnerable to flooding.
- Methods for assessing benefits in sectors not previously covered by MCM.
- Access to the rationale and background on the appraisal techniques, with links to the practical methods presented on a new web-based MCM (which now supersedes the previous 'Handbooks').
- A single compendium of a decade of research by the FHRC.

Importantly, the Manual provides approaches to implement the policies set out in the HM Treasury 'Green Book', and complements the *Flood and Coastal Erosion Management Appraisal Guidance* (FCERM-AG), published by the Environment Agency in March 2010. The Manual also complements Defra's *Making Space for Water* principles.

This Manual represents the results of intensive high-quality research to update completely the data and techniques in the previous 2005 version, which it now fully supersedes.

For the first time, this Manual is jointly published by the Flood Hazard Research Centre and the Environment Agency, representing the continuing close collaboration between us.

We hope you find this Manual and MCM-Online useful.

Jonathan Day
Head of Investment and Programmes
Environment Agency
August 2013

1 Introduction

The purpose and contents of this Manual

This chapter sets the scene for the appraisal of the benefits of flood and coastal erosion risk management in England and Wales. It discusses in Section 1.1 the purpose of this Manual, and locates this in the wider context of environmental assessment.

Section 1.2 is designed to guide the reader through the Manual, including 'How to get started' and 'Where to find what'. Reference is made to underpinning documents produced by the Treasury on public sector investment (HM Treasury, 2003) and by Defra on appraisal policy (Defra, 2009). In this respect some significant changes have occurred in the field of project appraisal and flood and coastal risk management since 2005, when our last Manual was produced (Penning-Rowsell *et al.*, 2005a), in terms of both approach and the use of different techniques (e.g. providing some financial data for local rather than national application).

SUPPORTING INFORMATION ON MCM-ONLINE

Throughout this Manual, readers will find reference to Supporting Information on MCM-Online. This is the supplementary material provided for many chapters. In addition to this Supporting Information, certain chapters herein are based on direct and indirect damage data contained previously on the Multi-Coloured CD (2010). These data are now replaced by those available on the new Multi-Coloured Manual website: http://www.mcm-online.co.uk.

Government policy for flood and coastal erosion risk management (FCERM) has also changed since the beginning of this century (Section 1.3). The Department for Environment, Food and Rural Affairs (Defra) is the government department responsible for flood and coastal erosion risk management (previously 'flood defence' and before that 'land drainage'). In the mid-2000s Defra developed a new policy framework in the form of 'Making Space for Water' (Defra, 2004a; 2005) which has had implications for project appraisal – e.g. balancing national and local priorities – and for government investment priorities. The Pitt Review (Pitt, 2008), following the 2007 floods, has also led to changes in flood management policy, enacted through the Flood and Water Management Act 2010 (HM Government, 2010a). This has primarily given new responsibilities to bodies at the local level to co-ordinate and participate in managing flood risk, within a new National Strategy developed by the Environment Agency (2011a). Defra has also developed an important new 'partnership funding' approach (Defra, 2011a) and at the same time delegated many responsibilities for project appraisal to the Environment Agency, and the Agency in turn has developed considerably more documentation on this subject than has ever existed before (Environment Agency, 2010a; see Table 1.1).

Section 1.4 outlines the aims and assumptions of benefit–cost analysis, before Section 1.5 discusses the difficulties of, and uncertainty in, predicting the future

Table 1.1 Sources of guidance on appraising flood and coastal erosion risk management schemes and plans

Source reference	Document	Purpose
HM Treasury 2003	The 'Green Book'	Identifies the preferred approach to public sector investment appraisal
Environment Agency 2010a	Flood and Coastal Erosion Risk Management Appraisal Guidance (FCERM-AG)	How a project appraisal and cost benefit analysis (CBA) should be completed for flood and coastal erosion risk management projects
Flood Hazard Research Centre and the Environment Agency 2013	The new 'Multi-Coloured Manual' (MCM)	This volume: gives details of relevant research and detailed guidance on benefit assessment methods and data
http://www.mcm-online.co.uk http://www.fhrc.mdx.ac.uk	Middlesex University FHRC MCM-Online	Provides data and other information (including questionnaires) for the support of flood and coastal erosion risk management project appraisals

for which risk management is planned in a 'non-stationary' world. More pragmatically, Section 1.6 lays out a framework for different levels of project appraisal – generalised and detailed – and Section 1.7 gives some guidance about the time and resources needed. There follow a few words on estimating scheme costs, and on how to update the data contained in this Manual by using its parallel web-based 'Multi-Coloured Manual Online' (termed herein 'MCM-Online'), containing the data and techniques covered by this Manual (Sections 1.8 and 1.9, respectively). Finally, we outline what is new in this volume in relation to our previous Manual publications, and provide a list of some of the papers that we have published in the last eight years from the Flood Hazard Research Centre.

1.1 THE RATIONALE, PURPOSE AND STRUCTURE OF THIS VOLUME: THE NEW 'MULTI-COLOURED MANUAL' ANALYSIS (MCA)

1.1.1 The rationale

The rationale of this Manual is to aid and improve decision-making. The relevant decisions are about investment in fluvial flood risk management policies, plans or schemes, and at the coast in policies, plans and schemes to manage the risks of both coastal flooding and the erosion of the land by the sea. These decisions should be seen in the context of the modern philosophy of an integrated and holistic approach to sustainable catchment and coastal management, with stakeholder engagement embedded in the decision process, and sustainability and adaptive management always foremost in our thinking and in the processes that we design to guide this thinking and our decisions.

The focus of this rationale should not be misinterpreted. Although much of this Manual is aimed at 'projects' or 'schemes', by which it is usually meant any plan, programme or engineering works for which appraisal and decision-making are required, this should by no means be the only or indeed the primary outcome from using this Manual. There are many alternative strategies which should be reviewed for catchments and for the coast within the modern approach, including the 'do nothing' option which would let flooding and erosion continue. In summary, the term 'scheme' here is not meant to imply an engineering scheme but includes both structural engineering ways to reduce flood or erosion risks and non-structural alternatives (flood warning, emergency response, land use planning, etc). The term 'scheme' is used hereinafter for the sake of simplicity alone.

Another caveat concerns the understandable tendency when evaluating investment projects to consider just the site in question. But it is clear that a regional perspective is necessary in the planning of coastal erosion risk management policies and works, and that a whole-catchment approach gives a significant new dimension to the planning of particular fluvial flood risk management schemes.

This arises because there is a measure of hydrographic interdependency between locations along a coastline, such that erosion in one place can result in protective deposition elsewhere. In addition, flooding in particular catchment areas can be a function of problems created elsewhere, such as accelerated runoff caused by the urbanisation of upland areas, or the encroachment on to floodplains by urban land uses which thereby are threatened with inundation. A modern regional and catchment perspective – first pioneered in the UK by Gardiner (1991) – allows consideration of these factors, which may be missed in any purely local analysis.

In addition, a range of alternative strategies needs to be reviewed and appraised. These can be structural engineering schemes or non-structural alternatives. The latter may take the form of stricter land use control in floodplain locations, or tighter designation of environmental 'no go areas', such as Sites of Special Scientific Interest, where no flood risk management schemes should be implemented. Equally valid may be insurance arrangements to spread the impact of floods, or warning systems to allow those at risk to reduce their vulnerability.

The modern integrated approach is needed, rather than one which looks at only local issues and uses a single-discipline approach to analysis and policy-making. A multi-functional approach is also needed in the management of river catchments, so that flood risk management plans and policies fit in with other water management objectives such as navigation, water supply and pollution control, in the spirit of the EU Water Framework Directive (Green, 2003). This is not easy, and makes many of the decisions on investment very complex and difficult. But only with a systematic approach to evaluation will decision-making improve, unintended consequences be minimised, and the nation's stock of economic and environmental resources along our rivers and at the coast be enhanced.

This Manual continues the designation as the 'Multi-Coloured Manual' because it embraces techniques contained in previous Middlesex University volumes in this series: the 'Blue Manual' (Penning-Rowsell and Chatterton, 1977), the 'Red Manual' (Parker *et al.*, 1987), and the 'Yellow Manual' (Penning-Rowsell *et al.*, 1992). We also first produced an early draft of this 'Multi-Coloured Manual' in 2003 (Penning-Rowsell *et al.*, 2003), which MCM (2005) and its associated Handbook and CD then replaced (Penning-Rowsell *et al.*, 2005a; 2005b). Eight years later this volume adds new results and is accompanied by the 'Multi-Coloured Manual Online' (MCM-Online), the web-based repository for all the data to which we here refer.

This Manual is therefore an updated and improved version of all of those previous manuals, dealing as they did, respectively, with flood risk management benefits (1977), indirect benefits (1987), coastal erosion risk management and sea defence benefits (1992), and all three combined (2003 and 2005). Although the user of this Manual is occasionally referred to sections of those manuals, it is intended that this volume stands alone and readers should not need to refer to the previous publications.

Both this Manual (MCM) and MCM-Online build on the definitive Treasury 'Green Book' guidance on investment in public sector projects (HM Treasury, 2003), including, for example, the use of weightings to correct for distributional impacts, optimism bias considerations when assessing project costs, and variable discount rates for projects with long lives (Table 1.1). They also have both built on Defra's original Project Appraisal Guidance series (e.g. MAFF, 1999: PAG3) and are complemented now by the Environment Agency's own appraisal guidance, issued in 2010 (Environment Agency, 2010a), and its additional guidance documents on its website (Table 1.2). In this respect, and to be clear, references here and in the Handbook on MCM-Online are therefore to 'Environment Agency (EA) appraisal guidance' (Environment Agency, 2010a) and to Defra's 'appraisal policy' (Defra, 2009).

1.1.2 The purposes of this Manual

This Manual has three main purposes, which interconnect and permeate the whole volume:

1 The primary aim is to present the user with a range of techniques and data that can be used in a practical way to assess the benefits of: (a) fluvial flood risk management schemes and policies; and (b) plans and schemes to alleviate the impact of

Table 1.2 Additional guidance on aspects of FCERM appraisal provided by the Environment Agency and Defra

Environment Agency FCERM Appraisal Guidance documents	Reference
Flood and Coastal Erosion Risk Management: Appraisal Guidance – Main Report (FCERM-AG)	Environment Agency (2010a)
Environment Agency supporting guidance for the flood and coastal erosion risk management appraisal guidance (FCERM-AG).	
Economic appraisal spreadsheets are available to assist economic analysis (Excel, 775KB)	Environment Agency (undated a)
Supporting document for the appraisal summary table (PDF, 396KB) – enables the clear presentation of negative and positive impacts to aid the consideration of wider issues within the appraisal	Environment Agency (2010b)
Guidance on applying the scoring and weighting methodology (PDF, 410KB) – enables the economic valuations of more intangible benefit types	Environment Agency (2010c)
Economic Valuation of Environmental Effects (PDF, 2.7MB)	Eftec (2010a)
Assessment of coastal erosion and landsliding for the funding of coastal risk management project guidance notes (PDF, 2.9MB)	Environment Agency (2010d)
Adapting to climate change: advice for flood and coastal erosion risk management authorities (PDF, 411KB) – this advice replaces the Defra 2006 guidance 'Treatment of climate change impacts' using the latest science from UK Climate Projections 2009, UKCP09. This applies to England only	Environment Agency (undated b)
Managing flood risk by working with natural processes	Environment Agency (2010e)
The following Defra Supplementary Notes support the use of the FCERM-AG	
Treatment of climate change impacts (PDF, 70KB) – following the publication of new climate change advice for England (above) this guidance now applies to Wales only. The Welsh Government is working with the Environment Agency to produce revised guidance using the latest science from UKCP09	Defra (2006)
Treatment of agricultural land – May 2008 (PDF, 104KB)	Defra (2008a)
Risk to people guidance – May 2008 (PDF, 170KB) – covers a method for the evaluation of the risk to life associated with flood risks	Defra (2008b)
Interim guidance note – July 2004 (PDF, 161KB) – which takes account of Defra's policy on socio-economic equity and appraisal of human-related impacts of flooding	Defra (2004b)
Treatment of risk (optimism bias) (PDF, 235KB) – guidance for applying optimism bias is covered in Revisions to economic appraisal procedures arising from the new HM Treasury Green Book March 2003	Defra (2003)

Source: All are accessed through the Environment Agency's website: http://www.environment-agency.gov.uk/research/planning/116707.aspx

erosion at the coast. The Manual is orientated to practising engineers, and river or coastal zone managers who wish to see their schemes represent good value for funders' money and a sustainable contribution to environmental management. In this way we aim to promote good designs through emphasising the need to evaluate the impacts of the plans or works that are proposed, in both economic and environmental terms. Indeed, this Manual may well be used to determine that plans or works to alleviate flooding or erosion are not appropriate, either because they are not cost-effective or because they will be so environmentally damaging that they should not proceed; or a mixture of both.

2 Second, the Manual describes the results of research at Middlesex University between 1999 and 2013 designed to update the data and improve the techniques that are contained in the previous manuals. We have attempted, in writing this Manual, to complement the production of the methods and techniques presented here with an explanation of the research undertaken to derive them and some of its limitations. This will give a context to the methods that we have produced and recommend. In certain areas of this Manual we present research results from previous investigations. Since we produced the 'Yellow Manual' in 1992 we have undertaken a number of project appraisals at Middlesex University which feed into this Manual. Particularly important has been our work on 'social issues', attempting to gain a better understanding of the 'intangible' benefits of managing flood and coastal erosion risks (e.g. Tapsell *et al.*, 2002).

In addition, we have also undertaken more conceptual development in the area of benefit–cost analysis, focusing particularly on choices and the way that choices conflict and also on the distributional consequences of flood risk management decisions (e.g. Penning-Rowsell and Pardoe, 2012a; 2012b; see Chapter 2). Some of this will be controversial and we do not pretend that any of this conceptual material necessarily makes benefit–cost analysis any easier to use, but we consider that it provides some enhanced understanding that project appraisal involves decisions about alternatives and their consequences, rather than just providing the justification of pre-determined policies, plans or schemes.

3 We explain, third, the limitations and complications of benefit–cost analysis to potential users. This is done so that they can use it in a thoughtful and critical way, to guide their decision-making on investment in river and coastal risk management schemes. Therefore, the users of this Manual will be those who are analysing problems of flooding from rivers and at the coast, or erosion in coastal situations, and who are considering the possibility of proposing some plan or works to lessen these risks or reduce their impacts. It should be noted, however, that the approach to project appraisal adopted throughout the Manual is generally the efficient investment of resources by the *nation* – not the Agency, local authorities or individuals – and users must take this important point into consideration in their use of its contents. The focus therefore is primarily still on benefit–cost analysis to aid the allocation of the nation's scarce economic resources, not the financial appraisal of investment for individual private or corporate gain.

Notwithstanding this point, we provide in parts of this Manual (in contrast to all the results in all previous manuals) some data on the *financial* consequences of floods and coastal erosion, because since the development of 'partnership funding' (Defra, 2011a) the locus of fundraising and consequentially of decision-making has somewhat changed. Instead of all the resources required for interventions in these fields being raised nationally, from national taxpayers, some of the resources are now being raised locally, for example via local authorities. This means that decisions may well be made on the grounds of the financial returns to the locality from these interventions, rather than economic returns to the nation. If this is the case, then decision-makers need to know of these financial returns in order to know whether or not to make their investments. Users of this Manual should be aware of the difference between economic and financial benefits and be careful not to confuse the two.

This Manual complements the guidance volumes provided by the Department for Environment, Food and Rural Affairs and the Environment Agency (Defra, 2009; Environment Agency, 2010a). It adopts a critical perspective about investment appraisal: the data and techniques do not aim to justify pre-determined designs, but to analyse the extent and character of a flooding or coastal erosion risk management risk problem prior to making decisions about whether to intervene or to do nothing. The main discussion about the critical use of economic appraisal is in Chapter 2. But throughout the Manual we stress that there are pitfalls and complications throughout the benefit assessment part of benefit–cost analysis, and that an understanding of its principles is essential to its appropriate use. Furthermore, benefit–cost analysis inherently involves some assumptions about the distribution of income and wealth in society (Mishan, 1971) – notably that the current distribution is optimal – and thus its use involves some moral decisions as to what is 'right' or 'wrong', as well as the purely technical considerations. Moreover, even at the technical level, incorrect understanding, incorrect assumptions and the wrong data can produce benefit–cost analysis results which are meaningless and indeed dangerous in that unwise decisions would result. We provide some guidance so that these pitfalls are minimised.

The first of these aims is facilitated by providing an electronic web-based version which contains all the data that are within the printed volume, plus extensive other datasets which previously were reproduced only in hard copy form or on the 2005 and 2010 CDs. By providing these datasets as readily accessible electronic versions we aim to facilitate the use of the data in all of their various forms, rather than the kind of limited use that follows when data are not easily accessible. We also produce, on the web-based MCM-Online, model versions of the questionnaires that we advocate are used to assess a range of benefits, so that they can easily be used in the appropriate circumstances.

The parallel MCM-Online has three elements. The most readily accessible is a free-of-charge part of that platform, which will contain a limited dataset so that all users can benefit from that set of basic data. The second is a companion to this manual: Providing Supporting Information. The third part of the platform has much more detail and considerably larger datasets, and this is available on a charged-for basis, accessible at http:www.mcm-online.co.uk or through the website of the Flood Hazard Research Centre at Middlesex University.

1.1.3 Benefit–cost analysis (BCA), environmental assessment, and multi-criteria analysis (MCA)

There is a range of techniques that can aid decision-making concerning investment appraisal, with benefit–cost analysis just one technique amongst many (Defra, 2009, 33). Similarly, there is a range of techniques that can be used to quantify the environmental impacts of a number of policy options, and this is an important component of the data that the decision-maker needs (see Chapters 2 and 10).

We see benefit–cost analysis as part of a more general procedure termed 'environmental assessment'. This is because BCA is concerned with a particular 'product' (the benefit–cost ratio or net present value as the measure of the economic return from that investment). Environmental assessment is first and foremost more concerned with a process of incorporating information on all environmental attributes, values and changes into the decision-making sequences.

On the other hand, 'multi-criteria analysis' attempts to quantify in some way all aspects of environmental and economic significance related to a particular decision, and weight them so that a simple range of indices can be developed that capture all adverse and beneficial results – or potential adverse and beneficial results – from an investment decision (Defra, 2009, 33).

So:

- Benefit–cost analysis can provide a sophisticated means of comparing very different investments and outcomes by reducing them all to a common monetary form. It is limited to consideration of those impacts to which a monetary value can be attached but it leads to a simple set of parameters on which choices can be made (e.g. the benefit–cost ratio).
- Multi-criteria analysis can be used to broaden the scope of analysis, but brings significant difficulties in terms of determining the appropriate weights to use or the different criteria involved.
- Environmental assessment is broader still, but lacks the precision (or apparent precision) of BCA, and can avoid the kind of discipline involved in quantification that both BCA and MCA bring to the decision-making process.

We do not recommend that decision-makers reject any vehicle which aids the elucidation of the decision that they are to make, but would always warn that excessive simplification brings dangers as well as advantages.

1.2 GUIDE TO CONTENTS AND THE USE OF THE MANUAL

This Manual is not intended to be read right through from start to finish (no doubt the typical reader will breathe a sigh of relief!). Rather, the component parts can be used in a 'standalone' fashion, and have been written with that in mind. However, the user should be wary of using parts of the Manual in a standalone fashion without considering the research base of the methods and data that are presented.

1.2.1 Advice to the reader as to how to get started

In summary, the user should therefore proceed as follows:

1 Read Section 1.4 on the aims and assumptions of benefit–cost analysis.
2 Review the contents of Table 1.3 to see where within the Manual are the parts that are relevant to the appraisal being prepared.
3 Read as much as possible of Section 1.3 on the development of sustainable, participative and catchment-based approaches to river and coastal management, as contained in agreements and conventions to which the UK Government has been a party.
4 Determine from Section 1.6 the scope of the analysis. Bear particularly in mind the need to take a broad perspective initially by evaluating the whole catchment as a context to the particular fluvial flooding problem being investigated and by evaluating the whole of the hydrographic cell in which the coastal problem is located.
5 Consider carefully the time required for any studies by consulting Section 1.7 concerning the phasing and timing of each part of the assessment and the resources needed for surveys and data processing. Many incomplete or flawed project appraisals result from an over-hasty approach, which leads to significant omissions in terms of the required data and the results obtained.
6 Be aware of the numerous 'health warnings' contained within the substantive chapters of this Manual, and therefore be aware of the limitations of the analysis you are undertaking.

Beyond this introductory chapter, Chapter 2 presents a critical analysis of the basis of benefit–cost analysis, and a number of warnings about its use. Thereafter, Chapters 3 to 10 analyse the different aspects of appraising river flood risk management and coastal erosion risk management, and give the approaches, techniques and datasets that we recommend are used.

To give a measure of consistency throughout this Manual, the user is given the following 'signposts':

- At the start of each chapter is a summary of information to be found in that chapter.
- In addition, within each chapter, there are sections which checklist important points, and some 'health warnings' about the problems of using the data and techniques that are recommended.
- We also provide at the end of each chapter several 'lessons from experience' which may help the user of this Manual to obtain results which are meaningful and appropriate.

1.2.2 The relationship with related plans and appraisal processes (e.g. CFMPs, SMPs, CHaMPs, WLMPs, etc.)

Since the production of our first Manual in 1977 the development of appraisal techniques and policies for rivers and the coast has continued apace. A range of policies and plans has been developed, and significantly more guidance is provided by central government through Defra.

In general, the appraisal policies provided by Defra (Defra, 2001a; 2009; etc.) and the Environment Agency (Environment Agency, 2010a) give an overview of project appraisals and step-by-step procedures for calculating costs and benefits (Table 1.1). The documents provide very little data with which to calibrate those analyses, but make reference to the Middlesex manuals and other important sources of data (e.g. the work on agricultural benefit–cost analysis at Cranfield University). In contrast, the Middlesex manuals give both theoretical

Table 1.3 Where to find what in the Manual

Topic	Theoretical perspectives	Appraisal methods	Data requirements	Data in the Manual	Lessons from experience
Flood risk management					
Assessing flood losses to residential property	4.2	4.9	4.9	4.3	4.10
Indirect flood losses for residential property	Table 4.1	4.9.1	Table 4.31		
'Intangible' flood losses to householders	4.8	Tables 4.26, 4.27			
Flood damage to park homes	4.4	4.43	4.4.3		
Vehicle damages in floods	4.5	4.5	Tables 4.12, 4.13, 4.14	4.53, 4.55	
The damage reducing effect of property-level protection (PLP) and flood warnings	4.6	4.6.1–4.6.3	Table 4.16	4.6.1–4.6.3	
The impact of evacuation during flooding	4.7	4.7.1	Tables 4.17–4.25	4.7.2, 4.7.3	
Direct flood damage data for non-residential properties	5.6	5.6.1–5.6.9		5.6.1–5.6.9	5.12
Road traffic disruption during floods	6.5.1	6.5.1	Tables 6.15–6.18	6.5.1	Boxes 6.5, 6.6
Potential losses owing to rail disruption	6.5.2	6.5.2	Tables 6.19–6.22	6.5.2	Boxes 6.7, 6.8
Disruption to public services	6.6	6.6	6.6	Tables 6.25–6.31	Boxes 6.1, 6.3
Emergency services costs	6.8	6.8.1, 6.8.2	Tables 6.26–6.28	6.8	
Agricultural benefits of flood risk management	9.2, 9.3, 9.5	Figure 9.4, Table 9.21	9.8, Table 9.11	Tables 9.4–9.6	9.12
Protection against erosion from the sea					
The nature of erosion benefits	7.3.1				
Methods of project appraisal: property losses	7.4.1	7.3	7.3.3, 7.4, 7.4.3	Tables 7.3, 7.4	
Methods of project appraisal: infrastructure losses	7.5.1	7.5.5, 7.6			7.7
Loss of agricultural land to the sea	9.3.8		Table 9.7		

Table 1.3 continued

Topic	Theoretical perspectives	Appraisal methods	Data requirements	Data in the Manual	Lessons from experience
Flood risk management and protection against erosion from the sea					
Recreational gains and losses: fluvial sites	8.2	8.3, Figure 8.1, 8.6.2	Table 8.1	Tables 8.5, 8.8, 8.9	8.8
Recreational gains and losses: coastal sites	8.6	8.3, Figure 8.1, 8.6.1	8.7	Tables 8.6, 8.7, 8.10	Case study: 8.7, 8.8
Environmental gains and losses	10.1–10.3	10.5		9.7 (agriculture), 10.7, Table 10.4	10.10

considerations and data to be used. Thus we see the two sources of information and guidance as complementary. As with all Middlesex manuals, this Manual is acknowledged as a source of user guidance by Defra but the research results remain the responsibility of Middlesex University and the Environment Agency.

In addition to the formal appraisal techniques, catchment and coastal planning agencies have developed a series of additional approaches and procedures. Catchment Flood Management Plans (CFMPs) bring together a strategic-level hydrological and hydraulic analysis with similarly broad-scale economic and environmental evaluations to provide a holistic view of the long-term measures that will provide the most appropriate management of flood risk at the catchment scale. The results represent a broad-brush analysis rather than a detailed evaluation of particular site situations. The same applies to shoreline management plans (SMPs), water level management plans (WLMPs) and coastal habitat and management plans (CHaMPs). Each contains data as to regional and local situations, and suggested indicative solutions to particular large-scale risk management problems.

Thus CFMPs give an overview of the flood risk across each river catchment. They recommend ways of managing those risks now and over the next 50–100 years. These CFMPs also consider all types of inland flooding, from rivers, ground water, surface water and tidal flooding, but not flooding directly from the sea (coastal flooding), which is covered in SMPs. The CFMPs also take into account the likely impacts of climate change, the effects of how we use and manage the land within a catchment, and how areas could be developed to meet our present-day needs without compromising the ability of future generations to meet their own needs.

An SMP is also a large-scale assessment of the risks associated with coastal processes and is intended to help reduce these risks to people and the developed, historic and natural environments. Coastal processes analysed include tidal patterns, wave height, wave direction and the movement of beach and seabed materials. The Environment Agency has a legal duty to protect and enhance the natural environment through its operations: the management of important nature conservation sites that are at risk of coastal erosion or flooding is therefore considered in CHaMPs.

WLMPs explain the way the water levels are operated within an area, and try to balance the needs of agricultural and conservation interests there. They were developed to provide a means by which the water-level requirements for a range of activities within a particular area, including agriculture, flood risk management and conservation, can be balanced and integrated. An action plan is produced which sets out the investment that Internal Drainage Boards and others need to make to address these problems, allowing the Boards, for example, to plan for Grant in Aid applications and understand their appraisal needs.

We see these manuals, therefore, as complementing both government and Agency advice and these local and regional plans. This Manual provides data and techniques appropriate to all levels of evaluation (from generalised to detailed) and the user should exercise judgement as to the appropriate level of analysis to be undertaken in support of any particular decision. Large-scale plans, such as SMPs and CFMPs, will normally provide the context for more detailed investigations within the areas covered.

1.3 FLOOD AND COASTAL EROSION RISK MANAGEMENT: STRATEGIES AND POLICIES

Underpinning flood risk management and coastal erosion risk management policies should be a number of key principles and strategies to ensure wise decision-making and beneficial outcomes. Generally these are codified in international agreements and conventions such as those discussed below but are also detailed in outputs having an international focus (e.g. Sayers *et al.*, 2013). International agreements are often transcribed into national legislation and regulations as appropriate, when accepted by the UK Government.

1.3.1 Sustainable flood risk management and coastal erosion risk management strategies

The United Nations Convention on the Protection and the Use of Trans-Boundary Watercourses and International Lakes recommended guidelines to 'prevent, control and reduce the adverse impacts of flood events on human health and safety, on valuable goods and property, and on the aquatic and terrestrial environment' (United Nations, 2000, i).

These guidelines recommend that there are at least seven principles and approaches regarding sustainable flood prevention. To implement them requires co-operation at all government levels and co-ordination of policies regarding environmental protection, land use planning, agriculture, transport and urban development. These principles are outlined in Box 1.1.

As with other strategies, the United Nations' guidelines on sustainable flood prevention emphasise a holistic approach, integration with other land use and water functions, and the development of adequate information systems so that project appraisals are soundly based. Whereas the guidelines are non-binding on member governments, they represent recommended measures and best practice in the development of sustainable policies.

BOX 1.1 SEVEN BASIC PRINCIPLES AND APPROACHES REGARDING SUSTAINABLE FLOOD PREVENTION

- Flood events are part of nature. They have always existed and will continue to exist.
- Human interference into the processes of nature has increased the threat of flooding. As far as possible, such interference should be reversed, compensated and, in the future, prevented.
- Structural measures will remain important elements of flood prevention and protection. However, these measures should primarily focus on the protection of human health and safety, and valuable goods and property.
- Requirements of nature conservation and landscape management should be taken into consideration.
- Everyone who may suffer from the consequences of flood events should also take his/her own precautions. To this end, an appropriate information and forecasting system should be established by the competent authority.
- Human uses of floodplains should be adapted to the existing hazards. Appropriate instruments and measures should be developed to reduce the risk of flooding.
- In flood-prone areas, preventive measures should be taken to reduce possible adverse effects of floods on aquatic and terrestrial ecosystems, such as water and soil pollution.

(United Nations, 2000)

1.3.2 A catchment focus and a participative approach: the Water Framework Directive and the Aarhus Convention

The Aarhus Convention (DETR, 2000a) and the Water Framework Directive (European Parliament, 1998) have fundamentally changed the nature of water management in Europe and, with that, the purpose and nature of project appraisal.

From the Dublin conference onwards (ACC/ISGWR, 1992), the holistic management of water and land across a catchment has been taken to be one of the defining criteria for the achievement of sustainable water management. The Water Framework Directive requires the introduction of such holistic catchment management within Europe. A second criterion that the Dublin conference established for sustainable water management was the involvement of the public in all levels of decision-making; the Aarhus Convention sets out the conditions for this to occur and was fully adopted by the UK Government in 2005.

The shift to stakeholder-driven decision-making requires a radical shift in the nature of project appraisal; the purpose of project appraisal is now to provide a rigorous analytical framework within which decisions can be argued, negotiated and debated by the stakeholders. It includes a key role in promoting communication between the stakeholders as well as in enabling them to assess and argue the consequences of adopting different alternatives (Green, 2002).

At the same time, the shift to holistic land and water management across catchments under the Water Framework Directive changes the nature of the options that will be considered. The implication is that there may cease to be single-purpose flood risk management schemes or wastewater treatment plants: actions will no longer be considered from a single functional perspective. Instead, the overall goal will be to improve the functioning of the catchment as a whole where one, and perhaps the primary, purpose of a possible set of interventions will be to improve the management of the variability of water quantity across the catchment and over time.

It is clearly intended that catchments should be managed as dynamic systems rather than just to address localised problems. As a result, the arbitrary differentiation between variations in water availability over time into droughts, water resources and floods may cease (Green, 2002). For example, in arid climates and regions, floods are the water resource. In turn, project appraisal techniques will also have to cover all aspects of the functioning of a catchment, and specifically the three fluxes that occur (of water, soil and pollutants). There is a similar requirement for coastal zone management (Green and Penning-Rowsell, 2011).

However, with the adoption of the Directive and the acceptance of the Convention it will not be enough to consider only the holistic functioning of the catchment or coastal zone. Actions within the catchment or at the coast will also have to mesh into national policies on, for example, urban and rural development (Green, 2002). In addition, the resources allocated towards improved catchment management will necessarily not be available for other purposes, such as health care and education.

1.3.3 The European Union's 'Floods Directive'

Under the Floods Directive (2007/60/EC) on the assessment and management of flood risks, which came into force on 26 November 2007, all EU countries are obliged to assess the risk and management of all types of flooding in their territories (European Parliament and the Council, 2007; European Commission, 2011, 1).

The Directive required member states to have carried out a broad-scale assessment by 2011 'to identify the river basins and associated coastal areas at risk of flooding', and to involve their government departments, agencies and other bodies in drawing up this preliminary flood risk assessment. This assessment needed to consider impacts on human health and life, the environment, cultural heritage and economic activity, with a completion date of December 2011. That information was then to be used to identify the areas at significant risk which would then be modelled in order to produce detailed flood hazard and flood risk maps.

These maps were to be in place by December 2013, and will include detail on the flood extent, depth and level for three risk scenarios (high, medium and low probability). Flood risk management plans then need to be produced to indicate to policy-makers, developers and the public the nature of the risk and the measures proposed to manage these risks, focusing on prevention, protection and preparedness. The plans are to be completed by December 2015, and in order to ensure community/stakeholder buy-in they will require input from interested parties during their development.

In England and Wales the Flood Risk Regulations 2009 (HM Government, 2009a) transposed the EU's Floods Directive into domestic law, to implement its provisions. In particular, the Regulations placed duties on the Environment Agency and local authorities to prepare these flood risk assessments, flood risk maps and flood risk management plans. But EU directives are interrelated: the flood risk management plans that result from the Floods Directive should take into account the relevant environmental objectives of Article 4 of the Water Framework Directive.

1.3.4 Defra policy priorities and high-level targets

The UK Government's flood risk management strategy has continued to evolve and broaden since 2005 (Environment Agency, 2010a). For example:

- The Climate Change Act 2008 (HM Government, 2008) requires a UK-wide climate change risk assessment every five years accompanied by a national adaptation programme that is also reviewed every five years. The Act has given the UK Government new powers to require public bodies and statutory organisations, such as water companies, to report on how they are adapting to climate change.
- *Future Water 2008* (HM Government/Defra, 2008), the UK Government's overall strategy for water, looks mainly at water supply and provision. However, it also reaffirms *Making Space for Water* as the basis for managing river and coastal flooding (see below). Furthermore, it sets out a vision for better management of surface water to address the dual pressures of climate change and housing development.
- The Pitt Review (Pitt, 2008), following the 2007 floods, made ninety-two recommendations, the government response to which culminated in the Flood and Water Management Act 2010. There is a particular emphasis in the Act's provisions on surface water and other local flood risk, a main cause of damage in the 2007 floods. The Act established lead local flood authorities and gave these and other local bodies, such as water companies, new powers and responsibilities to co-ordinate and participate in flood risk management. Among other things, it also gave the Environment Agency a duty to establish a national strategy for managing flood risk and coastal erosion in England. The aim of the national strategy, which builds on the earlier *Making Space for Water* (Defra, 2005), is shown in Box 1.2.

Sustainable schemes had been defined previously as 'projects which take account of the inter-relationships with other defences, developments and processes within a catchment or coastal sediment cell, and which avoid as far as possible tying future generations into inflexible and expensive options for defence' (MAFF/Welsh Office, 1993). The more recent elaboration of this is given in Box 1.3, being the guiding

BOX 1.2 THE AIM FOR FLOOD RISK MANAGEMENT IN UNDERSTANDING THE RISKS, EMPOWERING COMMUNITIES, BUILDING RESILIENCE – THE NATIONAL FLOOD AND COASTAL EROSION RISK MANAGEMENT STRATEGY FOR ENGLAND

The overall aim of the strategy is to ensure the risk of flooding and coastal erosion is properly managed by using the full range of options in a co-ordinated way.

Communities, individuals, voluntary groups and private and public sector organisations will work together to:

- manage the risk to people and their property;
- facilitate decision-making and action at the appropriate level – individual, community, or local authority, river catchment, coastal cell or national; and
- achieve environmental, social and economic benefits, consistent with the principles of sustainable development.

(Environment Agency, 2011a)

principles from the national strategy (Environment Agency, 2011a).

In terms of project appraisal, what is sought for sustainable flood and coastal risk management is the achievement of long-term stable benefits rather than schemes with higher benefit–cost ratios where the standard of protection reduces over time. There is an emphasis on whole-life costs and benefits, to get a truer comparison between options than simply looking at limited, short-term-scheme life costs and benefits. In addition, there should be a proper consideration of the 'do nothing' option. The assumption that 'something must be done' is to be avoided, as it can lead to the introduction of non-sustainable regimes of flood risk management and coastal erosion risk management work.

BOX 1.3 EXTRACTS FROM THE 'GUIDING PRINCIPLES' IN THE NATIONAL STRATEGY

Flood and coastal erosion management may mean that difficult decisions have to be taken on where risk management activities can and cannot be carried out at both national and local levels. These decisions, and the processes by which they are taken, will be guided by a number of high-level principles. These are outlined below (extracts only):

Community focus and partnership working

The risk management authorities should work in partnership with communities to understand the community perspective of flooding and coastal erosion, help communities understand and actively prepare for the risks, and encourage them to have direct involvement in decision-making and risk management actions. This includes giving communities a bigger say in what action is taken, greater responsibility for managing their own risks and decisions on local funding priorities, and as a result greater accountability for the level of safety and protection achieved and the way in which the risks are managed. The aim is to ensure that decision-making and ownership of risk management measures are as local as possible but within a catchment, coastal cell and national framework that ensures a fair allocation of funds and avoids the transfer of risk elsewhere without prior agreement.

A catchment- and coastal cell-based approach

In understanding and managing flood and coastal risks locally, it is essential to consider the impacts on other parts of the catchment or coast. Activities must seek to avoid passing risk on to others within the catchment or along the coast without prior agreement . . . Catchment flood management plans (CFMPs) and shoreline management plans (SMPs), or equivalent, provide an important building block for this co-ordination.

Sustainability

Flood risk and coastal erosion management authorities should support communities by managing risks in ways that take account of all impacts and the whole-life costs of investment in risk management. The risk management solutions should be forward-looking, taking account of potential risks that may arise in the future and being adaptable to climate change. They should also work with natural processes where possible and enhance the environment.

Proportionate, risk-based approaches

It is not technically, economically or environmentally feasible to prevent flooding and coastal erosion altogether. A risk-based management approach targets resources to those areas where they have greatest effect. Risk management measures consider both the probability over time of a flood or coastal erosion happening and the consequences that might arise if it did, for example by assessing the average annual damages that arise from floods or coastal erosion. To do this the sources, pathways, receptors and consequences of risk need to be understood and addressed as appropriate to manage all of the factors that combine to create risk. Further detail on this approach is available in the Guidelines for Environmental Risk Assessment and Management (DETR/EA/IEH, 2000).

Multiple benefits

As well as reducing the risks to people and property, FCERM can bring significant economic, environmental and social benefits. It can enhance and protect the built, rural and natural environments, cultural heritage and biodiversity by preventing loss and damage to habitats and heritage assets and reducing pollution, for example through the use of sustainable urban drainage systems (SUDS). It can contribute to regeneration and income generation, protect infrastructure and transport links, and contribute to economic growth.

Beneficiaries should be encouraged to invest in risk management

The benefits achieved when flood and coastal erosion risks are managed are in many cases localised and lead to personal or private gain through the protection of specific individuals, communities and businesses. They can also be public, through the reduction of future costs to society arising from incident recovery. The private as well as public nature of the benefits suggests that costs should not fall to the general taxpayer alone . . . In his review of the summer 2007 flooding, Sir Michael Pitt suggested that better aligning beneficiaries with those that pay would create a more efficient and responsive system. To do this he recommended that 'Government should develop a scheme that allows and encourages local communities to invest in flood risk management measures' [and] that developers, in potentially increasing local flood risk, should 'make a full contribution towards both the costs of building and maintaining the necessary defences'. In taking this recommendation forward, the Government has made clear that 'we cannot continue all of the work that the Environment Agency has historically done at the taxpayer's expense. Government investment in flood and coastal erosion risk management is significant, but we need to ensure that we get best value for money.'

(Environment Agency, 2011a)

Again, in terms of project appraisal, this is not easy, and the aim of delivering 'environmental, social and economic benefits, consistent with the principles of sustainable development' (Box 1.2) raises the question of how benefits are to be identified, defined and measured. However, these difficulties are recognised, and Defra has indicated that this 'will be complex as will ensuring a good fit between national and local priorities' (Defra, 2005, 10). In this respect, appraisers of FCERM schemes should also be aware of the types of risk management expenditure that Defra currently funds, not least because some benefits might not be supported in this way. Those benefits (e.g. major recreational benefits) might currently need the support of other funding streams where they are not central to the relevant scheme.

Defra has also developed a formal appraisal policy (Defra, 2009, 4). This 'sets out why appraisal is necessary, and the principles and policies that should guide this work'. In particular, it emphasises the need to ensure that appraisals for all activities (whether strategic-level plans or individual projects):

- give more consideration to 'risk management' and 'adaptation', as opposed to only 'protection' and 'defence';
- are undertaken consistently, transparently, with value for money in mind and in a way that complies with the Treasury guidance on appraisal and evaluation in central government (the 'Green Book');
- help achieve better social and environmental outcomes as part of sustainable development, both by considering a broader range of issues and by using a broader range of analysis techniques; and
- adopt a risk-based approach, whilst considering impacts within the whole of a catchment or shoreline process area.

In addition, five principles have been enunciated to 'ensure that flood and coastal erosion risk management contributes to sustainable development' (Defra, 2009, 11). These principles are those of:

- living within environmental limits;
- ensuring a strong, healthy and just society;
- achieving a sustainable economy;
- promoting good governance; and
- using sound science responsibly.

The approach to appraisal is set out by Defra (2009, 26), involving three key stages:

1. Define the issue and consider the case for government intervention. Set SMART objectives if there is a case (SMART = Specific, Measurable, Achievable, Realistic, Timely).
2. Develop a full range of possible options, describe the options, and then value the positive and negative impacts of each of the options.
3. Compare options in a systematic way and select the most effective and deliverable solution

To complement Defra's policy framework, as indicated above, it defines a series of Outcome Measures against which Environment Agency performance is measured and which are used to set Exchequer funding contributions to risk management activities (see next section). These Outcome Measures are set out in Table 1.4.

1.3.5 'Partnership funding': a new FCERM financing scheme

A significant change in the funding of FCERM schemes was promoted by Defra in 2011 to launch a system of local/national cost-sharing. The significance of this change is that it introduced a far greater and critical element of 'localism' into what previously was a highly centralised arrangement. The pre-existing block grant system remains (whereby Defra allocated a block of money to the Environment Agency), at a somewhat reduced level, but projects or schemes in many cases can proceed only if the national contribution is complemented by locally derived resources.

The new 'partnership funding' arrangement (Defra, 2011a; Environment Agency, 2012) operates on a formula basis to determine the Flood Defence Grant in Aid (FDGiA) – how the reduced Environment Agency block grant (provided by Defra) is to be allocated, scheme by scheme (Johns, 2011).

Equation 1.1

$$\pounds\text{FDGiA} = H + B + E$$

Where (see Table 1.4):

H is the value of qualifying Household benefits for that scheme, multiplied by the payment rate.
B is the value of Other Whole-life Benefits for that scheme, multiplied by their payment rate.
E is the number of Environmental Outcomes for that scheme, multiplied by their payment rates.

From this can be derived an Outcome Measure (OM) score (as a percentage) – being the £FDGiA sum divided by the scheme costs – and this metric can be used to prioritise decision-making.

Many schemes will have an OM value greater than 100 per cent, and then the full cost is available from the Defra-funded grant. In other cases there is a shortfall: the grant fails to cover the costs and this shortfall needs to be met from local contributions if

Table 1.4 The new payment rates for the partnership funding arrangement

OM no.	Outcome Measure definition	Benefits and outcomes qualifying for national funding	Payment rate	Examples of funding levels from Government
OM 1 OM 1a	Average benefit to cost ratio of schemes delivering OMs Present value of whole-life benefits per £1 of FDGiA	Under OM1, present value of whole-life benefits of the current investment, less benefits paid for or payments made under the other OMs	5.56p per £1 of qualifying benefit (i.e. seeking an 18 to 1 return from national investment)	These include avoidance of damages to e.g. business, agriculture, local government, communications, infrastructure, utilities and public health
OM 2	Households moved from one category of flood risk to a lower category Households must be at direct risk of flood damage and have been built or converted into housing before January 2012 to be counted	Under OM2, present value of direct damages to residential properties and their contents avoided, in the: 20% most deprived areas 21–40% most deprived areas 60% least deprived areas	45p per £1 30p per £1 20p per £1	Based on moving a single household from very significant risk to low risk for a duration of 50 years £15,399 per household protected £10,266 £6,844
OM 3	Households better protected against coastal erosion Households must be at direct risk of damage from coastal erosion and have been built or converted into housing before January 2012 to qualify	Under OM3, present value of the reduction in direct damages to residential properties, in the: 20% most deprived areas 21–40% most deprived areas 60% least deprived areas	45p per £1 30p per £1 20p per £1	Based on protecting a single household at risk of loss within 20 years, for a period of 50 years £35,601 per household protected £23,734 £15,822
OM 4	Statutory environmental obligations fully met through flood and coastal erosion risk management	Outcomes specifically funded under OM4:	£15,000 per hectare	
OM 4a	Hectares of water-dependent habitat created or improved to help meet the objectives of the Water Framework Directive	Water-dependent habitat created or improved		
OM 4b	Hectares of inter-tidal habitat created to help meet the objectives of the Water Framework Directive for areas protected under the EU Habitats or Birds Directive	Inter-tidal habitat created	£50,000 per hectare	

Table 1.4 continued

OM no.	Outcome Measure definition	Benefits and outcomes qualifying for national funding	Payment rate	Examples of funding levels from Government
OM 4c	Kilometres of river protected under the EU Habitats or Birds Directive improved to help meet the objectives of the Water Framework Directive	Protected rivers improved	£80,000 per km of river bed	

Source: Defra, 2011a

the scheme is to proceed. These contributions can come from a number of sources, with the three prime headings being the pre-existing 'local levy' raised by the Environment Agency, the public (e.g. local authorities) and the private sector (e.g. developers or industry).

In the expenditure programme for schemes active in the financial year 2012/13 there were 140 schemes (at an average cost of £4.078 million) for which a local contribution has been potentially forthcoming, out of a total of 398 schemes (i.e. 35 per cent), with that contribution varying from 0.3 to 95 per cent of total costs. Of the schemes with some local contribution, the total contributions as at early 2012 were £141 million out of total scheme costs of £585 million (i.e. an average of 24.1 per cent), representing a sizeable additional resource over and above that provided nationally. One might expect that these are the smaller schemes, which they are, but the schemes without a contribution average £5.403 million, or only 32 per cent larger by total cost than those with the local resources. Hence it is not simply the case that 'local' is small and 'national' is much larger.

One of the characteristics of this partnership funding is that any scheme delivering worthwhile benefits (as defined in the policy) can receive some level of central government funding, unlike the situation previously where, if the scheme did not warrant proceeding by the rules then extant (MAFF, 1999), the central government contribution was zero.

To be eligible for a grant, the partnership funding score (PF) must show that there are sufficient funds available: the PF score must be above 100 per cent (the score broadly is the maximum permitted grant for the scheme plus the local contribution divided by cost of scheme). To receive a grant in a particular year, a scheme must be successful in a process that prioritises funding against the PF scores and accommodates local choices.

Regarding the latter criterion, a stated objective of Defra's policy here is to enable more local choice in the solutions adopted to reduce flood risk, and Regional Flood and Coastal Committees now play a central part in ensuring that local choices are reflected in the programmes that they are now required to agree. One of the reasons why the contribution level for some schemes is so high (i.e. >90 per cent) is that previously they would probably have been funded entirely through the local levy but will now be attracting a modest level of government grant: the level of local contribution is not capped and can form any proportion of the total cost needed to implement the scheme.

In terms of rationale, Environment Minister Richard Benyon has indicated that:

> This new funding system means more flood defence schemes will benefit from [central] government money so we can protect more people and properties. Many schemes in areas at high risk will continue to receive full funding from government, whilst others will receive large contributions that will go a long way towards meeting the amount needed for the defence to be built . . . This will mean that local communities have a much greater say in how and where schemes are built and are no longer hampered in their ambitions by what [national] government alone can afford.
>
> (Defra, 2011b)

In this respect, the term 'defence' should be seen broadly, because the partnership funding scheme is certainly not meant to be restricted to engineering solutions to flood risk and coastal risk management problems.

There are significant implications of the move towards partnership funding for the contents of this Manual. Fundamentally, the national economic perspective that has pervaded all our manuals since 1977 is no longer fully appropriate for the schemes that are part funded by local communities. These communities will be motivated by financial (local) impacts from floods, rather than their – lesser – economic (national) effects. That is to say, a local community faced with shops losing business as a result of flooding will want to see this business restored or the losses mitigated. On the other hand, a national economic analysis would accept that losses to one shop would be recouped by sales increases elsewhere, rendering the national economic loss to be close to zero (see Chapter 5).

We have responded to this change by providing financial (local) flood damage figures on MCM-Online to complement the economic (national) figures that we have provided previously. On MCM-Online we have indicated very clearly which figures are economic values and which are financial values, and users should be careful not to confuse the two.

1.3.6 Environment Agency perspectives

The Environment Agency in England and Wales is the principal organisation implementing the UK Government's fluvial flood risk management plans and schemes, while local authorities are the principal organisations tackling 'surface water flooding' (the lead local authorities in England and all local authorities in Wales) and implementing coastal erosion risk management and coastal flood risk management schemes.

Here the most common forms of floods (Environment Agency, 2009a: England) are:

- River flooding that occurs when a watercourse cannot cope with the water draining into it from the surrounding land. This can happen, for example, when heavy rain falls on an already waterlogged catchment.
- Coastal flooding that results from a combination of high tides and stormy conditions. If low atmospheric pressure coincides with a high tide, a tidal surge may happen which can cause serious flooding.
- Surface water flooding which occurs when heavy rainfall overwhelms the drainage capacity of the local area. This is difficult to predict and pinpoint; much more so than river or coastal flooding.
- Sewer flooding that occurs when sewers are overwhelmed by heavy rainfall or when they become blocked. The likelihood of this flooding depends on the capacity of the local sewerage system. Land and property can be flooded with water contaminated with raw sewage as a result. Rivers can also become polluted by sewer overflows.
- Groundwater flooding that occurs when water levels in the ground rise above surface levels. It is most likely to occur in areas underlain by permeable rocks, called aquifers. These can be extensive, regional aquifers, such as chalk or sandstone, or may be more local sand or river gravels in valley bottoms underlain by less permeable rocks. This type of flooding is not significant in Wales (Environment Agency Wales, 2009).

1.3.7 Some key project appraisal differences between the coastal and the fluvial environment

In flood and coastal erosion risk management, there is a fundamental difference between alleviating flooding and alleviating erosion. Flood losses have the potential to be recurrent: there is a risk of flooding in a given low-lying area in any single year and it is possible that floods of different magnitudes may occur in any combination of years. Erosion is an ongoing process that can normally only be delayed.

Hence we generally assume in assessing the impacts of flooding that the same land uses will continue in that area and that they may repeatedly suffer flood damage.

However, where as a result of repeated flooding problems more land might become permanently inundated, or is flooded repeatedly with a frequency of, say, once every two years, then it is almost inevitable that the existing land use will change permanently. In these cases property will become abandoned and in the benefit assessments it is 'written off', just as where property lost to the sea is deemed to be lost for ever in the case of coastal erosion.

Losses from erosion, however, are very different from flood losses: erosion is a progressive or sequential process that is largely irreversible. Erosion can be countered by reclamation from the sea, but this, in effect, is merely a variant on the delay of erosion (i.e. it puts back the process of erosion still further than simply protecting the current line of the land/sea interface). Although the erosion process can be halted for a while, once land and property have been eroded they cannot usually be recovered and thus the losses are one-off and permanent.

In terms of this Manual, it should therefore be remembered that assessing the benefits of erosion risk management is somewhat different from assessing the benefits of flood risk management. In summary, the former assesses the benefit of *delaying* the erosion event, whereas the latter assesses the benefits of *reducing or eliminating* flood damages and losses.

The benefits of protecting agricultural land from both erosion and flooding are assessed as the changes in productivity of the land that would result from either the continuation or a reduction in risk, producing higher levels of output or lower levels of costs than in a 'do nothing' situation of existing cropping regimes and the returns that these produce (see Chapter 9).

1.4 THE AIMS AND ASSUMPTIONS OF BENEFIT–COST ANALYSIS

This section is designed to be a brief introduction to Chapter 2. The basis of that guidance is that the principal purpose of project appraisal methods is to identify and clarify the issues involved in a particular decision. The purpose is not simply to calculate the benefit–cost ratio for a particular scheme.

This means that the development and adoption of particular scheme options should be appraisal-led rather than design-led. Moreover, appraisal methods should always be used critically and, in a pluralist society where there are conflicts between the views of different individuals and groups, one of the functions of appraisal methods is to aid communication between these individuals and different groups within society.

All project appraisal methods involve the comparison of at least two options or alternatives; one is the baseline option, which is usually the 'do nothing' option. This is also, in effect, a comparison of the 'with scheme' and 'without scheme' scenarios that are fundamental to the application of benefit–cost analysis. The impacts of different options usually differ in:

- their nature;
- when they occur; and
- who is affected.

Any method of comparison or evaluation must encompass and allow for this and be able to highlight and analyse these differences.

The economic techniques used to make these comparisons involve the assessment of value – usually by methods related directly or indirectly to market prices – and also methods to bring all values to some common base date for analysis, using techniques such as discounting. Such comparisons also need to differentiate between financial and economic analysis, and an understanding of these differences is crucial to the proper appraisal of coastal erosion risk management and flood risk management schemes.

Market prices are only one basis of judging value to society, and are generally not useful when assessing non-use values such as the significance of environmental assets or heritage sites. For these and other reasons we do not therefore generally recommend the quantitative economic valuation of non-use values on a site-specific basis.

The fundamental stages within project appraisal involve the processes of:

- identification;
- quantification; and
- evaluation.

To apply economic analysis to compare a particular option to the baseline option, therefore, it is necessary to identify all the changes that will occur, determine when these changes will occur, and distinguish who will be affected by the changes.

A number of different techniques can be applied to the valuation of these changes and these are discussed in Chapter 2. Within this Manual we rely on market-based prices for assessing the benefits of flood risk management in terms of direct flood losses to property and land. In assessing the recreational benefits of coastal erosion risk management or river management, contingent valuation methods have become more prominent, so as to elicit from members of the public the value they put on the maintenance or enhancement of recreational activities and environmental values (their 'willingness to pay' for these goods – see Chapter 2).

The incorporation of risk and uncertainty into appraisal and the management of uncertainty are discussed in Chapter 2. We suggest that uncertainty is only important if its resolution would make a difference to which option is chosen: that is, whether the preferred option is 'robust' to remaining uncertainty.

Therefore, we recommend that appraisal should be an iterative process of progressively refining the estimate to the benefits and cost of the alternative options being considered. At each stage it is necessary to decide whether any reductions in the uncertainty concerning the estimates of the benefits of the option justify the costs of the work necessary to improve those estimates. In summary, project appraisal itself should be pursued only in so far as its benefits justify its costs.

1.5 PREDICTING THE FUTURE

Any project appraisal is concerned with predicting the future, and this is a risky business.

When it could be assumed that the hydrological record in England and Wales was characterised by random fluctuations about a constant and unchanging average, it was simple to calculate the annual average flood losses at a particular site and extrapolate these into the future to cover the design life of any particular scheme or plan.

With climate change, however, research at the Centre for Ecology and Hydrology (CEH) for the River Severn and the River Thames had indicated that increases of up to 20 per cent in peak flows for a given annual probability could be experienced under a changed climate regime within fifty years. But new advice is now available that replaces Defra's *Supplementary Note to Operating Authorities – Climate Change Impacts* (October 2006). It is provided as supplementary information to Defra's policy statement on *Appraisal of Flood and Coastal Erosion Risk Management* (Defra, 2009) and the Environment Agency's *Flood and Coastal Erosion Risk Management Appraisal Guidance* (FCERM-AG; Environment Agency, 2010a).

The advice is that to assess the potential impacts that climate change may have on extreme rainfall, river flood flows, sea level rise and storm surges, 'change factors' are provided such as are given in Table 1.5. The change factors quantify the potential change (as either mm or percentage increase, depending on the variable) to the baseline. It is recommended that scheme options are developed based on planning for the change factor covering the whole of the decision lifetime. However, rather than base options solely on the change factor, the upper- and lower-end estimates can be used to refine the options to prepare for a wider range of future change. Similar approaches are provided for sea level rise at the coast and for extreme rainfall.

In each of these situations of 'non-stationarity', the calculated benefits of flood risk management or coastal erosion risk management will be significantly affected by the predictions of secular change. In effect,

Table 1.5 Potential changes in peak flow for the Northumbria river basin district

	Total potential change anticipated for the 2020s	Total potential change anticipated for the 2050s	Total potential change anticipated for the 2080s
Northumbria			
Upper end estimate	25%	30%	50%
Change factor	10%	15%	20%
Lower end estimate	0%	0%	5%
Northumbria H++*	35%	45%	75%

Source: Environment Agency, undated b, 6

Notes: *The H++ scenario provides an estimate of sea level rise and river flood flow change beyond the likely range but within physical plausibility. It is useful for contingency planning to understand what might be required if climate change were to happen much more rapidly than expected. It is not possible to say how likely the H++ scenario is.

we are reducing the discount rate applied to the project appraisal results, and the impact that this will have on the calculated benefit–cost ratio will almost always be very significant. This means that close attention needs to be paid to the predicted changes in river discharge and sea levels into the future, and some sensitivity analysis applied within the project appraisal process.

1.6 GENERALISED AND DETAILED BENEFIT ASSESSMENTS

As indicated above, project appraisal is an iterative process where ideas go through a cycle of being developed, reviewed and refined. The overall framework is encapsulated in Figures 1.1 and 1.2, and it is advisable that appraisal techniques are matched to the appropriate level within this hierarchy.

1.6.1 National-level analysis and high-level benefit assessment methods

For effective policy development and programme planning (including national-level resource allocation), it is necessary to have a good understanding of the nature of the flood and coastal erosion problems experienced in England and Wales as a whole. This has been provided through the application of a range of high-level techniques, such as started with the project to define the National Appraisal of Assets at Risk (NAAR; Defra, 2001b) and was subsequently pursued with the National Flood Risk Assessment (NaFRA) in 2006 and 2008 (Willatt, 2006; Halcrow Group Limited, 2007; Environment Agency, 2009a). This project has used a series of datasets on average losses due to flooding and coastal erosion in relation to the extent of flooding, the existing standards of service provided by existing flood risk reduction measures, and the number of properties or other assets lying within areas subject to different general levels of flooding and erosion severity.

A decade ago the 2002–04 Foresight Future Flooding project (Evans *et al.*, 2004) took this work several steps further by modelling flood risk for the whole of England and Wales, based on four scenario 'futures', and also assessed the impact of a range of policy responses. The project showed that under some future scenarios annual average flood damage could increase twentyfold over the next eighty to a hundred years.

A general conclusion from the NAAR, Foresight and the NaFRA studies was that the then current level of flood and coastal risk management investment was at too low a level to prevent deterioration in flood risk management standards in the future. This evidence, and other information, has led these investment levels to be reviewed and increased in recent years, reaching a peak of over £600 million in 2010/11. It is not clear, with partnership funding providing local contributions, whether totals will continue at this level, but the indications in 2013 were that they would.

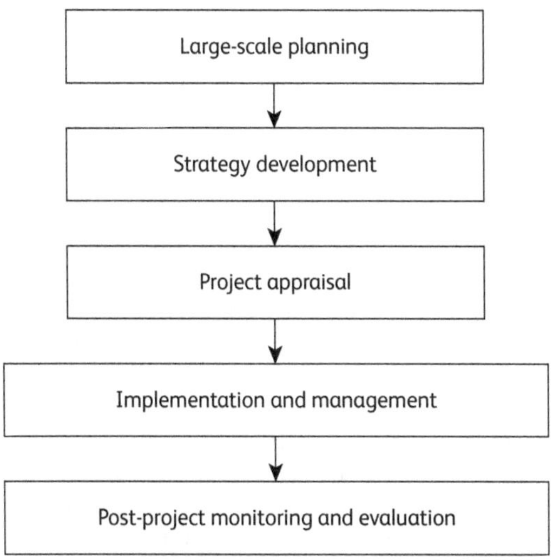

Figure 1.1 A hierarchy of planning and project appraisal

Source: Defra, 2001a

1.6.2 Meso-scale analysis and generalised assessment methods, as in catchment flood management plans, etc. (i.e. for 'large-scale planning' in Figure 1.1)

Within England and Wales CFMPs have quantified the potential impacts of flooding at a broad scale and thereby assist with the prioritising of policy and scheme development. In the fluvial situation, this requires more detailed hydrological and hydraulic inputs – and more detailed flood loss estimates – than could be handled at the national scale.

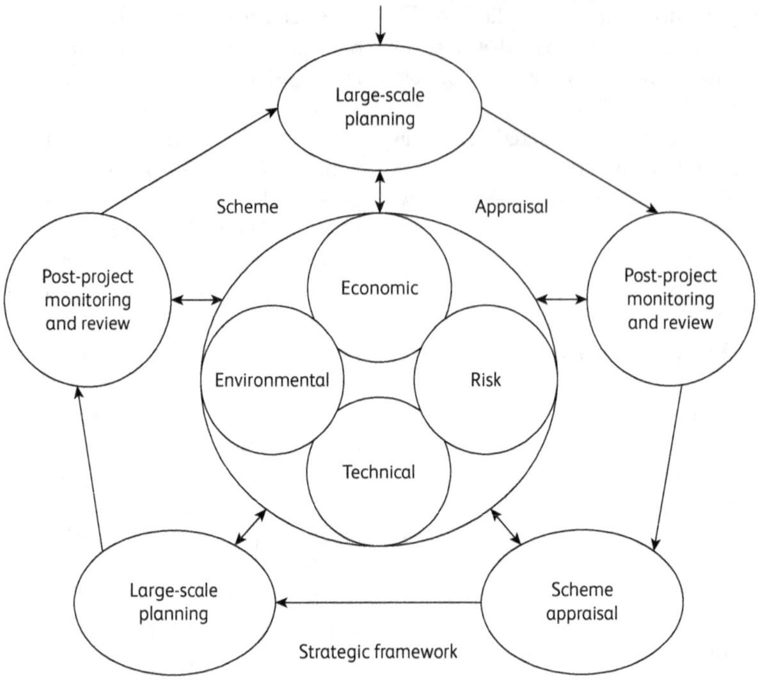

Figure 1.2 Stages of project planning and development

Source: Defra, 2001a

Commensurate with this greater detail, for example, a measure of social vulnerability has been introduced within the CFMP process, in addition to the potential for flood damage and losses, so as to pinpoint locations where the population may be particularly vulnerable to flooding owing to its social and economic character. The Intermediate Level Method in the RASP suite of programs from HR Wallingford (Gouldby *et al.*, 2008) uses measurements or model estimates of flood water levels, flood defence levels and ground elevation to generate better estimates of flood risk. It is being used to inform strategic decisions on flood risk management at this meso-scale.

1.6.3 More detailed appraisals (i.e. 'strategy development' and 'project appraisal' in Figure 1.1)

The broad-scale analysis as indicated above can pinpoint areas where flooding and coastal erosion are particularly problematic.

Proceeding down the hierarchy as indicated in Figure 1.1, 'strategies' are more detailed plans to tackle, generally, a series of linked problems at a subcatchment or sub-cell coastal scale. Project appraisal is then the detailed consideration of works or other activities to deliver a flood or coastal risk management strategy or a standalone site-specific solution.

At each stage we can use more and more detailed or complex economic methods, commensurate with the worthwhileness of this greater precision.

'Initial' and 'detailed' appraisal studies

It is useful here – and indeed throughout this Manual – to differentiate between 'initial' and 'detailed' appraisal studies. The objective of the 'initial' studies is to identify where more detailed investigations are required, using generalised data and methods, generally in a desk-based exercise. In relation to this Manual, this data includes the standard or average data on the value of flood losses (e.g. in Chapters 4 and 5), and at the coast it is possible, for example, to use average/standard data on a value of beach recreation (Chapter 8). Other chapters within this Manual also provide comparable average data for use in these initial studies.

Once 'hot spots' have been identified at the macro-scale, or through initial studies, more detailed investigations (previously termed 'feasibility' studies) will use all the detailed data contained in this Manual to provide as accurate as possible an assessment of the benefits of flood risk management or coastal erosion risk management.

These detailed studies will demand, in general, at least three times the resources of an initial study (and often much more), and will almost always involve field data collection, be it selected land use surveys of floodplain properties (for example, to identify properties with basements), site-specific damage surveys using the questionnaire (see MCM-Online Supporting Information for Chapter 5) or beach recreation surveys to assess the benefits of enhancing beaches or maintaining the status quo (although these last are expensive and need to be planned with care).

In the fluvial domain it will be necessary in detailed studies to assess accurately the extent and depths of flooding for events with different annual probabilities, both before the scheme is implemented (the baseline situation) and after the scheme or plan is completed. Such studies need to be closely related to the design process, which we advocate is led by the appraisal results, rather than vice versa.

Again, the appraisal process needs to be iterative, with appraisal results informing planning and design, which in turn inform the extent of post-scheme flooding or the extent of erosion delay. Those parameters, in turn, affect the net benefits of that particular scheme option, being the difference between the 'with scheme' and 'without scheme' situation. In many detailed studies the analysis of options is complex and time consuming if a major £multi-million scheme or other intervention is anticipated, although in some circumstances the preferred option is more self-evident and the appraisal process is therefore more straightforward.

1.6.4 Post-project evaluations ('post-project monitoring and evaluation' in Figure 1.1)

As levels of investment in flood and coastal risk management increase, there will be increasing concern to maximise its effectiveness. This can be assisted by 'hindsight' evaluations of what has been achieved in the past.

Such 'post-project appraisals' are an essential part of informing the process of scheme appraisal and design, since the performance of flood and coastal risk management schemes can be judged easily only after they are implemented and in relation to the flood and erosion events that they then experience.

In the flood and coastal risk management field the National Audit Office (2001) has recommended that the Environment Agency should collect more information on the success of its flood and coastal risk management schemes, and should be more systematic and objective in its post-project appraisals/evaluations. However, this is not easy, since a scheme designed to a 1 per cent per annum flood design standard may not experience such a flood within its design lifetime (which might be fifty years), and therefore its performance in this respect cannot always be determined. Other indirect methods of evaluation therefore need to be adopted instead (e.g. model-based extrapolation from the events that have been experienced).

However, whilst this Manual is not oriented specifically at post-project appraisals, research in the past at Middlesex University has investigated the performance of a number of flood risk management schemes (Thompson *et al.*, 1991). In addition, an assessment has been made of the damages caused by the severe autumn 2000 and summer 2007 flood events, as a post-incident investigation (Penning-Rowsell *et al.*, 2002; Chatterton *et al.*, 2010).

1.7 THE PHASING AND TIMING OF EACH PART OF AN ASSESSMENT

Careful timing and phasing of the various parts of a comprehensive benefit–cost appraisal for scheme assessment are vital. Different parts of a comprehensive assessment need to be started at a different time period ahead of the results being required, and there is a real danger that these different parts will be out of synchrony if careful attention is not given to this subject (Penning-Rowsell *et al.*, 1992).

1.7.1 Tasks needing at least a calendar year

These will involve assessing the potential environmental impacts of a proposed flood risk management or coastal erosion risk management scheme, for example to allow the existing status and potential impacts on flora to be investigated over a complete growing season, and wildfowl and wader importance to be assessed over a complete cycle of breeding, migration and wintering seasons.

In addition, any survey of enhanced recreational potential as a result of river management or coastal works should encompass a complete year (or at least the period between Easter and the end of October). It is important in this way to gauge the use of the water or coastal environment over the whole of this period, so as to permit valid grossing-up of site survey data to obtain annual use figures.

1.7.2 Tasks needing three to six months

These will include situations when site-specific surveys of flood damage locations are to be undertaken for non-residential property; it is likely that these will take some months to organise and implement. As a rule of thumb guide, it will take at least two (and probably three) person/days per site-specific survey to initiate, interview the appropriate personnel, analyse the results and complete each survey.

With both flood risk management and erosion studies, one of the tasks that consumes time is the survey of impacts on utility infrastructure within the benefit areas (see Chapter 6). Any interview surveys undertaken as part of this process will take time to arrange and complete, and for analysis of the data collected to be undertaken. The same applies to an assessment of the agricultural benefits of flood risk management and coastal erosion risk management. Farm interview surveys will be needed in detailed appraisal studies here, and these can take several days per interview to set up, conduct the interview and analyse the data thus collected (see Chapter 9).

Another item of benefit assessment that can take more time than is generally appreciated is the survey of traffic disruption caused by flooding and/or coastal erosion (again, see Chapter 6). Data need to be collected on the use of the communication routes concerned, and the diversions to be used at times of flooding or other dislocation. If these communication benefits are substantial, the analysis can be complex and time consuming, but yield very significant benefit totals. Thus, time needs to be reserved for this activity if it is shown in initial investigations or studies that these benefits will be significant in the subsequent decision-making.

1.7.3 Tasks that can be accomplished more quickly

Obviously the length of time needed to complete any analysis depends on the scale of the problem and the available resources. Nevertheless, tasks that can be accomplished within three months are simple land use surveys of flood risk benefit areas (often using web-based data and/or the Environment Agency's National Receptor Dataset (Environment Agency, undated c)) and a small number of site-specific interviews. Likewise, a basic survey of the impacts of erosion on cliff-tops can be undertaken within this time frame, assuming the data on property prices can be obtained reasonably quickly.

However, the number of appraisals that can be completed within a three-month period is likely to be very small indeed. This is because of the need, within the appraisal process, to link benefit assessment with hydrological and hydraulic calculations, as well as integrating this part of project appraisal with the general area of environmental impact assessment and obtaining full stakeholder involvement.

A clear implication of this is that in nearly all cases the users of this Manual should begin the preparation of studies for the benefit assessments required within benefit–cost analysis at least a full year in advance of the date when the results are required.

1.7.4 Resources for surveys and data processing

For many of the users of this Manual the types of surveys and data collection involved will be new or infrequently deployed. This means that in almost all cases the relevant personnel will require staff training and supervision.

The most skilled tasks required in the use of this Manual are the site-specific flood damage surveys of non-residential properties for which trained and experienced interviewers will be needed (Chapter 5). In addition, for the contingent valuation surveys of recreational benefits it will normally be necessary to recruit interviewers with experience of market research surveys, who should be properly briefed and trained for the purpose (Chapter 8).

The conduct of farm interviews to assess the agricultural benefits of flood risk management or coastal erosion risk management (Chapter 9) requires a high level of experience, and experts should be used from academic organisations (e.g. Cranfield University) or agricultural consultants.

On the other hand, land use data for areas liable to flooding can be assembled by relatively inexperienced technicians, using the Environment Agency's National Receptor Dataset, but these staff will need to be briefed with reference to the relevant chapters of this Manual (Chapters 4, 5 and 6) and the relevant Land Use Classifications (see MCM-Online Supporting Information for Chapter 3).

With regard to data processing, computerised data analysis is recommended for all but the simplest and quickest initial studies. Many consultants working for the Environment Agency have developed their own software packages based on spreadsheets or similar computational systems, so that this type of analysis has now become relatively routine. In addition, the Environment Agency (2010a) appraisal guidance includes spreadsheets which allow rapid calculations of the benefits of particular flood risk management options and likely Grant in Aid ('partnership funding') contributions. These developments have revolutionised the task of calculating the economic efficiency of particular schemes, and their computing power should be used to the full.

We recommend that the user decides at an early stage in the detailed study what systems are to be used for data processing, and that the data collection and survey systems are then geared to that decision.

1.8 ESTIMATING SCHEME COSTS

This Manual is predominantly concerned with the estimation of the benefits of flood risk management and coastal erosion risk management plans and schemes. However, the reliability of a benefit–cost comparison depends crucially on the accurate estimation of costs. All expected whole-life costs of works and management are required to be included in this, including capital, maintenance and operating costs. Maintenance costs should include the expected value of the costs of repairs as a result of flood and storm damage and, where significant, decommissioning costs. Environment Agency and Defra appraisal guidance (Environment Agency 2010a; Defra, 2009; MAFF, 1999) provides guidance on the assessment of costs, including what to include and what not to include and how to handle issues such as optimism bias in estimation.

1.9 UPDATING

Most of the data contained in this Manual are given at 2013/14 prices or as otherwise specified, in which case the relevant date is given with each data item.

The accompanying MCM-Online will have data updated on a regular basis, so that the user can turn there for assistance when up-to-date information is required. The exception is data on agriculture benefits, for which the user is referred to sources such as *Farm Management Pocketbook* (e.g. Nix, 2012, and subsequent editions) or to Cranfield University, but there will be few instances in the current Defra policy regime (Defra, 2005) when flood or coastal risk management schemes will depend on agricultural benefits alone (or even to a major extent).

However, as indicated, the MCM-Online will not have all data accessible free of charge. Therefore, if users wish not to use that resource, or for some other reason find it inaccessible, we recommend that within two or three years of 2013/14 updating can be undertaken simply using the most appropriate index series. In most cases this is the Consumer Price Index (CPI), available from the Office for National Statistics and its website.

1.10 WHAT IS NEW IN THIS NEW 'MULTI-COLOURED MANUAL'

1.10.1 What was new in the 2010 Multi-Coloured Handbook (Penning-Rowsell *et al.*, 2005b) and is now incorporated herein

The production of a series of manuals can lead to confusion. In between the 2005 edition of the Manual and this volume, published in 2013, came the 2010 'Multi-Coloured Handbook'. In that 2010 edition, changes since 2005 were noted, and they are reproduced here, in outline, so that the reader can appreciate what changes were made, and when:

- Major floods in 2007 had led to research that has altered our understanding of the costs of emergency services in flood incidents, and hence questioned the universality of the 10.7 per cent uplift factor recommended in 2005 (see also Chapter 6).
- The 2007 floods also led to substantial disruption of electricity and water supplies, and these need more emphasis now, especially their off-floodplain effects (again, see Chapter 6).
- The land use data available for benefit assessments continue to improve, with the Environment Agency's National Property Database (NPD) 3 and its 2010 successor, the National Receptor Dataset (NRD). This reduces the need for expensive, time-consuming, field-based surveys.
- Climate change impacts on flood frequency are more fully understood now than in 2005 and need to be factored in to assessments of the return periods of future floods (see Environment Agency, undated b).

Also new on the 2010 MCM CD, however, was:

- Better data on the damage to household inventory items as a result of more exhaustive analysis of the ownership and value of such items. This shows under-inflation increases in these values, reflecting the declining real cost of 'white goods' and other household contents.
- A complete reworking of the flood loss data for the building fabric of residential properties, using the same building surveyor and quantity surveyor as we have used for the last twenty-plus years. This shows over-inflation increases in these values, reflecting the labour-intensive nature of this work.
- Much better data on the costs of clean-up operations following floods, using staff from the National Flood School who, in turn, consulted widely in the relatively new clean-up and recovery 'industry'. Here we see lower values than in 2005, reflecting greater competition between the firms involved.
- Updated data on the damage to non-residential properties (NRPs), based on a range of updating indices produced by the Office for National Statistics (ONS). This updating is thus not as fundamental as that for flood damage to residential properties (Chapter 4) but has been completely superseded by the data researched for this volume (see Chapter 5).
- The values used to calculate road traffic disruption costs were updated according to the latest Department for Transport data.

In the chapter on agriculture for the 2010 'Multi-Coloured Handbook' we noted that recent concerns about global food security and the possible impacts of climate change had led to some renewed interest in maintaining the productivity of British agriculture. We noted also that almost 60 per cent of Grade 1 agricultural land in England is dependent on flood risk management and land drainage, although research on the 2007 floods (Chatterton *et al.*, 2010) showed that agricultural losses then amounted to less than 1.6 per cent of total losses.

1.10.2 What is new for 2013/14 (this volume)

This new volume of the 'Multi-Coloured Manual' is based on research commissioned between 2012 and 2013 by the Environment Agency to enhance the database of the Manual and provide greater accessibility to the data that it contains.

By far the biggest change has been to the data on non-residential property flood damage. These have been completely reworked with a different methodology from that existing previously, which had relied upon interviews with major non-residential property owners to assess potential damage. The new approach (Chapter 5) is more akin to that developed in the residential sector, where 'synthetic' depth/damage datasets are assembled from the accumulated knowledge of the impacts of floods upon the structure of the relevant building and its contents. This approach enlarges very considerably the sample size of the database thereby constructed, and adds significantly to its rigour. Whilst the changes have been substantial, this is an ongoing process, and data will continuously be updated from time to time and published on MCM-Online.

In addition, a more systematic approach has been taken to the assessment of flood losses described in Chapter 6. In particular, we have pursued more rigorously damage caused to roads and rail infrastructure, while also addressing the question of the spread of these effects off the floodplain into areas previously not analysed. This was the direct result of the impacts of the 2007 floods, where it was systematically noted that major infrastructure was at risk and the impacts of this, if flooded, would spread far beyond the locations of the infrastructure itself.

A new entry has been provided in the residential Chapter 4 on flood damage to park homes. These have not been considered previously, but unfortunately such residential developments can be located in flood risk areas, and at the coast, and are particularly vulnerable to flood damage owing to the relatively insubstantial nature of their construction. Nevertheless, as reported in Chapter 4, the losses as recorded here are relatively low, because the damage or losses are assessed in relation to the remaining value of the park home, rather than its value when new or its replacement cost.

In Chapter 9 we have produced a simplified procedure for assessing the benefits of managing flood and coastal risks with regard to agricultural land. This follows feedback on the methods we advocated in 2005, which stressed their complexity and resource-intensive characteristics. We believe that the simplified version we describe herein is no less rigorous than that which we provided previously, and we would welcome feedback on the changes we have made, even though they will be rarely used.

Whilst every chapter within this Manual has been thoroughly reviewed and updated, much of the material in Chapters 7 and 8 remains relevant and is unchanged as to its fundamentals in this volume compared with the one we published in 2005. Nevertheless, in both chapters we have referred to new research in the area of assessing recreational benefits of flood and coastal erosion risk management, and to documents related to the approach to evaluating intangible benefits in the form of 'ecosystems services'.

Users will be reassured to learn that the fundamental approach to assessing the benefits of flood and coastal risk management have not changed in the last decade, and therefore the contents of Chapters 2 and 3 remain essentially the same, although updated in many parts. What has changed is the extent to which the UK Government (Defra, 2009) and the Environment Agency (2010a) have provided far more sophisticated guidance to the user than previously, and this is referred to continuously throughout the chapters within this Manual.

1.10.3 Other relevant research at the Flood Hazard Research Centre since 2005

The Flood Hazard Research Centre at Middlesex University has continued to be active in the last decade, with research on a number of aspects of flood and coastal erosion risk management.

Table 1.6 gives an extended list of the publications that have resulted. To summarise, the following themes have been pursued in some detail:

- the effectiveness and cost effectiveness of flood warning systems;
- the better management of flooding incidents;
- the impacts of flood on human health;
- who gains and who loses from flood risk management;
- the integration of the social sciences and the environmental sciences when developing policies and schemes, particularly at the coast;
- the loss of life in floods;
- the communication of flood risks, both between professionals and with the public, including via flood maps;
- stakeholder engagement issues, and the development of better stakeholder engagement processes; and
- governance aspects of flood and coastal risk management.

In addition, the Flood Hazard Research Centre has been intimately involved in the major Foresight projects on both flooding and hazard-induced migration (e.g. Evans *et al.*, 2004; Penning-Rowsell *et al.*, 2012) and in the EU Floodsite project, our reports from which can be found at http://www.floodsite.net/html/publications2.asp.

We hope that you find all this research interesting and useful.

Table 1.6 Flood Hazard Research Centre publications (selected) 2005–13

2005
Green, C. (2005) 'The Social Construction of Floods', in Knight, D.W and Shamseldin, A.Y. (eds), *River Basin Modelling for Flood Risk Mitigation,*Taylor & Francis, London, pp519–536
Johnson, C.L., Tunstall, S.M. and Penning-Rowsell, E.C. (2005) 'Floods as catalysts for policy change: Historical lessons from England and Wales', *Water Resource Development,* vol 21, no 4, pp561–575
Parker, D.J. and Penning-Rowsell, E.C. (2005) 'Dealing with Disasters', in Hunt, J. (ed.), *London's Environment: Prospects for a Sustainable World City,* Imperial College Press, London, Chapter 13, pp175–201
Parker, D.J., Tunstall, S.M. and Wilson, T. (2005) 'Socio-economic benefits of flood forecasting and warning', International Conference on Innovation Advances and Implementation of Flood Forecasting Technology, 17–19 October 2005, Tromso, Norway
Penning-Rowsell, E.C., Floyd, P., Ramsbottom, D. and Surendran, S. (2005) 'Estimating injury and loss of life in floods: A deterministic framework', *Natural Hazards,* vol 36, pp43–64
Penning-Rowsell, E.C., Tapsell, S. and Wilson, T. (2005) 'Key policy implications of the health effects of floods', in Kirch, W., Menne, B. and Bertollini, R. (eds), *Extreme Weather and Climate Events: Risks to Human Health and Public Health Responses,* Springer-Verlag, Berlin/Heidelberg, pp207–223
Pottier, N., Penning-Rowsell, E.C., Tunstall, S.M. and Hubert, G. (2005) 'Land use and flood protection: Contrasting approaches and outcomes in France and in England and Wales', *Applied Geography,* vol 25, no 1, pp1–27
Priest, S.J., Clarke, M.J. and Treby, E.J. (2005) 'Flood insurance: The challenge of the uninsured', *Area,* vol 37, no 3, pp295–302
2006
Ebi, K.L., Hajat, S., Penning-Rowsell, E.C., Kovats, R.S. and Wilson, T. (2006) 'Floods and human health', in Menne, B. and Ebi, K.L. (eds), *Climate Change and Adaptation Strategies for Human Health,* World Health Organization Regional Office for Europe 2006, Steinkopff Darmstadt, pp99–127
Evans, E.P., Hall, J.W., Penning-Rowsell, E.C., Sayers, P.B., Thorne, C.R. and Watkinson, A. (2006) 'Future flood risk management in the UK', *Proceedings of the Institution of Civil Engineers: Water Management,* vol 159, pp53–61
McFadden, L.M., Nicholls, R.J. and Penning-Rowsell, E.C. (eds) (2006) *Managing Coastal Vulnerability,* Emerald Group Publishing Limited, Bingley
Penning-Rowsell, E.C., Johnson, C.L. and Tunstall, S.M. (2006) 'Signals from pre-crisis discourse: Lessons from UK flooding for global environmental policy change', *Global Envrionmental Change A,* vol 16, no 4, pp323–339
Priest, S.J., Clark, M.J. and Treby, E.J. (2006) 'Confronting flood risk: Implications for insurance and risk transfer', *Journal of Environmental Management,* vol 81, no 4, pp351–359
Tapsell, S.M. and Tunstall, S.M. (2006) 'The mental health aspects of floods: Evidence from England and Wales', in Few, R. and Matthies, F. (eds), *Flood Hazards and Health: Responding to Present and Future Risks,* Earthscan, London, pp89–110
Tunstall, S., Tapsell, S., Green, C., Floyd, P. and George, C. (2006) 'The health effects of flooding: Social research results from England and Wales', *Journal of Water and Health,* vol 4, no 3, pp365–380
Tunstall, S.M. and Eden, S. (2006) 'Ecological versus social restoration? How urban river restoration challenges but also fails to challenge the science-policy nexus in the United Kingdom', *Environment and Planning C: Government and Policy,* vol 24, pp661–680
2007
De Brujin, K., Green, C., Johnson, C. and McFadden, L. (2007) 'Evolving concepts in flood risk management: Searching for a common language', in Begum, S., Stive, M.J.F. and Hall, J.W. (eds), *Advances in Natural and Technological Hazards Research. Flood Risk Management in Europe: Innovation in Policy and Practice,* vol 25, pp61–75
Drobot, S. and Parker, D.J. (2007) 'Advances and challenges in flash flood warnings', *Environmental Hazards,* vol 7, pp173–178
Faulkner, H.F. and Ball, D. (2007) 'Environmental hazards and risk communication', *Environmental Hazards,* vol 7, no 2, pp71–78

Table 1.6 continued

Green, C. and McFadden, L. (2007) 'Coastal vulnerability as discourse about meaning and values', *Journal of Risk Research*, vol 10, no 8, pp1027–1045
Green, C., Donner, E., Faulkner, H., Garlick, H., Jones, H., Revitt, M. and Scholes, L. (2007) 'Water and health', in *Royal Society of Chemistry Sustainable Water: Chemical Science Priorities Report*, Royal Society of Chemistry, London, n.p.
Johnson, C.L., Parker, D.J. and Penning-Rowsell, E.C. (2007) 'Natural and imposed injustices: The challenges in implementing "fair" flood risk management policy in England', *The Geographical Journal*, vol 173, no 4, pp374–390
Johnson, C.L., Penning-Rowsell, E.C. and Tapsell, S.M. (2007) 'Aspirations and reality: Flood policy, economic damages and the appraisal process', *Area*, vol 39, no 2, pp214–223
McCarthy, S. (2007) 'Contextual influences on national level flood risk communication', *Environmental Hazards*, vol 7, no 2, pp128–140
McCarthy, S., Faulkner, H., Parker, D.J., Tunstall, S.M. and Howe, J. (2007) 'Risk communication in emergency response to a simulated extreme flood', *Environmental Hazards*, vol 7, no 3, pp179–192
McFadden, L. (2007) 'Governing coastal spaces: The case of disappearing science in Integrated Coastal Zone Management', *The Coastal Management Journal*, vol 35, no 4, pp429–443
McFadden, L. (2007) 'Vulnerability analysis in environmental management: Widening and deepening its approach', *Environmental Conservation*, vol 34, no 3, pp195–204
McFadden, L. and Green, C.H. (2007) 'Defining "vulnerability": Conflicts, complexities and implications for Coastal Zone Management', *Journal of Coastal Research*, special issue 50, pp120–124
McFadden, L., Nichols, R.J. and Spencer, T. (2007) 'Broad-scale modelling of coastal wetlands: What is required?', *Hydrobiologia*, vol 577, no 1, pp5–15
McFadden, L., Nicholls, R.J., Tol, R.S.J. and Vafesdis, A.T. (2007) 'A methodology for modelling coastal space for global assessment', *Journal of Coastal Research*, vol 23, no 4, pp911–920
Parker, D.J., Beven, K., Faulkner, H. and Green, C.H. (2007) 'Developing a translational discourse to communicate uncertainty in flood risk between science and the practitioner', *Ambio*, vol 36, no 7, pp692–704
Parker, D.J., Tunstall, S.M. and McCarthy, S. (2007) 'New Insights into the benefits of flood warnings: Results from a household survey in England and Wales', *Environmental Hazards*, vol 7, no 3, pp193–210
Tapsell, S.M. and Ball, D. (2007) 'Flood event management', in Thorne, C.R. and Evans, E.P. (eds), *Future Flooding and Coastal Erosion Risk*, Thomas Telford, London, pp340–359
Thorne, C., Evans E.P. and Penning-Rowsell, E.C. (eds) (2007) *Future Flooding and Coastal Erosion Risks*, Thomas Telford, London
2008
Harries, T. (2008) 'Feeling secure or being secure? Why it can seem better not to protect yourself against a natural hazard', *Health, Risk & Society*, vol 10, no 5, pp479–490
Johnson, C.L. and Priest, S.J. (2008) 'Flood risk management in England: A changing landscape of risk responsibility', *International Journal of Water Resources Development*, vol 24, no 4, pp523–525
Jonkman, S.N. and Penning-Rowsell, E.C. (2008) 'Human instability in flood flows', *Journal of the American Water Resources Association*, vol 44, no 4, pp1–11
Lonsdale, K.G., Downing, T.E., Nicholls, R.J., Parker, D., Vafeidis, A.T., Dawson, R. and Hall, J. (2008) 'Plausible responses to the threat of rapid sea-level rise in the Thames Estuary', *Climate Change*, vol 91, nos 1–2, pp145–169
McCarthy, S., Penning-Rowsell, E. and Tunstall, S. (2008) 'Public attitudes to "community based" small scale flood risk reduction measures in England: A case study in the Lower Thames catchment', in Bosher, L.S. (ed.), *Hazards and the Built Environment: Attaining Built-in Resilience*, Taylor & Francis, London, pp150–171
McFadden, L. (2008) 'Exploring the challenges of integrated coastal zone management and reflecting on contributions to "integration" from geographical thought', *The Geographical Journal*, vol 174, no 4, pp299–314

Table 1.6 continued

McFadden, L. and Green, C. (2008) 'Is stakeholder engagement too much like hard work?', *Proceedings of Solutions to Coastal Disasters 2008 Conference, April 13–16 2008*, pp968–979
Parker, D.J., Priest, S., Schildt, A. and Handmer, J. (2008) *Modelling the Damage Reducing Effects of Flood Warnings – Final Report*, FLOODsite Report No. T10-07-12
Sultana, P. and Abeyasekera, S. (2008) 'Effectiveness of participatory planning for community management of fisheries in Bangladesh', *Journal of Environmental Management*, vol 86, no 1, pp201–213
Sultana, P., Johnson, C. and Thompson, P. (2008) 'The impact of major floods on flood risk policy evolution: Insights from Bangladesh', *International Journal River Basin Management*, vol 6, no 4, pp339–348
Sultana, P., Thompson, P. and Green, C. (2008) 'Can England learn lessons from Bangladesh in introducing participatory floodplain management', *Water Resources Management*, vol 22, pp357–376
Tapsell, S. and Tunstall, S.M. (2008) 'I wish I had never heard of Banbury: The relationship between place and the health impacts from flooding', *Health and Place*, vol 14, no 2, pp133–154
Vafeidis, A.T., Nicholls, R.J., McFadden, L., Tol, R.S.J., Hinkel, J., Spencer, T., Grashoff, P.S., Boot, G. and Klein, R.J.T. (2008) 'A new global coastal database for impact and vulnerability analysis to sea-level rise', *Journal of Coastal Research*, vol 24, no 4, pp917–924
2009
Hutter, G. and McFadden, L. (2009) 'Strategic planning for long-term flood risk management: Findings from case studies in Dresden and London', in Samuels, P., Huntingdon, S., Allsop, W. and Harrop, J. (eds), *Flood Risk Management Research and Practice*, CRC Press, Netherlands, pp865–875
Johnson, C., Tunstall, S., Priest, S., McCarthy, S. and Penning-Rowsell, E. (2009) 'What's "fair" about flood and coastal erosion risk management? A case study evaluation of policies and attitudes in England', in Samuels, P., Huntingdon, S., Allsop, W. and Harrop, J. (eds), *Flood Risk Management Research and Practice*, CRC Press, Netherlands, pp1503–1512
McFadden, L., Penning-Rowsell, E. and Tapsell, S. (2009) 'Strategic coastal flood-risk management in practice: Actors perspectives on the integration of flood risk management in London and the Thames Estuary', *Ocean & Coastal Management*, vol 52, no 12, pp636–645
Parker, D.J., Priest, S.J. and Tapsell, S.M. (2009) 'Understanding and enhancing the public's behavioural response to flood warning information', *Meteorological Applications*, vol 16, no 1, pp103–114
Priest, S.J., Parker, D.J. and Tapsell, S. (2009) 'New developments in maximizing flood warning response and benefit strategies', in Samuels, P., Huntingdon, S., Allsop, W. and Harrop, J. (eds), *Flood Risk Management Research and Practice*, CRC Press, Netherlands, pp1695–1705
Steinführer, A., Kuhlicke, C., De Marchi, B., Scolobig, A., Tapsell, S. and Tunstall, S. (2009) 'Towards flood risk management with the people at risk: From scientific analysis to practice recommendations (and back)', in Samuels, P., Huntingdon, S., Allsop, W. and Harrop, J. (eds), *Flood Risk Management Research and Practice*, CRC Press, Netherlands, pp945–955
Tapsell, S.M., Priest, S.J., Wilson, T., Viavattene, C. and Penning-Rowsell, E.C. (2009) 'A new model to estimate risk to life for European flood events', in Samuels, P., Huntingdon, S., Allsop, W. and Harrop, J. (eds), *Flood Risk Management Research and Practice*, CRC Press, Netherlands, pp933–943
Twigger-Ross, C., Fernandez-Bilbao, A., Colbourne, L., Tapsell, S., Watson, N., Kashefi, E., Walker, G. and Medd, W. (2009) 'Putting people and places at the centre: Improving institutional and social responses to flooding', in Samuels, P., Huntingdon, S., Allsop, W. and Harrop, J. (eds), *Flood Risk Management Research and Practice*, CRC Press, Netherlands, pp849–854
Twigger-Ross, C., Fernandez-Bilbao, A., Walker, G., Deeming, H., Kasheri, E., Watson, N. and Tapsell, S. (2009) 'Flood warning in the UK: Shifting the focus', in Samuels, P., Huntingdon, S., Allsop, W. and Harrop, J. (eds), *Flood Risk Management Research and Practice*, CRC Press, Netherlands, pp1163–1171
Watson, N., Kashefi, E., Medd, W., Walker, G., Tapsell, S. and Twigger-Ross, C. (2009) 'Institutional and social responses to flooding from a resilience perspective', in Samuels, P., Huntingdon, S., Allsop, W. and Harrop, J. (eds), *Flood Risk Management Research and Practice*, CRC Press, Netherlands, pp911–917

Table 1.6 continued

2010
Faulkner, H. (2010) 'Muddy floods: Challenges for the future. Short communication', *Land Degradation & Development*, vol 21, no 4, pp406–411
Green, C. (2010) 'Towards sustainable flood risk management', *International Journal of Disaster Risk Science*, vol 1, no 1, pp33–43
McFadden, L. (2010) 'Coastal hazard, vulnerabilities and resilience', *Environmental Hazards*, vol 9, pp217–221
McFadden, L. (2010) 'Exploring system interactions for building resilience within coastal environments and communities', *Environmental Hazards*, vol 9, no 3, pp266–283
Steinführer, A,. Kuhlicke, C., De Marchi, B., Scolobig, A., Tapsell, S. and Tunstall, S. (2010) *Local Communities at Risk from Flooding Social Vulnerability, Resilience and Recommendations for Flood Risk Management in Europe*, FLOODsite Report, T11-10-01
Sultana, P. and Thompson, P. (2010) 'Local institutions for floodplain management in Bangladesh and the influence of the Flood Action Plan', *Environmental Hazards*, vol 9, no 3, pp26–42
2011
Harries, T. and Penning-Rowsell, E. (2011) 'Victim pressure, institutional inertia and climate change adaptation: The case of flood risk', *Global Environmental Change*, vol 21, pp188–197
Kuhlicke, C., Scolobig, A., Tapsell, S., Steinfuhrer, A. and De Marchi, B. (2011) 'Contextualizing social vulnerability: Findings from case studies across Europe', *Natural Hazards*, vol 58, no 2, pp789–810
Kuhlicke, C., Steinführer, A., Begg, C., Bianchizza, C., Brundl, M., Buchecker, M., De Marchi, B., Di Masso, M., Di Masso Tarditti, M., Hoppner, C., Blaz, K., Lemkowe, L., Luther, J., McCarthy, S., Pellizzoni, L., Renn, O., Scolobig, A. and Supramaniam, M. (2011) 'Perspectives on social capacity building for natural hazards: Outlining an emerging field of research and practice in Europe', *Environmental Science and Policy*, vol 14, pp804–814
Nye, M., Tapsell, S. and Twigger-Ross, C. (2011) 'New social directions in UK flood risk management: Moving towards flood risk citizenship?', *Journal of Flood Risk Management*, vol 4, no 4, pp288–297
Pardoe, J., Penning-Rowsell, E. and Tunstall, S. (2011) 'Floodplain conflicts: Regulation and negotiation', *Natural Hazards and Earth Systems Sciences*, vol 11, pp2889–2902
Parker, D.J., Priest, S. and McCarthy, S.S. (2011) 'Surface water flood warning requirements and potential in England and Wales', *Applied Geography*, vol 31, no 3, pp891–900
Priest, S., Pardoe, J., McCarthy, S. and Viavattene, C. (2011) 'Narrowing the focus: Public understanding of flood maps in the River Thames catchment', in Zenz, G. and Hornich, R. (eds), *Proceedings of the International Symposium 21st–23rd September 2011 – Graz, Austria. URFRIM – Urban Flood Risk Management Approaches to Enhance the Resilience of Communities*, pp309–320
Priest, S.J., Parker, D.J. and Tapsell, S.M. (2011) 'Modelling the potential damage-reducing benefits of flood warnings using European cases', *Environmental Hazards*, vol 10, pp101–120
Priest, S.J., Parker, D.J., Hurford, A.P., Walker, J. and Evans, K. (2011) 'Assessing options for the development of surface water flood warning in England and Wales', *Journal of Environmental Management*, vol 92, pp3038–3048
2012
Bubeck, P., Kreibich., H., Penning-Rowsell, E.C., Botzen, W.J.W., De Moel, H. and Klijn, F. (2012) 'Explaining differences in flood management approaches in Europe and the USA: A comparative analysis', in Klijn, F. and Schweckendiek, T. (eds), *Comprehensive Flood Risk Management, Research for Policy and Practice*, CRC Press, Netherlands, p302
Lavery, S., Penning-Rowsell, E.C. and Haigh, N. (2012) 'Protecting London from the sea: 21st century policies, plans and their implications', in Klijn, F. and Schweckendiek, T. (eds), *Comprehensive Flood Risk Management, Research for Policy and Practice*, CRC Press, Netherlands, p288
Hurford, A.P., Priest, S.J., Parker, D.J. and Lumbroso, D. (2012) 'The effectiveness of extreme rainfall alerts as surface water flood warnings in England and Wales', *International Journal of Climatology*, vol 32, no 11, pp1768–1774

Table 1.6 continued

Menoni, S., Molinari, D., Parker, D., Ballio, F. and Tapsell, S. (2012) 'Assessing multifaceted vulnerability and resilience in order to design risk-mitigation strategies', *Journal of the International Society for the Prevention and Mitigation of Natural Hazards*, vol 64, no 3, pp2057–2083
Meyer, V., Kuhlicke, C., Luther, J., Fuchs, S., Priest, S., Dorner, W., Serrhini, K., Pardoe, J., McCarthy, S., Seidel, J., Palka, G., Unnerstall, H., Viavattene, C. and Scheuer, S. (2012) 'Recommendations for the user-specific enhancement of flood maps', *Natural Hazards and Earth System Sciences*, vol 12, no 5, pp1701–1716
Meyer, V., Priest, S. and Kuhlicke, C. (2012) 'Economic evaluation of structural and non-structural flood risk management measures: Examples from the Mulde River', *Natural Hazards*, vol 62, pp301–324
Parker, D.J. and Priest, S.J. (2012) 'The fallibility of flood warning chains: Can Europe's flood warnings be effective?', *Water Resource Management*, vol 26, pp2927–2950
Penning-Rowsell, E.C. (2012) 'A "realist" approach to the extent of flood risk in England and Wales', in Klijn, F. and Schweckendiek, T. (eds), *Comprehensive Flood Risk Management, Research for Policy and Practice*, CRC Press, Netherlands, p162
Penning-Rowsell, E.C. and Pardoe, J. (2012) 'Who benefits and who loses from flood risk reduction?', *Government and Policy 2012*, vol 30, no 3, pp448–466
Penning-Rowsell, E.C. and Pardoe, J. (2012) 'Who loses if flood risk is reduced: Should we be concerned?', *Area*, vol 44, no 2, pp152–159
Penning-Rowsell, E.C., Haigh, N., Lavery, S. and McFadden, L. (2012) 'A threatened world city: The benefits of protecting London from the sea', *Natural Hazards*, vol 66, no 3, pp1383–1404
Penning-Rowsell, E.C., Sultana, P. and Thompson, P.M. (2012) 'The "last resort"? Population movement in response to climate-related hazards', *Environmental Science & Policy*, vol 27, no 1, ppS60–S75
Sayers, P., Galloway, G., Penning-Rowsell, E.C., Shen, Y.Y.F., Chen, Y., Wang, T., Le Quense, T., Wang, L. and Guan, Y. (2012) 'Strategic flood risk management: Comparing international challenges and developing common approaches', in Klijn, F. and Schweckendiek, T. (eds), *Comprehensive Flood Risk Management, Research for Policy and Practice*, CRC Press, Netherlands, pp230–231
Tseng, C.-P. and Penning-Rowsell, E.C. (2012) 'Micro-political and related barriers to stakeholder engagement in flood risk management', *The Geographical Journal*, vol 178, pp253–269
Viavattene, C. and Faulkner, H. (2012) 'An uncertainty index to measure the feasibility of whole-life cycle costing approach in flood risk management', *Journal of Flood Risk Management*, vol 5, no 3, pp215–225
Walker, M., Whittle, R., Med, W., Burningham, K., Moran-Ellis, J. and Tapsell, S. (2012) 'It came up to here: Learning from children's flood narratives', *Children's Geographies*, vol 10, no 2, pp135–150
2013
Aberg, U. and Tapsell, S.M. (2013) 'Revisiting the River Skerne: The long-term social benefits of river rehabilitation', *Landscape and Urban Planning*, vol 113, pp94–103
Alexander, M., Viavattene, C., Faulkner, H. and Priest, S. (2013) 'Translating the complexities of flood risk science using KEEPER: A knowledge exchange exploratory tool for professionals in emergency management', *Journal of Flood Risk Management*, http://onlinelibrary.wiley.com/doi/10.1111/jfr3.12042/full, accessed 13 June 2013
Penning-Rowsell, E.C. (2013) *Foresight Flooding: A Retrospective Analysis of Impact*, Foresight project report
Penning-Rowsell, E.C., Yanyan, W., Watkinson, A.R., Jiang, J. and Thorne, C. (2013) 'Socio-economic methodologies for the Taihu Basin, China', *Journal Flood Risk Management*, vol 6, no 1, pp23–32

2 Using appraisals to make better choices

This chapter presents key points on how to improve decisions through project appraisal, and is in part structured to consider six key questions covering the project appraisal process:

- What is project appraisal?
- Why do project appraisals?
- How to involve stakeholders?
- What is value?
- How to compare options?
- How to make the decision?

At the beginning, therefore, we need to ask ourselves the question: what is project appraisal? In this respect Defra's guidance (Defra, 2009) suggests that it is the process of identifying and then evaluating options in order to select the one that most likely satisfies the defined project objectives. The purpose of the project appraisal process is to improve decision-making towards making the 'best' choice. Good decisions and 'best' choices are most likely to result from considering all economic, social, environmental and technical issues for a full range of options, although the effort in doing so must always be proportionate to the likely gains in the understanding obtained.

The methods used in project appraisal are therefore aimed at:

1 Simplifying the complexity of choice.
2 Understanding what choice involves.
3 And enabling this understanding to be shared by stakeholders.

To ensure that project appraisal is not a mechanical exercise, appraisal-led design is essential. Appraisals should drive the design process, with the identification and specification of project options evolving through the appraisal process. To be useful, appraisal methods should ensure best value and hence the highest rate of return for pubic monies. They must also provide accountability, transparency of basis for choice, and result in a rational comparison of the available options and their consequences. Finally, the effort expended on appraisal must be proportionate to the task in hand: we should not extravagantly pursue details that will not affect the decision that we are about to take: an appraisal and its techniques should themselves be the subject of some sort of an analysis of costs and benefits.

2.1 INTRODUCTION

This chapter sets out the theoretical context for the other chapters. But the 'how to do it' questions raised here are logically consequent upon the answer to the 'why to do it' question.

So, this chapter begins by discussing the latter question and goes on to assert that there are three basic reasons for undertaking formal project appraisals. First, we want to make better choices than we might make otherwise. Choice is argued to be a learning process; we have to discover which course of action is the best to adopt. An important part of that learning process is the invention or discovery of different options. A key requirement for a project appraisal is thus that we have a better understanding of what the choice involves than we had before. The expectation is that consequently we will find a better option than those which are initially identified.

The choice we make in one instance restricts the choice that can be made in another instance; the money spent upon a flood risk management scheme in Yorkshire is consequently not available to be spent upon a flood risk management scheme in Sussex, or upon undertaking hip replacements in the National Health Service. So the second reason for adopting a formal project appraisal procedure is the need for accountability; we are spending other people's money, and a formal project appraisal technique both provides consistency within and between choices and establishes an audit trail. When other people read the project appraisal report, it should be clear to them both why the particular selected option was chosen and that this choice was made through a logical and rigorous process.

Third, there is a requirement to engage with the different stakeholders in making decisions (Cabinet Office, undated) and hence discussion, argument and negotiation as to what course of action should be adopted. An increasingly important function of project appraisal techniques is therefore to support and enable these debates by providing a means through which a common understanding of the nature of the choice that must be made can be shared or argued. Since it will never be possible for all the potential stakeholders to be engaged in these debates – particularly for all the taxpayers who will share the cost of the course of action adopted – a formal project appraisal process provides a way of incorporating their interests into the choice. Equally, it is in their interests that it is necessary to provide an audit trail.

Economics can be defined as the application of reason to choice; so economics seeks to develop a logical and rigorous framework with which to approach choices. The force of economics lies in the extent to which we can use economic methods to make 'better' choices.

This chapter therefore has three main purposes:

- To outline the nature of choice.
- To discuss what we may mean by a 'better' choice.
- To outline how economics conventionally approaches these two issues, including the limitations of the conventional economic approach.

A final purpose is to introduce a number of key terms of economic language; some words are used in economics in quite different ways to their everyday meanings.

Because monetary evaluation is the most widely used system of valuing all forms of resource use and widely different types of benefit, the accepted form of appraisal for most public investment situations is benefit–cost analysis (for decision support, not decision-making). This is recognised in HM Treasury guidance (HM Treasury, 2003) and is reflected in the Defra project appraisal policy (Defra, 2009), both of which stress important aspects of the form of this benefit–cost analysis.

However, it is recognised that there are situations where critical aspects of a decision cannot easily be valued in monetary terms, and in these cases other techniques may need to be employed to help reach a decision. Such techniques include multi-criteria analysis (scoring and weighting), and an example is given later in this Manual, in Chapter 10 (the TE2100 project).

2.2 WHY DO WE HAVE TO CHOOSE?

2.2.1 What are choices?

If we want to make better decisions – choices – in flood and coastal erosion risk management, then we need to start by understanding why we have to make the particular choices in the first place. Choices have a number of important characteristics:

- All choices are prospective; they are always attempts to choose the future, even when that future is the immediate future. In flood and coastal erosion risk management projects we are looking to the future over perhaps the next one hundred years.
- Choices involve power: who has the power to influence the outcome? In applying reason to choice, we are asserting that reason is the appropriate form of power to adopt.
- We want to make the 'right' choice in both senses of the word: a choice that is *both* fair, equitable or just *and* technically correct; to select that option which is most likely to be fair, equitable or just (see Johnson *et al.*, 2007) from a long-term perspective – it is that option which increases well-being.

- Choice is a learning process: we seek to discover which option should be preferred.
- Choices are always about what to do: which course of action to adopt.
- Choices are always comparative and relative: we have to choose between alternative courses of action.

In turn, for a choice to exist, there have to be at least two mutually exclusive and desirable options: we can imagine few circumstances where only one option is available (and if there were, it would still need appraising!). Hence the first condition for the existence of choice is conflict or disagreement as to what to do. The second condition for the existence of a choice is uncertainty: if all are agreed that one option should be preferred over all others then the choice has already been made. A choice is thus a process through which we seek to resolve the conflicts that make the choice necessary and to become confident that one option should be preferred to all others. In the simplest terms a choice is required when there is conflict and uncertainty about a course of action (Green, 2003).

In terms of 'fairness', this is not normally an explicit aim of appraisal, even if it is important to those making choices. Defra, however, has indicated that 'As now, the future will require us to spend public money in ways that are fair; fair to those that live and work in areas at risk and to the majority of taxpayers that do not' (Defra, 2009, Foreword). And 'social justice, fairness and equity should always be carefully considered in the appraisal process' (Defra, 2009, 12).

2.2.2 What do we need to choose?

If we are all agreed which is the best option (i.e. there is no conflict or uncertainty) then only in the most trivial sense is there a choice to be made. Instead, we make a real choice when the options are mutually exclusive (i.e. they are either in conflict or spending on one choice means we cannot afford a second). This may occur because the options are mutually exclusive in time and/or space (e.g. a wetland and drained agricultural land cannot co-exist in the same space) or because different stakeholders bring different objectives to the choice, which may themselves be mutually exclusive. Those stakeholders typically disagree about the relative importance that should be given to the achievement of the different objectives.

In addition to these conflicts, or disagreements, choice is a function of decision-maker uncertainty, whereby 'uncertain' is defined as 'the rational state of not knowing what to do' (Green *et al.*, 2000a, 73). One reason why decision-makers are uncertain is that they cannot resolve the conflict between objectives and/or they cannot resolve the social argument as to what the objectives should be. The other reasons are that they are uncertain about the state of the world now, and the state of the world in the future.

In practice, all decisions are made when we are uncertain, simply because we are choosing a future over both the short and longer term. This uncertainty is especially relevant for flood and coastal erosion risk management projects with a typical life of thirty to fifty years. In thirty years' time the economy will be radically different to the present. If we then take into account possible changes in social values and systems of governance, as in the Foresight exercise (Evans *et al.*, 2004), the result will be societies and economies that are dramatically different to those in which today's decisions are made.

In short, we must accept that we will always be uncertain about the future. However, this does not mean that we must be uncertain about which option to choose; where one option is preferable to all others, under all conditions, there is no decision-maker uncertainty. Instead, recognising that the future is inherently uncertain influences how we make choices and which options we adopt.

Having recognised that there is a choice to be made, the scarcity of financial resources then forms an external constraint. This is particularly important in flood and coastal erosion risk management, which must compete with other societal needs for its share of general taxation and other sources of revenue.

Thus, even if there were universal agreement about the policies that should be adopted, resource constraints will limit the options that can be chosen. In reality, however, such universal agreement is unlikely and even if we had infinite resources there would still be conflicts between objectives, and disagreements between people as to the importance that should be attached to each objective. Hence the need for 'hard choices' would remain.

2.2.3 Summary

So, to summarise, if we want to make better decisions in flood and coastal risk management then we need to start by understanding why we have to make the particular decision or choice in the first place. This commences with identifying the problem and defining the objectives (Green, 1997).

In the simplest terms, a choice is required when there is conflict (i.e. disagreement) and uncertainty about a course of action to meet the defined objective(s). Uncertainly arises because of initial limited knowledge of an option's pros and cons (benefits and costs), and whether the 'best' choice to be made will be the most sustainable.

Economic appraisal enables the comparison of widely differing options, with careful consideration applied to how options are appraised as to their 'value' to arrive at the 'best' choice.

2.3 MAKING THE 'RIGHT' DECISION: STAKEHOLDER INVOLVEMENT AS A 'GIVEN'

A better – or 'right' – decision is one that is both a 'just' decision and one that turns out to be 'correct' in the long run. For a decision to be 'just', it is not only the outcome that must be seen as equitable but so too must the process by which the decision is made (Thibaut and Walker, 1975; Johnson *et al.*, 2007). Critical to the achievement of a 'just' process and a 'right' decision is stakeholder involvement. Thus Defra (that part formerly the DETR) observed:

> Public participation in making decisions is vital. It brings benefits in making an individual decision and also for democracy more generally . . . It is also a moral duty. Public authorities work for the public. To do so in a way that the public want and to ensure that they know what the public needs, they must involve the public when they make decisions.
>
> (DETR, 2000a, 11)

More recently, Defra (2009, 36) has indicated that:

> Engagement with interested parties should aim to gain a sound understanding of local issues and an appreciation of the concerns of individuals, communities and businesses potentially affected, as well as more strategic and national perspectives. Community and stakeholder participation and information should be used to help identify and develop options for risk management and gain an understanding of local people's preferences.

In addition, the Environment Agency (2010a, 21) has stipulated unambiguously that 'Working with others (individuals, a group of individuals, communities, organisations or political entities) is critical to good appraisal and must be done from the start.'

In project appraisal, therefore, this involvement needs to be included in the definition of the problem through to the identification of options and the assessment of the relative importance that should be given to achieving the different objectives. To do so, project appraisal has two roles:

1 Stakeholders need informed involvement, with information available to all (DETR, 2000a; Environment Agency, 2010a, 21–22): the project appraisal technique itself can contribute to creating a shared knowledge base.
2 The project appraisal method must serve as a framework through which stakeholders can explore, argue and negotiate their concerns. This can apply equally to both benefit–cost analysis (BCA) and multi-criteria analysis (MCA), although it is not always so apparent in the former as in the latter (Flood Hazard Research Centre and RPA, 2001).

2.4 APPLYING REASON TO CHOICE

As a learning process, the project appraisal process has four key features:

1 The design process should be appraisal-led.
2 A key stage is to identify what parameters are critical in determining the choice between the available options.
3 It is an iterative process in which estimates of the values of these critical parameters are progressively refined.
4 A primary purpose is to aid the discovery or invention of new options, possibly through combining the most successful elements from options already being considered.

As indicated by the Environment Agency (2010a, 244), it is important to test the effect of uncertainties, and assumptions made about them, on option choices. Within economic appraisal, the purpose of sensitivity analysis and robustness testing is to determine whether, within the reasonable bounds of confidence:

- the project is economically worthwhile (benefits outweigh the costs); and
- the option choice is robust (where the option choice would not change to another option under reasonable changes to the assumptions made during the appraisal).

The identification of the critical four features as those listed above determines the choice between the options. *Sensitivity analysis* is a crucial step – exploring if particular parameter values affect the calculated results in a major way – and should be undertaken in spirit at least at the start of the appraisal process, not at the end as is commonly recommended (but undertaking both is also valuable). If the value of a parameter makes no difference to the choice between the options available then there is no point in considering that parameter in greater detail.

Attention needs to be focused on those parameters whose values will determine the choice between options. Depending upon the parameter, either *risk management* (Cabinet Office, 2002) strategies need to be adopted for these parameters or *better estimates* sought of the values of these parameters. In general, those parameters which are most likely to be critical are those which occur early in the life cycle of the project (e.g. capital costs; initial losses through erosion) or those which are frequent (i.e. frequent floods; annual operation and maintenance (O & M) costs). It is the relative contribution of any parameter to the total costs and benefits of the option that is important.

At the end of each cycle of the analysis, a *robustness analysis* should be undertaken; this is a series of tests to determine whether, given the inherent uncertainties, it is possible to be confident as to which option should be preferred. The logic is to take the critical parameters as identified in the sensitivity analysis and test how far each singularly and in combination has to change before the rank order of preference between the options changes: specifically, the rank order between the best 'do something' option and the baseline option. In this regard, it should be remembered that a benefit–cost ratio of 1.0 is the point of maximum doubt as to whether that option should be preferred to the baseline option; we can be more confident in our choice the further the benefit–cost ratio is away from 1.0.

If, in the early cycle of analysis, the benefit–cost ratio of the best 'do something' option is approximately 1.0:

- Unless a better option can be found, then the scheme is almost certainly marginal at best and there are likely to be schemes or other parts of the country which should be given higher priority.
- There is a high risk that spending more time and money on refining the estimates of the benefits and costs will simply make it clearer that the benefit–cost ratio really is close to 1.0 and hence that the scheme should be given a low priority.
- A potential exception to this rule is when there are significant benefits from the scheme which cannot be included into the benefit–cost analysis and which do not accrue to other schemes.

2.5 CHOICES AND VALUES

Choice is comparative and relative; we are choosing between alternative courses of action where those alternative courses of action can have very different consequences.

Conventionally, we compare the consequences of alternative courses of action against some *baseline option*, usually the '*do nothing*' option. We then assess the consequences of the alternative '*do something*' options by evaluating the changes that would incur in consequence against the baseline option.

In effect, the situation that occurs under the baseline option defines the zero point for measuring the changes. These changes can be very disparate; different people are often affected in quite different ways at quite different times in the future. The dimensions of choice are thus: *how?*, *who?* and *when?* Any choice involves making some implicit or explicit judgements as to how these different changes can be weighed against each other.

The first advantage of using some standard procedures to carry out these weightings is that consistency is then achieved both within an individual choice and between choices. The second advantage is that of making the basis of these assessments explicit and hence open to review.

Two possible baselines may be used:

1 A 'zero' point.
2 The continuation of the existing conditions.

In multi-criteria analysis it is often convenient to use some notional zero point against which to estimate the performance of the 'do something' options. In benefit–cost analysis, we use a 'do nothing' baseline '*walk away*' option (i.e. take no further action whatsoever and allow flood risk management assets to deteriorate), and the benefits and costs of the different options are then the changes from these conditions. Here, continuing with the existing maintenance and repair of a structure would be one obvious 'do something' option, as would be continuing with an existing flood warning system, or the existing arrangements for managing flood incidents.

A 'do nothing' or 'walk away' situation would be discontinuing the issuance of flood warnings, ceasing to maintain sea walls or river bypass channels, and the abandonment of arrangements for managing flood incidents; these would be the baselines against which alternative warning and incident management arrangements would be assessed. A '*do minimum*' is also often appraised, as another baseline, and is usually taken as the minimum amount of action (maintenance and minor repair) to retain some defences or protection works (Environment Agency, 2010a, 152) and comply with legislation (e.g. concerning health and safety). If major capital works would be required in the future (for example, to replace defences) then this would not be classed as 'do minimum'. This may mean that 'do minimum' would revert to 'do nothing' when it is no longer feasible to continue maintaining the defences.

When we compare the consequences of the different possible courses of action relative to the baseline we are commonly comparing apples with oranges. But, contrary to the common saying, this is the kind of comparison at which we are quite good. Offered an apple or an orange, few people say that it is impossible to compare the two and choose between them.

2.5.1 What is value?

In conventional economics, the 'how' question is addressed through evaluation. In economics, value is used in a particular and distinct way. Figure 2.1 is a typology of the potential ways in which the term 'value' can be used. Looking at the right-hand side of the diagram, there are three main categories of value. In everyday speech, when someone is asked what their values are, they are likely to reply in terms of those moral, ethical or religious principles that they believe should govern their lives and the lives of others. That is, value is an end, a principle that people consider should guide their lives.

Conversely, in conventional economics, value is solely instrumental (Coker and Richards, 1992): it is the contribution that some action makes to the attainment of one or more ends or objectives (Green, 2003). Where that value is determined by individual preference, all other forms of value are excluded from economic analysis. Although it is actions that have an instrumental value, it is more convenient to speak of things or *goods* having a value when the choice concerns whether to acquire that good.

The value of a good can be thought of as the potential actions that can be provided by that good; the value of a flood risk management scheme, for example, is the reduction in flood losses that result from it. As Coase (1991) pointed out, when we buy something, what we acquire are entitlements to perform certain actions. If one buys an apple, one acquires an entitlement to eat that apple or to give it to someone else; one has not acquired an entitlement to throw the apple at someone else or to throw it away in the street as litter.

In conventional economics, the individual is assumed to be the best judge of his/her own interests and so the value of any action is the contribution that action makes towards achieving their personal objectives. Thus economic value is ultimately subjective, reflecting individual preferences; the problem for the economist is to measure it rigorously, a problem which economics seeks to resolve by using money as a yardstick.

2.5.2 Consumption

Since economic theory developed around the demand for consumer goods, it is convention to speak of what the individual wants as *consumption*. This is a rather misleading term because anything that the individual wants is counted as consumption whether or not that good is used up in the process. A walk by a river

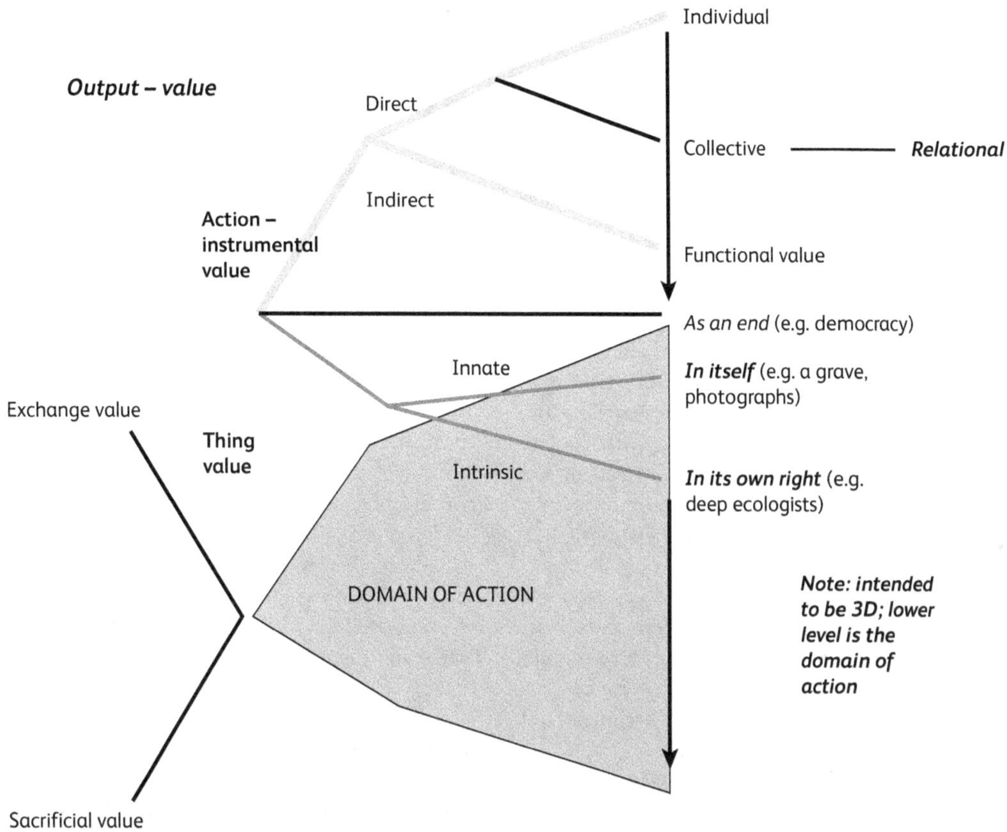

Figure 2.1 A typology of value

counts as consumption if the individual values walks by rivers, but the only thing used up by a walk is time. Since the value of an additional unit of some good is assumed to decline the more we have of that good, the *marginal value* of an additional unit of that good falls with the quantity that the individual has already consumed.

Some goods have an indirect instrumental value; rather than directly providing consumption, they provide a resource which is indirectly used in production. The most important examples of such goods are the different environmental assets. These often provide important *functional values* (de Groot, 1987) or *provisioning* and *supporting* services (Alcamo *et al.*, 2003). A wetland, for example, may scavenge pollutants (Mitsch and Gosselink, 2000), and/or it may store flood waters (Acreman, 2003). Such environmental resources provide *services* – usually termed 'ecosystem services' – at no cost which might otherwise have to be provided by other means: in the wetland example just given by constructing a wastewater treatment plant or some form of source control; or by constructing a flood storage reservoir.

As a resource, it is fairly straightforward to evaluate these services once they have been identified; the difficulty often lies in discovering the existence of the service. For example, it has been shown that migrating salmon are an important source of nutrients for the forests of the north-west USA and Canada; the decaying bodies of salmon killed by bears and birds, together with the waste products of those hunters, providing nitrogen and other minerals to the trees (Gende *et al.*, 2002). As the example of the salmon illustrates, it can be the linkages that are important rather than any single element.

In collective choices, objectives relating to interpersonal relations become engaged. Democracy, justice, liberty and most obviously equality are examples of objectives which refer to what should be the relations between us. As these examples illustrate, values as ends often refer to the relations that an individual believes ought to exist between people. There is arguably little scope for applying such objectives in private consumption choices – except in, for example, buying 'fair trade' coffee and tea as opposed to conventional coffee or tea.

But in collective choices, these objectives can become central. So, for example, those who argue most strongly for a reliance upon market-based approaches do so on the basis of claims as to personal liberty and freedom (Friedman, 1980). Moreover, collective choices can revolve around conflicts regarding the relative importance that should be given to each objective; for example, between equality of treatment and individual liberty. Those objectives, notably democracy and justice, concerning interpersonal relations cover the ways in which collective choices should be made. There is no way within benefit–cost analysis to handle such objectives, or conflicts between objectives, or disagreement as to the importance that should be attached to each objective. Multi-criteria analysis instead provides a framework in which argument and debate over objectives can take place.

The final main category of value (Figure 2.1) is when something has a value in and of itself. 'Deep ecologists' (Naess, 1993) argue that other species have an intrinsic value which exists over and above any value which humans may have for them. A second form of inherent value is when something has an innate value to an individual or group of people, but is of no value to anyone else. Wedding photographs and mementoes of the birth of a baby are two examples of such innate values. Some objects may come to form part of the individual's identity. They are of little or no value to anyone else but often of very great value to the particular individual so that if they are destroyed in a flood (see Chapter 4), their loss is experienced as more severe than damage to the replaceable contents of the home.

Framed in these terms, the problem for the individual or the community is to maximise the total value of consumption given the available *resources* that are required to provide or produce the good in question. The conventional economic assumption that value is given only by individual preference leads to the use of the *Pareto improvement* rule in collective choices: if everyone is left better off by a change then the change should be made. *Economic efficiency* is achieved when no one can be made better off without at least one person being left worse off. One condition for economic efficiency is *productive efficiency*, when the highest value of consumption is produced for the available resources.

Pareto improvements are rarely found in practice so the rule adopted instead is that of a *potential Pareto improvement*. This is embodied in the *benefit–cost ratio* and *net present value*; if the benefits of some action exceed the costs then, in principle, it would be possible for those who gain fully to compensate those who lose whilst themselves still gaining by the action.

2.5.3 Total economic value

Economists generally accept that an environmental asset has more value than the immediate use that may be made of that asset. For example, someone may want to bequeath a high-quality environment to their descendants. Thus, *Total Economic Value* is defined (Pearce and Turner, 1990) as:

Equation 2.1

$$\text{Total Economic Value} = \text{Use value} + \text{Non-use value}$$

Use value can be taken to include all values arising from a physical conjunction of the individual and the asset in question. So, it includes walks by a river as well as that river as a fishery. *Non-use value* is probably best defined by antithesis: it includes all elements of individual preference that do not require the conjunction of the individual and the asset.

A wide range of reasons can be proposed for why an individual may value an asset even when that individual is not currently making any use of it (Krutilla, 1967). But, critically, Total Economic Value refers only to *direct* individual value; it includes none of the other forms of value illustrated in Figure 2.1. In so far as an individual takes account of any other form of value, this, in principle, is excluded from Total Economic Value. But, in practice, when someone is asked to put a monetary value on some environmental asset, that person may take account of those other values so what the economist is actually measuring is something different from what is defined by economic theory. The person asked for a value may, for instance, include value as an end.

The right-hand side of Figure 2.1 is centred upon abstractions but the problem is what to do; what courses of action should be adopted to deliver those abstract conceptions of value?

If, for example, we decided that everyone should be exposed to no more than some tolerable level of risk from flooding, we would then have to decide what to do in individual cases. We cannot reduce risk other than through action; we can only take actions to reduce the frequency or severity of floods. It is choices of actions between which we have to decide.

The left-hand side of Figure 2.1 then forms the bridge to action. Here, there are two further categories of value:

- Sacrificial value: what the individual is prepared to give up in order to have the action or thing or end in question.
- Exchange value: what some other individual will require in order to exchange the action or thing in question.

2.5.4 Opportunity and other costs

The fundamental assumption in economics is that there are no 'free lunches'; if we want something, we will have to give up something else. What will we sacrifice? The sacrifices we are prepared to make are the *costs* of the proposed action. The fact that the cost of something is given by the value of what else we are prepared to give up in order to have the first thing means that value and cost are closely related. The cost of one thing is the value of the forgone alternative, and hence it is common to speak of the *opportunity cost* of some good.

The second branch on the left-hand side of Figure 2.1 is exchange value: how much someone else requires in order to give up the action, thing or end in question. It is assumed that the more they are prepared to make available, the higher the exchange value they will require; that the *marginal cost* will increase. It is the difference between one individual's sacrificial value and another individual's exchange value that makes markets possible and advantageous to both sides.

In a *perfect competitive market* the marginal sacrificial and exchange values of the good are in balance and the *price* of the good is determined by the market at this level. For a perfect competitive market to exist, it is necessary that none of the participants has any power to influence either the quantity bought and sold or the price at which it is bought and sold. Where a market in a good exists, the price of that good can be used to estimate the value or cost of that good. Since the market is seldom a perfect competitive market, corrections usually have to be made to the price in order to derive an estimate of the value or cost: a *shadow price*.

If there is no reliable method of estimating the value or cost of something, it has to be left as an *intangible*. This is problematic in that we do not know how important these intangibles are relative to the streams of all the benefits and costs included in the analysis. Therefore, we do not know whether the analysis includes the most important benefits and costs or only those that are simplest to evaluate and measure. Thus, we do not even know whether we have included the largest proportion of the true benefits and costs.

It is also problematic because the consequences of the project regarding other societal objectives – such as nature conservation – are often excluded from a benefit–cost analysis. By contrast, using multi-criteria analysis it is possible to include all the impacts in terms of all objectives. The problem is how to do so in a manner that is both coherent and consistent with the plethora of stakeholder views.

As with value, something does not need to have a price for it to have a cost. Rather, it just needs to have been possible to put that resource to an alternative, mutually exclusive use (e.g. building a hospital rather than adopting a flood risk management scheme). If there is no such alternative use then, in theory, the resource does not have a cost.

The cost of producing some good, whether it is a flood or coastal risk management project, has two main elements:

- A fixed cost that once committed will not vary, irrespective of the amount of the good that is actually produced (e.g. a site office).
- A variable cost that changes according to the amount that is actually produced (e.g. concrete revetment, per square metre of embankment). The incremental cost of producing an additional unit of some good is then its marginal cost.

It is the nature of their costs that serves as a primary reason for the public funding of most flood risk

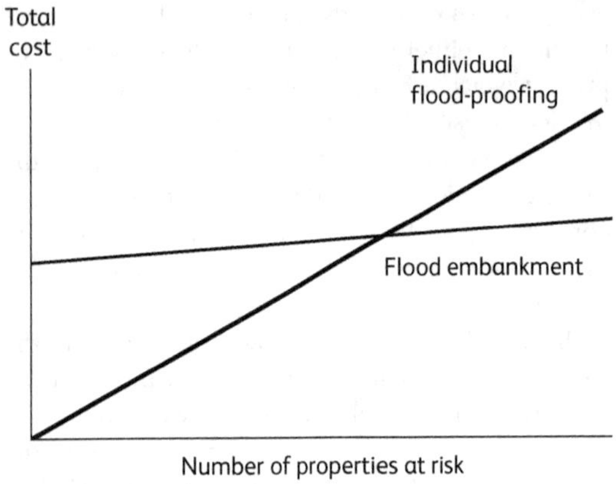

Figure 2.2 Flood and coastal risk management economies of scale

Source: Green *et al.*, 2000b

management and coastal protection schemes (although private companies can provide public goods). This is because the more that is provided, the lower the average cost of provision will be (i.e. there are *economies of scale*). Thus, the cost to the public purse of funding a flood embankment project is generally significantly lower than individual private flood-proofing of individual buildings (Figure 2.2). For example, Penning-Rowsell *et al.* (1987) found that the cost of providing local protection around a number of small urban centres was greater than the cost of undertaking one all-embracing project. Flood risk reduction is, in many cases, an example of a public good because within a scheme it is generally not possible to deny risk reduction to anyone included (Box 2.1).

2.5.5 Value and money

It is only when the concept of a market is introduced that it becomes necessary to discuss money. In economics, money is used in a number of ways. In a market it acts as a medium of exchange; it is simply easier to use money than to find someone with whom to barter what one wants to sell and who has what one wants to acquire. More importantly, in a market money then becomes a means of comparing the value and cost of different goods; it acts as a yardstick – or *numeraire* – of value. Since economic value is given by individual preference, something has a value whether or not it has a price; it is simply easier to estimate that value if the good has a price in a competitive market.

A further advantage of money is that it provides a generalised measure of sacrificial and exchange value; rather than having to look at each individual sacrifice that would have to be made to provide the good, the money cost of the action covers all possible combinations of possible potential sacrifices. Rather than thinking about the cost of a flood risk management scheme in terms of, say, forgone hip replacements – at roughly £4,000 per operation (National Audit Office, 2000) – the money cost of a

BOX 2.1 KEY DEFINITIONS: 'PRIVATE', 'PUBLIC', 'COLLECTIVE' AND 'INDIVIDUAL' GOODS

Those goods that are bought and consumed by individuals such that they are then not available to others are termed *private* goods. The assumption here is that individuals make their own purchasing decisions for their own purposes. This applies to most marketed goods, although some goods can be shared between individuals without being 'used up' (e.g. newspapers and books).

Public goods, by contrast, are those where the provision of a good by one individual necessarily means that it is also provided for others without diminishing its value. The assumption here is that there is no way for the provider of that good to exclude others from receiving the benefits of the good(s) provided (e.g. a lighthouse, or a ring flood embankment around a town).

There are some goods that any individual, given sufficient resources, can acquire for him/herself. These are termed *individual* goods (e.g. flood-proofing a house).

Collective goods, by contrast, can be provided only – or only efficiently – collectively (e.g. a public flood warning system).

scheme relates the cost to all other possible alternative sacrifices that could be made to fund the flood risk management scheme, be these a reduction in the number of pints of beer drunk or a reduction in the number of nurses. Economics looks only at the transfer of resources to consumption: the resource economy. It does not consider the money economy – the stocks, flows and distribution of money – so it has nothing to say about the distribution of money.

Although both economics and accountancy use money, that does not mean they are concerned with the same thing: economic and financial values are different. Accountancy is concerned only with those things that have a market price; economics is concerned with everything that has a *value*, whether or not it also has a price. Moreover, the market price of a good is not always a good indication of its economic value or cost (Section 2.7.2), but it is the financial value or cost of that good.

Financial analyses are prepared from the viewpoint of some individual or organisation, such as a farmer, a firm or a council, whereas an economic analysis is prepared on the basis of the country as a whole. The actions of one individual or organisation can have consequences for another individual or organisation without there being any counterbalancing payment. Those consequences are externalities of the action of the first individual or organisation. This is said to occur when some of the consequences of an individual's decision are borne by another without a corresponding transfer of money in the opposite direction.

Although an externality can be positive, this is not usually the case. Smoke from a factory chimney is an obvious example of such an externality: there is no financial implication for the factory owner of the emissions of smoke provided that the consequences are borne by others, and the factory owner is not required either to reduce smoke emissions or compensate others. Hence, those costs to others are excluded in a financial analysis but included in an economic analysis.

A comprehensive project appraisal needs to 'capture' these negative externalities. Consequently, the boundary of analysis is invariably wider than the geographical extent of the project or scheme. Thus, the way in which the boundaries are set has implications for what should, and what should not, be included in the analysis (see Chapter 3 for more detail on this point). For example, it is necessary to include in the evaluation of a coastal erosion risk management scheme that reduces cliff erosion the loss of the sediment generated by that erosion in beach building downdrift, but not if that beach is in another country (MAFF, 1999).

If consumption or resources are simply redistributed between individuals or organisations, there is no change in national resources and consumption. All that is involved is a *transfer payment*. Hence, from an economic perspective, those changes can be ignored: a conventional economic analysis is concerned only with the national totals of consumption and resources. But a financial analysis undertaken for an organisation that has experienced a gain or loss from such a redistribution will obviously take it into account. An example is a shop that is closed through flooding; if all its customers simply go to the shop down the road instead, there is no real national change. However, there is a real economic loss if those customers incur higher costs or have to make do with lower-quality goods, or if the second shop incurs higher costs in satisfying the increased demand (see Chapter 5 for how this applies to flood losses in the non-residential sector).

In summary, value is central to benefit–cost analysis, and in economics all values are subjective: the value of some 'good' is given by the individual and reflects his or her subjective preference for that 'good'. Value does not have to be measured in monetary terms only, although the Treasury's 'Green Book' (HM Treasury, 2003, 21) suggests that 'real or estimated market prices provide the first point of reference for the value of benefits', and that 'benefits should be valued unless it is clearly not practicable to do so'.

In this respect, the shorthand term 'good' is used to denote any commodity, resource or item which an individual prefers or desires (for example, a coastal risk reduction project, a flood risk management scheme, a beach, a river or a recreational experience). The values assigned to any such good then reflect the relative contribution that this good makes to an individual's *utility* (or well-being).

Value is also *sacrificial*. This means it quantifies or reflects the degree to which the individual would be willing to give up an amount of that 'good' in order to have more of another: more flood risk management means fewer hospitals. Values are, therefore, not absolute but reflect the basis upon which choices are made between enjoying these different goods (which, as we have said, the economist calls 'consumption').

2.5.6 Who is affected?

The 'who' is essentially ignored in conventional economics; only the net change to the nation as a whole is considered. This becomes more questionable with, on the one hand, the development of transnational governments, such as the European Union, and on the other hand with the increasing devolution of power to regions, such as, for example, Wales. Thus, it may sometimes be more appropriate for project appraisals to consider the effects on the people in the area who are paying for the project rather than the nation as a whole.

This has become more pertinent with the adoption by Defra of a new arrangement for flood and coastal erosion risk management, termed 'partnership funding' (Defra, 2011a). Under this arrangement the UK Government provides a certain element of funding, driven by a set formula, and local communities are invited to contribute the remainder of the cost of any scheme (see Chapter 1). For the local community, who benefits from and who pays for flood risk management become critical issues, and the metric by which these are likely to be decided is not the economic value to the nation of risk reduction, but the (financial) value to the people at risk and the cost to those paying.

The Treasury's 'Green Book' (HM Treasury, 2003) states that it is reasonable to suppose that a loss of £1,000 matters less to someone with an income of £100,000 a year than it would to someone with an income of £5,000 a year. Hence, the 'Green Book' recommends the use of income weightings. Critically, this form of weighting takes no account of differences in vulnerability, nor of wider considerations of equity (McCoubrey and White, 1993).

The effect is to increase the measured value of the benefits of flood risk management to low-income groups, and to reduce the value of the same benefits to those on higher incomes. One reason for this change is to negate the effect of unweighted assessments prioritising the provision of flood risk reduction measures for those who are the most affluent (and therefore have more assets to be damaged by flooding) as opposed to those of more modest means. In this regard the previous arrangement for unweighted benefit–cost analysis of flood and coastal risk management schemes ran the risk of reinforcing an unequal distribution of income by rewarding with investment and flood risk management those with higher incomes and more valuable assets at risk. Whether that regressive effect was ever a reality is open to debate.

2.5.7 When will the effect occur?

In assessing flood and coastal erosion risk management schemes we may have to consider effects that occur up to one hundred years in the future.

The first issue is how far into the future it is reasonable to seek to predict the impacts of the different options. HM Treasury (2003) recommends one hundred years. The purpose of taking such a long view is to ensure that we do not neglect impacts that would be harmful in the long term.

The second issue is to decide what weight to give a gain of, say, £20,000 in fifty years' time as compared to a cost now of £1,000. To bring the stream of costs and benefits over time to a common base, it is conventional to discount future costs and benefits back to a common base date.

There are two reasons for *discounting* future gains and losses, using money as a general analogy. If there is a choice between investing £1,000 to produce a return of £100 a year or investing the same sum to yield a return of £80 a year, then obviously we should prefer the former investment. This is the concept of the *opportunity cost of capital*: since capital invested in one way is not available to be invested in another way, in any one investment decision we have to take account of the highest return that could be obtained from a given investment.

But this is to assume that we are prepared to set aside some money now for investment rather than spending that money on consumption now. It is reasonable to expect both individuals and communities to have some preference for the distribution of consumption over time, which determines the rate at which they are prepared to sacrifice consumption now for consumption in the future: *time preference*. But it is more problematic to determine what this preference actually is (Friedman, 1957; Modigliani, 1966).

A cost or benefit that occurs in one year's time is thus treated as having a lower real value now than an identical benefit or cost today. The corresponding term used to calculate the present value of a flow of future benefits (and costs) is the *discount rate*. The formula is:

Equation 2.2

$$PV = \sum_{t=1}^{t=T} \frac{Xt}{(1 + r)_t}$$

Where:

- PV = the present value of the benefit or cost occurring in a future year;
- r = the discount rate;
- t = the number of years into the future the benefit or cost occurs after the base date of analysis;
- Xt = the benefit or cost in year t;
- T = the life of the scheme.

The Treasury's 'Green Book' now uses a discount rate which decreases with time (HM Treasury, 2003, Annex 6).

Discounting often causes discomfort; why should we be prepared to spend less to reduce the risk of a death occurring in fifty years' time than to reduce the risk of a death tomorrow? The key point, however, is to recognise that in any decision we are making some judgement which compares impacts now to impacts at some point in the future. The advantages of discounting are then that at least it enforces consistency and makes the assumptions explicit. However, there are limitations to both arguments for discounting.

First, the opportunity cost of capital argument is logical when applied to money capital (a stock) and money income (a flow). Income can be saved to create capital and capital generates income in the form of interest payments. In practice, however, we can be dealing with choices where the trade-off between a stock and a flow is not reversible and where the stock does not produce a monetary income, or where the stock itself has no market value. We may, for example, have to decide whether to preserve an attractive riverscape in order that those who walk by the river will continue to enjoy those walks.

Second, the individual time preference argument asserts that individuals have a preference for the

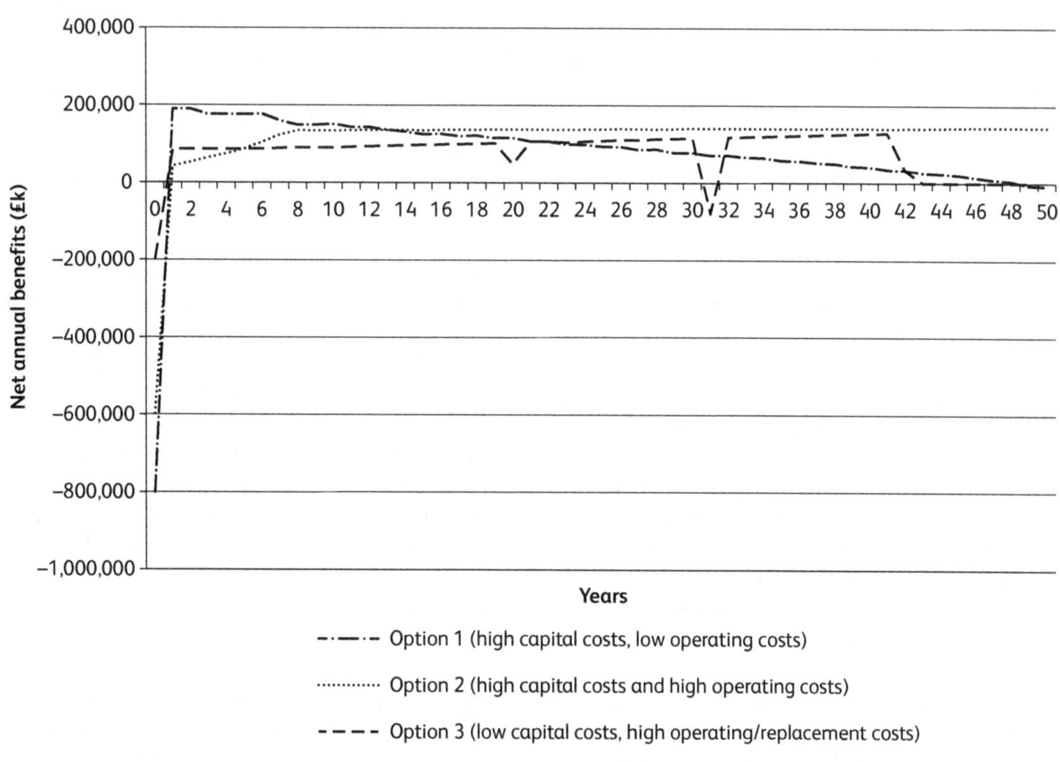

Figure 2.3 Comparison of options with different expenditure profiles

Source: FCDPAG3 (MAFF, 1999)

distribution over time of income or a wider distributional time preference. Discounting does not address such a distributional preference. If the net annual benefits from a project are plotted over time, all that discounting does is provide a weighted measure of the area under the curve, as a measure of total net benefit. However, for any one discount rate, there are an infinite number of curve shapes that would yield the same weighted area (Green, 2003).

The problem is that one curve can be preferred to another only if we assume that there is no preference about the shape of the curve. However, intergenerational equity is clearly a case where it is the shape of the curve that is of concern.

The three project options plotted in Figure 2.3 illustrate this issue. The three options have very different distributions of net annual benefits (i.e. the difference between expected annual benefits and costs for each year of the scheme's life) but very similar benefit–cost ratios and net present values. If we rely on the discounting rule then the choice between the options is dictated by the discount rate applied.

But since discounting does not address these preferences for the distribution of benefits and costs over time, all that can be done is to prepare plots of the form shown in Figure 2.3 and then prepare a reasoned argument for the choice of option based on *net present values*. Thus, there is no technical rule that can be provided to determine the choice; rather, it is again down to preferences and value judgements.

2.6 CHANGES IN RISKS AND VALUES OVER TIME

Flood and coastal management is about managing or changing risks. Thus, the aim of flood and coastal erosion risk management is to reduce the probability that a flood will occur and/or limit its damage, or to delay the onset of erosion. Where there is an existing flood risk management scheme then the objective of the project is to change the probabilities that the structure will fail over time. In most cases the probabilities of failure increase as the structure progressively decays.

The obvious question, therefore, is whether it is better to wait until the risk of failure has increased or erosion has progressed before renovating the structure. The problem is that there is only a weak scientific base for assessing the existing and future probabilities of failure. Instead, engineering judgement will have to be used.

In addition to the probability of failure changing over time, so too may the consequences of loss change over time. This will then alter the costs and benefits of the project. In project appraisals, future values are not adjusted to reflect anticipated inflation; instead, *real prices* are always used.

The problem is that these real prices may also change. They may be forced upwards by increases in the demand for goods as incomes rise (*income elasticity*): this appears to have been happening with house prices (discussed in Chapter 7). On the other hand, prices may be forced downwards by economies of scale and technological innovation. This is illustrated by the real price of consumer durables, such as computers and white goods, which have fallen sharply over time whilst their performance has improved (discussed in relation to flood damage to residential property in Chapter 4).

Such real changes in values can be incorporated into the analysis using *growth factors*. However, since to assume constant real prices for all streams of benefits and costs is generally a conservative approach, there should be a reluctance to use growth factors, not least because predicting what will happen to prices in twenty, fifty or a hundred years' time is often nonsensical. If they are used – and for FCERM schemes this would be most unusual – there are three basic rules:

1. If growth factors are applied to one stream of benefits or costs then they must be applied to all streams.
2. Growth rates should be measured relative to the assumptions built into setting the discount rate.
3. When the local economy is growing faster in one part of the country than the country as a whole, there is an argument for not using a growth factor at all but for increasing the discount rate that is being applied. The higher rate of growth in that area implies that, locally, the opportunity cost of capital is also higher.

Clearly one of the most important factors affecting change into the future is the possibility of climate change affecting both flood risk and coastal erosion risk management plans and schemes. This was discussed in Chapter 1, where the reader was referred to the latest advice provided by the Environment Agency.

2.7 MEASURING VALUES

Earlier, it was noted, first, that in conventional economics value is defined as the contribution of some action towards the achievement of some objective. Second, it was also assumed that the sole societal objective is the maximisation of the sum of individual utilities within the limit of available resources. In turn, it is assumed that the individual seeks to maximise his or her utility.

The individualistic base of conventional economics leads logically to a piecemeal approach to the evaluation of benefits and costs:

- Estimate the number of people or properties who experience some impact.
- Estimate a unit value for that impact.

2.7.1 Avoiding the pitfalls in piecemeal analysis

It is usually easiest to carry out economic analysis on a piecemeal basis. For example, in estimating the benefits of flood risk management, we generally determine how many properties of each different type will be flooded, calculate the losses that each will suffer, and then sum the results.

We need to be careful here, as the danger in undertaking this type of piecemeal analysis is that some effects will be omitted or the whole is more than the sum of its parts. As an alternative, we can adopt a more holistic technique. For example, ECLAC (2000) report the estimated impacts of a number of different disasters on the GDPs, debt levels and government revenues of the countries affected. Similarly, input–output analysis (van der Veen *et al.*, 2003) can, in principle, be used to track how some loss would impact on the total economy. However, practical consideration of scale (the average car contains 35,000 components from a dedicated supply chain) and the need to be able to define each and every input–output relationship limit the practical use of such techniques. Similarly, economists have tended to assume that one site of environmental importance is a potential substitute for another such site. But ecologists see all such sites as part of a system and argue that what must be evaluated is the change in the behaviour of the system as a whole.

Even when taking a piecemeal approach, it is necessary to think holistically because all the significant effects have to be identified before they can be included. Easy to omit are the 'knock-on' or indirect effects that arise from the interrelations that characterise a developed economy.

In this respect a flood is a shock which results in an immediate loss of well-being by those most directly affected. That shock then can propagate through the socio-economic systems, creating a further loss of well-being until, over time, well-being recovers. Figure 2.4 illustrates the losses from a flood disaster as the difference in the areas under the two main trajectories, between the 'without disaster' and the individual response curves. In this example, we would obviously prefer trajectory A to all others (recovery is quickest) and least desire trajectory B (recovery does not occur). We might seek to intervene by reducing the initial shock, including by making changes which, whilst reducing the gradient of the trajectory of well-being before the disaster, reduces the proportional magnitude of the shock (i.e. invest to mitigate impacts). Or we could invest to enhance recovery from the shock. Importantly, direct damages do not of themselves measure the initial shock. In fact, direct damages are relevant only in so far as they influence either the initial shock or the path of subsequent recovery.

Another form of indirect effect arises from the dependency of the economy on the environment. This relationship is unpriced. But when we make changes to the environment we can cause price changes within the economy as a result of the losses of these unpriced services provided by the environment.

2.7.2 Strategies for deriving information on values

There are four general strategies for deriving values, and hence also opportunity costs, for use in benefit–costs analysis:

- using market prices;
- using inferential methods;
- using expressed preference methods; and
- benefit transfer.

These are generally used in different circumstances, such that inferential and expressed preference methods are reserved for detailed appraisals rather than initial investigations, given that they tend to be

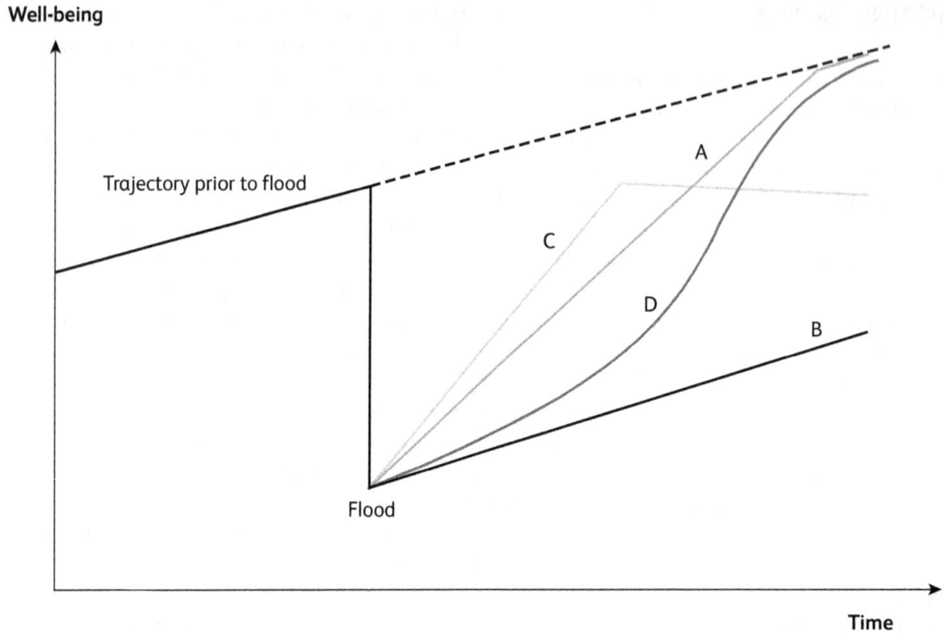

Figure 2.4 Shock of and recovery from a flood

somewhat expensive, and benefit transfer is used in both but is particularly useful for initial studies.

Market prices

For a priced good sold in a perfectly competitive market, the prevailing price would exactly equal both the marginal value and the marginal cost (Samuelson, 1970). However, the conditions that define a perfectly competitive market are so restrictive (Henderson and Quandt, 1971) that this provides only a starting point for estimating the value or cost of anything. In addition, even if a market were to be perfectly competitive, it has been shown (Coase, 1991; Stiglitz, 2008; Williamson, 1981) that the result may not be the optimal allocation of resources.

In the real world, markets are commonly imperfect or distorted in some way. Hence, it is often necessary to make some corrections to the observed market price – to derive its shadow price – in order to estimate the economic value or opportunity cost of the good or service in question.

In assessing the benefits of flood risk management schemes, one important correction is that for second-hand goods, the prices of which are depressed below their true value for many reasons (Akerlof, 1970). Therefore, in evaluating the loss in a flood of, say, a five-year-old television, the economic value of that television before the flood is *greater* than the market price of an apparently identical second-hand television. Equally, the economic value of the television is clearly *lower* than the cost of a replacement new television, since that new television would have a longer expected life. Therefore, the pre-flood values given in Chapter 4, as the baseline for assessing the flood losses to building contents, are based upon the use of straight-line depreciation, rather than second-hand values.

It is also necessary to remove direct and indirect subsidies within prices. For example, in much of the world, irrigation water is heavily subsidised either directly or indirectly (Garrido, 1999). If an irrigated crop is lost as a result of a flood, the real loss is given by calculating the market value of the crop less the costs of the inputs, including the real costs of providing the irrigation water. Again, agricultural outputs are commonly either subsidised or supported. Here, too, the value of a lost crop needs to be adjusted downwards by removing the subsidy or price support element (see Chapter 9).

In turn, where inputs or outputs are subsidised in this way, the capital value of the asset that produces the outputs can also be artificially inflated. Thus, when assessing the value of agricultural land that may be lost to erosion, the market price of that land must also be corrected downwards to the level which would exist were there no subsidies on inputs or outputs (MAFF, 1999). There is, however, an exception to the general rule for subsidy removal: where a subsidy is intended to represent a positive externality. Some agricultural subsidies are in this form; for example, those intended to promote environmentally friendly farming.

Indirect taxes must also be removed as they are simply taxes intended to generate revenue rather than real opportunity costs. However, 'green taxes' – those that are intended to reflect the externalities caused by the use of a particular good – are real economic costs and for that reason they should not be removed. For example, any tax on pesticides intended to reflect the damage caused to the environment, or the costs of removing pesticide residues from potable water, is a real economic cost. The practical problem in any analysis is that it is necessary to determine whether any indirect tax is solely for revenue-raising purposes or if it is designed to incorporate the costs of externalities into the cost of using the product (many so-called 'green taxes' combine both functions).

Where a resource would be put to no other use if it were not applied to the project then its opportunity cost is zero, whatever the price actually paid for it. This situation is most likely to be found when looking at labour inputs. Where there is a high level of unemployment or under-employment, the wage rates paid to those building an engineering scheme to reduce flood probabilities will overstate the opportunity costs of that labour and a *shadow wage rate* needs to be estimated (Squire and van der Tak, 1975).

At times, where something is not priced, there may exist a near-perfect substitute that is priced, and this can then be used to set an upper bound on the value of the good we need to evaluate. However, the value of the good in question can be no greater than the cost of this *least cost alternative*. Thus, for example, the contribution to the benefits of a flood storage reservoir, in terms of reducing the losses when the local telephone exchange floods, can be no greater than the cost of flood-proofing that exchange.

One limitation of this approach is that the alternative must either not yet have been undertaken or all the costs can be recovered if it has already been adopted. Any irreversible expenditure is termed a *sunk cost*. For example, where the occupiers of a floodplain have flood-proofed their property, in estimating the benefits of constructing a flood risk management scheme for the whole town, the costs of these walls or stop-gates must be ignored since the costs of their construction are irrecoverable in the project appraisal. This said, the benefits that are gained from their construction *should* be included; we should not assume that they are not there when assessing flood extent or depth.

Inferential methods

The use of market prices is often referred to as the *revealed preference technique* but, as can be seen from the above discussion, it is often necessary to make quite a large number of corrections to the observed prices in order to calculate economic values and costs. The second group of techniques may be termed *inferential methods* and these use statistical techniques to infer the value of something that does not have an observable price.

The *travel cost method* (Clawson, 1959) is a way of putting a value on a recreational resource by determining the distances that different visitors have travelled to a site and the costs they have thereby incurred. By regression analysis, a relationship between the distance travelled and the probability of making a visit is deduced and the resulting function is then used to estimate the value of a visit to the site.

One problem is that any function can be fitted more or less badly to any set of data and we have no theoretical reason for expecting one functional relationship rather than another. Depending upon what relationship is actually applied, the values derived may then vary by a factor of perhaps ten. A second problem is that the assumptions about recreational choices made in the model seem to be inconsistent with recreational behaviour (Green *et al.*, 1990). We therefore do not recommend this technique.

The *hedonic price technique* (Rosen, 1974) is intended to determine the influence of the area in question's characteristics – such as a sea view – on the market price of the land or, more usually, houses. A number of studies have been undertaken of the effects of flood risks on house prices (e.g. Donnelly, 1989). Using this technique, the reasonable assumption is made that the market price of a particular house is the result of a series of parameters, including flooding or erosion threat, that make the house more or less desirable/valuable.

In practice, we do not know exactly which are the parameters that most influence house prices. We cannot measure all of them, have to use surrogate variables in some cases and do not know the functional relationship between them but have to infer this relationship in order to undertake the statistical analysis.

Ducci (quoted in Ardila *et al.*, 1998, 17) refers rather despairingly to the practical problems of obtaining results of any real value from this method, referring to the time spent on 'torturing the data to provide a decent coefficient' and claiming the hedonic price method is 'more unreliable [than the contingent valuation method] with respect to whether you will be able to get a usable result'. Again, therefore, we do not recommend this technique.

Expressed preference methods

Neo-classical economists long resisted the use of the remaining approach, *expressed preference techniques* (conjoint analysis and contingent valuation), which employ social survey techniques, because these involve asking people what choices they would make. The basis of the criticism is that what people say they will do in a situation is not necessarily as reliable as observing what they actually do in such situations, were we able to observe their behaviour. Likewise, there is a problem in that people are being asked to respond to hypothetical circumstances.

Conjoint analysis was developed in market research (Green and Srinivasan, 1978) and then adopted in transport research under the name 'stated preference' (Hensher, 1994). Using this method, respondents are asked to make a series of choices between different combinations and different levels of those characteristics believed to be most important, including the characteristic in which we are interested (e.g. flood or erosion risk).

For example, if we were interested in how flood risk affected house prices, we would first determine the critical parameters that determine the amounts individual households are prepared to pay for their dwellings. One of the characteristics then included in the study is price. By statistical analysis it is hoped to determine how changes in the quantity of the characteristic of interest affects the amount people are prepared to pay for the good. Conjoint analysis is thus the expressed preference equivalent of the hedonic price method, and is subject to some of the same problems. However, its merits are demonstrated by its widespread use by consumer companies when seeking to determine whether there is a market for a potential new product, and to identify who will buy it and at what price.

The *contingent valuation method* (CVM) is a rather simpler technique. Respondents to the interview survey are asked directly how much they are prepared to pay for a change in the availability of the good in question (e.g. coast protection of beaches; see Chapter 8). As this is essentially a social survey approach, it is a virtue of the CVM, as opposed to the travel cost and hedonic price methods, that validity and reliability tests can be applied. As with conjoint analysis, it is essential to follow good social survey practice in terms of question design, sampling and fieldwork (Mitchell and Carson, 1989).

Unfortunately, the influential NOAA Panel (NOAA, 1993) recommended that CVM studies should be conducted using the 'referendum' format: each respondent is asked whether or not they would be prepared to pay a fixed amount, with this amount varying between respondents. Using this approach, we would not know precisely what any respondent would be prepared to pay and can only estimate, by fitting a statistical relationship to the data, the sample mean. As with the hedonic price and travel cost methods, the resulting estimate then depends upon what functional form is fitted to the data. Ardila *et al.* (1998) have, therefore, argued that the approach of simply asking respondents how much they would be prepared to pay should be reassessed.

Nevertheless, both techniques have a couple of virtues:

- The first is that they are experimental techniques. Hence, we can test whether or not there is a particular relationship rather than being forced to assume it.
- The second is that they involve listening to the public, both through the exploratory qualitative studies (Krueger, 1988) that invariably should precede the quantitative interview survey and in the quantitative study itself.

It was argued at the beginning of this chapter that one primary goal of using project appraisal methods is to understand what the choice involves. The use of social survey techniques, therefore, is potentially a good way of learning how members of the public, or of more specialised groups, interpret the choice and the issues it raises.

Benefit transfer

Benefit transfer (Podar *et al.*, 2000; Rosenberger and Loomis, 2001) – usually now termed *value transfer* (Eftec, 2010a, 6fn) – is simply the use of an estimate of the unit value of some benefit (or cost) derived in one context within a specific and different benefit–cost analysis. All benefit–cost analyses use some form of benefit transfer: an obvious example is the use of unit rates in estimating the capital costs of any engineering works or flood warning system. The alternative is to develop estimates of every stream of benefit or cost that is specific to the site and scheme in question.

There are four basic methods of benefit transfer:

1 using a specific value derived in another context;
2 using standard or average values;
3 using site-specific values within a predictive equation derived in other conditions; and
4 meta-analyses: the statistical analysis of values derived for different sites and the use of site-specific values within that equation.

In the initial stages of the iterative project appraisal process, it is appropriate to make use of whatever information is available; benefit transfer data is therefore useful. Only if that particular parameter proves to be critical is it logical to seek to improve estimates of that parameter's value. Much of the data associated with this Manual are standard or average values.

The central issues in benefit transfer are:

- What is the variance in the value between apparently similar contexts?
- What are the causes of that variance?

This is something of a chicken-and-egg situation because until we know how much variance there is, we are unlikely to have a good idea of the causes of that variance. For standard or average values to be appropriate, we have to have good knowledge of the causes of variance and also its extent. If we use a predictive equation derived in one condition for another context (e.g. an equation which predicts the number of visits to a site as a function of the nature of the population and the costs of visiting a site) then we are assuming that the pattern of visiting between sites is determined by the same factors as explain visits to the single site.

Meta-analyses (Smith, 1992) are a means of evaluating those types of benefits and costs which are otherwise expensive to measure for a particular scheme. A meta-analysis is a statistical analysis of the results obtained of the estimates of the value of a particular good in a large number of different studies in different contexts of different sites. The obvious advantage of such an approach is that we can make the best use of what we have learned.

Such meta-analyses have been carried out for the value of wetlands (Brouwer *et al.*, 1999) and the recreational value of coastal sites (Polome, 2003). It is hoped that the explanatory variables included in the statistical analysis will provide a valid explanation of the differences in the values obtained in the different studies. If this is the case then the value of that good for a new and different site can be derived simply by inserting the site-specific values in each of the explanatory variables. That value should be expected to be some function of site characteristics and population characteristics.

How valid and reliable those estimates are then depend upon:

- The degree to which the individual studies are a random and representative sample of the population of all possible sites of that nature.
- The size of that sample of sites.
- The inclusion of the relevant explanatory variables (site characteristics and population characteristics).
- The quality of the original studies.

For example, there is a wide variety of different types of wetland (Mitsch and Gosselink, 2000) and the value of a specific form of wetland should be expected to vary according to its relation to wider systems. Thus, the value of a wetland for flood risk management purposes will depend upon whether it is groundwater or surface water fed, and its location within the catchment.

We need to have a good theoretical understanding of the relevant variables that should be expected to explain differences between sites, and data as to the value for each parameter at each site.

The original studies have to be of adequate quality and ideally each should have adopted exactly the same methodological approach. Unfortunately, meta-analyses tend to discover that a major form of explained variance is a result of methodological differences (e.g. Brouwer *et al.*, 1999).

Whilst meta-analyses are promising in the longer run, they remain somewhat experimental at present. But certainly, for the initial stages of an analysis, some form of benefit transfer should be used as it is the only method of trying to assess at an early stage in an investigation whether or not a particular parameter is critical to the choice between options.

2.8 MAKING THE DECISION

The purpose of an appraisal is decision-support, to indicate:

- If there are better ways of achieving a given objective (e.g. flood risk reduction).
- If those resources could be better used elsewhere (e.g. building a hospital).

Also, a good appraisal will indicate how confident we can be that one option is superior to all others.

2.8.1 Decision criteria

Economic constraints are no longer the only criterion used in making flood and coastal erosion risk management decisions, although they remain important. The two economic criteria to be used in comparing different options are:

- benefit–cost ratio: the *ratio* of the present value of all of the streams of benefits over the present value of all of the streams of costs; and
- net present values: the *difference* between the present value of all of the streams of benefits and the present value of all of the streams of costs.

Brent (1990) provides the following rules for the use of the benefit–cost ratio and net present values:

1 *When the options are mutually exclusive*: e.g. in a flood risk management project with options with, say, different standards, select the project with the highest net present value; and
2 *When there is a budget constraint*: e.g. when the issue is which projects to include in a national programme, select the projects that maximise the benefit–cost ratio of the programme as a whole.

Unfortunately, these rules were developed at a time when the costs of a scheme were only the capital together with operational and maintenance costs. As externalities have come to be included as well, two problems have emerged concerning the benefit–cost ratio. First, it is sensitive to how both positive and negative externalities are treated. Hence, the following rule has been adopted (MAFF, 1999):

- *Desirable changes* are benefits.
- *Undesirable changes* are costs.

For example, if gravel extracted during the construction of a bypass channel can be sold, then this should be treated as a benefit rather than the income being deducted from the costs of the project. Net present value is not sensitive to such issues.

Second, the rationale behind the use of the benefit–cost ratio is that it assumes we maximise the returns on the resource costs. Once we include negative externalities as costs, the benefit–cost ratio no longer measures returns on resource costs. However, there is no obvious replacement decision criterion so the benefit–cost ratio should continue to be used, albeit with caution.

2.8.2 Robustness analysis

At the beginning of the analysis, those parameters to which the choice is most sensitive will have been identified. Having refined the estimates of those parameters during the course of the analysis, there will still be significant uncertainties. The problem at this stage is to determine whether those uncertainties are sufficient to influence the choice of option.

Two of the changes in the 'Green Book' (HM Treasury, 2003) have the result of increasing uncertainties. First, the time frame of analysis has been increased; uncertainty will tend to increase the further into the future we extend our predictions. Second, the reduction in the discount rate since the 1990s gives somewhat greater weight to impacts occurring in the medium to long term.

Over the time frame of the analysis, any risk management option is likely to have to be renewed or replaced several times. If the proposed option at the coast is a 'big beach' then that beach will probably have to be replenished on several occasions over the next hundred years. But each occasion on which major investment will be required is an opportunity to decide whether or not to extend the life of the scheme. On each of these occasions, other options include abandoning the scheme and opting for managed realignment, or adopting some other option, perhaps a portfolio of measures. The opportunity to adopt a radically different approach some time during the time frame may be a positive advantage as it offers the opportunity to adapt to changing circumstances: *adaptive risk management*. However, the consequence of adopting a shorter-life option is that the future 'explodes' with uncertainty.

We do not want the choice between the options to be highly dependent upon the time frame adopted for the analysis. Therefore, it is helpful to plot the net present value as this changes with an increased time frame of analysis. An example is shown in Figure 2.5. In this instance, option 1 is superior to either of the two alternatives being considered over all time frames of analysis. Neither of the other two options is economically viable, although for a time frame greater than approximately eighty years option 2 should be preferred to option 3.

This is a methodological uncertainty; so, similarly, is the issue discussed earlier concerning the trajectory of the net annual benefit over time. In neither case is there some simple rule which can dictate which option should be adopted.

Sensitivity analysis, on the other hand, can reveal the implications of uncertainty concerning the parameters. The benefit–cost ratio itself contains a great deal of information. For example, a benefit–cost ratio of 1.5 means that the costs could have been underestimated by a third or, similarly, the benefits could have been overestimated by 50 per cent and the scheme would still seem to be worthwhile: a robust situation. Thus, the benefit–cost ratio is, therefore, a good guide to how confident we can be that one or another 'do something' option is justified. With experience, and hence the knowledge as to those parameters upon which the result is highly sensitive, the benefit–cost ratio itself can be a sufficient guide.

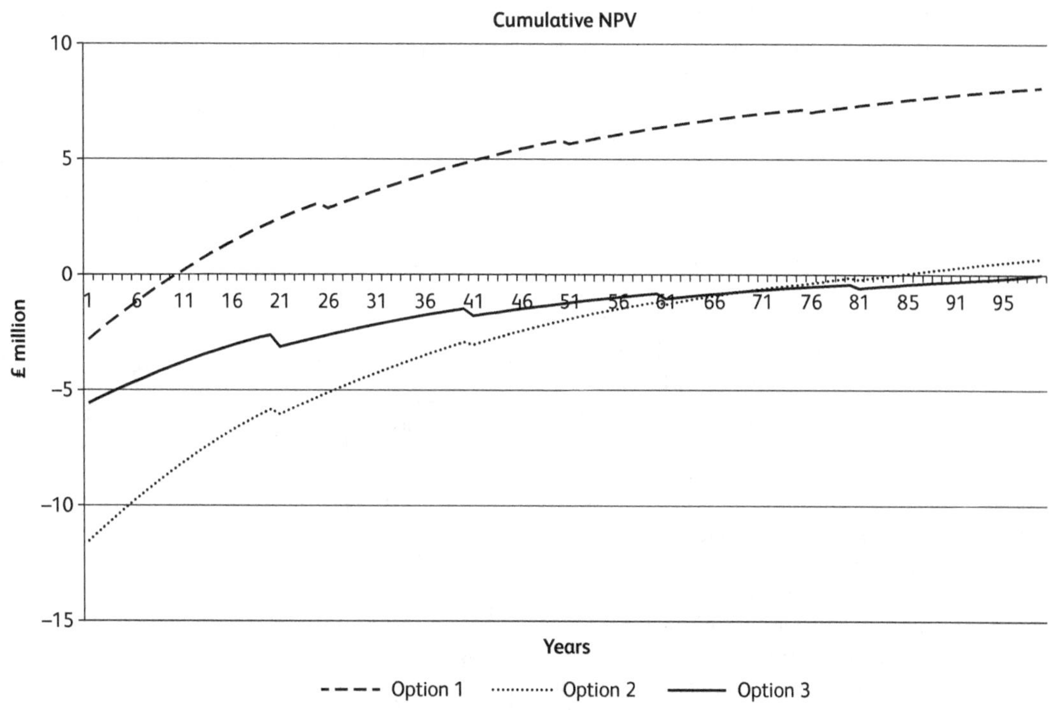

Figure 2.5 The relation between net present value (NPV) and the time frame of analysis: three examples

More generally, however, we need to explore how robust the conclusion is that a particular option is preferable to the baseline option, and also to the other options. Fortunately, in deciding how confident we can be that one option is preferable to all others, we need only examine our rank order of preference (not the detailed numbers). We need to determine the *robustness* of our order of preference across the 'do something' and baseline options as against all the inherent uncertainties of the world in the future.

The problem, then, is assessing how far our estimates of the different parameters could be wrong before the rank order of preference for the options changes (Green *et al.*, 2000a). One approach, recommended in the Treasury's 'Green Book' (HM Treasury, 2003), involves adopting probability distributions about the point estimate of the value for each of the critical parameters, and then undertaking a *Monte Carlo analysis*. The result is a probability distribution for the benefit–cost ratio or net present value. The appropriate techniques are well documented (ADB, 2002; Greenley-Polhemus Group, 1992) and specialist software is readily available.

However, this approach has a number of drawbacks and dangers which can be illustrated using the example of the approach given in the 'Green Book' (HM Treasury, 2003). The example relates to the construction of a small building. For each major construction phase, a triangular probability distribution is assumed around the point estimate of the cost of that construction phase. The first element is foundation and ground works. Amongst the reasons why the costs of this phase might overrun are the unexpected: the discovery of archaeological remains, rare species, made-up ground, contaminated soils, abandoned wells or mineshafts, or unexploded bombs. One consequence is that the probability distribution could well be very skewed. More importantly, seeking to identify what could increase costs leads directly to risk management (Cabinet Office, 2002) and deciding whether action should be taken to reduce that risk. One cannot but conclude that it is more important to seek to identify potential problems and take action than to apply probability distributions.

Second, some decision has to be taken about whether the uncertainties around any one parameter are correlated with those for any other parameter. For example, if the foundation works cost more than expected as a result of delay, is it plausible that painting and decoration could cost less than expected? Some parameters are necessarily correlated; for example, if there is uncertainty about the extent of the five-year return period flood, there will probably be even greater uncertainty about the extent of the hundred-year return period flood.

Finally, the result of the Monte Carlo analysis is likely to show that the benefit–cost ratio and net present value could be anywhere within a quite large range. We should expect this; we cannot expect to predict the next hundred years with any precision. All we need to determine is whether we can be confident that one option should be preferred to all others.

Instead, plausible but extreme values of the critical parameters are used in a robustness analysis, and the effect on the rank order of preference across the variables is determined (see Table 2.1). Table 2.1 also illustrates that the tendency to express concern about possible cost overruns of, say, 10–20 per cent is to worry about entirely the wrong parameter. There are almost invariably other parameters which have much greater effects on the benefit–cost ratio and about which there is much greater uncertainty.

When, even after a detailed analysis, it is found that the benefit–cost ratios of all of the 'do something'

Table 2.1 Robustness analysis for flood alleviation options for Hubei and Hunan provinces, China (US$million)

Case	HUBEI			Hunan
	Jianan	Wuhan	Babu	
Base	11	4.9	10.2	2.7
Delay benefits by two years	8.4	3	5.9	1.8
Probability of failure by existing dikes is lower	1.7	2.3	6.3	1.5
New dikes are not properly maintained	9.1	5	9.4	2.3
Bank protection works required every five years	8	4.9	10.1	1.2

Source: Green, 2001

Table 2.2 Critical parameters for some hypothetical 'do something' options

	Options				
	A	B	C	D	E
	Source control	Flood storage	Channel improvement	Flood warning	Dikes
Probability of failure on demand	●			●	●
Capital costs		●			●
Maintenance costs	●		●	●	
Effective scheme life	●				
Rate of increase in runoff					●

Source: Adapted from Green *et al.*, 2000b

options are close to 1.0, this illustrates that some decisions are truly marginal. In this case, for it to be worthwhile to undertake a project, either a better option needs to be found or the decision must be based on considerations not included in the benefit–cost analysis. The alternative is to do nothing, and one of the main purposes of using project appraisal is to reject some projects altogether.

The above methods apply when the 'do something' options are all variations on a theme. That is, the set of critical parameters is the same for all options. The analysis becomes much more complicated when the 'do something' options differ in kind and hence the different options have different critical parameters (Table 2.2). In such cases the robustness analysis is almost certain to show a different rank order of preference under different assumptions. In such cases a reasoned argument is needed in favour of the option that is preferred (i.e. applying reason to choice without further economic analysis).

2.8.3 Summary

According to the Treasury's 'Green Book' (HM Treasury, 2003), the purpose of an appraisal is to indicate that no policy, programme or scheme is adopted without first clarifying that there are no better ways of achieving a given objective (e.g. reduced flood or coastal erosion risk) and whether those resources could be put to better use (e.g. building a hospital). An appraisal should also explore how confident we can be that one option is better than a range of other options.

Generally, projects are judged to be economically viable only if the benefits exceed the costs (i.e. the ratio of benefits to costs is greater than 1.0). Where benefits marginally exceed costs, there is often high uncertainty as to whether an option is justified, because only a small change or error in either the benefits or the costs would tilt the balance the other way. So, when comparing a 'do something' option to the baseline option, confidence is needed to be content that the 'do something' option is clearly preferable.

In this regard, the decision process explores whether the best value for money is provided while achieving the most appropriate standard of flood or coastal risk management. This can be undertaken by assessing the incremental benefit–cost ratio of each economically viable option. The mechanics of this decision process are illustrated in Section 3.7.1, but Environment Agency, Defra and Treasury guidance also – in one way or another – all explicitly and usefully note that the decision process should be modified as necessary to take account of factors that cannot be taken fully into account in the economic analysis.

2.9 SOME LESSONS FROM EXPERIENCE

- The increasingly central role of stakeholder engagement (e.g. in the EU's Floods Directive) means that the role of economic analysis is to support the stakeholders in deciding which is the best option.
- Do not look for a mechanical means of making choices. What project appraisal can provide is greater understanding of what the choice involves: decision-support, not decision-making. In a real sense, choice is a learning process through which we seek to discover what choice we should make.
- We are seeking to make 'better' choices. A key issue is therefore to establish what we mean by 'better'. One of the central conflicts in the choice may therefore be different understandings concerning what is a 'better' choice.
- In turn, this means that the process of designing flood risk management measures (engineering or otherwise) should be appraisal-led.
- As a learning process, we should start by identifying the critical parameters as these affect the choice between options and concentrate our attention upon those parameters.
- It also means that we should hope that project appraisal will help us to invent or discover new options during the course of the appraisal process. For example, once we have discovered the advantages and disadvantages of two different options, this may suggest that a combination of some aspects of those two options, in a portfolio of measures, could be superior to both of the two original options.
- There is no universally superior project appraisal technique; the choice of technique has to be matched to the reasons why the choice is necessary.
- Do not expect all choices to be clear-cut; some will be truly marginal.

Even though economic analysis is a central component of the appraisal process, it should always be used critically and only as an aid to decision-making; it is not an end in itself. This is because many choices will remain subjective, irrespective of the economic analysis conducted. For example, core societal values, often left as intangibles in a benefit–cost analysis, are arguably more important than the tangible values that can be priced and are, therefore, easier to handle. In any case, economic values are themselves subjective, not absolute.

3 Flood risk management benefits

Theory and practice

This chapter contains a summary of the latest information and techniques concerned with gauging the benefits of flood risk management, both for fluvial environments and at the coast.

In this respect, Section 3.2 gives information about the types of flood damage that can be averted by flood risk management schemes,[1] and Section 3.3 outlines the stages involved in the calculation of annual average flood damages, and the nature of the loss-probability relationship that is central to this calculation. Sections 3.4 and 3.5 provide a guide to the data required for flood risk management appraisals, including the use of flood damage data, the impact of flood warnings on flood damage reduction, and the whole question of managing data quality.

Section 3.6 identifies some key loss-probability curve issues related to practical problems of gauging the benefits of relatively low design standard schemes, where these lead to considerable residual flooding (as is often now the case). Important in this is the discussion of Above Design Standard (ADS) benefits, since these have often been neglected in the past but can be crucial to scheme worthwhileness.

Section 3.7 briefly reviews the parameters surrounding decision options, including the use of benefit–cost ratios and net present value statistics and the issues of write-off and 'capping'. Finally, Section 3.8 highlights some particularly important lessons from experience, which we hope will guide the users of this Manual to efficient scheme appraisals.

3.1 INTRODUCTION

Inevitably, in this relatively abbreviated but self-contained Manual it has not been possible to incorporate a discussion of all the complexities of fluvial and coastal flood risk management project appraisals.

The majority of users will find this volume sufficient for their needs, but others who wish to understand some of the background discussion are therefore referred to previous manuals in this series. These include the 'Blue Manual' (Penning-Rowsell and Chatterton, 1977), which remains useful as an overview, and the 'Red Manual' (Parker *et al.*, 1987), which describes in particular detail the indirect effects of flooding. The reader is also referred to the 'Yellow Manual' (Penning-Rowsell *et al.*, 1992), which discusses flood alleviation only at the coast, but does present some information which remains useful. Finally, the 'Multi-Coloured Manual' (Penning-Rowsell *et al.*, 2005a) brought all this material together. However, like this volume, it is necessarily abbreviated in parts.

In addition, the reader is referred to material in Penning-Rowsell and Green (2000a and 2000b), who review some of the lessons learned in this research field since 1992 and forms the basis of some of the sub-sections below.

In this respect we would like to repeat the point made in Chapter 1, that this Manual is not a completely new volume but rather a continuation of the process of updating and improvement that has been undertaken throughout our series of volumes, starting in 1977. This process is one of incremental improvement, not one of complete change.

3.2 TYPES OF FLOOD DAMAGE, FLOOD LOSS AND LAND USE ENHANCEMENT

3.2.1 Direct and indirect flood damages: tangible and intangible

The benefits of flood alleviation within flood risk management comprise the flood damage averted in the future as a result of plans and schemes to reduce the frequency of flooding or reduce the impact of that flooding on the property and economic activity affected, or a combination of both. Hence we need to understand the nature of flood damage losses,[2] and also the impact of flood risk reduction and land drainage on agricultural productivity in areas liable to be affected.

It is customary to categorise flood damage and losses as direct or indirect, and by whether they are tangible or intangible (Parker *et al.*, 1987, 2). Direct damages result from the physical contact of flood water with damageable property. The magnitude of the damage may be taken as the cost of restoration of the property to its condition before the flood event, or its loss in market value if restoration is not worthwhile. Indirect flood losses are losses caused by disruption of physical and economic linkages of the economy, and the extra costs of emergency and other actions taken to prevent flood damage and other losses. These indirect effects of floods can be suffered by productive activities and emergency services both within and beyond the area of immediate, direct physical flood impact.

Many items of flood damage loss are a function of the nature and extent of the flooding, including its duration, velocity and the contamination of the flood waters by sewage and other contaminants. All these affect damages and losses, and the location of the flood will affect the networks and social activities disrupted, causing indirect losses.

This situation is summarised in Table 3.1, and different appraisal techniques are necessary to gauge accurately the magnitude and extent of the flood losses categorised there. What is generally important, however, in each case is to ensure that for the purposes of benefit–cost analysis we assess only the national economic losses caused by floods and their indirect consequences, rather than the financial losses to the individuals and organisations which are affected.

Table 3.1 Direct, indirect, tangible and intangible flood impacts, with examples

		Measurement	
		Tangible	Intangible
Form of loss	Direct	Damage to building and contents	Loss of an archaeological site
	Indirect	Loss of industrial production	Inconvenience of post-flood recovery

Source: Parker *et al.*, 1987

The exception here, as discussed in Chapter 1, is when the funders of flood risk reduction measures are restricted to local communities, or local communities make a significant contribution to total costs. Here it may be more important to look at the financial losses incurred in these communities in order to gauge the financial worthwhileness of them making their contributions towards these costs. The policy move towards 'partnership funding' (Defra, 2011a) makes this approach more likely than has been the case in the past.

3.2.2 Flood losses and land use

The land use of a flood-prone area profoundly affects its likely damage characteristics. Houses are affected differently from offices and shops, which in turn suffer different kinds of damage from that experienced in industrial or commercial premises. Areas of recreational land use such as golf courses and playing fields are likely to suffer little damage, as are certain agricultural land uses.

The effect of land use on flood damage and losses has determined the structure of all the manuals that we have produced since 1977, and this remains the case for this volume. Therefore, Chapter 4 discusses damage and loss in residential properties, Chapter 5 discusses flood losses in non-residential properties (NRPs), and Chapter 6 examines other losses, related both to land use and independent of land use. Chapter 9 discusses the situation with regard to agricultural land.

The rationale for this approach is that the type of use of a particular area of the floodplain affects both its likely direct damage potential and the magnitude of losses caused by an interruption of the economic activities that are carried out there (i.e. the indirect flood losses). The type of land use will also affect the magnitude of any necessary emergency services and works required to protect or alleviate the damage to these areas during times of flooding. The type of land use on the floodplain will also determine, to some extent, its environmental value, which could be enhanced by flood risk management schemes or harmed if such schemes are implemented without due consideration for those environmental values. Hence the term 'land use' encapsulates a number of characteristics of a floodplain area, and should not be confused with 'property' or 'urban development'.

3.2.3 Secondary indirect losses

As benefit–cost analysis has become more sophisticated, and project appraisers have sought to identify and quantify more and more benefits, so attention has been given to the effect of floods on the wider economy, beyond the immediate floodplain. The idea is that floods affect not just the property directly affected and the related indirect economic transactions, but that the impact spreads further into the neighbouring economic units affected by a general disruption to the regional or sub-regional economy: this is termed a *secondary indirect effect*.

Quantifying any secondary indirect flood effects has always been elusive and problematic (Higgins, 1981), yet a recognition of the increasing interconnectedness of modern industry and society leads to suggestions that losses in one location could spread to other economic activities well away from the flood-affected area. Adopting a proportional approach to the assessment of secondary indirect effects is recommended: focusing on where significant additional losses are most likely to occur.

In this vein, some research evaluated the effects on the wider local and regional economy of loss of trade in the flood-affected area (Flood Hazard Research Centre, 1992). The hypothesis was that a major flood in the Thames Valley, UK, would have a serious effect on that local economy and that this effect would spread far into the economy of south-east England.

The research used a model of the economic structure of the sub-region and the degree of interaction (as inter-trading) between businesses located there. Economic multipliers (Parker *et al.*, 1987) were used to quantify the linkages between the floodplain economy (the businesses affected by the flooding) and the economy in, first, the local flood-free area (within 15–20km of the floodplain) and, second, the sub-regional economy (up to 50km from the floodplain). The stronger the linkages, the larger would be any adverse economic effects of the floods, in the form of lost income in the adjacent economy owing to lack of sales to the flood-affected businesses.

To quantify these impacts, we needed to know what the effect of any loss of income by the flooded businesses would be on the income of the other businesses in the local economy and the more distant sub-region through reductions in their demand for inputs to their businesses when they are disrupted owing to the floods. Using multipliers calculated from surveys of how and what the flood-affected businesses buy from the local area and the sub-region, we found that every £1 million of income generated by businesses located in the floodplain generates a further £60,000 and £90,000 of business income in the local area and sub-region, respectively. Losses of £1 million, owing to the flooding, will have the reverse effect.

Applying this multiplier to the indirect losses calculated for the flood-affected businesses gives the results in Table 3.2. This reveals that while the indirect losses are small in relation to the direct losses (*c.* 3.4 per cent), the secondary indirect effects are even smaller in the locality (0.2 per cent) and in the much larger economy of the sub-region (0.33 per cent). Smaller still is the effect on the local and sub-regional economy from reduced purchases owing to loss of wage income by employees laid off during the disruption of the flooded businesses: this was assessed at only £68,000 at the sub-regional level for the 0.5 per cent probability flood (0.08 per cent of direct damages).

The reason for these low values is that the economy of this area (and most floodplains adjacent to large centres of population in modern market-based economies, such as Britain) is comprised largely of services, rather than manufacturing, for which there are low indirect flood losses because demand for these services can easily be transferred elsewhere: the economy is very 'open'.

Thus, while the economy of an area seriously affected by floods may appear at first sight to be 'hard

Table 3.2 Secondary indirect effects of floods: the case of the lower Thames area (1992 prices)

Loss category		Damage and losses caused by a major flood (0.5 annual probability)		Size of economy: input linkages (£m) (purchases from floodplain businesses)	Secondary indirect losses as percentage of input linkages
		Potential losses*	As percentage of direct damages		
Direct flood damage		£85,404,000	100.00 %		
Indirect flood losses		£2,866,041	3.36 %		
Secondary indirect effect:	Locality	£171,962	0.20 %	86.73	0.20 %
	Sub-region	£278,006	0.33 %	140.25	0.20 %
Loss of income from wages loss:	Sub-region	£68,000	0.08 %	140.25	0.05 %

Note: * It is recognised that these figures are given in a form that is far too precise; this is done for illustrative purposes only. For definitions, see text.

Source: After FHRC, 1992 in Penning-Rowsell and Green, 2000c

hit', the impacts are largely financial (i.e. affecting only the firms flooded) rather than economic (affecting the nation). Therefore, the national significance is small, given that the trade lost by the affected premises simply forms extra income for those located elsewhere.

This research – albeit restricted in scope – suggests that the vast majority of the full economic effects of floods in terms of damage and losses are located in the floodplain and the nearby adjacent areas, and that to pursue these losses further afield is not likely to be worthwhile. Financial losses, however, will be much higher and therefore this area may have added significance from a partnership funding perspective.

The exception might be situations where the floodplain comprises a high proportion of the country concerned. Thus, in Hungary the country's floodplains constitute 22.3 per cent of the nation's area, compared with the UK's 4.8 per cent (Evans *et al.*, 2000, 02.4.2). Appraisals results for Hungary (Evans *et al.*, 2000) showed that the potential one-hundred-year event losses could be as high as 8 per cent of GDP, or 35 per cent of the country's annual public expenditure budget: very high figures, and a threat to that country's whole economy. The losses in the 2007 floods in England (Chatterton *et al.*, 2010) were only approximately 0.36 per cent of national GDP (at *c.*£1,400 billion in 2007) and therefore perhaps only some 0.6 per cent of public expenditure totals. So the potential impact of such flooding in England – whilst serious – is less than might be possible in Hungary and not a threat to economic activities far removed from the flood-affected area.

3.2.4 Land enhancement with flood risk management

Flood risk reduction may result in enhanced land values in both urban and rural areas. In each case this may result from the same land use becoming more profitable, as a result of the reduced hazard, measured either in terms of the greater production sustained or in terms of the rent that the land commands. Alternatively, a new land use may be possible – in the urban context perhaps allowing residential development where only recreational land existed before, and in agricultural areas allowing cereal or root crops where previously rough grazing predominated.

In the urban context, the Treasury has not allowed as a benefit of flood risk management the enhanced value of land which could be converted to more productive economic activity (i.e. the move from recreational land to residential development, indicated above). This is because to do so would provide a subsidy to that development, through flood risk management being provided by the state, based on counting the private gain to the developer from the protection thus afforded. However, once a flood risk management scheme has been constructed an 'escalator effect' may occur whereby further flood management may be required – and can be justified under the Treasury rules – to protect against increasing flood damage potential owing to post-scheme development (Parker, 1995).

In the agricultural context, however, the opposite has been the case. Here the calculation of the benefits of flood risk reduction (and land drainage) may include the enhanced future productivity of the land in terms of its production of crops of higher value or greater volume (Chapter 9). Although the costs of agricultural land drainage and agricultural flood risk management have, in the past, been substantially met from the public purse, the gains in producing more crops have been interpreted as benefits to the nation as a whole, particularly at times when national self-sufficiency in food was a primary agricultural objective (Tunstall *et al.*, 2004). Nevertheless, those appraising flood risk management schemes across both the rural and the urban sectors should note the potential for the different treatments.

3.3 CALCULATING ANNUAL AVERAGE DAMAGES

The methodology for assessing the benefits of flood risk management is well known and relatively uncontroversial (Penning-Rowsell *et al.*, 1992, 1994). In essence, it combines a hazard assessment, in terms

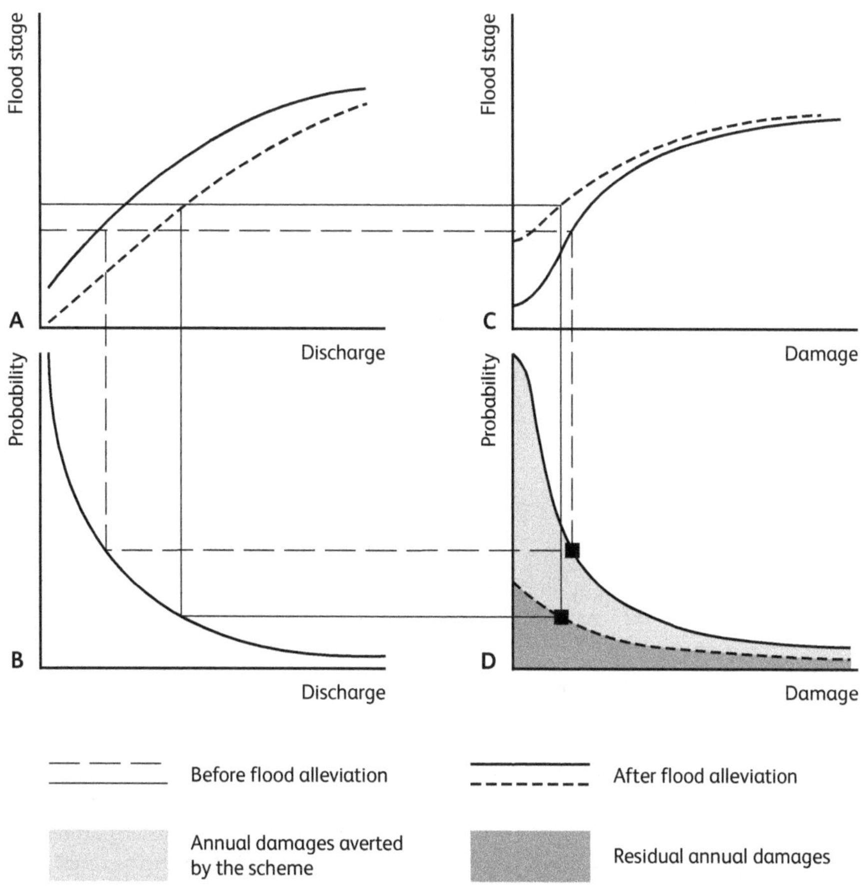

Figure 3.1 The classic four-part diagram summarising the calculation of annual average flood losses

of the probability of future flood events to be averted, and a vulnerability assessment in terms of the damage that would be caused by those floods and therefore the economic saving to be gained by their reduction.

3.3.1 Urban flood risk management schemes

Figure 3.1 provides the classic four-part diagram summarising the interrelation of hydrology, hydraulics and economics as the basis of calculating the benefits of flood risk management.

Those benefits must always be calculated with the 'with and without' criterion in mind: benefit–cost analysis looks at the difference in the nation's resources with and without the investment that is being appraised (Chapter 2). Thus, in Figure 3.1 (D) it is the difference between the relationship between probability and damage with and without the flood risk management scheme that comprises the benefits of that scheme. Therefore, if a scheme does not prevent all the flooding that might occur in the future, but residual overtopping of fluvial or coastal defences is likely or a failure of the scheme is possible, then the damage from such continuing or possible additional future flooding in these situations should be appraised. The current relationship between flooding and damage is the 'do nothing' situation, and the reduction in flood damages at a range of flood probabilities represents the result of the necessary investment in reducing the frequency of flooding or the impact of that flooding on the land use affected.

In terms of calculating the benefits of flood risk management, therefore, assessing the relationship between probability and damage (the loss-probability relationship) is crucial. Figure 3.2 gives a simplified

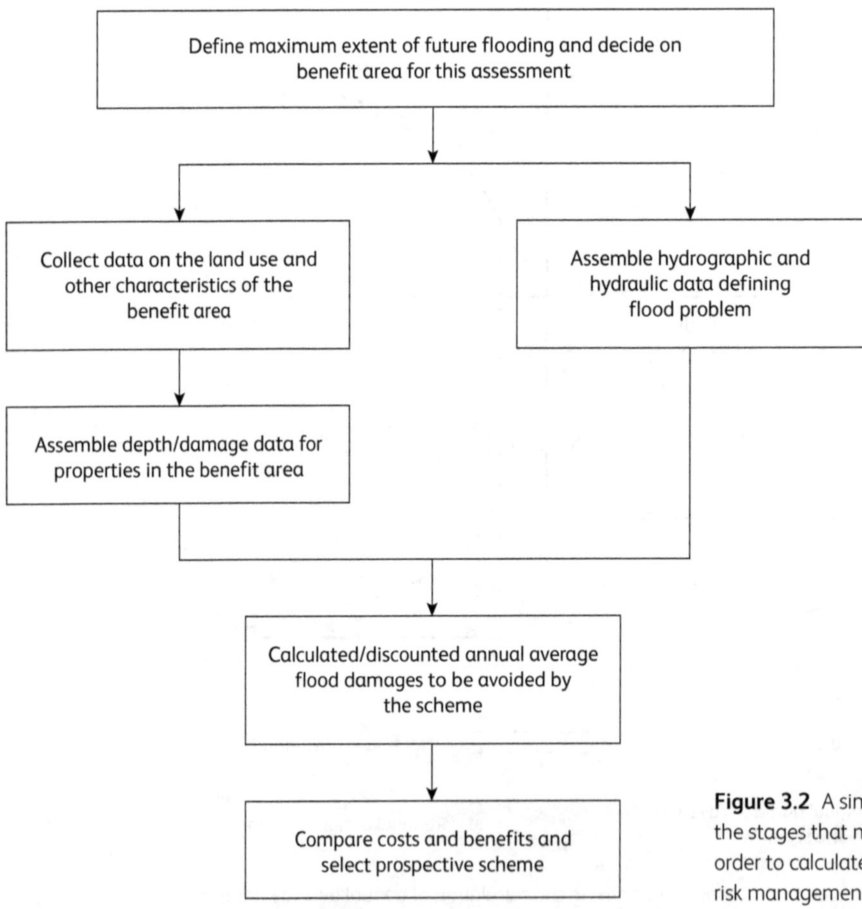

Figure 3.2 A simplified flow chart of the stages that need to be followed in order to calculate the benefits of flood risk management

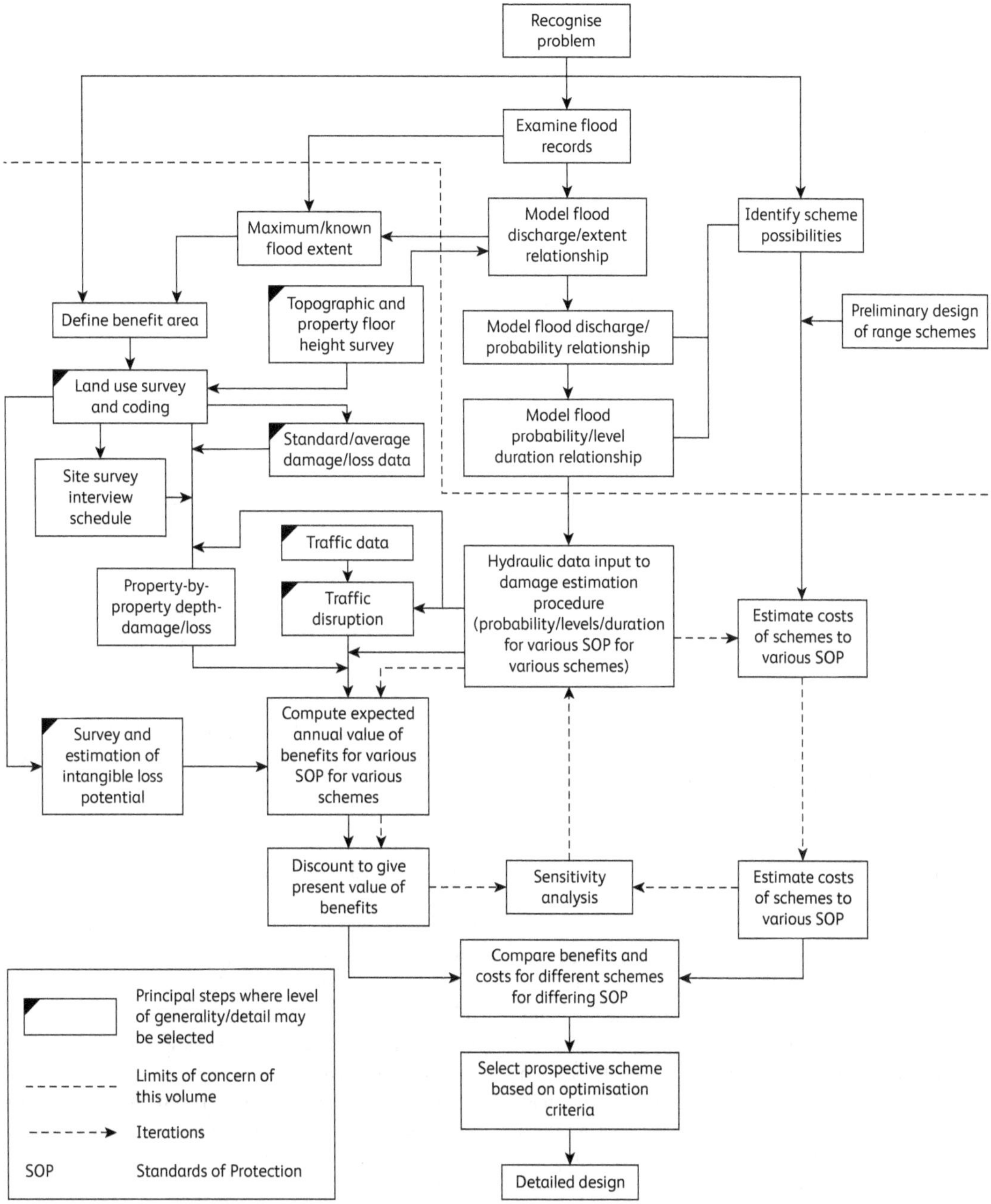

Figure 3.3 A more detailed flow chart of the various stages that have to be completed so that flood risk management benefits can be compared with the costs

flow chart of the stages that need to be followed in order to calculate the benefits of flood risk management (or, put another way, the stages for calculating the capitalised value of flood losses that will occur in the future if a 'do nothing' option is adopted). Figure 3.3 gives a more detailed flow chart, identifying the various stages that have to be completed before the benefits can be compared with the costs and the option selected that best passes this test of economic efficiency.

The annual average flood damage is the area under the graph of flood losses plotted against exceedance probability (the reciprocal of the return period in years). This is the area under the curve in Figure 3.1 (D), and this curve should be derived from an analysis of several future floods with a range of severities, or the results will be unreliable (at least three and preferably six flood events should be used). It is important, in particular, to determine the annual probability of that flood at which damage begins, assuming that some floods cause no measurable damage. It will also be important in the coastal situation to gauge accurately the timing and annual probability of the failure of any existing defences, if that is the mechanism of flooding to be alleviated. Indeed, assessing the probability of failure of existing defences remains one of the most difficult fields in the area of project appraisal for flood risk management expenditure, since there is still a very sparse research record on which to base the assessment of such probabilities (HR Wallingford, 2005a; Buijs *et al.*, 2005).

3.3.2 Agricultural flood risk management and land drainage schemes

This subject is dealt with more comprehensively in Chapter 9, but some general principles need elaboration here.

First, the 'with and without' criterion must still apply, namely the benefits of agricultural flood risk management and land drainage are the difference between the 'do nothing' approach and the economic change brought about by the investment in arterial and field drainage. The calculation of the benefits of agricultural flood risk management and land drainage are assessed as the value of the enhanced agricultural productivity, net of subsidies, discounted to present values. In addition, there may be an element of crop damage reduction inherent in reducing the frequency or severity of flooding on agricultural land, and this can be added to the results of the benefit calculations for enhanced land productivity.

In the agricultural scene it is not customary simply to calculate the annual average enhancement to land productivity as the benefit of flood risk management and land drainage. This is because those benefits accrue some years after a scheme has been developed, as the farmers take up the enhanced agricultural circumstances and enjoy the rewards of different cropping regimes or reduced costs.

In this respect it should be noted that few agricultural schemes are now being pursued within the capital programme for flood risk management in England and Wales. In the most part the maintenance of flood risk management and land drainage schemes is currently concentrated on high-risk urban systems provided the benefit–cost ratio calculated in the system asset management plans is substantial. Low-risk systems of primarily agricultural land use will attract only minimum needs maintenance expenditure, for example to guarantee legal environmental objectives. Any assessment of the worthwhileness of such maintenance regimes needs to be based on an assessment of the benefits of this investment using methods described in Chapter 9.

3.4 DATA INPUTS FOR FLOOD RISK MANAGEMENT APPRAISALS

Many data inputs are required for the accurate appraisal of flood risk management schemes, and use of these techniques over several years has identified the key priorities discussed below.

3.4.1 Defining the benefit area

The benefit area is the starting point for assessing the benefits of flood risk management; it is the area affected by the flood problem, both directly and indirectly. Usually the benefit area will be the maximum known extent of flooding in the area or catchment involved.

In the UK this will often be based on the indicative floodplain (IFP) maps, as collated by the Environment Agency. But the limitations of the current IFP maps should be appreciated. Although much improved from their predecessors, they are still not intended to identify localised areas where flooding may occur due to blocked or inadequate drainage, nor flooding associated with many minor watercourses or other causes, such as overland flow or high groundwater levels. The Environment Agency is continually improving flood mapping in these areas, including 'surface water flooding' maps (Environment Agency,

2013a), but some of these are in development and require further work whereas the basic fluvial and coastal floodplain map is now the result of concerted effort over many years, including local inputs to seek the maximum possible accuracy.

However, it may also be necessary to extend the benefit area beyond the floodplain as conventionally defined by, say, the 1 or 0.1 per cent annual probability event. This is because the calculation of Above Design Standard benefits (see Section 3.6.2) generally requires the assessment of the impacts of reducing more extreme flood events beyond any anticipated 'design flood'.

An assessment of the indirect effects of flooding can also extend well beyond the floodplain (see Chapter 6). Telecommunications, road and rail traffic disruption can occur many kilometres from the floodplain, as a flood can cause disruption to those communication and economic linkages and that disruption extends to communication links not themselves flooded. Emergency services costs will be incurred by local authorities, police authorities and fire authorities located well beyond the floodplain (see Chapter 6).

In the coastal situation it will generally be necessary to assess the floodplain as the area subject to flooding if current defences are breached. This is not an easy matter, because breaching can result in either widespread flooding or a fairly small flood event, geographically, depending on whether the breach is sealed either slowly or rapidly.

3.4.2 Assessing vulnerability to flooding: land use in the benefit area

As indicated above (Section 3.2.2), the approach to assessing the benefits of flood risk management is through investigating the potential damage to a variety of land uses in the areas to be affected.

As in previous manuals, we provide a Classification of Land Use (see MCM-Online Supporting Information for Chapter 3). Whilst this is very detailed, it is not matched by the potential flood damage data as researched at Middlesex University and presented here in Chapters 4, 5 and 6. Indeed, in some respects such detail is inappropriate, in that flood risk management investment is generally being appraised over a life of a scheme lasting perhaps fifty years or more. Therefore, one particular property on the floodplain which is now, say, a butcher's shop, could well not have these characteristics in fifty years' time; by then it could have become a residential property or indeed a small factory. It could even have been demolished and replaced by a warehouse.

However, it is customary within benefit–cost analysis of flood risk management investment to consider only the land use as currently existing (except where the future flood regime is likely to make current use untenable and property is assumed to be 'written off' or subject to change of use, or agricultural land becomes suitable only for less productive uses). Lessons from experience suggest that the level of detail as contained in the classification can be excessive, and the land use classification provided on MCM-Online is for guidance only.

The philosophy underlying the presented land use classification (see MCM-Online Supporting Information for Chapter 3) is one of a hierarchy: if flood damage data are not available at the level of the 'sub-category', then the 'category' or 'type' data should be taken instead. In general we now believe that the focus should be on the 'category' or 'type' of land use, rather than the 'sub-category' (see above).

For a fully comprehensive and detailed assessment of benefits it will be necessary to determine:

- the geo-reference of the property (the grid reference);
- the altitude of the threshold of flooding at that property; and
- the area of the property in square metres if the property is non-residential (in this Manual, as in previous manuals, the residential house is taken simply as a house rather than as a number of square metres of residential property).

A number of properties can be recorded at the same location, and properties can be separated by reach or sub-area if this is desired.

Research evidence indicates that the social grade of the occupants of residential properties is a good indicator of damage potential and these differences are reflected in the standard damage tables provided with this Manual. Clearly for some initial or high-level strategy studies it will not be appropriate to obtain the detailed information required for such social grade banding, though this is increasingly available from standard sources (e.g. http://www.statistics.gov.uk). In these cases it is reasonable to use average values for each property type. However, in any detailed studies where decisions depend upon realistic damage assessments, it will be necessary to use the relevant damage tables by social grade of the residents and then consider use of the recommended distributional adjustments (see HM Treasury, 2003, and Chapter 4 herein).

It should be noted that if average/sector depth/damage data are used then the use of distributional adjustments is no longer appropriate. Nevertheless, it is a common mistake, we have learned, that this is done (and it has the effect of increasing benefits). The reason it should not be done is that the averages are weighted by the social grade of the occupants of the houses concerned, so the distributional adjustment is built into that average data and should not be applied twice.

3.4.3 Flood damage data and its application: our general approach

The general approach to assessing the benefits of protecting individual properties from flooding adopted in this and previous manuals encapsulates the following principles:

- It is necessary to assess the potential damage in the future from a range of severities of flooding, resulting from different depths of flood waters within the property. Only in this way will the shape of the loss-probability curve be accurately determined (see above).
- Much of the flood damage data presented here, as in previous manuals, is 'synthetic'. It is therefore not directly derived only from an analysis of properties which have been flooded in the recent past, because evidence suggests that post-flood surveys can either exaggerate flood losses (in the immediate aftermath of the flood) or incorrectly miss significant items of flood damage (where the survey is done before the full impact on the building has had time to emerge). (The philosophy underlying this approach is discussed in more detail in Penning-Rowsell and Chatterton, 1977, Chapters 2 and 3, and in Chapter 4 herein.)
- The losses to individual properties for use in benefit–cost analysis must generally represent national economic losses (see Section 3.2.1). Therefore, the assessment of damage to property components (i.e. inventory items) is based on their assumed pre-flood value – their depreciated value – rather than the cost of their replacement with new items at current market prices. This has serious implications for a comparison between our data and that produced by loss adjusters, who tend to assess damage as the cost of replacements with new items, because 'new for old' insurance policies now dominate the insurance marketplace.
- In addition to the above point, it is necessary to subtract any taxation element within potential flood losses, because these are transfer payments within the economy rather than real resource costs. Therefore, the VAT element in repair costs is not counted towards the benefits of flood risk management, significantly depressing the value of flood losses as assessed here in comparison, again, with loss adjuster calculations and other data derived from insurance claims. However, except for VAT adjustments, it is rare to adjust flood intervention scheme costs for other transfer payments (e.g. fuel duty or National Insurance payments).
- In the area of indirect flood losses, it is necessary strictly to separate financial and economic losses by not including within the benefits of flood risk management the loss of income in one particular retail shop if the trade this represents is likely to be deferred in time or transfer to another retail outlet.

We recognise the assumptions given above are not the same as those under which other flood damage datasets have been produced. Therefore, the Dundee flood damage data (Black and Evans, 1999) is a collation of data from insurance companies, inclusive of VAT and not counting flood damages as loss of depreciated values. That data should therefore not be used in scheme appraisals. A set of financial flood damage data will be made available by the Flood Hazard Research Centre (MCM-Online), and as a generalisation such damage values are approximately twice those derived when taking a strictly economic approach, as presented in Chapters 4 and 5.

3.4.4 The impact of flood warnings on flood damages

The flood damage data provided on MCM-Online follow the tradition established by Penning-Rowsell and Chatterton (1977) (and by Parker *et al.*, 1987) in representing the maximum potential damage, ignoring the damage reducing effects of action taken after flood warnings.

But we know that effective and timely warnings can and do result in lower levels of damage and loss. Since the late 1970s, research has continued on flood warnings and their effects, summarised by Parker (1991). Particularly important have been surveys of those to whom flood warnings have been issued, designed to determine the actions that they were able to take, the damage that was thereby averted and the factors that affect this (including flood warning lead-time, the availability of assistance with moving vulnerable household goods, etc). A total of over 1,200 interviews were undertaken in the 1980s and 1990s (e.g. Parker and Tunstall, 1991) and more have been undertaken since (see Chapters 4 and 5).

This research generally shows that the early estimates of what damage and loss could be saved with warnings were far too high; we know now that far less can be – and is – saved, perhaps because the recent research has been better but also perhaps partly because people now no longer see damage-saving as a high priority on receipt of a flood warning (given that many household and commercial premises' contents are insured on a 'new for old' basis), with preventing injury and loss of life the key priorities after a flood warning.

This in turn means that the maximum potential damage figures associated with this Manual are likely to be nearer actual damage totals than otherwise would be the case. However, scheme appraisal should recognise that warnings can reduce damages, and therefore that as warning systems become more sophisticated, and the resultant damage savings become more routine and secure, the economic justification for additional flood risk management measures will decline, because their benefits will be lower than would have been the case were there no warnings at all.

3.4.5 Topographic and flood surface data

Experience with many project appraisals has indicated that one of the most important inputs to benefit assessments is the topographic data describing the floodplain and the accuracy of the hydraulic profiles that intersect this surface. In Britain, many floods are relatively shallow, slow-moving and represent water accumulating towards the lower end of catchments. In these circumstances, accurate delineation of the area liable to flooding and the precise depth of flood waters on that floodplain are both essential in the production of accurate hazard and benefit assessments.

A good example is provided in the Lower Thames in southern England, where the floor levels of a very large number of properties were surveyed by field levelling (some 8,000 in total out of approximately 13,000 properties in the floodplain). Table 3.3 shows the build-up of damage with flood probability, as modelled by a combination of the Halcrow ONDA/ISIS hydraulic model and the Middlesex University ESTDAM damage assessment model (Thames Region *et al.*, 1992).

In the last two decades there has been a major improvement in terrain and hydraulic models such as ISIS, MIKE 11 and Jflow, allowing analysts to convert data describing the floodplain terrain and flood discharge into flood surfaces, using as many nodes as can be accommodated within the computer systems being used. Several thousand nodes can be employed, giving the point altitude of flood waters for different annual probabilities at each node. Many consultants have produced damage models which can extrapolate between these nodes to provide a three-dimensional flood surface.

The River Thames in this area is not out of bank until approximately the 25 per cent probability event but both the number of properties affected and the

Table 3.3 The relationship in 1992 between flood probability and impacts for the Datchet to Teddington River Thames floodplain, UK, prior to Jubilee River

Annual probability of flood event	Number of properties affected	Event damage (£000s)*
20%	440	1,200
11%	909	2,800
4%	5,200	13,000
1.70%	8,600	31,000
1%	10,400	49,000
0.50%	12,400	85,000

Note: * 1992 values.

Source: FHRC, 1992

damage caused rise sharply for events with annual probabilities smaller than around 10 per cent.

However, Table 3.4 shows the extreme sensitivity of the relationship between calculated damages and terrain and flood surface data. Whereas the best estimate for the 1 per cent probability flood event shows 10,400 properties being affected by flooding (including up to 30cm below threshold floor level), adding 10cm to the flood surface – or conversely subtracting 10cm from all the threshold heights – raises this to 11,600 properties, an increase of 10.8 per cent. The damage suffered in those properties in these circumstances increases by 30.4 per cent and the capital sum representing the effects of floods of all probabilities up to 1 per cent annually increases by 50.6 per cent (because the floods of greater probability are also increased by 10cm in depth in this simulation). Caution should be applied to interpreting these results as the calculations for the 0.5 per cent flood did not take account of properties outside the indicative floodplain that would probably be affected by an event of this magnitude.

The point of this example is to show that flood depths are therefore clearly crucial for accurate flood hazard appraisals (and benefit assessments). In this respect it appears inherently less likely that the water surface level will be inaccurately estimated by up to 10cm than that the topographic ground surface will be subject to errors, simply because the extra discharge required to increase flood volumes and therefore flood depth by 10cm is huge in areas such as the Lower Thames, whereas it is quite possible to have inaccuracies of this order of magnitude in the heights of property thresholds as a result of sparse benchmarks, inaccurate interpolations between known heights, or even inaccurate levelling and GPS errors.

There are, of course, counteracting factors. It is unlikely that all topographic data will be the subject of this type of systematic error. Rather, errors will tend to cancel each other out. The most serious impacts will be where the properties affected by the high-probability flood events are incorrectly surveyed, since these contribute proportionately more to annual average damages.

The implication of these results is that considerable attention should be given to this area of data input to project appraisals. Inevitably this will be costly, but our experience suggests that in large areas the levelling of the floor or doorstep levels of the properties in the floodplain – or similar GPS-based surveys – should not cost more than approximately £3 to £4 per property if the properties are relatively close together and modern levelling equipment is used. The use of helicopter-mounted light detection and ranging (LIDAR) for obtaining good topographic data is to be encouraged, notwithstanding its difficulties in dense urban areas where the necessary ground floor property heights can be confused with building-top heights. However, LIDAR methods are improving all the time, and the helicopter-based results are far better than the earlier plane-based systems, so this

Table 3.4 The impact of increasing flood depths by 100mm for the Datchet to Teddington River Thames floodplain*

		A + 10 cm	B Best estimate	C A/B (%)
1 % annual probability flood	Capital sum (£m)	34.8	23.1	50.6
	Event damage (£m)	64.3	49.3	30.4
	Properties affected	11,600	10,400	10.8
0.5 % annual probability flood	Capital sum (£m)	41.4	30.2	37.1
	Event damage (£m)	103.8	85.4	21.6
	Properties affected	13,600	12,400	9.2
PMF**	Capital sum (£m)	51.1	38.1	34.1

Notes: * All at 1992 values; ** Probable Maximum Flood.

Source: FHRC, 1992

whole area needs attention as the accuracy of the data derived continues to improve.

3.4.6 Flood probability data

The estimation of the probability of flood events contributing to appraisals is critical. This is illustrated in the case of the River Irwell in Manchester, north-west England.[3] A history of flooding over the last thirty years shows a concentration of this flooding in heavily urbanised areas with a mix of residential and industrial property (Penning-Rowsell *et al.*, 1994; Table 3.5).

An alteration of the flood probability by 10 per cent in each direction produces an 11.1 per cent and 9.1 per cent change in calculated benefits, other things being equal (Table 3.5D and E). Downward revisions of the probability of the lowest probability flood analysed (Table 3.5C) – i.e. the highest return period event – has a somewhat perverse effect in reducing calculated benefits (in this case by 9.1 per cent, with a change from the 0.66 per cent probability flood to the 0.57 per cent flood). This is because, with this change, the size of the area under the loss-probability curve is reduced.

The most significant effect of flood probability on calculated benefits shown in Table 3.5 is the change in the threshold of flooding (Table 3.5B). Merely changing from the 11.1 per cent probability flood to the 20 per cent probability threshold flood, being the flood that just causes no damage, all other data remaining the same, increases calculated benefits by up to 22.1 per cent. This emphasises the crucial importance of accurately assessing the annual probability of this threshold of flood damage to the assessment of the benefits of flood risk assessment, yet this parameter is often neglected and much more emphasis is unwisely given to much larger events, such as a 1 per cent annual probability flood (which in fact generally has very little impact on the value of calculated annual average benefits).

3.4.7 Emergency costs

Flood risk management policy has shifted away from simplistic flood defence towards 'living with floods' and 'making space for water' (Defra, 2004a), thereby accepting that significant flood damage and disruption will continue into the future. This highlights the need for efficient emergency response to flood events, yet until recently we have known very little about the resources and costs that this involves.

Research that is reported in more detail in Chapter 6 has evaluated the severe flooding in the UK in autumn 2000 and summer 2007, and shows that these costs are widely distributed, both geographically and institutionally. Geographically, they broadly match the pattern of property flooding, but they are much higher than we had hitherto appreciated. The significant implication is that we should take this topic more seriously in the future, and not see emergency response to floods and other disasters as a low-cost or even a cost-free option.

The research on the 2000 event showed, therefore, that the costs of tackling a flood emergency are more substantial than previously appreciated and these costs form part of the benefits of flood risk management. As reported in Chapter 6, the research from the 2000 event has shown that these emergency costs can add some 10.7 per cent to property damages, and that it is therefore legitimate to multiply all other potential damages in a benefit assessment by 1.107 to account for these costs.

The same approach was adopted for assessing the total emergency costs and recovery during the summer 2007 floods in England (Chatterton *et al.*, 2010). The total emergency costs were then £110 million – that is, 5.6 per cent of the total economic property losses.

The difference in terms of percentage between the 2000 and 2007 floods may be explained by an effect of economy of scale. Indeed, the summer 2007 floods affected a greater number of properties (up to 73,000 versus *c.*10,000) but fewer local authorities claimed under the Bellwin scheme (38 versus 87).

We suggest the figure obtained from autumn 2000 reflects dispersed flood-affected communities whereas the figure obtained from the summer 2007 floods reflects more densely populated areas. The total property damage calculated in project appraisals of flood risk management schemes should therefore be multiplied by a factor ranging between 1.107 and 1.056 to allow for the emergency and recovery costs that can be justified as real economic costs, not counted elsewhere in the benefit assessments. This figure should be applied for floods at all annual probabilities and for all scales of flood risk management scheme, in the absence of better information.

Table 3.5 The impact of alterations to the return periods of appraised floods (River Irwell, Manchester, UK)

A		
Baseline case		
Annual probability	Event damages*	Annual benefits**
11.10%	£0	
10%	£7,405,794	£41,143
4%	£8,425,382	£516,078
2%	£13,567,411	£736,005
1%	£54,917,490	£1,078,429
0.66%	£63,929,788	£1,276,507
0.40%	£94,547,480	£1,487,810
Rectangle		£1,865,999

B		
Higher probability: threshold flood		
Annual probability	Annual benefits	% of baseline
20%	£0	
10%	£370,290	
4%	£845,225	
2%	£1,065,153	
1%	£1,407,578	
0.66%	£1,605,657	
0.40%	£1,816,960	122.10%
	£2,195,149	117.60%

C		
Higher probabilities: maximum flood		
Annual probability	Annual benefits	% of baseline
11.10%	£0	
10%	£41,143	
4%	£516,078	
2%	£736,005	
1%	£1,078,429	
0.66%	£1,276,507	
0.57%	£1,351,972	90.90%
Rectangle	£1,892,246	101.40%

D		
Higher probabilities overall: +10%		
Annual probability	Annual benefits	% of baseline
12.30%	£0	
11%	£45,715	
4.40%	£573,421	
2.20%	£817,785	
1.10%	£1,198,257	
0.70%	£1,418,345	
0.44%	£1,653,126	111.10%
	£2,073,336	111.10%

E		
Lower probabilities overall: –10%		
Annual probability	Annual benefits	% of baseline
10.10%	£0	
9.10%	£37,403	
3.60%	£469,162	
1.80%	£669,096	
0.90%	£980,391	
0.60%	£1,160,463	
0.36%	£1,352,557	90.90%
Rectangle	£1,696,366	90.90%

Notes: * These and comparable figures here are recognised to be over-precise, but are used for illustrative purposes; ** This is the area under the loss-probability curve at the extremes of flood probabilities (see Parker *et al.*, 1987).

Source: Penning-Rowsell *et al.*, 1994

We recommend that the lower factor should be applied in urban areas to reflect the relative economy of scale in the work of the emergency services in such locations.

3.5 'FILTERING' AS AN EXPERIMENTAL AID TO TARGETED DATA QUALITY ENHANCEMENT

3.5.1 Introduction

An important source of data on the land use in floodplain areas is available through the Environment Agency's National Receptor Dataset (NRD) (Environment Agency, undated c). Use of this data reduces the need for extensive field surveys. However, allocating MCM land use codes (see MCM-Online Supporting Information, Chapter 3) and equivalent depth/damage to each property or land use item using the NRD is not an absolute science. Likewise, the accuracy of the assessment of property thresholds and 'footprint' areas (all fundamental to accurate evaluation of potential flood damages) depends on the techniques employed and the resources available. Too often hydraulic or terrain modelling 'absorbs' large proportions of scheme appraisal budgets at the expense of thorough damage and benefit assessments. Thus appraisal data requirements may be 'squeezed' and the resultant present value of damages (PVd) may become volatile or simply fallacious.

Errors in damage assessment can be categorised into two types:

- systematic error; and
- measurement error.

Systematic error relates to a misunderstanding of the necessary appraisal processes, for example:

- A property is given the wrong land use code and therefore incorrect depth/damage data are assigned, leading to an under- or over-estimate of damages.
- The benefit area is misunderstood, i.e. Flood Zone 3 is used instead of the area defined with more accurate hydro-dynamic modelling techniques, which again may seriously under- or over-estimate the number of properties within the floodplain benefit area and therefore under- or over-estimate potential damages and hence the calculated PVd.

Systematic errors can be reduced or eliminated through sound initial conceptualising within the project team, and a consistent process of audit and review followed throughout the study period.

Measurement error relates to errors in physical characteristics, such as:

- inaccurate threshold levels of property in relation to the flood profile surface; and
- inaccurate property footprint areas.

These errors can compound the systematic errors. Measurement errors are now less problematic as footprints may be easily extracted property-by-property using geographic information systems (GIS) tools. Threshold field levelling surveys are generally the norm for detailed project appraisals, whilst at a strategic level digital elevation modelling (DEM) can be an appropriate substitute if error ranges can be minimised.

3.5.2 Objectives

The object of the 'filtering' process is to minimise all potential sources of error through iterative reviews. This procedure was reported in 2005 but it remains experimental and is not yet standard practice; it is based on experience that indicates that the different data elements have different qualities. It is recognised that data often have imperfections, so an important task is to recognise these imperfections and react to them.

Thus all data sources, in this procedure, should be given a Data Quality Score (or DQS) (Posford Duvivier *et al.*, 2002) and successive iterations within the benefit assessment should seek to obtain more accurate results by focusing on those data sources which: (a) are most critical to the final results; and (b) have the lowest data quality.

The recommended objective is to improve the quality of the data that make most contribution to calculated benefits, using a system that is transparent and auditable. The description below is for calculating the benefits of flood risk management but the principles can apply equally to other areas of benefit assessment, including for coastal erosion risk management appraisals. We urge appraisers to continue to trial this approach and provide feedback.

3.5.3 Data assembly and DQSs

Appraisers need to assemble the following for each property in the benefit area. (The National Receptor Dataset (NRD) is a useful source of land use data.)

1 The land use category.
2 The floor area (NRPs only: see Chapter 5).
3 The threshold height of the property.
4 The most appropriate level of detail for depth/damage data (from MCM-Online).
5 The hydrologic/hydraulic profile data (or similar) for each return period analysed.

Appraisers should assign a DQS for each of the five elements of the dataset above: 1 = good; 4 = poor (see Table 3.6). The best achievable score is 5 and the worst is 20.

Table 3.6 The system of Data Quality Scores

DQS	Description	Explanation
1	'Best of Breed'	No better available; unlikely to be improved on in near future
2	Data with known deficiencies	To be replaced as soon as third parties reissue
3	Gross assumptions	Not invented but deduced by the project team from experience or related literature/data sources
4	Heroic assumptions	No data sources available or yet found; data based on educated guesses

Source: Posford Duvivier *et al.*, 2002

3.5.4 Procedure

Proceed as follows with the benefit assessment:

1 Calculate the present value of flood damages (PVd) for each property in the benefit area, then rank each property in order of the contribution it makes to the total PVd.
2 'Cap' PVd at each property's market value and rank again. Market value data sources include:

- Residential: Land Registry website, etc., for the property's postcode.
- Non-residential: Valuation Office Agency from http://www.voa.gov.uk (rateable value).

3 Highlight properties contributing individually more than a pre-determined percentage of total PVd (usually 1 per cent for schemes with up to 250 properties; 0.1 per cent with more properties).
4 If the DQS for either the threshold or hydraulic/hydrology data is 3 or more, this is unacceptable. If the total DQS for the threshold and hydraulic/hydrology data is 4 or more, this is probably unacceptable.
5 If the total DQS for any of the highlighted properties is more than 10 then revisit their data sources and improve the data items with the highest DQSs.
6 Recalculate PVds and re-rank the capped PVd list.
7 Repeat steps 2 to 5 as a second iteration.
8 Repeat iterations until total PVd stabilises (i.e. marginal changes in data quality make no significant difference to total PVd values).

3.5.5 An example from a large study in the Thames catchment

This process is illustrated in an example from the Lower Thames study, a complex assessment of 4,000-plus properties in the Datchet to Teddington reach of the River Thames. Note that at that stage of its development the DQS system did not extend to the hydrologic/hydraulic data, as indicated above.

The stages of this process (Figure 3.4 and Table 3.7) were as follows:

1 Iteration 1 applied preliminary water surface data for only three return periods, with property thresholds 'guesstimated' using LIDAR data, which was then notoriously poor in providing an accurate estimate of levels in complex urban environments. The land use from the National Property Database (NPD) was used 'raw' without field checking, and property footprint areas were taken without

refinement as the mean values for each property type from Modelling and Decision Support Framework (MDSF) datasets.

2 Iteration 2 introduced improved water surface data for 103 co-ordinates along the Thames for an additional four return periods. LIDAR was rejected in favour of the mean ground surveyed thresholds for clusters of properties within twenty sub-areas. No field checks were done to improve the quality and quantity of land use data, although footprint areas were now derived more accurately from OS Mastermap sources.
3 Iterations 3 and 4 introduced extensive field checks, thereby eliminating both erroneous and duplicate land use records from within the NPD.
4 Iteration 5 followed Iteration 4 in all respects except the weighted mean of all MCM NRP depth/damage data was used irrespective of property type (i.e. public house, office, shop, etc.). The very small difference between the two runs illustrates the insensitivity in the large Lower Thames study of utilising detailed NRP categories and sub-categories as opposed to average data.
5 Iteration 6 introduced further improvements in the flood water surface modelling resulting in the doubling of PVd.
6 Iteration 7 reflected continued improvements to property threshold data.
7 Iteration 8 introduced yet further improvements in threshold estimates using advancements by Halcrow in LIDAR calibration.
8 Iteration 9 eliminated all spurious grid references which – erroneously – located some property outside the floodplain. The co-ordinate system used to derive the water surfaces at the river was improved using 819 node points spread across the floodplain.
9 For the final run (Iteration 10) seven sub-areas were defined, with one of these sub-areas using the assumption of no flooding for all flood return period events (i.e. taking it out of the floodplain altogether), while in another no flooding was assumed for both the five- and ten-year events (i.e. altering the threshold of flooding). This change to the data used was designed to reflect known flood extents in an improved way, and thus determined more accurately the return period at which damage begins.

Iteration 10 stabilised the PVd of the 'do nothing' situation at £132 million. The number of properties affected by flooding for Iteration 10 stabilised at 246

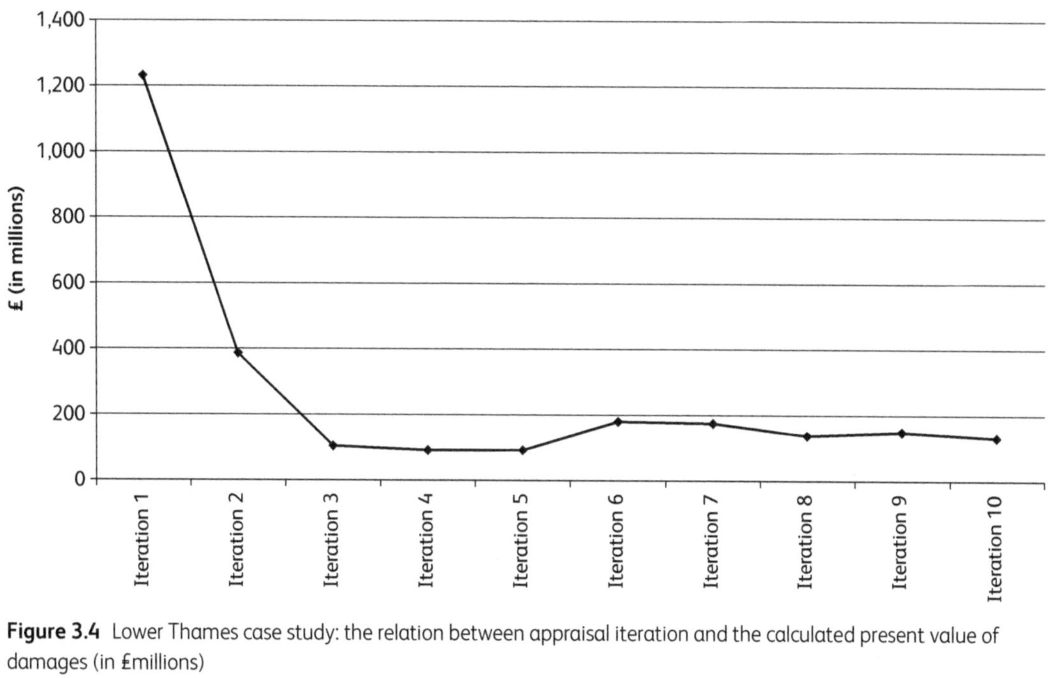

Figure 3.4 Lower Thames case study: the relation between appraisal iteration and the calculated present value of damages (in £millions)

Table 3.7 The stages in the benefit assessment process for the Lower Thames study, with the progressive and iterative refinement of the data used and the consequent reduction in total Data Quality Score (2004 prices)

Iteration number	TOTAL DQS	PVd £millions	Data Characteristics								
			Water profile surface	Property threshold	DQS	Footprint area	DQS	Land use code	DQS	Depth/ damage data	DQS
1	>9	1,229	2 water levels for 3 return periods U/s level linked to D/s with no intermediate	Lidar	4	Areas from Modelling and Decision Support Framework (MDSF) mean values	3	NPD, no field checks	2	NPD Appx C	Various
2	>6	387	103 co-ordinates for 7 return periods	Mean of levelled thresholds within each of 20 sub-areas	3	OS Mastermap	1	NPD, no field checks	2	NPD Appx C	Various
3	>6	104	103 co-ordinates for 7 return periods	Mean of levelled thresholds within each of 20 sub-areas	3	OS Mastermap	1	First field checks	1 & 2	NPD Appx C	Various
4	>6	90.2	103 co-ordinates for 7 return periods	Mean of levelled thresholds within each of 20 sub-areas	3	OS Mastermap	1	Further field checks, reducing properties confirming land use	1 & 2	NPD Appx C	Various
5	8	91.6	103 co-ordinates for 7 return periods	Mean of levelled thresholds within each of 20 sub-areas	3	OS Mastermap	1	As iteration 4	1 & 2	Weighted mean of all NRP D/D data in Appx C	2

6	>5	181	110 co-ordinates for 7 return periods with increased water levels	Mean of levelled thresholds within each of 20 sub-areas	2	OS Mastermap	1	As iteration 4	1 & 2	NPD Appx C	Various
7	>5	175	110 co-ordinates for 7 return periods with increased water levels	Mean of levelled thresholds from increased sample	2	OS Mastermap	1	As iteration 4	1 & 2	NPD Appx C	Various
8	>4–5	139	110 co-ordinates for 7 return periods with increased water levels	Threshold adjustments from Halcrow Lidar analysis	1 & 2	OS Mastermap	1	As iteration 4	1 & 2	NPD Appx C	Various
9	>4–5	147	819 water level points for 7 return periods	Threshold adjustments from Halcrow Lidar analysis with corrections to spurious georeferences	1 & 2	OS Mastermap	1	As iteration 4	1 & 2	NPD Appx C	Various
10	>4–5	132	819 water level points for 7 return periods. 7 sub-areas and water profile adjustments		1 & 2	OS Mastermap	1	As iteration 4	1 & 2	NPD Appx C	Various

for a five-year (20 per cent annual probability) flood event, rising to 4,422 for the 200-year (0.05 per cent) event.

This Lower Thames example is typical of a complex appraisal (more than 5,000 properties within the modelled floodplain), with up to ten iterations necessary to stabilise the PVd. For less sophisticated appraisals, it is envisaged that some three to four iterations are necessary to reduce total DQSs sensibly, so as to provide relative consistency and confidence in the calculated PVd.

3.5.6 Assessment

This filtering process should help to manage complexity in most medium and large appraisals, to develop consistency in the evaluation process, avoid unnecessary work and avoid erroneous conclusions. It does this by:

- Determining likely high contributors to total flood impacts – which will have implications on investment decisions under partnership funding (see Chapter 1).
- Quantifying the characteristics of these high contributors effectively, and assessing the quality of the data used.
- Ensuring, during the quantification process, that data quality is appropriate to the scale of analysis and the volatility of the results.
- Providing an audit trail.

3.6 LOSS-PROBABILITY CURVE ISSUES

As indicated above, the loss-probability curve describes the relationship between estimated flood losses and flood probability (Figure 3.1). Floods with a high probability of occurrence usually inundate a smaller area than those of lower probability, and to less depth, and hence low-probability floods generally affect more property and generate higher flood losses than those occurring more frequently.

3.6.1 Residual flooding and dis-benefits

The Environment Agency's FCERM-AG decision rules on flood risk management scheme design standards (Environment Agency, 2010a) encourage those promoting schemes to seek the lowest standard of protection commensurate with maximising the difference between costs and benefits. One consequence of this is that schemes may be chosen with lower design standards than was conventional in the past, and that these schemes therefore do not protect wholly or even significantly against the more major floods.

This is obviously easiest to understand when applied to an engineering scheme, which has a finite design standard which may be quite low. But the principle applies equally to other flood risk management measures, such that, for example, a flood warning system may reduce some flood losses but not all, and therefore damage will continue into the future, albeit at a lesser level than hitherto. In each case this leaves residual flooding after the scheme has been implemented, and this damage from residual flooding should not be counted towards the benefits of the scheme.

To assess these residual dis-benefits requires the assessment of the impact and damage of the major floods not avoided, and hence the modelling of the extent of these floods on the floodplain and the assessment of their indirect disruption costs (Penning-Rowsell *et al.*, 1994).

Such assessments will often not be easy, particularly for very low-probability floods because they can extend over a large area. Nevertheless, assessing the damages from the residual flooding can inform judgements about flood risk management measures ultimately for those floods, for example by providing a new or enhanced flood warning system to lessen some of the residual damage were a structural engineering scheme to be implemented at a relatively low standard. What is important, therefore, is to assess the full range of flooding at any particular location, irrespective of the design standard or flood risk management measures to be chosen. The residual flooding must not be ignored in any choice of policies or schemes, but these benefits should not be included to compare with the costs of the low-standard option if that is chosen.

3.6.2 Above Design Standard benefits

Above Design Standard (ADS) benefits accrue where engineered flood risk management schemes result in water levels changing for the whole range of floods experienced on a floodplain, not just the events with annual probabilities up to and including some design event (Penning-Rowsell, 1996). The situation where these ADS benefits will be most important is where there is much urban development at the outer edges of the floodplain, only affected by the rarest and most substantial floods, and where modest schemes can reduce flood water levels and therefore extents at these locations, even if only marginally.

However, only certain types of scheme have this hydraulic effect; for example, levée (raised embankments or walls) schemes do not. The most obvious schemes where ADS benefits accrue are bypass channels and, in most circumstances, flood storage reservoirs. In both cases water is taken off the floodplain during flood events and routed either into and through extra channels (the bypass channel) or into designated storage areas (the storage reservoir). The same principles apply to channel enlargement schemes; the flood waters stay within bank for flood events with lower probability or longer return periods, and the water surfaces are lowered for events with all return periods.

Calculating ADS benefits inevitably means modelling the nature and impact of floods that have a lower annual probability than the anticipated design standards to be provided (although that standard itself should not be decided in advance of the modelling but should be determined as a result of the economic analysis, following the appraisal-led design philosophy explained in Chapter 2).

Thus the full range of floods at a particular floodplain – not just the minor events and any prospective design flood – all need hydrological and hydraulic analysis. In addition, the land use survey used within the benefit assessment part of the benefit–cost comparison will need to be sufficiently extensive to allow for the extent of the largest flood to be analysed.

Thereafter the full impacts of all these floods will need to be assessed, thereby calibrating the relationship between flood probability and damage/loss for the pre-scheme 'do nothing' situation. Since the difference between the pre-scheme and the post-scheme situations is the basis for calculating the benefit total, a full analysis of ADS benefits requires the evaluation of the impact of the full range of floods after the implementation of the scheme or schemes being appraised, not just the pre-scheme situation.

This can be illustrated with respect to the standard relationship between flood damages and probability: the loss-probability curve, the area under which equals the annual average flood damage at that site (Figure 3.5). Here, area A in Figure 3.5 is the area up to the exceedance probability of the design flood, and this has often traditionally been taken as representing the total benefit. But total benefits are represented by A + B, and the proportion accounted for by the ADS benefits is therefore B/(A + B). In practice, the number of points on these loss-probability curves may be low, owing to considerations of the cost of the appraisal itself, but the very minimum number of points should be such as to fix points a, b, c and d, so that area B can be estimated.

Results are available for ADS benefits for three locations appraised on two parts of the River Thames (the Maidenhead, Windsor and Eton reach, and the Datchet, Wraysbury, Staines and Chertsey reach, to the west of London), and one scheme on the River Irwell, near Manchester.[4] All three cases show that ADS benefits can be substantial and, indeed, in certain cases, crucial to the cost-effectiveness of the proposed schemes (Penning-Rowsell and Green, 2000a; 2000b).

In the case of the Maidenhead bypass channel (the Jubilee River), in one of the final benefit–cost analyses the total benefits of protecting against all floods up to the 0.5 per cent annual probability flood amount to £66.48 million (Thames Region *et al.*, 1992). The

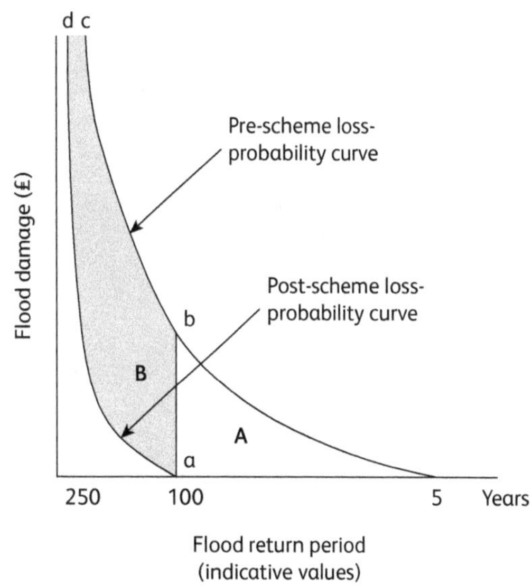

Figure 3.5 The derivation of ADS benefits

anticipated design standard of the bypass channel is an annual probability of 1.7 per cent, or approximately the annual probability of the 1947 event on the Thames (Table 3.8). The benefits of flood risk management for protecting against floods up to just that standard is £42.7 million for a scheme of life of fifty years and with the discount rate of 6 per cent then current. However, the bypass channel has an effect on all floods with lower annual probabilities, including the 0.5 per cent annual probability flood. Using this information, and netting out the effect of residual flooding in the post-scheme situation, yields the results shown in Table 3.8. The ADS benefits contribute 34.2 per cent of the £64.1 million of realisable benefits.

In the case of the Datchet, Wraysbury, Staines and Chertsey (DWSC) reach of the Thames the comparable results (Table 3.8) show that the ADS benefits amount to some 31.5 per cent of total benefits. The benefits of protecting the development on the floodplain in this area with bypass channels are some £33.7 million, net of the residual flooding, for a 1.7 per cent annual probability design standard, preventing damage to some 12,000 properties. Post-scheme residual flood damage would be substantial, however, given the extensive flat floodplain and the substantial impact of the larger flood events.

The extent of this post-scheme residual flooding amounts to some £1.13 million (capital sum) with the 0.5 per cent probability event still causing £9.2

Table 3.8 Above Design Standard (ADS) benefit results from appraisal of flood hazard problems on the River Thames, UK (1992 values)

River Thames reaches	Total annual average damages (to 0.5 % annual probability flood)	Total residual flood damage post-scheme (1.7 % annual probability standard)	Total ADS benefits (with 1.7 % annual probability scheme standard)	ADS benefits as % of total realisable benefits
Maidenhead, Windsor and Eton	£66.47m	£2.4m	£21.89m	34.18
Datchet, Wraysbury, Staines and Chertsey	£34.84m	£1.1m	£10.61m	31.48

Table 3.9 Above Design Standard (ADS) benefit results for the River Irwell, Salford, Manchester, UK (1994 values)

Scheme	Project description	Total 'do nothing' annual average damages (£m)	Total residual flood damages (£m)	Total ADS benefits (£m)	ADS benefits as % of total realisable benefits
A	Two-basin scheme (design standard 0.74% annual probability flood)	27.16	0.71	6.76	25.6
B	One-basin scheme (design standard 1% annual probability flood)	27.16	1.51	5.96	23.3
C	Channel improvements only (design standard 2% annual probability flood)	27.16	6.88	6.84	33.7

million of damage (but this is small compared with the £83.8 million in the pre-scheme situation). In capital sum terms the ADS benefits contribute £10.6 million to the £33.7 million of total benefits.

The Irwell situation is different in that flood storage was anticipated on the floodplain within large meander loops (Penning-Rowsell *et al.*, 1994). A number of scheme options were appraised, ranging from a two-basin flood storage option, a one-basin option at a lower design standard, and a much more modest river channel improvement scheme to retain the 2 per cent annual probability flood within bank. In each case the residual flooding is obviously different (being greater with the lesser design standards), and the amount of ADS benefits also vary with the design standard, mainly because of this different degree of residual flooding.

Table 3.9 gives the benefit calculation results. Despite the very different design standards of the proposed schemes, the ADS benefits lie in a relatively narrow range from 25.6 per cent (the 0.7 per cent annual probability flood standard two-basin Scheme A) to 33.7 per cent (the 2 per cent annual probability standard river maintenance works only Scheme C). Unlike the River Thames schemes, the justification of none of the schemes rested on these ADS benefits, since the benefit–cost ratio of even the most ambitious Scheme A was 5.7:1. Nevertheless, the ADS benefits are large and influence the choice between the three schemes because the marginal benefit–cost ratio is affected by these parts of the benefit totals (MAFF, 1999).

3.7 DECISION RULES AND OPTIONS

Flood risk managers and government policy-makers have begun to think more carefully about different types of plans and schemes for flood risk management, including non-structural alternatives, and about different standards of flood risk management or risk reduction. These choices are difficult ones, and need to be guided by careful analysis of their relative merits. The broader scope of such approaches should be reflected in the economic analysis which, when properly applied, can be a very useful aid to decision-making between very different options (see Chapter 2).

3.7.1 The choice of design standards: decision rules (2005)

The situation in 2005 was set out in our previous 'Multi-Coloured Manual' (Penning-Rowsell *et al.*, 2005a, 51–52). The Defra Project Appraisal Guidance Notes (FCDPAG3; MAFF, 1999), applicable in England and Wales, had provided a logical decision-making approach with regard to the standards of flood risk management to be implemented.

In summary, this approach identified the standard of flood protection which has a positive incremental benefit–cost ratio, compared with a lower standard, while also targeting 'indicative' minimum standards of protection for particular land uses. (These standards were given as ranges to reflect the range of conditions within the land use bands, which went from 'High density urban' through to 'Low density or rural communities' and to various intensities of agricultural use.)

The approach was designed to provide a significant improvement on a simple process of maximising benefit–cost ratios or net present values which might well tend to favour relatively low-cost solutions which might eliminate much of the economic damage from frequent events but would not be socially acceptable in densely populated urban areas.

3.7.2 The choice of design standards: a hypothetical example

We illustrated the broad situation with regard to Figure 3.6, which remains useful, if somewhat simplistic. What it raises is the role of the incremental benefit–cost ratio (IBCR).

That diagram gives results for a small hypothetical flood risk management scheme for a medium-sized low-density urban community. On strictly economic grounds the scheme to choose should be Scheme C because it has the highest overall benefit–cost ratio commensurate with an IBCR above unity and a 2 per cent annual probability protection standard (i.e. above the 4 per cent indicative standard).

But there are potential pitfalls here, as Scheme D could be chosen if Scheme C is never appraised. Scheme D has the desired positive IBCR (if Scheme C is omitted from the analysis) and more than meets the indicative protection standard of service. But it involves spending some £1 million more to generate

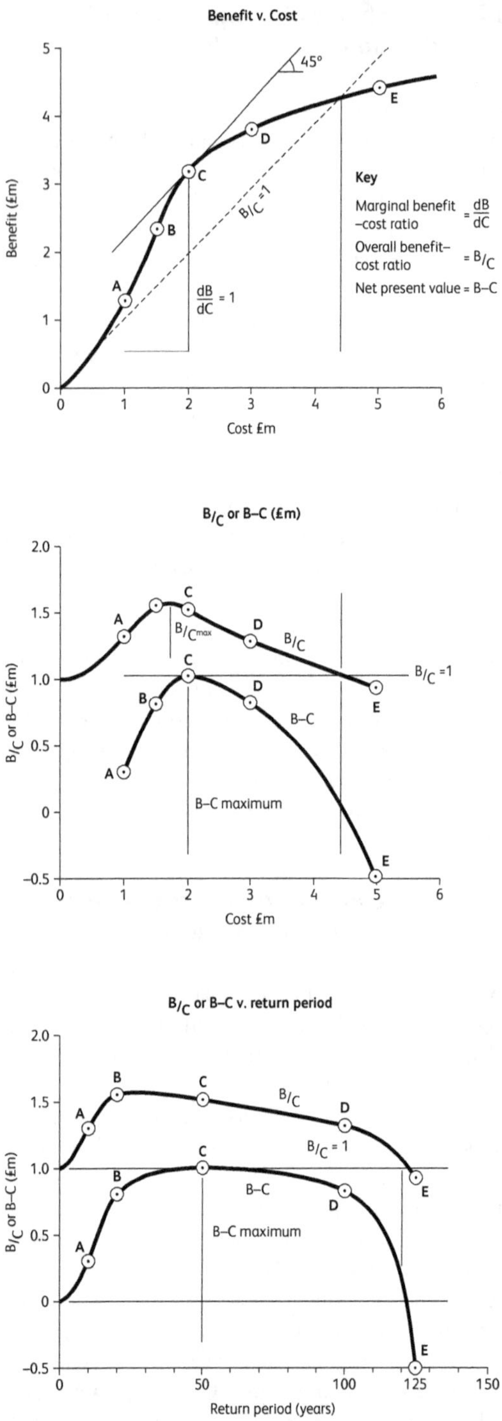

Figure 3.6 Benefit–cost ratio and benefit minus cost comparisons

Source: After Tim Yates

only approximately £500,000 extra benefits compared to Scheme C.

3.7.3 Decision rules and options: new arrangements (2010)

The Environment Agency (2010a) takes this matter several steps further, and gives, first, procedures for decision-making when cost-effectiveness analysis (CEA) is the appropriate appraisal tool (e.g. where legislative considerations override a purely economic approach and the least-cost scheme that delivers the proper legally acceptable outcome is required). Second, very slightly different procedures for benefit–cost analysis are provided for schemes in Wales as opposed to those for England.

In each case a step-by-step procedure is required (see also Figure 3.7), with key questions at each stage:

Stage 1 Is the average benefit–cost ratio greater than 1?
Stage 2 Identification of a leading option by ranking the options in one of two ways:

- by reducing probability of flooding; or
- by average benefit–cost ratio (where options cannot be organised by reducing probability of flooding).

Stage 3 Does the option choice change when you take contributions to the costs into account?
Stage 4 Does the option choice change when you take uncertainty into account?
Stage 5 Does the option choice change when you consider other factors (such as the extent to which wider objectives and/or providing enhancements are delivered)?

At Stage 2, it is necessary to calculate the incremental benefit–cost ratio for each option (this will be automatically calculated if using the spreadsheets provided in Environment Agency, 2010a, 254). Users should move through the options (one by one) to assess whether the IBCR is sufficient to allow them to identify a different leading option. They should then compare the leading option with the IBCR of the next option. If the IBCR of the next option (the next option is defined as the option that provides the next-lowest annual exceedance probability (AEP) compared with the leading option) is greater than the thresholds

Calculation of economic decision-making criteria (England)

- ***(average) benefit-cost ratio:*** divide the total PV benefits by the total PV costs (ensure that the total costs include optimism bias). The same approach is used for the implied benefit-cost ratio where scoring and weighting has been undertaken (in this case, the implied benefit-cost ratio should be used in addition to the average benefit-cost ratio)

  State 1: $\dfrac{\text{PV benefits Option 1}}{\text{PV costs Option 1}}$

  State 3: $\dfrac{\text{PV benefits Option 1}}{\text{PV costs}_{\text{FCERM}}\text{ Option 1}}$

  Where $\text{costs}_{\text{FCERM}}$ = whole life costs minus contribution from non-FCERM funding sources

  The average benefit-cost ratios of solutions taken further in the decision-making process must be greater than one (or the non-monetised benefits must outweight the costs) for the option to be implemented.

- ***incremental benefit-cost ratio:*** For the do-nothing option and those options with average benefit-cost ratios greater than one (see above), arrange the options in either increasing order of benefit or increasing order of cost. For each pair of neighbouring options in this "chain", subtract the PV benefits of Option "1" from those of Option "2" and divide these by the difference in PV costs of Options 1 and 2 (subtracting any contributions from non-FCERM funding sources in Stage 3). (As with the average benefit-cost ratio, the approach to calculation of the implied incremental benefit-cost ratio is the same as that for the incremental benefit-cost ratio):

  State 2: $\dfrac{\text{PV benefits Option 2} - \text{PV benefits Option 1}}{\text{PV costs Option 2} - \text{PV costs Option 1}}$

  State 3: $\dfrac{\text{PV benefits Option 2} - \text{PV benefits Option 1}}{\text{PV costs}_{\text{FCERM}}\text{ Option 2} - \text{PV costs}_{\text{FCERM}}\text{ Option 1}}$

  Where $\text{costs}_{\text{FCERM}}$ = whole life costs minus contribution from non-FCERM funding sources

  The incremental benfit-cost ratio is used to identify how 'much' can be delivered. Any option Different thresholds for the required IBCR are used when deciding which option should form the preferred solution (see the need for IBCR thresholds).

- ***net present value:*** total benefits minus total costs

  PV benefits Option 1 – PV $\text{costs}_{\text{FCERM}}$ Option1

  Where $\text{costs}_{\text{FCERM}}$ = whole life costs – contribution from non-FCERM funding sources

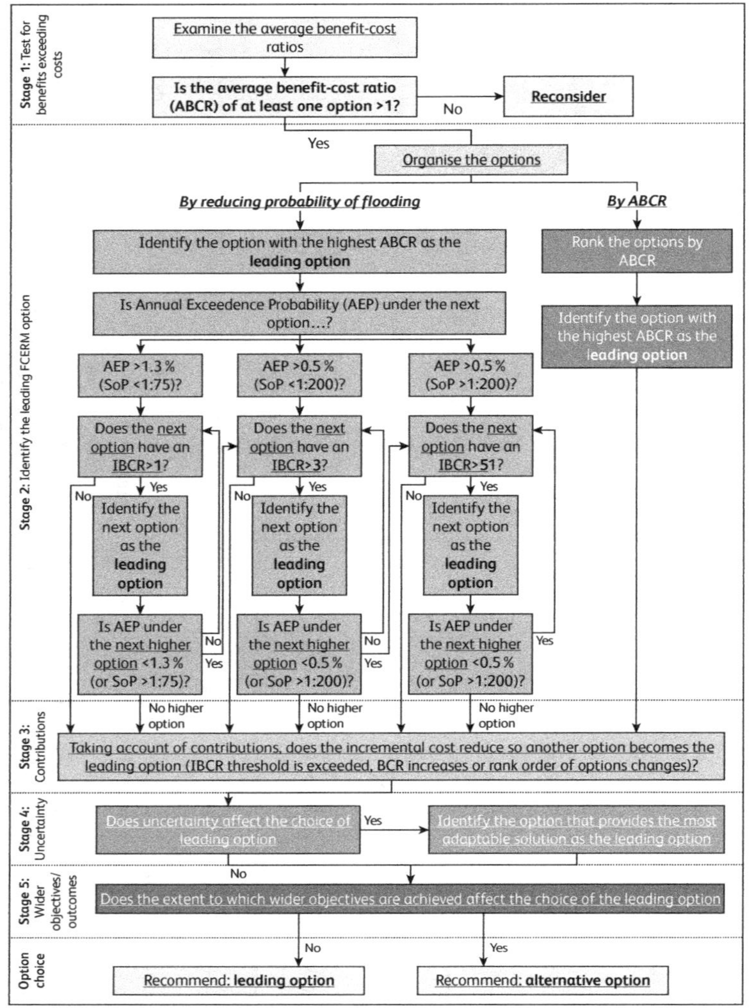

Figure 3.7 The 'optimisation' of option selection using a stepwise response to marginal benefit–cost ratios

Source: Environment Agency, 2010a, 273

Table 3.10 Incremental benefit–cost ratios

Option type/ risk level	Minimum requirement for option to be preferred
Options with existing AEP greater than 1.3 % (or Standard of Protection (SoP) <1:75 year)	IBCR > 1
Options with existing AEP less than 1.3 % but greater than 0.5 % (or SoP between 1:75 year and 1:200 year)	IBCR > 3
Options with existing AEP less than 0.5 % (or SoP >1:200 year)	IBCR > 5

shown in Table 3.10, then the next option becomes the leading option.

The thresholds set out in Table 3.10 identify the IBCR that is required at different levels of AEP to change the leading option. Users should then consider whether the next-highest option is viable (defined as the option with the next-lowest AEP compared with the previous next option, now the leading option).

It is important to remember that one can move from one option to the next only if the IBCR exceeds the threshold relevant to the level of AEP offered by the next option. Users should not 'jump over' options that have an IBCR that is lower than the thresholds. Once the user has identified the leading option using the IBCR thresholds, Stage 3 takes on board possible contributions (reducing costs to the FCERM budget), Stage 4 then tests for volatility in relation to uncertainty and Stage 5 asks if a broader view is necessary. If the answer here is 'No' then the last-considered option is the one that should be chosen.

3.7.4 Write-offs and 'capping'

Properties that are projected to be flooded on average more than once every three years are usually considered to be written off unless they are flood resilient or water compatible (Environment Agency, 2010a, 212–213). This is because there will not usually be sufficient time for the property to be repaired and return to full use following the previous flood before the next flood occurs. As a result, repairing the property would be a waste of money. Write-off values are taken as the risk-free market value of an asset because the actual market value of the at-risk property could be lower. (Where the risk is known, there may be lower demand for the property or higher insurance costs such that the market value is reduced.)

'Capping' is different to write-off. Care should be exercised where the total present value of projected flood losses exceeds the risk-free market value of the asset. In the case of residential or commercial property, appraisers should assume that the long-term economic loss cannot exceed the current capital value of the property and 'cap' the damages if this is likely. Capping will apply to any property if the PVd over the lifetime of the proposed scheme is greater than the market value.

Write-off and capping both use the risk-free regional average market value to ensure that the risks are not already reflected in the market value of the property (see Horsfield, 2012, for a map of the government regions). For a non-residential property it may be necessary to use its rateable value multiplied by a factor that reflects the added value or percentage rental yield from that property. The average rental yield (residential) in September 2012 ranged from 5.6 per cent to 6.5 per cent (Rightmove, 2012), so with current yields of around 6 per cent, a multiplier of 16.7 would be appropriate (we assume a 6% return annually on capital here; hence it takes 167 years to recoup the capital).

Appraisers need to proceed as follows:

- Where assets (such as properties, infrastructure, land) are flooded more frequently than once every three years – or eroded – they are written off.
- Where assets (such as properties, infrastructure, land) are flooded less frequently than once every three years, it is assumed that damages are incurred on each flood up to the point where the total PVd equals the risk-free market value of the asset and capped at that value. Once the damages have been capped, no further damages are estimated even though flooding may still continue.
- Where assets (such as properties, infrastructure, land) are flooded occasionally over the first part of the appraisal period and are written off at a later date as the frequency of flooding increases (as is usual under the 'do nothing' scenario), the approach is to determine when properties might be abandoned (i.e. flooded so frequently that their whole value would be lost) and discount their

write-off value. Add to this the damages that would occur in terms of average annual damages until the time of write-off. It may be necessary to cap the total damages where they exceed the market value of the property.

To give an example of the last case (Environment Agency, 2010a, 213), a property is currently flooded occasionally (say, on average every five years). However, by year 20 it is expected that – owing to the deterioration of its flood defences and climate change – the property will be flooded once every three years. The property would be written off in year 20 within the appraisal calculations. In practical terms this means taking the average annual damages for the property over the first twenty years and then writing off the value of the property (discounted from year 20 to the present day). The cash flow streams of damages are a complex mix of write-off capping and annual average damage (AAD) combinations of properties flooded at different stages changing through the time horizon of the appraisal.

In the case of other assets, such as roads, railway lines, pipelines or cables, some very large values can be generated for long-term disruption. Appraisers should assume that the maximum allowable economic benefit derived from flood risk reduction is equal to the economic cost – depreciated to allow for the age of the existing asset – of reconstructing an equivalent facility at a higher level or on an alternative alignment which avoids the flood risk. Care is needed when capping the damages of infrastructure to make sure that adequate account has been taken of the service that the infrastructure provides; otherwise, there is a risk of significantly underestimating the real effects of loss of the infrastructure.

For properties or infrastructure which do not have an obvious market value (such as ancient monuments, public lavatories, gas pipelines and electricity distribution pylons), a surrogate value based on the cost of moving, replacing or diverting the asset can be used.

It is important that the capping value used is the risk-free value (where the property value does not reflect the risk of flooding or erosion). This can be achieved by using county or regional averages, as discussed above.

Appraisers will know if they have a serious capping problem if the estimated damages of 'do minimum' exceed the damages of 'do nothing' (except where the intervention leads to significant environmental or other benefits). Less significant capping problems can occur where flood damage to a small proportion of the total number of properties affected exceeds their market value. Careful description of the impacts here, accompanied by clear and auditable spreadsheets to estimate the damages in money terms, should help to ensure that capping is undertaken wherever it is needed.

3.7.5 Decision rules and options: issues remaining

What the above ideas encapsulate is the need to optimise investment levels in relation to levels of risk and residual risk in order to obtain 'best value for money'. As such they are a major advance. However, as yet, these arrangements cannot optimise the level of investment when a portfolio of measures is needed, yet these are likely to be the best flood risk management arrangements in the future (see Evans *et al.*, 2004).

If a combination of flood risk management measures being appraised includes some conventional engineering works to reduce the probability of flooding, plus a better flood warning system and a more sophisticated arrangement for evacuation and other aspects of emergency response, together with a tighter spatial planning regime designed to prevent the build-up of flood damage potential into the future, then the optimisation of the different elements of the portfolio is in no way helped by the step-by-step procedure identified above, yet it is essential. What have to be decided here are not just the 'standards of protection' for each element of the portfolio but – crucially – 'how much' of each element is to be included within the portfolio. This area is fraught with difficulty, and is therefore a topic for further research.

3.7.6 The choice of flood risk management design standards: public perceptions and attitudes

The economic appraisal of flood risk management schemes has always assumed that the most efficient scheme, as far as the public to be protected is concerned, is that which gives the maximum protection at the minimum cost. The standard of protection to design for has always been assumed to be the highest that could thereby be justified or afforded. But research shows that this is not the complete picture: the public's view is that a higher standard of protection is not necessarily 'better' if it brings unwanted features.

Figure 3.8 summarises this research, undertaken in the Lower Thames catchment, by showing that a proportion of the population at risk from floods of different probabilities is prepared to live with this risk (Fordham, 1992; Penning-Rowsell and Fordham, 1994, Figure 7.2). Indeed, at the riverside itself, where the hazard is greatest, approximately one-third of those interviewed indicated that they were prepared to live with a 20 per cent annual probability flood in exchange for their riverside location being left undisturbed by river channel works and other flood risk management engineering structures necessary to give that level of protection.

At the other extreme, 94 per cent of this population was willing to live with the risk of a 0.5 per cent annual probability flood rather than suffer – as they saw it – the negative environmental and amenity consequences of a scheme to lessen the risk and the impacts that such a flood would bring. In general the proportion of the riverside population willing to take – or live with – this risk was between one-and-a-quarter times and double the proportion in areas away from the river (where the perceived amenity losses from river works were much less).

There therefore appears to be a trade-off in the public's mind between flood protection – as one good – and other associated goods or impacts: people are willing to trade reduction of risk against amenity loss. This reinforces the point that scheme appraisal methods need to appraise the total assemblage of features of flood risk management – 'good' and 'bad' – rather than simplistically assume that the public necessarily wishes to have the highest standard of protection that the benefit–cost comparison identifies. In this case a comprehensive multi-criteria appraisal would take account of perceived benefits and dis-

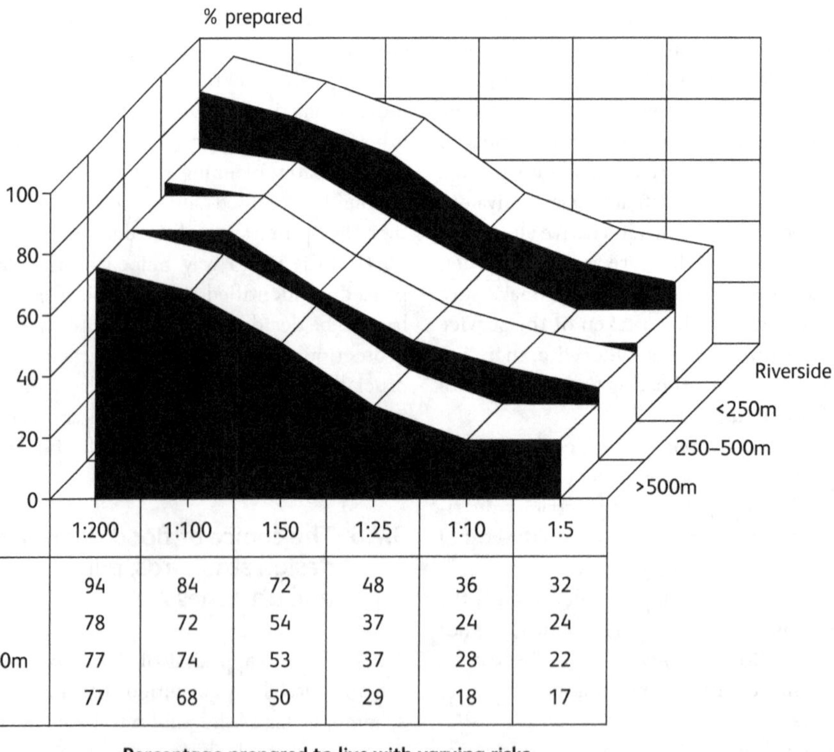

	1:200	1:100	1:50	1:25	1:10	1:5
Riverside	94	84	72	48	36	32
<250m	78	72	54	37	24	24
250m–500m	77	74	53	37	28	22
>500m	77	68	50	29	18	17

Percentage prepared to live with varying risks

Figure 3.8 Preparedness to live with flood risk: research results showing the trade-off between risk and flood plain location

Source: Fordham, 2000

benefits, and these should influence the chosen solution, even if they cannot all be valued in monetary terms.

3.8 LESSONS FROM EXPERIENCE

This chapter has covered many aspects of the theory and practicalities of appraising flood risk management schemes, and it is sometimes difficult 'to see the wood for the trees'. We would suggest prioritising the following in detailed appraisal studies, assuming that these follow on from more generalised initial investigations:

- For schemes resourced from public funds, the damages averted by flood risk management schemes should be assessed as national economic losses, not the financial losses to the individuals and organisations that suffer damage and loss. The possible exception to this, as explained in this chapter, is where local contributions form a significant part of total costs. Here there is a case for using financial analysis to judge whether those local contributions are worthwhile for the local community in terms of the financial benefits they obtain from their expenditure. In this regard, where possible, financial benefit figures are provided on MCM-Online.
- Close attention should be given to accurate determination of the area potentially affected by flooding (the floodplain). Within that exercise considerable effort should be devoted to determining the extent and annual probabilities of the lesser floods and the flood at which damage begins.
- The different quality of different data inputs needs recognition, using the DQS systems, if appropriate, to manage the process of benefit refinement targeted at those data inputs that are of poorest quality yet which contribute most to the variation in benefit totals.
- In locations where there is an efficient flood warning system which results in significantly lower damage and loss values (e.g. from the kind of sandbagging operations reported in Chapter 6), the assessment of the flood damages to be averted by further investment in flood risk management must reflect those lower damage and loss values.
- For major schemes involving considerable investment in low-lying areas (i.e. not steep catchments), close attention should be paid to the topographic data that define the thresholds of property flooding.
- Sufficient potential floods should be appraised so that an accurate picture can be developed of the shape of the loss-probability curve, including, where appropriate, such events needed to define and quantify any ADS benefits.
- Appraisers should not assume that the public necessarily wants the standard of flood protection that is identified as being optimal by the benefit–cost analyses that are undertaken.

NOTES

1 The term 'scheme' here and elsewhere in this Manual is used as shorthand to include all flood risk management policies (structural or non-structural), plans and engineering works to alleviate flooding. It could be a non-structural arrangement, such as a warning system, as well as a structural flood risk management option, such as river works, an embankment system or a dam.

2 The term 'losses' here is used to separate out the damages that result from contact with flood water and the economic losses that are not related to damaged items.

3 The figures here were derived from an initial study and do not represent final results. They are included for illustrative purposes.

4 The figures here were derived from initial studies of all three areas and do not represent final results. They are included for illustrative purposes.

4 Flood damage to residential properties and related social impacts

In our experience, residential flood damage alone rarely justifies major engineering works. Nevertheless, this should not deter the evaluation and subsequent implementation of other types of flood risk management measures to protect houses from flooding or manage that risk, not least because flood damage to residential properties is significant in almost all cases of serious flooding in the UK, and flooding in houses remains an area of public and governmental concern. This includes both direct damages and indirect losses, measured as the tangible and intangible impacts of flooding on residential properties and householders.

The majority of this chapter is dedicated to the appraisal of the direct damages and tangible impacts of flood waters on household inventory and building fabric items. In addition, important information is provided for the damage reducing effects of flood warnings and for incorporating new government guidance on the appraisal of the indirect and intangible impacts of flooding. This chapter also includes (for the first time) information on flood damage to vehicles and 'park homes'. New research has been conducted on the cost of drying and cleaning residential properties following a flood and this has been incorporated into the residential damage data provided on MCM-Online. All other depth/damage data and methods for residential properties remain unchanged from the previous 'Multi-Coloured Handbook' (Penning-Rowsell *et al.*, 2010), having been updated only to reflect the change in prices since then using the Consumer Price Index (CPI).

4.1 SUMMARY OF INFORMATION

Residential flood damage is significant in almost all cases of serious flooding in the UK, and flooding in houses remains an area of public and governmental concern. This includes both direct damages and indirect losses, measured as the tangible and intangible impacts of flooding on residential properties and householders (Table 4.1).

The majority of this chapter is dedicated to the appraisal of the direct damages and tangible impacts of flood waters on household inventory and building fabric items. In addition, important information is provided for the damage reducing effects of property-level protection measures, for flood warnings and for incorporating new government guidance on the appraisal of the indirect and intangible impacts of flooding. This chapter also includes for the first time information on flood damage to park homes, the one-storey residential properties which in theory are 'mobile' – as they are on wheels – and therefore may potentially be moved out of harm's way in the event of a flood, but which in practice are permanently sited on authorised sites and indeed are 'plumbed in' there. In the USA these would be known as 'trailers' or 'house trailers'.

The Manual does not recommend the use of detailed surveys within houses as a method for residential flood damage assessment. Rather, the assessment of direct residential property flood damage potential should utilise the standard data provided on MCM-Online. The most detailed standard data provided is for:

Table 4.1 The range of possible flood impacts on households (not exhaustive or necessarily mutually exclusive)

Direct tangible losses for flooded households	Intangible losses for flooded households	Indirect losses for flooded households	Indirect losses for non-flooded households
Damage to building fabric	Worry about future flooding	Permanent evacuation from area	Increased travel costs
Damage to household inventory items	Loss of memorabilia and irreplaceable items and pets	Disruption to household due to flood damage	Loss of income/earnings
Clean-up costs	Damage to physical and/or mental health, death or injury	Temporary evacuation costs	Loss of utility services
	Loss of community	Disruption due to flood warnings or alarms	Loss of other services
	Loss of confidence in authorities and services	Loss of utility services	Loss of leisure and recreational opportunities
		Loss of income/earnings	Increased cost of shopping and recreational opportunities
		Loss of leisure and recreational opportunities	
		Additional communication costs	
		Loss of services	
		Increased travel costs	
		Increased cost of shopping and recreational opportunities	

- five house types;
- seven building periods; and
- four different approximated social grades of the dwellings' occupants.

This gives 140 datasets in total. For each of these, information is available for:

- the basic datasets;
- the components of damage;
- long- and short-duration flooding;
- salt, wave, water quality and flood warning scenarios
- total damage per square metre.

4.1.1 Changes, 1988 to 2003/05

In comparison with the basic datasets of the Middlesex Manual published more than two decades ago (Suleman *et al.*, 1988), the sector average figures published in 2005 and given there in Appendix 4.1 (Penning-Rowsell *et al.*, 2005a), and now available on MCM-Online, increased on average by a factor of four times at a flood depth of 0.9m for building fabric items and six times at 0.9m for inventory items. In addition, changes to the susceptibility of building fabric and inventory items resulted in significantly increased damages at lower depths of flooding. These changes resulted from methodological developments, improvements in the data available and real increased costs. The primary methodological changes between 1988 and 2005 were:

- altered susceptibility of inventory items as these are now 'written off' at significantly lower levels of damage, and hence at lower depths of flooding, than was previously found to be the case;
- altered susceptibility of building fabric items due to changes in the materials used;
- the inclusion of kitchen furniture as a building fabric item, reflecting the fact that fitted kitchens are now deemed to be an integral structural component;
- the inclusion of enhanced repair costs for household electrical fittings which were previously undervalued;
- a greater understanding of the costs associated with domestic clean-up which were previously included as homeowner labour costs only;
- a re-evaluation of the ground floor plans of existing house types; and
- the inclusion of a new house type reflecting the last decade of building construction changes.

The primary data changes were:

- increased unit costs for building fabric materials, inventory items and labour;
- changes to building regulations resulting in the inclusion of additional items such as anti-fungal brickwork treatment and insulation under screed;
- the inclusion of new building fabric items such as paths and splayed skirting;
- the addition of new inventory items reflecting the increased affluence of households;
- the change in ownership of household items by social grade as reflected in market research publications; and
- increased damages for houses with cornicing, as it is now recognised that cornicing is significantly more expensive than coving. This was a new distinction.

4.1.2 Main changes, 2003/05 to 2010

For the production of the 2010 'Multi-Coloured Manual and Handbook', a comparison between the data from both 1977 and from 2003/05 with the

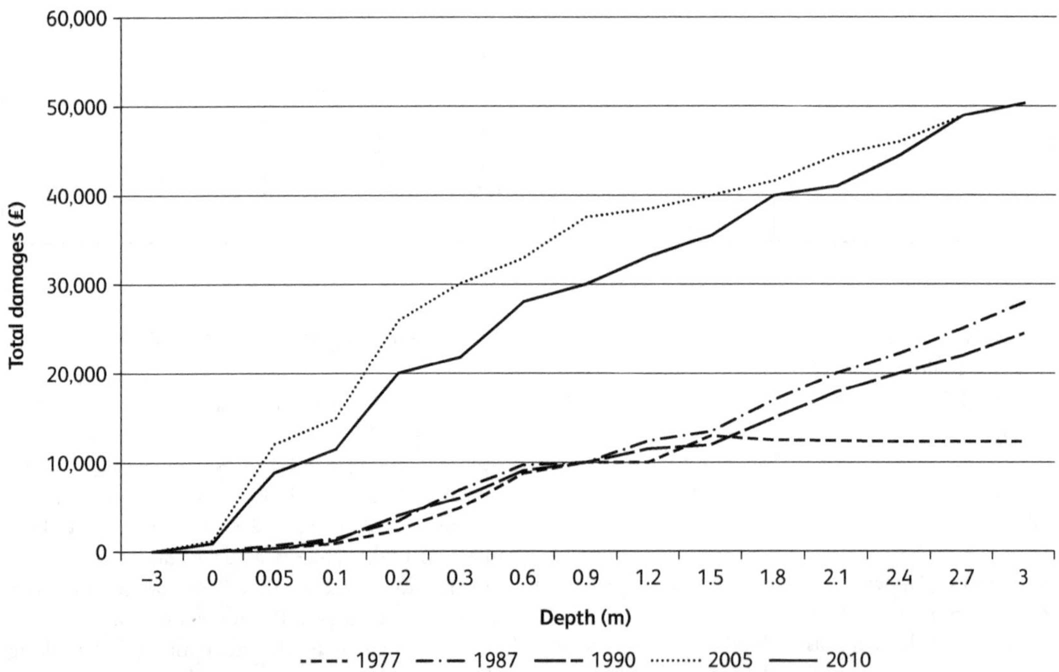

Figure 4.1 Residential property flood losses: 1977 to 2010

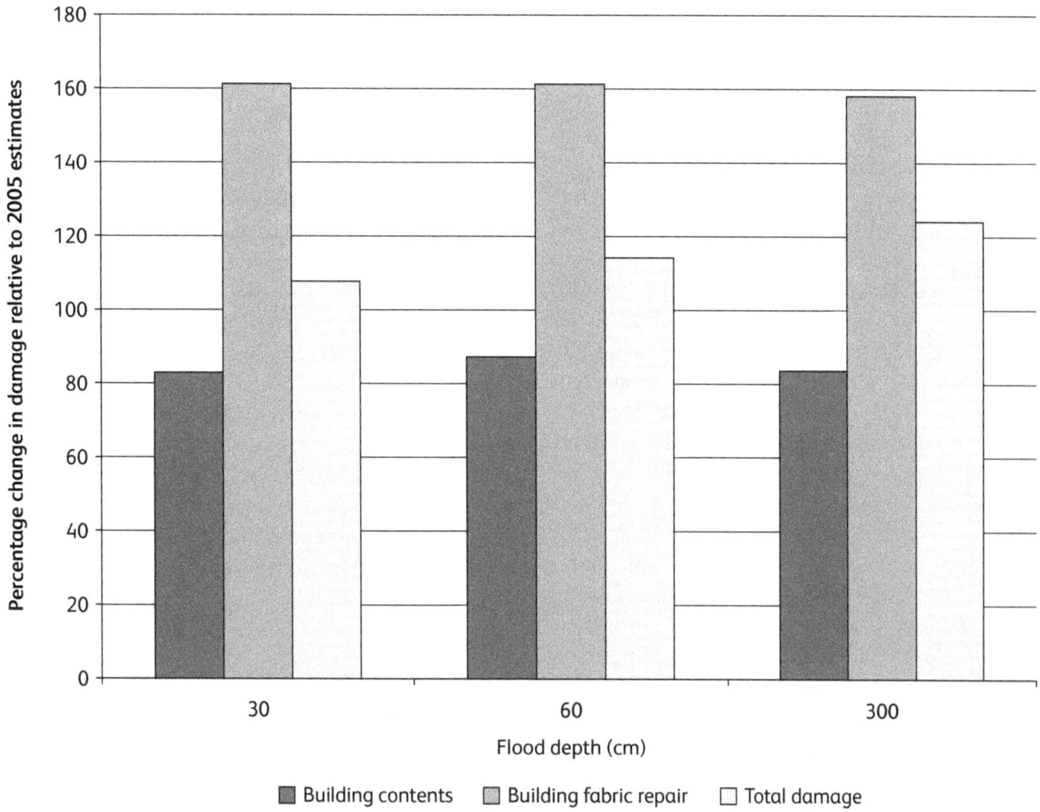

Figure 4.2 Changes in residential flood damages between 2005 and 2009 split between building fabric repair items, building contents and net residential damages for a range of typical flood depths

2009/10 data was undertaken (Penning-Rowsell *et al.*, 2010). Inevitably, the changes have been substantial (Figure 4.1).

The 2009/10 results show an increase in expected flood damage values associated with the cost of building fabric repair items (Figure 4.2). For example, at a typical flood depth of 0.60m, damage values for short-duration flooding (<12 hours) increased from *c.*£10,200 to *c.*£16,500, an increase of more than 60 per cent. The main driver of this increase has been the real rise in labour costs of building fabric repair items (fuelled by the 2003–07 housing 'boom' and its effect on builders' tender prices).

Conversely, our estimate of the potential flood damage for building contents has decreased as the cost of imported 'white goods', electronic household items and other domestic consumables has fallen. As a result, the total net expected damage is moderated, but it is still increasing across the range of flood depths. Therefore, we saw in this period a real (i.e. inflation adjusted) increase in flood losses for houses of between 8 per cent and 23 per cent across the range of possible flood depths. We have not undertaken a comparable analysis between 2010 and 2013, but it is likely that the real trends we saw before have continued, with flood damaged building fabric items increasing in repair costs at a faster rate than flood damaged building contents/inventory values.

4.1.3 Other changes since 2003

Since the first version of the 'Multi-Coloured Manual' (Penning-Rowsell *et al.*, 2003) was published, there have been two important changes in government guidance. First, appraisers must now incorporate social equity considerations (see HM Government,

2010b) in their appraisals through the use of distributional impacts (DI) where this is considered 'necessary' and 'practical' (HM Treasury, 2003). This requires an adjustment to the values of potential damages avoided by the application of weighted factors (Sections 4.9.3 and 4.9.4).

Second, appraisers are now required to include the intangible health benefits associated with flood risk management improvements by applying a risk reduction matrix. At its simplest, this requires appraisers to apply a benefit figure of £200 per household per year for these intangible impacts (2003 values). More specifically, it requires appraisers to consider how the level of exposure to household flood risk varies with, and without, the scheme (Defra, 2004b). The appraiser can then apply a risk reduction matrix to determine the intangible benefits before and after an intervention (Section 4.9).

Third, recent research has resulted in an altered approach to the assessment of the damage reducing effects of flood warning lead-times. The primary change in this regard is that the value of potential damages avoided is now derived from a combination of the damage data available with previous manuals and findings from flooded households surveyed in the research for the MCM (2005). In addition, this and other data have been synthesised in research since (Wicks *et al.*, 2012), and the results are reported in this volume (Section 4.6).

4.1.4 Chapter structure

The data used to generate the residential direct flood damage datasets are based on economic not financial values. Having set out a rationale for this distinction, the chapter then reviews the underlying assumptions in the compilation of the datasets on MCM-Online. We continue with sections on flood damage to park homes and vehicles and then provide a detailed evaluation of the approaches we recommend for the assessment of residential benefits, including allowing for property-level protection and flood warnings. This is followed by a section on the benefits of mitigating evacuation during flooding, then the importance of the intangible impacts of flooding is highlighted and the process for incorporating recent government guidance is explained. We conclude the chapter by highlighting some of the remaining issues, lessons from experience and 'health warnings'.

4.2 UNDERLYING ASSUMPTIONS

4.2.1 Economic and financial flood damages

Our residential flood damage data for household inventory and building fabric items are based on economic values, not financial values. This is important if the benefits of a particular flood risk management project are to reflect the real economic losses to the nation as a whole rather than the financial losses to individual householders. The rationale for this distinction is given in Table 4.2 and expanded on below.

The difference between financial and economic flood damage values has been illustrated by research at Dundee University (Black and Evans, 1999). An analysis of floods at seven locations in the UK yielded data on 2,600 insurance claims from residential properties, with average financial values derived from a range of 1,223 claims in Llandudno (flooded 1993) to

Table 4.2 Financial and economic damages related to household flood losses

Financial
Takes the standpoint of the individual household involved
Uses the actual money transfer involved to evaluate the loss or gain (e.g. if a household has a new-for-old insurance policy and they claim for a ten-year-old television, the loss is counted as the market price of a new television)
VAT is included, as are other indirect taxes, as they affect the individual household involved
Economic
Takes the standpoint of the nation as a whole – one person's loss can be another person's gain
Corrects the actual money transfer in order to calculate the real opportunity cost (e.g. in the case of the ten-year-old television, the real loss to the country is a ten-year-old television; the depreciated value of that ten-year-old television is taken as the loss)
VAT is excluded, as are other indirect taxes, because they are money transfers within the economy rather than real losses or gains

54 from Devon (flooded 1997). Averages from these data show damage values at far higher (financial) levels than shown by the Middlesex University (economic) flood damage data available in the 1990s; that is, up to £10,000 higher per property.

In general, this is as expected, but the Dundee University data have led to suggestions that the Middlesex University data, used in this Manual, underestimate the true economic effects of floods. In fact, a comparison of the Dundee University data with those derived in the EUROflood project (again, based on actual flood events) shows that structural losses/building fabric repair costs are comparable in the two datasets, but contents losses are higher in the Dundee University data than anything obtained at that date by Middlesex University (Ketteridge and Green, 1994).

This difference between the Dundee University and the Middlesex University flood damage information is due to the former being financial data and the latter being economic data. In accordance with the theoretical underpinnings of this Manual (see Chapter 2) and the Treasury's 'Green Book' (HM Treasury, 2003), economic data is the preferred method for the development of the standard datasets on MCM-Online, although we have also provided some financial data, for reasons explained in Chapters 2 and 3.

4.2.2 The susceptibility of residential properties to flooding

The direct flood damage potential of residential properties is estimated using details of the effects that flood waters of varying depths and durations would have on each of the components that make up a residential property. These components can conveniently be subdivided into building fabric and inventory items (Table 4.3). This section describes the susceptibility assumptions used when compiling the standard flood damage data on MCM-Online.

For both of these items, the expected damage is evaluated for long-duration (>12 hours) and short-duration (<12 hours) flooding, for a range of flood depths and as a percentage of the item's pre-flood (depreciated) value. Data on the susceptibility of building fabric and inventory items have been updated from Penning-Rowsell and Chatterton (1977) and Suleman *et al.* (1988) using three focus group studies with flood victims in Banbury, Ruthin and Bewdley and in consultation with loss adjusters, flood damage repair experts and quantity surveyors. The assumptions that have been used to compile these data are illustrated in Table 4.4 for building fabric items and Table 4.5 for inventory items.

Table 4.3 Residential property: building fabric and inventory components

Building fabric
Fabric of building, main and outbuildings (e.g. garage, shed) including decorations
Electric light and power fittings but not appliances
Fitted kitchens
Plumbing installation and normal fittings
Heating installation, including firing unit
Power/gas supply to cooker but not the unit itself
Boundary walls, gates and fences, landscape constructions but not horticultural layout
Inventory
Domestic appliances, heating equipment and electrical appliances (e.g. hi-fi equipment, microwave oven)
Furniture and soft furnishings
Personal effects (including books, clothes, etc.)

4.3 THE RATIONALE FOR THE COMPILATION OF THE STANDARD DEPTH/DAMAGE DATA

Flood damage potential has been estimated by collecting details of the effect of flood waters of varying depths and durations on the components of a dwelling's inventory and building fabric. Fifteen depths and two durations have been chosen (Table 4.6). This data does not take into account flood velocities, sediment loads or the pressure differentials highlighted by Kelman (2002). This is because velocities are often highly localised, sediment loads are not regarded as significant and Kelman's analysis was undertaken using only modern properties.

The standard residential data on MCM-Online was compiled from ground floor plans, the susceptibility of building fabric features for each of these properties, and the susceptibility of inventory items for each

Table 4.4 Residential property: building fabric susceptibility assumptions

Paths and paved areas, boundary fences, etc.
It is unlikely that any significant damage will occur to the main residential unit if the surrounding ground is submerged to a level below the air vents
Swollen sub-soils may require repairs to paths and paved areas
Fences, sheds and greenhouses may become displaced
Joint ownership of boundary fences should be ignored
Repairs to lawns, gardens, etc. are excluded
External main building
Brickwork is unlikely to be damaged unless accompanied by severe frost
Damage to mortar will increase with the duration of flooding
Damage to drains will increase with the duration of flooding
If water penetrates the sub-floor via air vents, floors will need to be lifted and disinfectant sprayed. The cost will be dependent on the number of sleeper walls
The extent of damage to basements and cellars depends on their use. If for storage, drying and two coats of limewhite may be required. If for living, replastering and redecoration will be required
Any external wall decoration will suffer damage under short- or long- (and extra-long-) duration floods
Depth and duration will influence damage to external doors and architraves
Apart from properties with basements, windows are unaffected until water reaches the sill at about 0.9m; short-duration floods will not affect timbers; long duration will
Internal plasterwork
Plasterwork absorbs water and the extent of damage depends on the depth and duration of flooding
Modern building methods employ dry lining (gyproc). This is highly susceptible to flood damage and any contact with water involves complete replacement
Stud partition walls absorb water and will need to be replaced
Brief immersion of plasterwork should not cause severe damage if the pre-flood condition was sound but prolonged immersion will saturate and renewal will be required
Contaminated water will often necessitate replastering
Old lime plaster would be affected by short- and long-duration flooding
For minor incidents, the area to be renewed might be two or three times the area exposed to contact; this will reduce the greater the flood depth
Stud partition walls will be damaged for short- and long-duration flooding but the damage will be less extensive for short-duration flooding

Table 4.4 continued

Floors
Floor joist damage is unlikely to upper floors
Floorboards should not be extensively affected on upper floors
Short-duration flooding will not affect softwood boards until 0.2m depth; greater depths will increase warping and sanding will be required
Long-duration flooding from 0.5m depth will affect soft- and hardboards; the greater the depth of flooding the greater the anticipated damage
Polished hardwood floors will increase resistance to penetration for several hours. Prolonged immersion would require sanding for soft- and hardwood floors
Damage to concrete floors should be no more than the cost of cleaning plus the possible application of a surface sealer. Cement and sand screeds and the insulation beneath them will be damaged by short- and long-term flooding
Asphalt and composition floors are susceptible to flood damage as with concrete
Damage to stone floors is likely to be related to the method of fixing and their pre-flood security
Thermoplastic tiles will be seriously affected from 0.05m of flood depth in long-duration flooding
Long-duration flooding will affect skirting from 0.05m of flooding
Joinery
Door frames, architraves, doors and window sills are unlikely to suffer permanent damage from short-duration flooding
Long-duration flooding on old doors will require repairs at flood depth levels below 1.2m and probably renewal at greater depths. For new doors flooding to 0.6m for a long-duration flood will probably involve subsequent replacement
Old doors would not require any repairs until 0.3m after short-duration flooding and would not require replacement until above 0.9m
After short-duration flooding new doors may require renewal at a flood depth of 0.3m
Except in the case of the shortest duration, floor level fitted furniture is quickly affected beyond the economical cost of repair. For example, kitchen cupboard base units, either on legs or with chipboard partitions built down to floor level, are usually a complete write-off. If the chipboard does not expand immediately, it will in due course
Internal decorations
Redecoration cannot normally be undertaken to part walls or even part rooms
Normal domestic maintenance costs are likely to be insufficient to cover flood damage because more extensive preparation is required. This additional cost would apply to damaged parts only. Above water levels no more work would be required than for normal maintenance
Lining paper is recommended where new plasterwork joints may be visible
After short-duration flooding up to 0.3m oil paint will only need cleaning

Table 4.4 continued

After short-duration flooding above 0.3m and long-duration flooding deeper than 0.05m walls will need repainting
Window sills and frames would only need cleaning after flooding to depths of 0.9m but above this level redecoration will be required
Plumbing, central heating and electrical installations
Pipes are not likely to be impaired but lagging will need to be renewed beneath floor levels
Flooding to a depth less than 1.2m will not affect sanitary fittings
Most modern central heating boilers will be above the level reached by most floods but any boiler immersed in water whilst hot is likely to be destroyed
Laggings are not easily renewed and relatively little damage is required before the level of uneconomical repair is reached
If tests indicate that rewiring is necessary, plugs, junction boxes and the like are often fit for reuse
Damage to electrical installations is inevitable when flood water penetrates the consumer unit which is often positioned at relatively low levels. Complete rewiring may then be necessary

Table 4.5 Residential property: assumptions concerning inventory susceptibility

Domestic appliances
Any domestic appliance in contact with water, no matter for how short a period, immediately presents a health and safety hazard. This is particularly true for refrigerators and deep freezers. Where items may theoretically be repaired, as is the case of electric cookers and ovens, high labour costs mean that it is often cheaper to replace these items. Therefore, all domestic appliances once in significant contact with water are assumed to be written off.
Electrical goods
Any electrical goods in contact with water, no matter for how short a period, immediately present a health and safety hazard. All electrical items will have to be replaced on contact with water.
Furniture
Bedroom furniture, often made of chipboard, is usually a complete write-off. Upholstered furniture such as sofas and armchairs usually soak up water through the fabric or valance. This furniture will need to be replaced on contact with water. Most polished furniture such as dining room chairs and tables will quickly become defective once in contact with water.
Floor coverings, curtains and personal effects
Carpets once wet have very little chance of renewal. Vinyl sheet flooring cannot be successfully taken up and reused. Thermoplastic tiles need to be removed as they will trap water in the floor screed and slow down the drying process. Both thermoplastic tiles and ceramic floor tiles are likely to lose their adhesion if under water for any length of time unless they have been laid with a waterproof adhesive. It is likely that soft furnishings, linen and clothes will be a total loss upon direct contact with flood waters.
Heating equipment
Radiators, gas fires and storage heaters can be repaired after contact with water but high labour charges will generally necessitate replacement. The electric pumps associated with gas and oil central heating systems will cut out when affected with flood water and any repair is labour intensive. Central heating systems will be affected, as outlined in Table 4.4.

Table 4.6 Depth and duration of flooding for standard depth/damage information

Two durations		
Short	Above ground floor level for less than 12 hours	
Long	Above ground floor level for more than 12 hours	
Fifteen depths above and below ground floor level		
1	−0.3m	To include damage to sub-floor areas
2	0.0m	Ground floor level to include damage to floors
3	0.05m	To include damage to carpets and floor coverings
4	0.1m	To include superficial damage to both internal fabric and inventory items
5 and 6	0.2m and 0.3m	To include superficial damage to both internal building fabric and inventory items
7–15	0.6m to 3.0m	In incremental steps of 0.3m to include progressively more items of damage

property. In this way, the standard data represent the accumulated damage to each component of each property. The specific compilation of the depth/duration/damage matrices for building fabric and inventory items which led to the standard data on MCM-Online is examined in more detail below.

4.3.1 Compilation of standard building fabric data

Ground floor plans of twenty-eight typical dwelling types were updated from Penning-Rowsell and Chatterton (1977) and supplemented from discussions with loss adjusters and quantity surveyors. These plans represent the likely dimensions and layouts of properties common throughout the UK, not differentiated by social grade, and are illustrated on MCM-Online (see MCM-Online Supporting Information for Chapter 4).

The plans were used as a reference for the calculation of potential damage to building fabric items. For each plan, expert judgement with specialists influenced the decisions made as to the most likely decorations and building fabric (e.g. flooring and plasterwork) in each dwelling. These decisions were then entered into a matrix for each house type under both short- and long-duration flood conditions. Data on MCM-Online provides examples of such a matrix for short- and long-duration flooding for a pre-1919 detached house occupied by a social grade AB household.

In each of these matrices, information is provided on the expected unit area of each building fabric item, the depth/damage susceptibilities for each building fabric type and the cost per unit of each item in 2013 prices. This then enables building fabric damage costs to be calculated for each flood duration and each house type. Whilst the susceptibility curves and unit costs of garages and conservatories are included in these tables, they are not included in the standard data found on MCM-Online. Instead, garages and conservatories/extensions are considered as special items. Once their occurrence is established, appraisers can then utilise the unit prices and susceptibility curves outlined on MCM-Online.

The susceptibility percentages illustrate the critical depths at which flood waters are expected to influence the damage potential. Although depth is clearly an important factor in these susceptibility figures, damage to flooring, joinery, plasterwork and decorations is often less clearly related to depth – as whole walls may have to be decorated or re-plastered when only a small proportion is affected. Likewise, doors, fitted cupboards and other pre-fabricated joinery often cannot be partially repaired. The assumptions used in the development of these matrices are provided in Table 4.7.

4.3.2 Compilation of standard inventory data

The presence or absence of household possessions has an obvious effect on flood damage potential. The standard inventory data needed to establish inventory damage was compiled from secondary sources (Table 4.10), to which the standard susceptibility data were applied, taking into account the average remaining value of each item.

Every dwelling in Great Britain can be classified using three basic variables: house type, age and the

Table 4.7 Assumptions in the development of depth/duration/damage matrices for building fabric items

Damages represent contractors' prices for repair work without regard for pre-flood conditions. No attempt has been made to reduce these figures to account for betterment. Thus we judge that the full cost of repairs reflects true damage.
Where redecoration is required the full cost is not included as decorations are naturally replaced from time to time. Of the redecoration costs, 50 per cent have been taken to represent an average true cost of flood damage.
The quality of replaced or repaired building fabric items is standard. No distinction between the cost of different quality items is included because this would not significantly affect the overall damage estimates. Labour costs remain the same irrespective of material costs.
The costs are averages and do not reflect regional variation. For the average price percentage difference in each region, appraisers are directed to apply the latest figures from the Office for National Statistics. For example, in 2010 (the latest release), the relative prices in London and the south-east were 7.9 and 2.3 per cent higher, respectively (ONS, 2011).
Standard sizes of building fabric items such as doors and windows have been assumed.
All the values exclude VAT.

occupants' social grade. The type and age of the house correspond to those used in the compilation of the ground floor plans (see MCM-Online Supporting Information for Chapter 4), with the social grade categories provided in Table 4.8. The total number of properties of each type, age and social grade of occupant has been compiled using the Survey of English Housing (ONS, 2001), the DTLR's *Local Housing Statistics* for 2001 (DTLR, 2001a) and the Office for National Statistics' *Family Spending* report (ONS, 2008).

In compiling the standard inventory data, the total inventory damage is dependent on the average remaining value (ARV), the house type, social grade and the ownership of household items for each social grade. The results of this analysis are provided on MCM-Online, with an explanation of the rationale provided below.

Table 4.8 Approximated social grade categorisation by occupation

Social Grade AB	Upper middle and middle class: higher and intermediate managerial, administrative or professional
Social Grade C1	Lower middle class: supervisory or clerical and junior managerial, administrative or professional
Social Grade C2	Skilled working class: skilled manual workers
Social Grade DE	Working class and those at the lowest level of subsistence: semi-skilled and unskilled manual workers unemployed and those with no other earnings (e.g. state pensioners)

Average remaining value

The average remaining value of inventory items has been used to ensure that the calculated flood damage figures reflect the pre-flood value of the inventory stock and not the cost of replacing damaged old goods with pristine new ones. This is important because the actual damage caused to household items is not equivalent to the new-for-old payments common in many insurance policies. Rather, the damage is the difference between the remaining value of the item before flooding and the residual value after flooding. Thus, the ARV is that proportion of the average new cost, or manufacturers' list price, which represents the average pre-flood value of the item.

Damage can also be assessed in relation to repair costs, as long as the average repair costs do not exceed the ARV. However, in our increasingly 'throwaway society', some, if not many, items are now rarely repaired. This was illustrated by the experience of flood victims in the three focus group studies in Ruthin, Banbury and Bewdley, where 94 per cent of items from the houses of thirty flood victims interviewed were replaced rather than repaired. This has also influenced our judgements on the susceptibility of household items (Table 4.5).

The percentage ARV is a function of the age of the household item, which can be treated simply, but justifiably, as a straight-line depreciation function. All

Table 4.9 Inventory items for which a 50 per cent ARV would not be suitable

Carpets and rugs, for example, have the potential for 'secondary use' and thus the percentage ARV is set at more than 50 per cent
Where exchanges are often made before the end of an item's effective life, through fashion changes, an average depreciation at halfway through an item's life may be 60 per cent to even 80 per cent of its current replacement value: only 40 per cent to 20 per cent is then allowable as the percentage ARV
The percentage ARV of books and pianos is highly variable and is more closely related to quality than for other items
The percentage ARV for items recently entering the market has been set higher than 50 per cent due to these items being, in general, less than halfway through their lives

items are assumed to be – on average – exactly halfway through their lives: a percentage ARV of 50 per cent is therefore commonly ascribed. Table 4.9 illustrates those items for which a 50 per cent ARV would not be suitable.

Inventory, ownership and social grade

The secondary data sources used to compile the inventory list, inventory prices and ownership figures by social grade are illustrated in Table 4.10. Ownership figures by social grade are not available for collective household items such as books, toys, clothes, etc. In order to establish the potential values of these items susceptible to flood damage, the stock values for food and toiletries/cosmetics are based on monthly expenditure, clothes on yearly expenditure and books and toys on five-yearly expenditure. Details of this expenditure were obtained from the Family Expenditure Survey (ONS, 2008). Recognising the importance of the 50 per cent ARV, one year's expenditure on clothes, for example, is equivalent to two years' worth of stock.

For all inventory items, the estimated ownership was multiplied by the pre-flood value of the item (present value excluding VAT multiplied by the percentage ARV). The resultant figure was further multiplied by the estimated susceptibility to damage at each depth increment to produce successive damage estimates for the two basic flood durations. For more detailed project appraisals, this information is provided as a function of the type and age of the dwelling and the social grade of the occupants. This then produces standard estimates for any combination of dwelling type, age and occupants' social grade (MCM-Online).

The ownership of household durables varies between social grades, and is an important variable in estimating damage potential. In addition, the quality of household items varies between households and social grades. Thus, the standard inventory damage data have been compiled using social grades as a measure of inventory quality and hence value. The social grade definitions are directly related to neither income nor wealth and, as such, are regarded as a more appropriate determinant of the quality of the contents of a person's house than income. Thus, 'good quality' has been ascribed to the contents of AB social grade dwellings, 'medium quality' to C1 and C2 social

Table 4.10 Secondary data sources

Source	Type of data
Office of National Statistics (2008)	Expenditure on: books, toys, clothes, toiletries/cosmetics
Office of National Statistics (2002a) Social Trends	Type and age of property by social grade
Focus group questionnaire results (Bewdley, Ruthin and Banbury)	Susceptibility of household inventory items to different depths and durations of flood
Store catalogues and websites	Inventory lists and prices of household goods
Office of National Statistics (2002b) Survey of English Housing 2001	Housing stock statistics
DTLR (2001a)	Housing stock statistics
Mintel (2003–2006)	Ownership figures by social grade – various household goods

grade dwellings, and 'poor quality' to DE social grade dwellings. This assumption was validated prior to the publication of our MCM (2005) using data from the British Household Panel Survey (1991–2000) for a range of consumer durables (Penning-Rowsell *et al.*, 2003).

4.3.3 Drying-out and clean-up costs

Context

The clean-up data here have been assembled by the National Flood School, which has over twenty years' experience in property restoration due to fire and flood damage. The data are based on the School's in-house knowledge and clean-up expertise. Our work with the School has taught us that the drying of buildings is more important than their cleaning, not least because the latter cannot commence until the building is relatively dry. And in terms of the duration of the effort, it is the drying period that is by far the longer, compared with the period required for cleaning. This is a new conclusion compared with our earlier assessments in previous manuals. The new data can be used to estimate the clean-up costs for five different house types of average quality (not including historic or listed buildings) and for five categories of flood water: major clean/grey; minor black; major flood/storm; major flood including sewage; and major flood 'contaminated' (Table 4.11).

Drying and cleaning flooded residential properties

There are various methods for drying and then cleaning a building. These include: open drying (naturally airing a property; this can generally be employed only in summer in the UK); refrigerant dehumidification (the most common method, refrigerating moisture out of the air where the ambient temperature should be above 15°C; this method cannot be used if the building's heating system is not working); and desiccant dehumidification (absorption of moisture using a desiccant material).

The actual methods employed for drying and cleaning a property will depend on a variety of circumstances: for example, the type of premises/room(s) in question (size, age, building materials, contents, etc.); the level of saturation (flood depth and more importantly the 'dwell time' of the water in the building); the ambient temperature; weather conditions; and any health considerations of the occupant

Table 4.11 Categories of flood water

Category of water	Description
Major clean/grey (IICRC Category 2)	Water contains significant contamination and can contain potentially unsafe levels of micro-organisms or nutrients for micro-organisms, as well as other organic or inorganic matter: commonly discharge from washing machines, dishwashers or toilet overflows (not including faeces).
Minor black (IICRC Category 3)	Water is grossly contaminated: as 'major clean/grey', but includes sewage backflow scenarios from an internal source where water may contain faeces, urine and other waste through toilet discharge system.
Major flood/storm (IICRC Category 3)	Water is grossly contaminated: this is the most common category for a typical fluvial, surface water or coastal flood scenario. Water may contain: organic matter, pesticides, heavy metals or toxic organic substances.
Major flood including sewage (IICRC Category 3)	Water is grossly contaminated: as 'major flood/storm', but with the inclusion of animal and human waste materials.
Major flood 'contaminated' (IICRC Special situations)	Water may contain regulated hazardous waste (as per Technical Guidance WM2, see: http://www.environment-agency.gov.uk/business/topics/waste/32200.aspx), including (but not limited to): asbestos, heavy metals, pesticides, solvents, caustic chemicals, etc.

Note: Adapted from IICRC S500, 2006, p14.

(namely, respiratory conditions). In order to establish average cost values, these clean-up data are based on a typical clean-up scenario at any time of the year.

The weather and outside temperature can have an impact on the time it takes to dry a property and on the costs associated with it. It is often less expensive to dry a building in the spring/summer than it is in the autumn/winter as it can take 20–30 per cent longer to dry a property during the colder months. It is worth noting, however, that warm, dry weather can promote mould growth if water is not removed efficiently from materials. This can increase overall clean-up costs. For clean-up and restoration, water depth is an important factor, but not as important as the flood duration or dwell time. A short-duration 100cm flood may cost less in clean-up costs than a long-duration 30cm flood. This can make estimating flood damage costs per square metre difficult.

Different clean-up contractors will use different methods; some will strip all contents following a 10cm flood whereas others will strip contents only after a 100cm flood. There is a mixed response across the industry.

'Major clean/grey' water (Table 4.11) will not exceed a depth of 0.1m as this category concerns internal household floods, usually involving a dishwasher or washing machine leak, or similar. Approximately 70 per cent of insurance claims involve this and the 'minor' category (not necessary for inclusion in this Manual). Approximately 5 per cent of insurance claims per annum will feature the category 'minor black' (usually involving sewage contamination).

A typical clean will involve the removal of foreign debris and silt; this can sometimes require a 'controlled re-flood' where small fragments are removed using a fine spray of water. An anti-microbial chemical will then be applied, and moisture removed. In order to establish the type and number of dehumidifiers required, it is necessary to determine the amount of moisture in the air. This is done by calculating the volume of a room and measuring the relative humidity within it (see the British Standard BSI PAS 64 for further information on the recommended methods for cleaning and drying a property in the UK).

Drying and clean-up costs

There are three ways of establishing cleaning costs: time and materials (labour costs plus number of machines for number of days); price per square metre (where the price tends to change only with the category of water); and a 'room rate' price (a fixed price, regardless of room size). Previous MCM clean-up costs (Penning-Rowsell *et al.*, 2005a) were based on the 'room rate' as this was the industry standard at that time. This room rate was then multiplied for the typical number of rooms for each type of housing, producing a total cost per building.

The new data for 2013 are based on the cost of clean-up per square metre (£/m^2). This cost for all property types has been calculated by establishing the typical clean-up costs for a standard 1946–79 bungalow. These costs have been divided by an average room size of 12m^2 and divided again to obtain the £/m^2 value. All other building types and ages are based on this figure, with percentage adjustments for the different housing types and the age of the property. Using the expertise of the National Flood School, a separate dataset for short-duration and long-duration floods has been produced. The long-duration data feature a 20 per cent uplift across all categories based on empirical clean-up and drying experience.

It has been assumed that a refrigerant dehumidification method of drying will be employed; this method is used in 90–95 per cent of all cases in the UK, using one dehumidifier and two fans per room. These data also apply to brick-built housing, which accounts for the vast majority of residential buildings in the UK. Other types of building materials could require a site-specific appraisal.

The prices are based on 2012 values, uplifted to 2013/14 prices, with no VAT included and no regional cost variation. Insurers tend to set prices in three–five-year cycles so these should not change drastically in the short term. The actual cost of on-site clean-up tends to decline at each cycle as the technology used becomes more efficient so less time is spent on the drying process. However, the price of fuel is increasing (by approximately 25 per cent between 2008 and 2013) and with it the transportation cost of machinery and workers to the clean-up site. This may lead to an overall increase in clean-up rates for the next cycle.

The new clean-up data provided are an average for the UK. However, the regional variation in housing stock – size, local style of building, cost of labour, etc. – may be a significant factor in determining the total clean-up cost.

Thus, in the 2007 floods, properties in Hull were flooded, as were properties in Gloucestershire. The events were a few weeks apart but both inland in nature. The housing stock in Hull was generally two-up–two-down terraced properties, whereas the Gloucestershire floods mainly affected large detached houses. Clean-up and drying data on 136 properties in Hull and 122 in Gloucestershire were analysed. The average costs were: Hull £3,299; Gloucestershire £4,575 (these data were recorded in the time and materials manner and not the room rate; there should be a smaller variation in the room rate charge as it is the number of rooms that drives the change in cost rather than the housing stock or flood depth). Also, as another regional effect, it is not uncommon for cleaning contractors to apply a 10 per cent price increase when working on cases inside the M25 motorway ring.

The National Flood School has verified the data we have provided here by calibrating the figures with completed clean-up jobs, and the data provide a very good match. We also presented summaries of these results to a conference of specialists in this area at the Belfry, Warwickshire, in 2012, and there was no objection to the levels of cost indicated. We can therefore be more confident than previously that these figures provide a very solid representation of the current clean-up market, for the different types of housing and flood scenarios.

4.4 FLOOD DAMAGE TO 'PARK HOMES'

4.4.1 What is a 'park home'?

'Park home' is the currently accepted name for a residential mobile home, installed on a site or 'home park' (ODPM, 2002). Although park homes are technically 'mobile' (i.e. built on a chassis with wheels), it is rare for the home to be moved once sited on a 'pitch'. Park homes are usually transported as complete units from the place of manufacture to the home park on a flat-back lorry. Most will remain in the same location for the duration of their lifetime, typically forty to fifty years if well maintained.

Approximately 160,000 people live in 84,000 park homes on 1,950 sites in England. These sites vary in size, with the average park housing approximately forty park homes. The vast majority of park home sites are privately owned, with a small number owned by local authorities (DCLG, 2012a). Government research (from 2002) states that between 1999 and 2000 the average value was £34,500 for a new single unit, and £13,500 for a second-hand unit (ODPM, 2002), equivalent to approximately £45,000 and £17,500 in 2013, respectively. The sale price of park homes varies dramatically if purchased on site. It is influenced by land value; the desirability of location; the park facilities; and the local amenities. Park homes depreciate over time, but because new parks are rarely given planning permission, the trade-in of once-new properties is brisk and occupancy levels are high (*This is Money*, 2010). In addition to the purchase price, park home owners usually pay the site owner a rent for the pitch and for maintenance of the communal areas (Housing Care, 2012). Pitch fees average around £1,000 per annum (excluding utilities) (PAS, 2010).

4.4.2 Specification and locations

Park homes vary in size and specification, but a typical size is 11m × 6m or 12m × 6m (Homeseeker Homes, 2012). The vast majority have a timber structure (usually plywood), which sits on a steel chassis. The outer material is durable and weatherproof (either acrylic or wood finish) and the inside is fully insulated (Stately Albion, 2012).

Homes are raised above ground level, usually at a height of 60cm (Towergate Insurance, 2011; IPHAS, 2012), although this may be increased if the site is prone to flooding. Some manufacturers offer a built-in flotation device as an additional extra. It is not uncommon for park homes to be bricked around the base, giving the appearance of a more solid, bungalow-like, permanent fixture.

The distinction between caravans and park homes is important because the former are deemed to be moveable if threatened by flooding (and therefore not to be counted against flood risk management costs). A park home could be classed as a caravan, which is defined by the Caravan Sites and Control of Development Act 1960 (HM Government, 1960) as 'any structure designed or adapted for human habitation which is capable of being moved from one place to another (whether by being towed, or by being transported on a motor vehicle or trailer)'. The key difference, however, is determined by building standard guidelines and the period of occupancy.

Caravans and other holiday homes are typically built to British Safety Standard BS EN 1647, which specifies health and safety requirements for various aspects of the design: ventilation, fire precautions, fitted furniture, etc. Park homes are usually built to British Safety Standard BS 3632. This specifies a higher standard for insulation and weather protection. Most new park homes are built to this standard and it provides a useful way of distinguishing between the two types of 'mobile home'.

The site licence, issued by the local authority, will specify the period of permitted occupancy throughout the year. Most holiday sites – often located on floodplains – will not be inhabited during winter months, whereas park home sites are inhabited all year round. Whereas caravan holiday homes tend to be located in 'beauty spots' alongside rivers or at the coast, park home sites can be found in various locations: in urban areas among conventional housing or in standalone rural locations throughout England and Wales.

Some park home sites are known to be located in flood-prone areas; one example is Redhill and Wheatplot Home Sites along the River Stour, north of Bournemouth (Dorset), for which flood warnings were issued several times in 2012. Some preliminary research using the Environment Agency's Flood Map reveals that several other sites are at risk but only during an extreme flood event.

4.4.3 Susceptibility and vulnerability to flooding

Planning Policy Statement 25 notes that park homes intended for permanent residential use are highly vulnerable to flood risk (DCLG, 2010). It is a little ambiguous whether this is due to location or durability.

As a result of the building materials involved in park home construction, if flood water reaches floor level (generally approximately 60cm above ground level) it is likely the home will be written off as a total loss (Homeseeker Homes, 2012). Indeed, Coast Insurance (2012), which specialises in caravan and park home insurance, have indicated that annual flood claims for park homes are very low, but where claims are made over half result in a total loss. For our purposes, we need a benefit figure, and Environment Agency Flood risk management (FRM) guidelines stipulate that 'in assessing economic damages, consider . . . mobile homes, chalets or other temporary buildings or structures as depreciating assets worth, on average, only half their replacement cost' (Environment Agency, 2008, 7).

In addition, the occupants are likely to have above-average vulnerability: approximately 80 per cent of park home residents are in or at retirement age (UK Parks, 2012). It is reported that two-thirds of park operators place age restrictions on their sites, with the most common minimum age limit being set at fifty or over (ODPM, 2002); these properties are not for family occupation. The 2001 Census showed that approximately 31 per cent of all people living in caravans or other mobile temporary structures reported a long-term limiting illness, compared with the national average of 17 per cent (DCLG, 2012b).

Park home site managers are required to consult and implement the recommendations in the Model Standards 2008 for Caravan Sites in England (or regional equivalents). Failure to do so may result in planning permission being withdrawn or refused by the local authority. The Model Standards stipulate that park site mangers shall establish whether the site is at risk from flooding and seek advice on flood depths and likely occurrence rates (DCLG, 2008).

4.4.4 Park homes: conclusions

Although park homes are technically mobile, it is unusual for them to be moved once sited on a pitch. Unlike most other types of caravan, a substantial warning period would be required to ensure homes can be moved. This is likely to be unfeasible in most cases as a flat-back lorry is required and interior furniture must be removed or at least secured before transportation. There is an absence of data relating to flood events at park home sites. It must be considered that this could be due to their rare occurrence. Field research at sites identified as prone to flooding may be necessary to make further progress in this area.

Individual park homes are undoubtedly at risk from flood water, particularly at depths of 60cm and above. The materials used in park home construction are not designed to prevent ingress from standing water, and even at floor level damage sustained is likely to be extensive, resulting in a total loss. The benefit figures that should generally be used, therefore (at 60cm flood depth at the site), are the total loss figures, being half the replacement costs.

Park home residents tend to be more vulnerable than the average, being generally at retirement age. In addition, our research indicates that pre-existing medical conditions among residents are not uncommon. This naturally raises the issue of resident vulnerability during flood events, specifically during evacuation following a flood warning. It would be useful to know if specific contingency plans are in place for such events on flood-prone sites, but this would require scheme-specific local research.

4.5 VEHICLE DAMAGES IN FLOODS

To assess the damage to road vehicles in floods, we need to know their value, their location and the effect of flood waters on their condition.

The Society of Motor Manufacturers and Traders (SMMT) produces an annual report detailing the state of the motoring industry in the UK. This report provides a wealth of information on various topics, such as vehicle manufacturing output, CO_2 emissions, the number of vehicles on the road and the average vehicle age.

SMMT calculates the average vehicle on UK roads to be 7.44 years of age (SMMT, 2012). For our purposes, we will base our figures on a more rounded 7.5 years.

Using SMMT data, we can see that four particular models have consistently appeared in the annual top-five bestselling vehicles list for the past eight years: Ford Focus, Ford Fiesta, Vauxhall Astra and Vauxhall Corsa. The Ford Focus is the most common vehicle on UK roads (1.4 million), closely followed by the Ford Fiesta (1.3 million) (DfT, 2012a).

To get a reasonably good indication of the average vehicle's value, we can look at the current sales price for these four bestselling vehicle types; specifically models aged 7.5 years. Used Car Expert provides the typical sales prices for 2005 models (Used Car Expert, 2012). Averaging the sales prices for the four bestselling vehicles gives a result of £3,600. It is worth pointing out that sale price is different from market value, but at least it provides an indication of value.

Although we can use this as a starting point, the price of the average vehicle is not the same as the average price of all vehicles on the road – which is the figure we require for this Manual – so we must now take this forward.

4.5.1 Average vehicle value

Finding the definitive value of all vehicles on UK roads is almost impossible due to the varying degree at which different vehicles depreciate in value. According to the Automobile Association (AA), most vehicles will have lost around 40 per cent of their original value by the end of their first year, but others will lose as little as 10 per cent (AA, 2012). The depreciation rate is determined by several factors: the production volume of a particular model; the condition and mileage of the vehicle; fuel consumption rate; etc. Even the colour of the vehicle can be a factor.

Using both the *What Car?* Depreciation Calculator (*What Car?*, 2012) and the *Parkers* Valuation Tool (*Parkers*, 2012), we can determine an average depreciation rate for all vehicles, which should provide a strong indication of average vehicle value. A large selection of vehicles was compared – slightly over-represented by the most common on the road, using figures from SMMT and the Department for Transport (DfT) – and the average depreciation rate was deduced (see Table 4.12).

The largest depreciation rate occurs in the first year, slows rather rapidly after the third year and virtually stops after seven years. The AA notes that vehicles aged eight years have largely completed their depreciation cycle and will retain whatever value remains.

Table 4.12 Average vehicle depreciation rate

Year following sale	1st	2nd	3rd	4th	5th	6th	7th	8th
Depreciation as a percentage of sale price (rounded)	43	52	61	65	70	75	80	82

Table 4.13 Residual value of the average vehicle

Residual value after each year (rounded)									
Sale price	1	2	3	4	5	6	7	7.5	8
£16,400	£9,350	£7,900	£6,400	£5,750	£4,900	£4,100	£3,300	£3,100	£2,950

One method for assessing value is to find the average sale price of vehicles today and depreciate this by 7.5 years to find the value of the average vehicle. However, depreciating into the future does not reflect present-day circumstances. It is more logical to seek the average sale price of vehicles during 2005 (7.5 years ago, to account for the average age of vehicles on the road in 2013) and use the information in Table 4.12 to depreciate this figure. Data provided by *Parkers*, used in conjunction with 2005 list-price catalogues from Honda (Honda, 2005) and Vauxhall (Vauxhall, 2005) (Ford and other prominent manufacturers do not supply historic sales data), allows us to calculate a rough average sale price for new vehicles when originally sold in 2005. This figure is £16,400.

Table 4.13 takes this aged sales figure and depreciates it for each year of the vehicle's life using the data in Table 4.12, providing us with a residual value of £3,100 per vehicle.

4.5.2 Vehicle location

The Department for Transport (DfT) provides the number of vehicles registered in each English region and in Wales. However, pinpointing the exact number of vehicles in any given location at any specific moment is problematic. The DfT also provides us with traffic volume for specific road types, but this information is not particularly useful in ascertaining the potential direct damages sustained when conducting a flood benefit appraisal as vehicles are not always in motion. One way of estimating vehicle damages is by finding out the ownership rate per residential property and assuming that vehicles will be parked within the vicinity of that property.

4.5.3 Vehicles per household

According to the latest statistics from the DfT, the average number of cars per household is 1.15 (DfT, 2011a). This figure varies between regions, particularly in London, where the average number of cars per household falls to 0.76, compared to 1.21 as the UK average excluding London. Urban areas have a lower average number of cars per property, while rural areas have a higher average number of cars. The averages for rural and metropolitan areas provide a high and low value, respectively, alongside our nationwide average value. Table 4.14 shows these average figures.

Table 4.14 Average value of vehicles per household

Property type	High (rural)	Medium (average)	Low (metropolitan)
Number of vehicles per household	1.61	1.16	0.99
Value of vehicles per household	£4,991	£3,596	£3,069

Using these figures, we can determine the value of vehicles per household (Table 4.14). The average value of vehicles per household is therefore approximately £3,600.

Where flood depth for a property is known, this damage figure can be applied to all properties at risk from a flood depth above 35cm from ground level. For more specific regional car ownership rates, use ONS data based on the 2011 Census, which can be accessed at: http://www.ons.gov.uk/ons/interactive/census-map-2-1---car-ownership/index.html.

4.5.4 Moving vehicles following a flood warning

Five post-event surveys, conducted between 1998 and 2007 (BRMB, 1999; 2001; Environment Agency, 2005; 2006; 2007), inform us that 25 per cent of respondents reported moving their vehicles to a safe place following a flood warning. This should be factored into any potential damage calculations based on the number of vehicles per household. We need to assume, therefore, that a quarter of all properties in the benefit area will not have any vehicles present.

4.5.5 Vehicle damage in relation to property damage

Detailed property damage figures tend to be collected for all major flood events, whereas vehicle damage figures do not. Fortunately, the ABI (2012) has collected such figures for the floods of 2007 and June–July 2012. In 2007, 120,000 residential property claims, 27,000 commercial property claims and 18,000 vehicle claims were made. In June–July 2012, 9,100 residential property claims, 3,300 commercial claims and 2,000 vehicle damage claims were made (further claims came later that year, too). These ABI figures cannot be assumed to apply to unique properties, as each flood-affected property owner may make two or more separate claims (for contents, structure, vehicles, etc.). To account for this, we recommend that the ABI's figures be decreased by 50 per cent, giving approximate individual property claims of 73,500 in 2007 and 6,200 in June–July 2012.

These refined figures mean that the percentage of vehicle claims to property claims was approximately 25 per cent in 2007 and 32 per cent in 2012. The average of these two events would be around 28 per cent. An estimate of total vehicle damages could therefore be ascertained by assuming that the number of vehicles damaged is approximately 28 per cent of the number of residential and commercial properties at risk in the benefit area and multiplying this figure by £3,100 (the value *per vehicle*, not the value of *vehicles per household*).

The insurance claim figures could be further refined by taking into consideration the prevalence of uninsured homes and vehicles. Research conducted after the 2007 floods found that approximately 25 per cent of UK residential properties are uninsured (GFK, 2008), so it would be reasonable to increase these residential insurance claim figures by 33 per cent to account for all properties (insured and uninsured) which may have been affected by the floods. The number of uninsured vehicles in the UK is estimated at 1.2 million (Motor Insurers' Bureau, 2013), approximately 4 per cent of all vehicles on the road.

4.5.6 Vehicle susceptibility to flood damage

As vehicles differ greatly by make and model, it is difficult to accurately predict how any particular vehicle will be affected by flood waters and how much damage will be caused. However, we can make some general assumptions by determining how flood waters can impact upon vehicles and then assessing how this might affect the average vehicle.

Most cars are surprisingly susceptible to flood water, and it takes a relatively shallow body of water to cause significant vehicle damage. Admiral Insurance advises that once flood water has entered a vehicle's interior cabin, it is highly likely that the vehicle will be written off as a total loss (Admiral, 2012). This is partly due to the amount of electrical components situated around the driver and passenger areas – damage can take months to present itself as corrosion causes mechanical parts to fail gradually (Esure, 2012) – but the main reasons for the write-off are the health and safety implications associated with water ingress, such as legionnaire's disease (RAC, 2012).

The Research Institute for Consumer Affairs (RICA) has collected sill height data for over 900 vehicles of various makes and models driven in the UK. The average height from ground level to sill for all sampled vehicles is 39cm. When we remove data for 4×4s and MPVs (multi-purpose vehicles) – which have an average sill height of 46cm and 40cm, respectively, and are therefore less susceptible to flood damage – the average height from ground to sill is 37cm (Ricability, 2012). This is significant, because at this level it is highly possible for water to penetrate door seals and other waterproofing measures which are not designed to resist water ingress during total immersion.

The door sill is usually the same height as the centre of the wheel, which is considered a critical height for other damage – particularly since moving vehicles will create a bow wave, raising water level by several centimetres – notably water entering the engine through the air intake and/or the exhaust. Air intakes are increasingly situated low down behind a vehicle's front grille, which is deemed to enhance performance but has the adverse effect of making vehicles more vulnerable to flood water. Both Zurich Insurance (2012) and the AA (2011) warn that as little as 25ml of water in the combustion chamber of an engine can cause serious and expensive damage. Motor industry writers at the AOL Autoblog (2009) stress that water-damaged parts generally need to be replaced to guarantee their pre-flood quality and this is another reason why insurers tend to declare the vehicle a total loss.

Email correspondence with the RAC (2012) provides us with a good understanding of which electrical components can be affected as water level rises:

> The lowest components would include engine sensors and connections, and as the level increases it is possible to contaminate more critical circuits and components such as charging systems, fuse boxes, ignition, fuel delivery etc. If the level rises to bumper height, this can then reach lighting circuits and a host of sensors that may be fitted by a manufacturer.

It is worth pointing out that a flood depth of between 30cm and 50cm, depending on water velocity, is enough to float a vehicle (Zurich Insurance, 2012; Smart Driving, 2012; HR Wallingford, 2005b; EMRFRA, 2009).

It is reasonable to surmise from the above that water depths above 35cm are almost certain to lead to a vehicle being written off. Even lower levels can cause critical damage as a result of the waves created by moving vehicles. Research indicates that drivers, especially males, tend to underestimate the risks involved with driving through flood water (Jonkman and Kelman, 2005) and may not make a connection between low water depths and high damage costs.

4.5.7 Summary

The average loss value that we recommend for project appraisals is £3,600 per residential property in the benefit area, for floods greater than 0.35m above ground level (not above property threshold level) at the location of the house in question. It should be assumed that 25 per cent of these residential properties will have no vehicles present if a warning has been issued. We cannot, however, anticipate the number of vehicles present when flooding occurs. If flooding begins during the night, this assumption is likely to be accurate. During daylight hours, it would be reasonable to assume that a proportion of these vehicles will not be parked at or near a residential property. In order to allow for this, we propose an alternative method, which assumes that the total number of vehicles likely to be damaged during a flood occurring at any time of day will equate to 28 per cent of the total number of flooded properties (from a flood of any depth). Once the number of likely vehicles has been ascertained, appraisers can multiply this by £3,100 (the value per vehicle, not the value of vehicles per household) to calculate the total estimated vehicle damage. This second method does not require an assumption to be made on the presumed location of vehicles when a flood occurs.

4.6 THE DAMAGE REDUCING EFFECT OF PROPERTY-LEVEL PROTECTION (PLP) AND FLOOD WARNINGS

Examples of the use of sandbags and individually installed flood-proofing devices for residences have been common for many years and have met with varying degrees of success. However, there are now dozens of examples of property-level protection (PLP) schemes in England and Wales, as well as others in Western European countries, often focused on residential properties (HR Wallingford, 2012). PLP schemes have become more common in England and Wales since the Pitt Report into the 2007 floods (Pitt, 2008) and have been encouraged by various grant schemes (May and Chatterton, 2012). These schemes install professionally designed and fitted flood protection products to residences.

The following kinds of measure are most commonly used:

- barriers for doorways and airbrick covers;
- non-return valves for domestic and foul drainage systems;
- de-watering pumps;
- waterproofing and sealants;
- internal rearrangements for electrical outlets and wiring; and
- replacement of floors and doors with materials which have a comparatively low damage-susceptibility (e.g. replacement of timber floors with concrete ones).

These measures have become more common as 'Kitemarked' flood resistance products have become available. A much lower capital investment is required to implement these measures compared with large-scale engineering projects, and they provide property owners and tenants with the opportunity to manage the consequences of flooding where costly flood risk management schemes cannot be justified. Data from Germany following the 2002 Elbe floods reveal fewer than 5 per cent of 1,200 flooded private residences had adapted their property (e.g. by adapting interior fittings, purchasing water barriers, moving utility services upstairs or sealing cellars) prior to the floods, but this increased to 20–40 per cent (depending on the measure taken) after the floods. For those residences where water barriers were available, the mean damage ratio (i.e. the fraction of damage in relation to total building value) was reduced by 29 per cent (Kreibich *et al.*, 2005; 2011). Clearly, experience of flooding encourages residents to adopt PLP measures, and Tunstall *et al.* (2007) found that the more frequently flooded householders were, the more likely they were to adapt their property to reduce flood damage.

De-mountable door and window guards, airbrick covers, sump pumps and external wall waterproofing, which are designed to prevent the entry and build-up of flood water within a property, are examples of flood resistance measures. They may be manually or automatically operated. Some are premised and dependent on a timely flood warning being received.

On the other hand, flood resilience measures treat mainly the internal components of properties to make them less susceptible to damage when flood water enters. Some measures, such as replacing timber floors with concrete ones, are permanent and do not require a flood warning to be implemented. However, property occupants may be able to remove damageable contents to flood-free locations or to an upper floor so that flood damages are saved if a timely flood warning is available.

4.6.1 Resistance and resilience measures

We can only estimate how common PLP measures are within residential properties in England and Wales. HR Wallingford (2012) refer to 646 resistance installations made between June 2009 and March 2010. May and Chatterton (2012) refer to approximately fifty locations in England and Wales where PLP schemes have installed resistance measures in 1,109 residential properties. An evaluation of residential take-up of flood resistance measures in England and Wales undertaken in 2011 and 2012 as part of the SC090039 research project for the Environment Agency (Wicks *et al.*, 2012) revealed that an estimated 8.1 per cent of 'ground floor' properties in the NaFRA 'significant risk' category (i.e. 23,247 residential properties) had installed flood resistance measures (Environment Agency, undated d).

Here 'significant risk' is where the chance of flooding in any year is greater than 1.3 per cent (i.e. >1:75 years), which is the broad flood probability range in which the costs of installing resistance measures are most likely to be cost-effective. A detailed analysis by Royal Haskoning for the Committee on Climate Change (2012) found that, from an economic perspective, the number of properties in England and Wales that would find it worthwhile to install the most cost-beneficial package comprising manual resistance measures would rise to between 344,000 and 537,000 by the 2080s due to climate change alone.

Research undertaken for the Environment Agency between 2010 and 2012 led to the development of a benefit assessment tool called ANSR (Appraisal of Non-Structural Responses), which is based on research on flood warnings and the development of the FWRBP (Flood Warnings Response Benefit Pathways) model as part of the FLOODsite research project (Parker *et al.*, 2008). ANSR specifies simple formulae and parameter values for estimating the benefits of PLP measures and flood warnings as well as other non-structural measures (Wicks *et al.*, 2012).

4.6.2 Resistance measures

Two approaches for estimating the benefits of resistance measures are available. The first involves increasing the ground floor residential property threshold level in a benefit area property database by 0.6m for those properties known to have resistance measures installed. This applies to the number of residential properties with resistance measures at each return period considered in the appraisal. Wicks *et al.* (2012) concluded that warning-related resistance measures are, on average, only approximately 75 per cent effective (see also Committee on Climate

Change, 2012) and so flood damage savings will be overestimated if this is not considered. This means that the estimated flood damage savings at each flood return period need to be factored by 0.75 before they can be used.

With this first approach, it is preferable to be able to identify which residential properties in the benefit area have resistance measures installed. Records providing this information may be available where grant schemes have led to PLP measures being installed; or, alternatively, a field survey may need to be undertaken. For an initial study, if the number of properties with resistance measures is small (say, 5 per cent of properties or less), it is probably not worth taking account of the effect of resistance measures in an appraisal. Otherwise (e.g. at the project appraisal level) the estimated damage reducing effect of resistance measures must be taken into account in the appraisal. This is because PLP measures reduce damage at the more frequent flood return periods and will, therefore, have a significant effect on estimated annual average damages.

The second approach involves either using formulae in which national average parameter values are used where local parameter values (e.g. for uptake of resistance measures) are unavailable or introducing some local parameter values (where available) and using national average values as default data values where a local parameter value is unknown. Here an assumption is made that where flooding probability is less than 1.3 per cent (i.e. <1:75 years), the likelihood of flood resistance measures being deliberately installed is zero. This is a conservative estimate because May and Chatterton (2012) concluded that the evidence for resistance measures being cost-effective related to properties where the chance of flooding was greater than 2.5 per cent (i.e. >1:40 years). Of the 8.1 per cent uptake value (UP) taken from Wicks *et al.* (2012) for England and Wales, the uptake of these measures was 4.8 per cent for warning-dependent resistance measures (WDRM) and 3.2 per cent for warning-independent resistance measures (WIRM), some of which are automatic.

As previously explained, other than UP (3.2 per cent), the effectiveness (EF) of resistance measures also influences the impact of WIRM on damage reduction. Resistance measures require maintenance, may be misplaced, installed incorrectly and have a relatively short life (e.g. twenty years). Wicks *et al.*'s (2012) value of 75 per cent (or 0.75) for EF is based on the work of Thurston *et al.* (2008). Resistance measures are also effective only in relatively shallow flooding; between 0.6m and 1.0m flood depths they may begin to leak, and they are inadvisable where flood depths are likely to exceed 1.0m. Because of these factors, we assume here that the maximum height to which resistance measures are effective is 0.75m. The following formula provides a means of predicting the frequency (FREQ1) of adoption of WIRM in NRPs based on the above parameter values dated 2011.

Equation 4.1

$$\text{FREQ1} = \text{UP } (0.032) \times \text{EF } (0.75) = 0.024 \text{ (i.e. 2.4 per cent)}$$

Where there is sound local knowledge of the degree of uptake of resistance measures in residential properties, a local value may be substituted for 0.032 in Equation 4.1. Local Environment Agency offices may have knowledge of properties in which measures have been installed.

For WDRM, we also need to take into account that they depend upon property owners or tenants receiving, and being available to receive, flood warnings – termed 'reliability and availability' (RA). RA depends, in turn, on: (a) coverage of the flood warning service; (b) uptake of this warning service; (c) effectiveness of the warning service; and (d) the availability of potential warning recipients. Wicks *et al.* (2012) analysed a range of evidence and concluded that, for England and Wales in 2011, RA was between 22.69 per cent and 30.02 per cent, although the higher of these figures was taken as most likely (here we round this value to 30 per cent). Furthermore, these measures also depend on the proportion of property owners operating the measures (OP), which Wicks *et al.* (2012) estimate as 63 per cent. For WDRM, therefore, prediction of the frequency (FREQ2) of adoption is provided by the following formula.

Equation 4.2

$$\text{FREQ2} = \text{RA } (0.30) \times \text{UP } (0.048) \times \text{OP } (0.63) \times \text{EF } (0.75) = 0.068$$

Table 4.15 Benefit breakdown for a detached house (2 per cent annual exceedance probability: fifty-year flooding threshold)

Benefit (i.e. damage reduction) category	Automatic resistance (£)	Manual resistance (£)
Building fabric	1,850	1,000
Building inventory	450	200
Clean-up	800	600
Temporary accommodation	725	200
Absence from work	125	50
Health/stress	125	50
Total	**4,075**	**2,100**

Source: May and Chatterton, 2012

Again, where there is sound local knowledge of any of these parameter values, they may be substituted for the relevant value in the above equation.

May and Chatterton (2012) provide benefit (i.e. damage reduction) figures for a detached residential property of $62m^2$ ground floor size (see MCM-Online Supporting Information for Chapter 4). Automatic resistance measures therefore have the potential to generate damage savings of £65.7 per m^2. The equivalent damage saving for manual resistance measures – £33.87 per m^2 – is significantly less.

The above data may be used to estimate an approximate value for the damage reducing effects of resistance measures for residential properties, assuming that a flood warning system is operational in the case of manual measures:

Step 1: Calculate the number and then the total ground floor size (m^2) of residential properties at risk at each return period in the benefit area up to the 1:75-year flood probability threshold where flooding is not expected to be greater than 1.0m (TGA).

Step 2: The following formulae may then be used to estimate the total damage reduction (i.e benefit – WIRMB) owing to residential property WIRM.

Equation 4.3

$$\text{WIRMB£} = \text{TGA} \times \text{DR} \times \text{UP} \times \text{EF}$$

Where:

WIRMB£ = estimated damage reduction (i.e. benefit) by employing WIRM
TGA = total ground floor area of residential properties located in benefit area within 1:75 year flood risk area and where flooding is not likely to be greater than 1m
DR = damage reduction: £65.7 per m^2 at 2012 £ values
UP = uptake of WIRM factor: 0.032
EF = effectiveness factor: 0.75

For WDRM, the equivalent formula is given below.

Equation 4.4

$$\text{WDRMB£} = \text{TGA} \times \text{RA} \times \text{DR} \times \text{UP} \times \text{OP} \times \text{EF}$$

Where:

WDRMB£ = estimated damage reduction (i.e. benefit) by employing WDRM
TGA = total ground floor area of residential properties located in benefit area within 1:75 year flood risk area and where flooding is not likely to be greater than 1.0m
RA = reliability and availability: 0.30
DR = damage reduction: £33.87 per m^2 at 2012 £ values
UP = uptake of WDRM factor: 0.048
OP = operated: 0.63
EF = effectiveness factor: 0.75

To summarise, Equations 4.3 and 4.4 are only indicative guides to the value of residential property damage reduction through use of resistance measures for the following principal reasons:

- the uptake factor (UP) applies to the number of properties although here we apply it to TGA, which is a surrogate measure for property numbers; and
- DR in £/m^2 is derived from research into detached houses only.

The equations are, to a degree, progressively more reliable as sound local parameter values are substituted for the national average values.

4.6.3 Resilience measures

For England and Wales, Wicks *et al.* (2012) estimate that the uptake (UP) of resilience measures in residential properties in benefit areas up to the 1:75 year flood probability in England and Wales is 8 per cent, although, owing to lack of research evidence, this estimate is fairly crude. The effectiveness (EF) of these measures is considered to be less than for resistance measures and is estimated at 50 per cent.

May and Chatterton (2012) provide two sets of benefit (i.e. damage reduction) figures for a detached residential property of $62m^2$ ground floor size, which may be averaged and used to derive a surrogate damage reduction value. These two values are for resilience measures with and without resilient flooring, the average for which is £4,200 or £67.74 per m^2 (comprising building fabric, building inventory, clean-up, temporary accommodation, absence from work and health/stress impacts).

A modified form of the equations above may be used to estimate a guide value for the damage reducing effects of resilience measures on residential properties.

Equation 4.5

$$RISMB£ = TGA \times DR \times UP \times EF$$

Where:

RISMB£ = estimated damage reduction (i.e. benefit) by employing resilience measures
TGA = total ground floor area of residential properties located in benefit area within 1:75 year flood risk area and where flooding is not likely to be greater than 1m
DR = damage reduction: £67.74 per m^2 at 2012 £ values
UP = uptake factor: 0.08
EF = effectiveness factor: 0.50

Very similar simplifying assumptions to those applicable to the procedure for estimating the damage reducing effects of resistance measures are also applicable to resilience measures. Thus:

- resilience measures are considered only where the annual flood probability is less than 1.3 per cent (i.e. <1:75 year);
- the DR value is derived by considering only a detached residence of $62m^2$ ground floor area; and
- UP and EF values are derived by Wicks *et al.* (2012) and these are national values: they may therefore be substituted by local values where sound local evidence is available.

A further simplifying assumption which can affect the reliability of these estimates is that in some cases resistance and resilience measures may be used in combination. The uptake factor (UP) may improve in the future and/or there may be sound local evidence of uptake of resistance and resilience measures, in which case the uptake value can be substituted to match more closely with evidence.

The parameter values used in the above equations (i.e. for calculating WIRMB£, WDRMB£ and RISMB£) are derived from a national study of England and Wales and they are mean values. They can be substituted by local values where there is sound evidence to support a more customised local appraisal. For example, if it is known that, say, 10 per cent of properties have WIRM in a particular benefit area, then the UP parameter value may be altered from 0.032 to 0.10 and so on with other parameter values.

In practice, when undertaking a benefit assessment, a decision will have to be made about (a) whether or not to take into account the damage reducing effects of resistance and resilience measures (if they exist and are ignored, benefits may be exaggerated) or (b) making allowance for them, possibly by using the above equations. At the project appraisal level, PLP measures must be taken into account unless the number of relevant properties is a very small proportion of the total (say, less than1 per cent).

4.6.4 Potential reduction in flood damages by moving moveable content inventories

A great deal of investment has been put into flood forecasting and warning systems in England and Wales, as well as elsewhere (Parker and Priest, 2012).

Although flood warning systems will never be perfect, appraisers must consider the potential impact of flood warnings for reducing damage estimates in benefit assessments. The most up-to-date figures available result from an examination of the residential damage datasets to be found on MCM-Online and the application of the flood warning research described below.

For each house type, building period and social grade, the inventory value as a percentage of total damages is 52 per cent, of which 41 per cent has the potential to be moved during the warning period prior to a flood. Therefore, only 21 per cent of the total potential damages can be influenced by the provision of a flood warning. Of this 21 per cent, Parker *et al.*'s (2007) empirical research evidence from England and Wales indicates that only 38 per cent of householders receive flood warnings and respond (RA) with varying degrees of effectiveness.

Taking into account the research of Parker *et al.* (2007), Wicks *et al.* (2012) examined the latest evidence surrounding this RA parameter value (which takes into account: (a) coverage of the flood warning service; (b) uptake of the flood warning service; (c) effectiveness of the service; and (d) availability of residents to receive a warning). They concluded that, of the two parameter values which emerged (22.69 per cent and 30.02 per cent), the higher should now be used (rounded here to 30 per cent). Of those in receipt of a flood warning less than eight hours prior to being flooded, an average of 55 per cent of their inventory value could be saved (due to a combination of their ability, knowledge and/or extent of action). For those provided with more than an eight-hour warning lead-time, 71 per cent of their inventory value could be saved.

Applying these parameter values to a theoretical residential property in which the total potential damage is £60,000 gives the potential damage values in Table 4.16. The figures to which these percentages should be applied would then be dependent on the type and scale of the appraisal being conducted. For initial, strategy-level, meso-scale appraisals using weighted AADs, the damages saved by warning lead-times should be applied as a percentage of the total damages (i.e. 3.58 per cent or 4.54 per cent), as shown in Table 4.16.

For all other appraisals, the data on MCM-Online are aggregated by inventory items, building fabric items and total damage figures. Therefore, where the appraiser knows the depth of flooding, the average inventory value as a percentage of total damage should be ascertained from the datasets on MCM-Online.

Table 4.16 The potential damage reduction in residential properties arising from flood warnings which trigger the movement of contents inventory items beyond the reach of flood waters

Parameter	Description	Parameter value % (p)	Example £ (x)	Calculation
TPD (A)	Total potential damage	100	60,000	
PID (B)	Potential inventory damage	52	31,200	Bp*Ax
MID (C)	Moveable inventory damage	41	12,792	Cp*Bx
RA (D)	Households in receipt of warning	30		
PHE	Effectiveness of:			
(E)	< 8 hours' warning lead-time	56		
(F)	> 8 hours' warning lead-time	71		
	TPD saved by:			
	< 8 hours' warning lead-time	3.58	2,149	Ap*Bp*Cp*Dp*Ep
	> 8 hours' warning lead-time	4.54	2,725	Ap*Bp*Cp*Dp*Fp
	PID saved by:			
	< 8 hours' warning lead-time	6.89	2,147	Cx*Dp*Ep
	> 8 hours' warning lead-time	8.73	2,724	Cx*Dp*Fp

This new inventory figure should then substitute the figure of 52 per cent (Table 4.16), resulting in a re-evaluation of the potential inventory items saved by warning lead-time. For example, if this figure were to rise to 55 per cent, the potential total damage saved by a warning of more than eight hours' lead-time would be 4.8 per cent.

Where the depth of flooding is unknown, the average inventory value (52 per cent) should be applied across all depths, depending on the level of appraisal. At the initial appraisal level, this would be the average inventory value from the 'residential sector average' tables to be found on MCM-Online. For all other studies, it would be averaged across all house types and ages, with the most detailed investigation inclusive of social grade.

4.7 THE BENEFITS OF MITIGATING EVACUATION DURING FLOODING

Evacuation of flood-affected properties is often considered in terms of a short-term emergency response to flooding – to limit loss of life, injury and the stress caused by the flood event – and indeed it is a sensible measure to have in place. In previous versions of the 'Multi-Coloured Manual' (Penning-Rowsell *et al.*, 2003; 2005a), the costs of emergency response and recovery have been developed to be included in appraisal calculations.

In this respect, where properties are affected by flooding, evacuation from the property may also be necessary to allow flood damage to be repaired. In such cases evacuation requires temporary or alternative accommodation for households affected and this incurs additional costs. To date, these additional household costs have not yet been developed into appraisal techniques within the previous 'Multi-Coloured Manuals'.

Our research has assessed whether these evacuation costs are relevant to benefit appraisals and, if so, we have aimed to develop data and methods for their incorporation into benefit calculations. In this respect, in 2008, we researched the non-emergency service/local authority costs associated with evacuation in the Lower Thames area. The work aimed to scope a methodology for assessing these costs and, in doing so, highlighted the potential costs of evacuation as a significant component of the costs of flooding.

Indeed, our research found potential total evacuation costs to average between £7,800 and £20,100 per evacuated property (2008 prices). These costs comprise a large proportion of residential damage costs and, as such, it was deemed valuable to develop the methodology and data further to enable their incorporation into flood risk management benefits assessments.

4.7.1 Methodology

Evacuation of a flooded property involves several stages (see Figure 4.3). The first two stages relate to the planning and preparation of an emergency response prior to the occurrence of a flood event. The third stage refers to the evacuation during an event, with stages four and five describing post-event activities. Finally, stage six describes the long-term effects, such as long-term health impacts.

The stages of evacuation that we are principally concerned with here are stages three and four. For

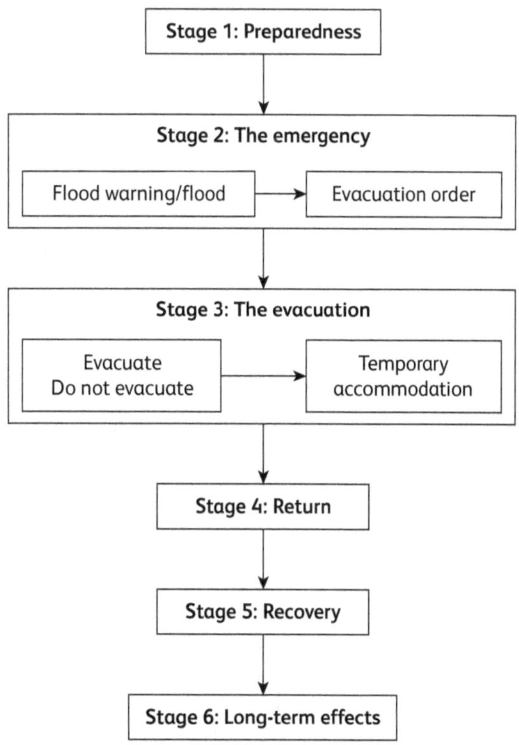

Figure 4.3 The evacuation process

Source: Adapted from Ketteridge *et al.*, 1996

Table 4.17 Components of evacuation costs

Component	Description
Temporary accommodation	Short-term accommodation for households in the early stages of evacuation, until flood waters recede and property damage can be assessed (e.g. rest centres, hotels/B&Bs, friends and family).
Alternative accommodation	Where accommodation is required for the duration of the repair and recovery phase of a flood event. Our methodology uses the evacuated property's rental value as a proxy for alternative accommodation costs.
Food	Assesses the additional costs of food associated with temporary accommodation.
Travel and time	For temporary accommodation, additional travel and time costs are associated with travelling to temporary accommodation in the process of first evacuating a flooded property and, second, costs are incurred in moving to alternative accommodation.
	For alternative accommodation, additional travel and time costs are associated with travelling additional distances to work and to school, etc. Furthermore, additional costs are incurred in travelling between alternative accommodation and the flooded property.
Loss of earnings	Flooding of residential properties may require households to take time off work to attend to the evacuation and organising repairs. This time off work is reflected as a loss of earnings.
Other costs	Relating to temporary accommodation, other costs incurred cover additional expenses such as medical supplies, toiletries and clothing, etc. that may be required as a consequence of evacuation.

these stages, we have identified the components contributing most to evacuation costs (Table 4.17) and which are clearly a net cost to society. These are components that are not already accounted for elsewhere in our appraisal techniques, but are important contributors to evacuation costs, and we discuss below the method developed and demonstrate some of the results we obtained.

Flood evacuation costs depend on many variables. However, from other research on flood 'intangibles' (RPA/FHRC, 2004), a link was revealed between flood depth and evacuation duration. The direct correlation here is particularly useful as other residential damage costs are also based on a relationship with flood depth. As such, the methodology we have adopted uses flood depth to provide a proxy for evacuation duration and it is duration on which the data and remainder of the method are based.

Whitehead (2003) developed a model for assessing the costs of evacuation. Adapted by Johnson (2008) to take into account the direct relationship between flood depth and duration, the model upon which the evacuation costs can be calculated is given below.

Equation 4.6

$$E\ (\text{Cost}) = \Pr\ (\text{evacuation}/d)c(dn)$$

Where:

E (Cost) = the total cost of evacuation
Pr = the probability
(evacuation/d) = the propensity to evacuate at a given flood depth
c = the cost associated with evacuation
dn = the duration of the evacuation

The findings from RPA/FHRC (2004) give the results in Table 4.18 as a guide to evacuation duration in relation to depth. This forms the basis of our calculations but the probability of evacuation is also important. From this it is possible to deduce the likely number of properties evacuated in a given flood, and this can be used to scale up the evacuation cost values which will be given here as a per household figure.

In addition to considering the number of households evacuated, it is important to account for how the

Table 4.18 The probability of evacuation and duration in relation to flood depth

Maximum depth in house (cm)	% who evacuated	Mean duration of evacuation in weeks
0	23	11
1–10	41	12
10–20	55	18
20–30	59	18
30–60	69	21
60–100	76	23
100+	87	33

Source: RPA/FHRC, 2004

numbers of people evacuated decrease as properties are repaired and people return home. The research by RPA/FHRC (2004) and Tunstall *et al.* (2007) found a mean evacuation of around twenty-three weeks (Table 4.18), although some households were evacuated for longer periods of up to (and more) than one year. The duration of evacuation has a major impact on total costs which are accrued over the time period from evacuation to return to the property (stages three and four in Figure 4.3). However, to assume that, of the properties originally evacuated, all remain evacuated for the longest duration (i.e. over one year) could result in a greatly overestimated cost figure. Instead, households will return over time and only a small percentage (around 8 per cent) are likely to remain in alternative accommodation for more than one year. The data and method used to calculate evacuation costs take this into account. Where duration is not known, an average figure of twenty-three weeks may be used. Otherwise, the average duration for the given depth has been used to calculate the costs.

4.7.2 Findings

Temporary accommodation

Temporary accommodation is required in the immediate instance of an evacuation until either the property is safe to return to or alternative accommodation is found for the duration of the flood and of the repairs to a flood damaged property.

The different types of temporary accommodation considered are rest centres, hotels/bed and breakfast (B&B) accommodation and staying with friends or family. Data from Doncaster Council (2008) have been used to determine the likely percentages of the population that will seek temporary accommodation for each of the different types (Table 4.19). As rest centres are included in the Bellwin returns (Chapter 6), these costs are not counted here. Furthermore, staying with friends and family or other temporary accommodation is deemed to represent no additional cost to society. As such, the focus for temporary accommodation costs is on those staying in hotels/B&Bs.

To assess the costs associated with temporary accommodation, as defined above, the costs of hotel and B&B accommodation have been calculated per household, taking into account the percentage of households utilising such accommodation. The costs of temporary accommodation are provided as averages per household, with high, mid and low values given for different durations. The durations are capped at eight weeks, as it is assumed that affected households would either have returned to their property or be residing in alternative accommodation by week 8. Recognising that the percentage of evacuated households remaining in temporary accommodation decreases over time, the costs are adjusted to reflect this, as shown in Table 4.20.

The figures in Table 4.20 provide an average figure which can be multiplied by the total number of properties evacuated to provide a total cost of temporary accommodation for a given area/scheme.

Alternative accommodation

Alternative accommodation presents a major component of evacuation costs as these costs accumulate

Table 4.19 The proportion of flooded households that evacuate to different types of temporary accommodation

Temporary accommodation	Percentage
Rest centres	42
Friends and family	38
Hotels/B&Bs	9
Other	11

Source: Adapted from Doncaster Council, 2008

Table 4.20 The average costs of temporary accommodation per household evacuated

	Remaining in accommodation (%)	Average potential cost per household evacuated (£)		
		Low	Mid	High
Up to 1 week	100	40	46	61
Up to 2 weeks	85	73	84	113
Up to 3 weeks	50	93	107	144
Up to 4 weeks	35	107	123	166
Up to 8 weeks	10	123	141	190

Sources: Adapted from RPA/FHRC, 2004; Tunstall *et al.*, 2007

over potentially long periods of time. As highlighted by RPA/FHRC (2004) and Tunstall *et al.* (2007), on average, properties were evacuated for twenty-three weeks, but some evacuees remain out of their properties for more than a year.

The alternative accommodation costs are calculated on the basis of the rental value of the property evacuated. The reasoning for this calculation is that, where better data are not available, it is assumed that the evacuees are paying for accommodation from which they are gaining no value – the property that they are not occupying – hence this represents a net cost to society (Johnson, 2008). As better data become available, the cost of alternative accommodation can be modelled again to include the difference between the rental equivalent rates and the actual alternative accommodation costs. Until then, however, the cost of rental equivalence is deemed acceptable (Johnson, 2008).

Calculating the rental equivalents, Table 4.21 provides average data for the costs of alternative accommodation by property type and duration of evacuation. The costs take into account the decreasing percentage of evacuated households over time and provide figures at high, mid and low ranges.

Table 4.21 The average costs of alternative accommodation per household evacuated

Duration	% of all evacuees	Detached (£)			Semi-detached (£)			Terraced (£)			Flat		
		Low	Mid	High	Low	Mid	High	Low	Mid	High	Low	Mid	High
Up to 1 week	100	155	215	274	127	159	190	118	148	178	96	126	155
Up to 4 weeks	96	602	833	1,063	492	616	739	459	574	689	374	488	601
Up to 12 weeks	83	1,633	2,258	2,883	1,335	1,669	2,003	1,244	1,557	1,869	1,013	1,322	1,631
Up to 26 weeks	75	3,262	4,511	5,760	2,668	3,335	4,002	2,486	3,111	3,735	2,024	2,641	3,258
Up to 39 weeks	45	4,170	5,767	7,363	3,410	4,263	5,116	3,178	3,976	4,775	2,588	3,376	4,165
Up to 52 weeks	15	4,473	6,185	7,898	3,658	4,572	5,487	3,409	4,265	5,121	2,776	3,621	4,467
Over 52 weeks	8	4,485	6,202	7,920	3,668	4,585	5,502	3,418	4,277	5,135	2,783	3,631	4,480

Source: Adapted from VOA, 2012

Other extra costs (food)

Our research on evacuation costs has found that when living in temporary accommodation, expenditure on food was higher as meals could not be prepared at low cost but had to be bought at restaurants or cafés. As such, the additional costs of food are included in evacuation costs for those living in hotels/B&Bs. In addition, those staying with friends and family, or in 'other' accommodation, were found to incur additional food costs. However, these were found to be lower than the costs for those in hotels/B&Bs, at 48 per cent for those staying with friends and family and 18.5 per cent for those staying in 'other' accommodation. For evacuees staying at rest centres, the costs of additional food are covered by Bellwin returns (Chapter 6) and are thus not considered here.

For each of the different types of accommodation, low, mid and high extra food costs have been determined, taking into account the proportion of households evacuated to each type of accommodation and the variations in additional food costs, as outlined above. Different sets of data tables have been produced and Table 4.22 gives an example of these for hotels/ B&Bs.

Travel and time costs

Additional travel and time costs accrue to evacuees partly through the process of evacuating but also during their stay in temporary and/or alternative accommodation. As such, two tables have been produced. The first provides data for the travel and time costs associated with the evacuation process (temporary accommodation), and Table 4.23 shows a sample of this data for hotel/B&B accommodation.

An additional set of data and tables has been developed for the travel and time costs associated with the duration of the evacuation when evacuees are staying in alternative accommodation. The travel and time costs associated with the duration of the evacuation (alternative accommodation) are seen as an additional distance travelled to schools, work and

Table 4.22 The average extra food costs per household evacuated to hotels/B&B

Time period	Percentage of hotel/B&B evacuees remaining in accommodation	Potential cost per household evacuated to hotel/B&B (£)			Cost for actual percentage of households evacuated to hotel/B&B (9%)		
		Low	Mid	High	Low	Mid	High
Up to 1 week	100	168	337	562	15	30	51
Up to 2 weeks	85	311	623	1,039	28	56	94
Up to 3 weeks	50	395	792	1,320	36	71	119
Up to 4 weeks	35	454	910	1,517	41	82	136
Up to 8 weeks	10	521	1,045	1,741	47	94	157

Table 4.23 The extra travel costs associated with temporary accommodation per household evacuated

Distance to temporary accommodation (km)		% of evacuees	Travel cost (£)			Time cost (£)		
			Low	Mid	High	Low	Mid	High
Hotel/B&B		9						
Low	5		1.06	1.56	3.55	0.32	0.51	1.26
Mid	30		6.34	9.33	21.27	1.89	3.03	7.58
High	80		16.92	24.88	56.73	5.05	8.08	20.20
Average per household evacuated			0.73	1.07	2.45	0.22	0.35	0.87

Table 4.24 The extra travel costs associated with alternative accommodation per household evacuated

Nature of visit – additional distances involved	Distance (km)	Average number of visits per week	Travel cost (£)			Time cost (£)*		
			Low	Mid	High	Low (km/h)	Mid (km/h)	High (km/h)
Return to home		4						
Low	5		4.23	6.22	14.18	1.26	2.02	5.05
Mid	10		8.46	12.44	28.36	2.53	4.04	10.10
High	30		25.38	37.32	85.09	7.58	12.12	30.30
Average per household evacuated			**12.69**	**18.66**	**42.54**	**3.79**	**6.06**	**15.15**
School		5						
Low	5		5.29	7.78	17.73	1.58	2.53	6.31
Mid	10		10.57	15.55	35.45	3.16	5.05	12.63
High	30		31.72	46.65	106.36	9.47	15.15	37.88
Average per household evacuated			**15.86**	**23.33**	**53.18**	**4.73**	**7.58**	**18.94**
Work		5						
Low	5		5.29	7.78	17.73	1.58	2.53	6.31
Mid	10		10.57	15.55	35.45	3.16	5.05	12.63
High	30		31.72	46.65	106.36	9.47	15.15	37.88
Average per household evacuated			**15.86**	**23.33**	**53.18**	**4.73**	**7.58**	**18.94**
Shopping		4						
Low	5		4.23	6.22	14.18	1.26	2.02	5.05
Mid	10		8.46	12.44	28.36	2.53	4.04	10.10
High	30		25.38	37.32	85.09	7.58	12.12	30.30
Average per household			**12.69**	**18.66**	**42.54**	**3.79**	**6.06**	**15.15**

Note: * £10.10 average hourly wage × 50%.

for shopping. In addition, trips to visit the flooded property are included. A sample of this data is given in Table 4.24. This data is then combined to provide total travel costs along with time costs, as calculated from national average hourly wage rates, as shown in Table 4.25.

Table 4.25 Time and travel costs associated with alternative accommodation per household evacuated

Time period	% Remaining in accommodation	Travel cost (£)			Time cost (£)*		
		Low	Mid	High	Low	Mid	High
Up to 1 weeks	100	57.10	83.97	191.45	17.04	27.27	68.18
Up to 4 weeks	96	221.55	325.80	742.83	66.13	105.81	264.52
Up to 12 weeks	83	600.69	883.36	2,014.07	179.30	286.88	717.20
Up to 26 weeks	75	1,200.23	1,765.05	4,024.31	358.26	573.22	1,433.04
Up to 39 weeks	45	1,534.27	2,256.27	5,144.30	457.97	732.74	1,831.86
Up to 52 weeks	15	1,645.61	2,420.02	5,517.64	491.20	785.92	1,964.80

Note: * £10.10 average hourly wage × 50%.

Loss of earnings

The very process of evacuation disrupts normal daily life for those affected. As such, it is not uncommon for evacuees to take time off work to organise the repair and recovery of their property.

According to RPA/FHRC (2004) and Tunstall *et al.* (2007), of those employed, 61 per cent took time off work as a result of their evacuation. Using 2013 values to derive the percentage of people employed and applying the 61 per cent – from the RPA/FHRC (2004) data – to this, the percentage taking time off work was derived. Adding this to the average length of time taken off (thirteen days), the average lost earnings per household were found to range from £140 (low), £270 (medium) and £395 (high).

4.7.3 Total evacuation costs

The data provided in the tables above are useful as general guides for evacuation costs. Pulling the figures together for each component should provide an average cost per property which can then be applied to the number of properties evacuated.

To utilise this detailed data and facilitate easier, more accurate appraisals, the data developed in this work are available on MCM-Online. In addition to facilitating easier cost allocations, the website data also provide opportunities to tailor the calculations when, for example, evacuation durations are known or the costs of alternative accommodation are locally identifiable and there is therefore no need to rely on national averages for these elements of cost to calculate the necessary totals. Where assumptions have been made based on local surveys and research, users can modify the assumed values where they have access to more suitable information such that costs can be tailored to specific circumstances and schemes.

4.8 THE 'INTANGIBLE' EFFECTS OF FLOODING

The 'intangible' effects of flooding are now recognised to be significant. These were originally investigated by Middlesex University researchers at Uphill, Avon, in relation to coastal flooding (Green *et al.*, 1985) but since then they have been studied on a number of occasions and in contrasting circumstances (Green and Penning-Rowsell, 1989; Tapsell *et al.*, 1999; Defra/Environment Agency, 2004).

4.8.1 Health effects of flooding

Several qualitative studies have been carried out since the flooding of Easter 1998, June 2000 and autumn 2000, investigating the health and social impacts of these flood events (Tapsell *et al.*, 1999; Tapsell, 2000; Tapsell and Tunstall, 2001; Tapsell *et al.*, 2002). Although limited in nature, these studies have illustrated that the magnitude of the intangible impacts – particularly the human health consequences of flooding – are complex and may be far reaching. Likewise, the factors influencing these impacts are also now considered to be extensive (Table 4.26).

The effects on human health as a consequence of flooding are illustrated in Table 4.27, which shows both the range and severity of these 'intangible' impacts. These range from risks to life – hypothermia and injuries during, or immediately after, the flood – to more long-term physical and psychological health effects during the weeks or months following a flood. Follow-up qualitative research in Banbury and Kidlington found that psychological effects were still evident up to eighteen months after the flood, suggesting that the effects of flooding on people's health and general well-being can continue for many months, or even years, after the flood event.

More recently, Defra and the Environment Agency have funded a large research project which sought to establish an economic valuation of the 'intangible' health impacts of flooding for project appraisal (Defra/Environment Agency, 2004). This research confirmed the significance of the health impacts of flooding and led to the publication of interim guidance (Defra, 2004b). This guidance is the recommended approach for establishing potential 'intangible' benefit values in this Manual, the details of which are provided in Section 4.9.

4.8.2 Risk to life from flood events

Fatalities and injuries from flooding in England and Wales are fortunately few when compared with some other countries. However, in some situations in England and Wales there is high potential for risk to life, an assessment of which could be included within a project appraisal.

Identifying whether it is proportionate to include an assessment of the potential for risk to life within an appraisal is related to an understanding of how fatalities from flooding in England and Wales occur. In research for the Environment Agency's Flood Incident Management Investment Strategy (Environment Agency, 2013b), we have investigated the circumstances in which members of the public have been killed or severely injured during flooding. This research produced a typology of events/situations

Table 4.26 Factors contributing to the consequences of flooding on human health

Factors
Characteristics of the flood event (depth, velocity, duration, timing, etc.)
Type of property, e.g. single storey, two storey, etc.
The amount and type of property damage and losses
Whether flood warnings were received and acted upon
Previous flood experience and awareness of risk
Any coping strategies developed following previous flooding
Having to leave home and live in temporary accommodation
The clean-up and recovery process and associated household disruption
Frustration and anxiety dealing with insurance companies, loss adjusters, builders and contractors
Pre-existing health conditions and susceptibility
Increased anxiety over the possible recurrence of the event
A loss in the level of confidence in the authorities perceived to be responsible for providing flood protection and warnings
Financial worries (especially for those not insured)
A loss of the sense of security in the home
An undermining of people's place, identity and their sense of self (e.g. through loss of memorabilia)
Disruption of community life

Table 4.27 Physical and psychological health effects of flooding

Physical and other health effects reported during, or immediately after, Easter 1998 and summer and autumn 2000 floods	Physical health effects reported in the weeks or months after Easter 1998 and summer and autumn 2000 floods	Psychological health effects reported in the weeks or months after Easter 1998 and summer and autumn 2000 floods
Injuries from being knocked over by flood waters or thrown against hard objects, or from being struck by moving objects	Gastro-intestinal illnesses	Anxiety (e.g. during heavy rainfall)
Injuries from over-exertion during the flood (e.g. sprains)	Cardiac problems	Panic attacks
Hypothermia	Respiratory problems (e.g. asthma, chest infections, pleurisy)	Increased stress levels
Fear of electric shocks (although none was reported)	Lacerations, abrasions and contusions	Mild, moderate and severe depression
Cold, coughs, flu	Sprains and strains	Lethargy/lack of energy
Headaches	Skin irritations (e.g. rashes, dermatitis, etc.)	Feelings of isolation
Sore throats or throat infections	High blood pressure	Sleeping problems
Skin irritations (e.g. rashes)	Kidney or other infections	Nightmares
Shock	Stiffness in joints	Flashbacks to the flooding
	Muscle cramps	Increased use of alcohol or prescription (or other) drugs
	Insect or animal bites	Anger/tantrums
	Erratic blood sugar levels (diabetics)	Mood swings/bad moods
	Weight loss or gain	Increased tensions in relationships (e.g. more arguing)
	Allergies (e.g. to mould spores)	Difficulty concentrating on everyday tasks
		Thoughts of suicide

which are indicative of those situations in England and Wales where lives are likely to be threatened or where fatalities have occurred in the past (Table 4.28).

This typology highlights that there are a number of situations and different elements which lead to the potential of fatalities from flooding. In recent years, the majority of the deaths in England and Wales have not been related to the characteristics of flood itself, but have been caused by behavioural responses or actions of the victims (i.e. driving or walking through flood waters or trying to clear drains in flood conditions). Such fatalities are very difficult to account for when assessing the potential for risk to life from flooding and to include within a project appraisal.

There are circumstances, however, whereby the nature of the flood risk may lead to a higher potential for risk to life (such as fast-onset, high-velocity or very deep floods) or the circumstances of the flood (e.g. with an onset at night or an unexpected flood). These situations have occurred in the past (such as the Lynmouth floods in 1952 and the east coast floods in 1953) and, although there have been significant improvements in flood warning and emergency

Table 4.28 A typology of reasons why risk to life is threatened from flooding

	Category	Reason why risk to life is threatened	Examples
Hazard	Areas of extreme flood depth or velocity	The severity of the flood hazard: e.g. high flood depth or flood velocity.	Inundation of the east coast in particular areas behind defences
	Rapid response catchments	Severity of the flood hazard /coupled with limited time to undertake incident management activities.	Boscastle
Vulnerability	Vulnerable locations	Locations which will provide little shelter from flood waters or buildings which would suffer extensive damage, thereby exposing people to flood waters.	Caravan and camping parks; single-storey buildings
	Vulnerable people	People who may require assistance to leave their properties at the time of flooding or who are unfamiliar with the flood risk in the area and therefore may require warning separately or other additional assistance.	This may include the elderly or disabled or those transient to the area
Behavioural	People are exposed to flood risk through their actions	These situations relate to those circumstances whereby people place themselves in risky situations through their actions: either through ignoring instructions not to cross rivers or misjudging the dangers of the flood waters.	Including activities like driving or walking through flooded areas
Other	Other flood-related activities	This includes a range of other situations whereby people may be injured or lose their lives via other means.	Exposure to fumes from drying equipment; falling when cleaning

Source: Adapted from Environment Agency, 2013b

planning, severe flooding might still create the potential for such fatalities in the future. The other situations are when there are high numbers of vulnerable people (e.g. in care homes or schools) or people are located in vulnerable locations (e.g. caravan/camp sites) where there is little shelter.

Appraisal of risk to life should therefore be considered in those situations which are shaded on Table 4.28 and include (or are a combination of):

- Where the expected flood has high-velocity or deep flood waters (or a combination of both).
- Where the speed of onset is rapid, permitting little time for warning the public or evacuating areas at risk.
- Where there are large numbers of highly vulnerable people (e.g. elderly or schoolchildren).
- Where people may be vulnerable due to lack of shelter (e.g. caravan/camp sites).

From an appraisal perspective, an assessment to risk to life will be most efficiently undertaken when the economic efficiency of an FRM option is marginal (i.e. as an additional benefit) or it is considered that the impacts of flooding on the likelihood of loss of life is disproportionate to the expected property damage.

This might include a situation, for instance, where there is a small number of permanent properties located in a benefit area and therefore property damage will be low, but risk to life may be high due to rapid onset of flooding affecting residents or caravan/camp sites.

In particular, there may be circumstances where one or more of the categories are located within a benefit appraisal area (e.g. a caravan site located in an area of rapid-onset flooding) and in these situations an appraisal should clearly be undertaken. Appraisal also should not be undertaken without consideration of the wider flood management measures as the presence or absence and effectiveness of flood incident management activities (in particular flood warnings) will have mitigating impacts in many circumstances.

Assessing risk to life from flooding

Several approaches have been developed to calculate the potential risk to life from flood events (Jonkman *et al.*, 2002). Table 4.29 summaries some of the existing models, highlighting their basis and key characteristics. These approaches have tended to centre on deriving mortality fractions based on empirical observations which have then been applied to the exposed population according to flood severity and other parameters, such as the presence of flood warning and/or awareness. However, availability of quantitative data from historical floods is in general very limited and therefore these functions are based on datasets of inadequate size and quality. There is also an unequal focus on dam-burst or flood-defence-failure type events, which are of limited use in the context of England and Wales.

'Risk to people' methodology (HR Wallingford; 2003; 2005b; Defra, 2008b)

One approach that developed different models to predict loss of life was the 'flood risk to people' method (HR Wallingford, 2003; 2005b). This was developed for application in the UK and differs from the other approaches as fatalities are estimated as a function of injuries, which are calculated according to the flood, area and population characteristics, rather than applying a uniform mortality fraction to the exposed population. This approach was used in the Catchment Flood Management Plans by the Environment Agency.

Priest *et al.* (2007) undertook a sensitivity analysis on the approach for the Boscastle event and highlighted that potential fatalities ranged from 0.67 to 109.2, depending upon the input values. This assessment highlighted the significance of the data input to the model, and the fact that a range of values might be a more appropriate output rather than an absolute prediction of deaths. The Boscastle analysis also illustrates the impact of such factors as the timing of the flood, any unofficial warning systems and evacuation, showing that these are major components that are missing or under-represented in this approach.

The risk to people method was developed and calibrated using a limited number of events in England and Wales, and the model has been properly tested and calibrated on only a small number of case studies. Many of these cases occurred a number of years ago (e.g. 1912, 1952, 2000, 2005), since when there have been major advances in flood warning, communication technologies and changes to search and rescue practices. Moreover, the case studies used to calibrate and test the approach are instances of quite severe flooding; the approach has not been tested on more minor flooding cases in the UK.

The risk to people approach was also found to contain several structural weaknesses, including the need to divide benefit areas into risk zones wherein estimations of hazard, people characteristics and the characteristics of the area were constant. This 'zoning approach' limits the flexibility of the model to account for differences in population vulnerability and the type of properties impacted. The approach therefore has the potential to overestimate numbers of fatalities severely.

A 'threshold' approach to assessing risk to life

Priest *et al.* (2007) developed a new methodology to estimate loss of life from flood events utilising a semi-qualitative 'threshold' model which combines hazard and exposure thresholds.

The model is based on the premise that not all people are equally vulnerable at the same depths and velocities of flood waters, with vulnerability varying depending upon individual and situational characteristics. However, there is value in identifying the broad

Table 4.29 Existing approaches for estimating the risk to life from flooding

Study	Flood context/ scenario	Empirical basis	Characteristics of interest	Critique
Waarts (1992)	Large-scale coastal flood – flood defence failure	Based on data from the 1953 coastal flood in the Netherlands	Water depth Population at risk – calculates the fraction of the population that drowned based on water depth	Its applicability to other areas is limited.
Jonkman (2001), discussed in Jonkman (2007)	Large-scale coastal flood – flood defence failure	Based on data from the 1953 coastal flood in the Netherlands Incorporates Abt's *et al.* (1989) work on human stability into the Waarts formula	Evacuation time, depth and velocity are all a function of the location in the floodplain	Based on limited data and therefore transferability is limited.
Vrouwenvelder and Steenhuis (1997)	Large-scale coastal flood – flood defence failure	Based on data from the 1953 coastal flood in the Netherlands	Population at risk Building collapse Dyke breaches	Its applicability to other areas is limited.
Graham (1999)	Dam failure	Forty historical dam breaks	Flood severity, warning lead-times, the number of people occupying the at-risk area and the population's understanding of flood severity	Dissimilar to Brown and Graham model (Brown and Graham, 1988), this method attempts to take into account the response time to flood warnings by including the 'flood severity understanding' parameter and provides a range of potential fatalities.
Jonkman (2003)	Large-scale coastal flood	Based on data from the 1953 coastal flood in the Netherlands	Probability of drowning statistically related to hydraulic characteristics of the flood such as water depth, velocity and rate of rise of flood waters	Although based on one flood event this method does relate the mortality to three different types of event: drowning due to rapidly rising water, high flow velocities and deaths due to other causes.
Zhai *et al.* (2006)	Various	A total of 269 historical flood events that occurred in Japan during the period 1947–2001	The authors propose that the number of flooded residences is an indicator of: flood severity, size of the at-risk population and number of flood-related injuries/fatalities	An interesting feature of this research is that when the historical data was analysed at decadal intervals, it was found that fatality/injury coefficients decreased with time. This may be due to improved institutional arrangements such as flood warnings and emergency response.

Table 4.29 continued

Study	Flood context/ scenario	Empirical basis	Characteristics of interest	Critique
Jonkman (2007)	Large-scale floods that occur in low-lying areas due to flood defence failure	Eleven historical flood events, with a total of 165 separate locations	Based on the factors of flood depth, flood water velocity, rate of rise (speed of onset) and the arrival time of flood waters at a location following a breach Model also considers the effect that evacuation, available shelter, rescue and escape (self-evacuation) has on numbers exposed	As stated above, Jonkman's model is designed for large-scale floods caused by flood defence failure in low-lying areas which are characteristic of deltaic regions. However, flood events in many other parts of Europe are quite different from the situation in the Netherlands, thus Jonkman's model may not be applicable in these different situations.

characteristics of both people and areas where different levels of risk to flood waters exist. Figure 4.4 illustrates the sequencing of the components which are considered within the model.

Full details of the approach and the model can be found in Priest *et al.* (2007) and there is a summary in Tapsell *et al.* (2008). This approach draws principally on existing data concerning the stability of humans during flooding (Jonkman and Penning-Rowsell, 2008) and the likelihood that they will be killed if they are both inside properties and outside during flood events. The foundation of the approach is the depth–velocity function, and utilising the human and building stability work undertaken by Abt *et al.* (1989), Jonkman and Penning-Rowsell (2008) and Karvonen *et al.* (2000), these functions identify when different people and buildings become vulnerable to flood waters. Table 4.30 illustrates these depth–velocity thresholds and the main factors that may lead to fatalities from flooding.

The approach then adds mitigating variables, such as the nature of the area (i.e. the availability of shelter and the integrity of that shelter under different flood conditions) and the presence or absence of mitigating procedures (e.g. flood warning and evacuation). Notably, in addition to the more commonly used variables, the approach takes account of the types of property situated in the area of interest and the likelihood that they will suffer building collapse (a significant cause of fatalities) as well as factors which might in turn reduce the risk of fatality.

When combined with estimates of the numbers of people vulnerable (i.e. through census information), this approach can be used as a tool to allow appraisers

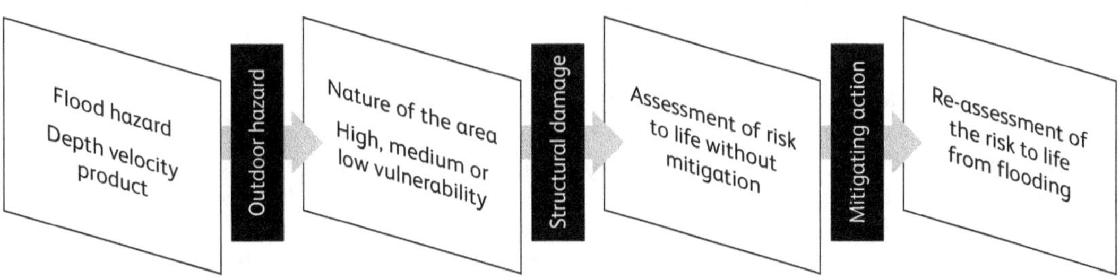

Figure 4.4 Sequence and combination of the different components of the threshold modelling approach

Table 4.30 Main factors leading to fatalities from flooding

Depth–velocity thresholds (m^2/s)	Nature of area categories	Main factor leading to fatalities	Description
<0.25	All	Low risk	There is low risk to people from the flood waters.
0.25–0.50	All	People vulnerability dominated – some behaviour related	The fatalities are likely to be concentrated amongst the vulnerable people, e.g. children either playing in or near flood waters, or elderly people (often trapped in their properties).
0.50–1.10	Low and medium vulnerability	Behaviour dominated	In most circumstances people will be able to find shelter away from the floods. However, deaths and injuries may still occur if people undertake risky activities such as driving through the flood waters or taking unnecessary risks in the waters.
0.75–1.75	High vulnerability	Hazard dominated	In these situations, fatalities are likely to occur from direct contact with the flood waters.
1.75–7.00	Low and medium vulnerability		
1.75–7.00	High vulnerability	Hazard and building collapse dominated	Fatalities will occur if people are in direct contact with the flood waters or if caught in buildings that are structurally compromised by the flood waters.
>7.00	All		

Source: Tapsell *et al.*, 2008

to make general and comparative assessments of risk to life under a 'do nothing' scenario and once a flood risk management measure is implemented, thereby establishing the benefits of lowering the potential risk to life. The approach permits the mapping of risk to life, and thus the assessment of this risk in different areas is possible (Priest *et al.*, 2007).

This approach presents a semi-qualitative approach to assessing risk, rather than fully quantifying that risk in terms of numbers of fatalities. It is possible to monetise the potential for risk to life through utilising figures presented in HM Treasury's 'Green Book' (HM Treasury, 2003). This provides a figure of £1.145 million per fatal casualty prevented (at 2000 values), which was originally developed by the Department for Transport for deaths in road accidents. This may therefore be applied to the estimated number of fatalities. However, although it is possible to monetise this risk, we ask appraisers to think carefully about whether it is appropriate to do so. Due to the often small numbers of fatalities that would be expected in many flood situations, placing an economic valuation on this risk may trivialise the value of life for known individuals when it is compared to other losses, such as those from property damages.

An appraiser also needs to be aware that the circumstances leading to the highest potential for risk to life are likely to occur in the most extreme flood scenarios. Therefore, the impact of flood risk management measures on these very severe floods needs to be carefully examined. In such an examination it needs to be gauged whether *any* flood risk management activity is likely to reduce the potential for fatalities due to floods of the highest magnitudes, and whether any other parallel flood management measures would also need to be in place and operating (e.g. flood warning) for the appraised measure to be fully effective.

4.9 RECOMMENDED APPROACHES TO RESIDENTIAL BENEFIT ASSESSMENT

As emphasised in Chapter 2, benefit–cost analysis is an iterative process whereby the most significant benefits and costs are first identified and then estimates of the values of each of these streams of benefits and costs are progressively refined.

This progressive refining of estimates is a central feature of the benefit assessment procedure. Hence, the recommended approaches for assessing residential benefits in this Manual are similarly progressive in detail and refinement – strategic, initial and full detailed project appraisal (Table 4.31).

At the heart of any benefit assessment is the loss-probability curve, which is derived using estimates of

Table 4.31 Assessing the potential damage to residential properties and households

Level of analysis	Guidance	Data requirements	Direct/tangible method of assessment	Indirect/intangible method of assessment
Overview	For rapid desktop appraisals. This provides first approximations which are useful to identify the areas where more detailed work is required	Number of properties in benefit area	Annual average damages	Intangible benefits
		Annual average direct damage	Sector average	
		Sector average direct damage		
		Weighted average value of £200 per property per year for intangibles		
Initial	For more detailed project appraisal information where further assessment of household loss potential is considered warranted	Number, type and age of houses in benefit area	Generalised standard residential depth/damage data	Intangible benefits
		Generalised standard data for type and age of houses		Vulnerability analysis where feasible
		Standard of protection (pre- and post-scheme) for intangible values		
Full-Scale Detailed	For the greatest accuracy and detailed study of potential benefits using the most detailed of the standard datasets	Number, type, age and social grade of houses and householders in benefit area	Detailed standard residential depth/damage data	Intangible benefits
		Detailed standard data for type, age and social grade of houses and householders	Distributional impact analysis	Vulnerability analysis
		Standard of protection (pre- and post-scheme) for intangible values		
		Government weighting factors for distributional impact analysis		

flood losses for flood events of differing return periods (with a minimum of three flood events – the threshold event, one above the anticipated standard of protection, and one mid-way between). In order to estimate this loss-probability curve, two items of data are required for each return period event considered:

- the number of properties of each land use class that are flooded; and
- the depth of flooding in each of those properties.

At an early stage in the benefit assessment process, the appraiser will be required to determine a 'ball-park' estimate of potential damages. At this stage, it is unlikely that detailed information on the depth of flooding will be available. Likewise, it is unlikely that detailed information will be available on the type and age of properties in the benefit area or the social grade of their occupants. However, as an absolute minimum, the number of properties in the benefit area must be known. This enables the appraiser to conduct an initial appraisal using weighted annual average damages (WAADs). Where the depth of flooding is known, the appraiser can further refine this estimate by applying the sector average figures for each flood depth (Section 4.9.1).

Where the results of this initial benefit assessment suggest that further analysis is justifiable, the appraiser is then required to gather improved information in three important ways:

- the number of flood events and the return periods of those events for which losses are estimated;
- the number of properties of different land use types (type, age and social grade) that are flooded in each of those flood events; and
- the depth of flooding of each of these properties in each of the events.

The Environment Agency (2010a) provides more detailed guidance on the most efficient means of reducing the uncertainty about the estimates of flood risk management benefits, but in general it is necessary to concentrate on the high-frequency events. In addition, flood losses are highly sensitive to relatively small changes in the depth of flooding at shallow depths. A first stage refinement is thus to calculate the depths of flooding from two or three high-frequency events. Once some estimates of the depths of flooding are available, it is then possible to use the depth/damage curves provided alongside this Manual (on MCM-Online) in a progressively more detailed manner.

The benefit assessment methods recommended in this Manual have been categorised according to the level of appraisal being considered (Table 4.31). As stated above, this scale is dependent on the extent to which the estimates have been refined:

- Overview or 'generalised' project appraisals: very much a 'ball-park' estimate to establish if further analysis is justifiable. This is the highest level of analysis available for rapid desktop appraisals. It is intended to provide first approximations and is designed to identify the areas where more detailed work should be undertaken. It is also designed to identify those variables which are critical to the final appraisal result.
- Initial project appraisals: for more detailed assessments which take account of the age and type of houses in the benefit area when assessing direct damages. As this kind of information is not available in the National Receptor Dataset (NRD), a site survey or the use of Google Street View (or any equivalent) will be necessary. It also provides a more in-depth analysis of the intangible benefits and considers the vulnerability of the households in the benefit area.
- Full-scale detailed analysis project appraisals: for the most detailed study of potential benefits. This differs from an initial appraisal in that social grade is included in the analysis of direct benefits. In addition to refining the direct damage estimates, this enables appraisers to account for distributional impacts through the use of weighted factors (but *not* if social grade averages are used: see below).

Irrespective of the level of analysis, it is important for appraisers to note that government guidance requires any expected value of damage not to exceed the market value of the property. These market values can be obtained from the government's Land Registry website (http://www.landregistry.gov.uk). However, this 'capping' at market value creates regional distribution issues (e.g. houses within the M25 are significantly more expensive than comparable houses in the north of England) for which there is, at present, no official counter-mechanism.

It is also important to note that the figures given in this Manual do not include damage from saltwater.

Where saltwater damage is expected, flood damage repair costs to building fabric items are estimated to increase by around 10 per cent (Penning-Rowsell, 1978) for still or low-velocity saltwater inundation, typical of estuarine locations (see Penning-Rowsell *et al.*, 1992, 98–102, for further details). Where properties are subjected to wave or shingle damages, this uplift factor may be substantially increased (see Section 5.6.7 for further details).

4.9.1 Overview or generalised project appraisals

This desktop analysis is intended to provide a rough approximation of potential benefits (as in the Modelling and Decision Support Framework 2) and to identify where more detailed analysis is required. Its primary purpose is to provide an estimate of the likely magnitude of the benefits in order to determine whether it is worth spending money on additional studies.

At this stage, there is unlikely to be information about the depth of flooding in several flood events. So, a preliminary estimate of the direct damages can be made only by making assumptions. Where only the number of properties in the benefit area is known, an approximate estimate of the order of flood risk management benefits can be derived by making some assumptions about the depth of flooding to be expected from different return period floods. To do so, we recommend using weighted AAD figures as rough estimates of potential direct damages (see 'Weighted annual average damages (WAADs)' below).

Where there is a basic understanding of the depths of flooding, appraisers can use the data provided on MCM-Online. However, during an overview-level study, only the sector average figures should be used (see below).

For assessing the indirect and intangible benefits, we recommend using surrogate values for indirect losses (i.e. the rental costs of the house vacated in evacuation circumstances, rather than the cost of the accommodation actually used – see Section 4.7.2 – and the value of £225 per property per year for intangible health benefits.

To employ these methods, the appraiser needs to determine the size of the benefit area, the number of properties and, where available, the depth of flooding:

- The size of the benefit area is determined by the characteristics of the scheme being appraised.
- The number of properties can be obtained from the National Receptor Dataset, available from the Environment Agency.
- The depth of flooding is determined from the ground level data and the results of the hydraulic modelling (if any).

This high-level analysis does not differentiate any variation between houses, using average values across all the properties. However, for a more informed overview study, knowledge of the types of houses in the benefit area is recommended. Much of this can be obtained by using Google StreetView (or equivalent), with the remainder determined during field visits.

Weighted annual average damages (WAADs)

A number of approaches have been developed over the last decade or so to provide annual average direct damage data for overview studies. Foremost of these is the WAAD approach by warning lead-time and standard of protection. This approach uses the sector average damage figures, compiled by Middlesex University, as a means of calculating mean damage for a range of flood return periods (Table 4.32). This is the recommended method to be used where the appraiser has little, or no, understanding of the potential flood depths and return periods associated with their benefit assessment.

Using the updated 2010 calculations in Table 4.32, the WAAD to the average house with no flood warning and no flood protection is £4,728. This figure has decreased since 2010, due to the reduced costs of cleaning and drying a flooded property. Table 4.33 gives the reduced values provided by different standards of protection and different levels of flood warning, such that a property already provided with a 100-year standard of flood protection, but without warning, has an annual average potential damage figure of £65. This average potential damage value is progressively reduced with warning such that with eight hours' warning it becomes £62.

Our recommendation is to use the figures in Tables 4.32 and 4.33 for the assessment of WAAD, because the underlying analysis of this is based on accurate hydraulic modelling of flood depths and extents, rather than relying on more arbitrary flood loss thresholds and probabilities.

Table 4.32 Weighted annual average damage (WAAD) calculations: residential property with no protection (2013 prices)

Flood frequency	Distribution of flood depths		Damage (£)	Weighted damage (£)
	M	%		
5 years	<0.1	81	6,577	5,327
	0.1–0.3	7	17,697	1,239
	0.3–0.6	11	24,133	2,655
	0.6–0.9	1	27,864	279
	0.9–1.2	0	30,583	0
	>1.2	0	32,106	0
	Total weighted damage			**9,500**
10 years	<0.1	50	9,690	4,845
	0.1–0.3	31	22,545	6,989
	0.3–0.6	10	29,588	2,959
	0.6–0.9	6	32,922	1,975
	0.9–1.2	2	35,511	710
	>1.2	1	36,860	369
	Total weighted damage			**17,847**
25 years	<0.1	45	9,690	4,361
	0.1–0.3	24	22,545	5,411
	0.3–0.6	22	29,588	6,509
	0.6–0.9	5	32,922	1,646
	0.9–1.2	4	35,511	1,420
	>1.2	1	36,860	369
	Total weighted damage			**19,716**
50 years	<0.1	32	12,803	4,097
	0.1–0.3	20	27,392	5,478
	0.3–0.6	21	24,133	5,068
	0.6–0.9	21	27,864	5,851
	0.9–1.2	4	40,440	1,618
	>1.2	3	41,614	1,248
	Total weighted damage			**23,360**

Table 4.32 continued

Flood frequency	Distribution of flood depths		Damage (£)	Weighted damage (£)
	M	%		
100 years	<0.1	22	12,803	2,817
	0.1–0.3	16	27,392	4,383
	0.3–0.6	26	24,133	6,275
	0.6–0.9	19	27,864	5,294
	0.9–1.2	12	40,440	4,853
	>1.2	6	41,614	2,497
	Total weighted damage			**26,119**

Return period (years)	Exceedence probability	Damage (£)	Probability of flood in interval	Mean damage (£)	Annual interval damage (£)
2	0.5	0			
			0.3	4,750	1,425
5	0.2	9,500			
			0.1	13,673	1,367
10	0.1	17,847			
			0.06	18,781	1,127
25	0.04	19,716			
			0.02	21,538	431
50	0.02	23,360			
			0.01	24,739	247
100	0.01	26,119			
			0.005	26,118	131
200	0.005	26,119			
			Weighted annual average damage		**4,728**

Notes: >0.1m = all sector residential damage figures at 0.05m. Due to rounding, percentages will not sum to 100.

However, where this value is used in strategy studies, as the WAAD per residential property within a defined benefit area (say, 1-in-200-year floodplain), the number of properties affected by successively more frequent return period floods could be reduced as follows (OST, 2004):

- One-hundred-year property numbers
  93 per cent of 200-year floodplain AADs
- Fifty-year property numbers
  80 per cent of 200-year floodplain AADs
- Twenty-five-year property numbers
  25 per cent of 200-year floodplain AADs
- Ten-year property numbers
  10 per cent of 200-year floodplain AADs
- Five-year property numbers
  5 per cent of 200-year floodplain AADs

Table 4.33 Weighted annual average damages (WAAD) (2013 prices) assuming variable threshold standards of protection (SoP)

Existing SoP	No warning (£)	<8 hour warning (£)	>8 hour warning (£)
No protection	4,728	4,559	4,513
2 years	4,728	4,559	4,513
5 years	2,828	2,727	2,700
10 years	1,400	1,350	1,336
25 years	612	590	584
50 years	261	252	249
100 years	65	63	62
200 years	33	32	32

Note: Using AAD figures from Table 4.32.

Failure to moderate property numbers to account for flood frequency will dramatically overstate the WAAD.

As outlined above, the WAAD method is designed to assess the risk to any particular property from a range of flood probabilities and severities. It is an important summary statistic. The WAAD dataset, described above, was developed in the 1980s from an analysis of several flood alleviation project appraisals in the English Midlands. The analysis was comprehensive but based on a sample of probably somewhat atypical schemes.

New research for this volume sought to improve on this dataset, using as far as possible a standard set of flood return periods (5, 10, 20/25, 50, 100 years), taking data from the Environment Agency's Project Appraisal Reports and their 'do nothing'/'do minimum' flood depth/return period data. We used this data for the following locations:

1 Riverside Newport (Tidal, Wales).
2 Ouse to Seaford Head (Coastal, south-east England).
3 Teignmouth (Tidal, south-west England).
4 Kempsey (Fluvial, West Midlands, England).
5 Glynbridge Gardens (Fluvial, south-west, England).
6 Ashbourne (Fluvial, East Midlands, England).

To increase the number of schemes, we have included three others in our calculations (Exeter, Lower Thames and Chertsey – all fluvial).

The total weighted damage per return period and the resulting WAAD value have been calculated for each scheme using our automated WAAD calculator for the 'do nothing' and 'do minimum' scenarios. The 'do nothing' scenarios include overtopping, breaching, failure or removal of embankment scenarios and their representativeness for other schemes to which the WAAD data might be applied can be questioned. The 'do minimum' scenarios include different standards of protection (SoP). The considered return periods are also variable (i.e. not the standard pattern 5, 10, 20/25, 50, 100, as above). The WAAD values (Figure 4.5) are therefore of different ranges, some of them not strictly comparable.

The new average WAAD value, calculated as the average total weighted damage per return period, is £7,709, assuming no damage for floods more frequent than one in two years (0.5 annual probability). In comparison, the 2010 MCM WAAD value was £5,393 (Penning-Rowsell *et al.*, 2010, Table 4.4).

What is apparent from Figure 4.5, however, is that the spread of WAAD values is very large indeed, even allowing for the different types of location considered (coastal/fluvial; 'do nothing'/'do minimum'). This makes the aforementioned average of dubious value, and we do not believe that we have produced a new set of data that is fit for purpose, as yet. More research is clearly needed, and the previous WAAD value should perhaps be used until a better result is obtained and posted on MCM-Online.

Sector average damages

To provide a more refined estimate of direct damages, the depth of flooding across a range of flood events must be known. The absolute minimum number of flood events that can be considered is three:

- the threshold flood event (the most extreme flood event which does not cause any losses);
- an intermediate mid-way flood; and
- a flood larger than the anticipated standard of protection.

Ideally, the appraiser should calculate the depth of flooding from two or three high-frequency events. This will demonstrate the sensitivity of potential flood damages to floods of shallow depths.

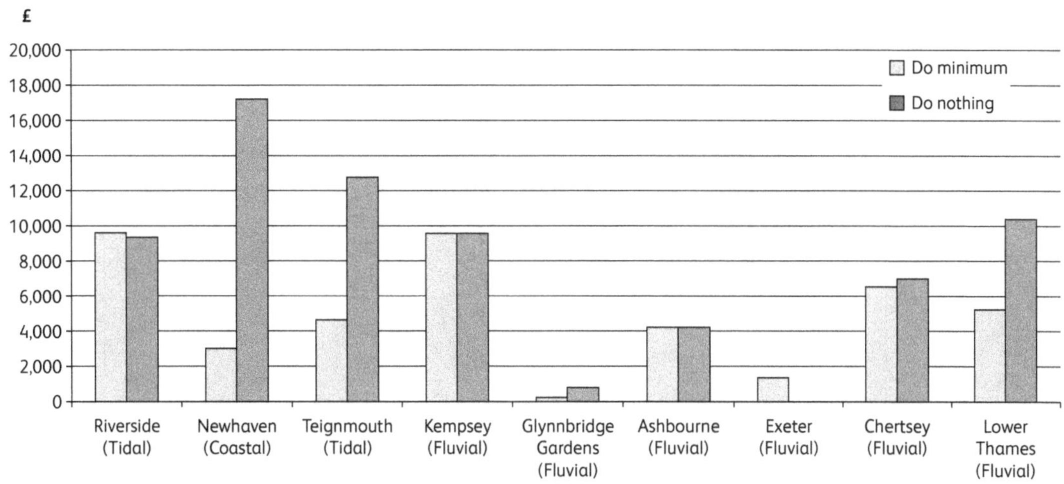

Figure 4.5 WAAD values for the different schemes (2013 results)

Where this flood depth information is available, it is then possible to use the depth/damage curves on MCM-Online. The highest-level data available is the sector average for all residential properties, given as data for long and short flood durations. For example, at a 0.1m flood depth the sector average damage figure for flooding of less than twelve hours is £10,487, whilst for floods of more than twelve hours' duration the corresponding figure is £19,865. This data is available across a range of flood depths which can be extracted by users and plotted as depth/damage curves. Where the appraiser's calculated flood depths do not correspond with the tabulated depth/damage figures on MCM-Online, these curves should be interpolated linearly.

In addition to the sector average figures for all residential properties, data is also available for categories of building fabric items and inventory items.

4.9.2 Initial project appraisals

For an initial project appraisal, but where a more refined estimate of potential benefits is considered warranted, an initial project appraisal is recommended. This requires the appraiser to have information on a range of flood depths for each event considered, and to have a more detailed understanding of the properties in the benefit area. In particular, the appraiser will need to know the following:

- The depth of flooding across a range of flood events: this should be obtained from ground level data and the results of hydraulic modelling.
- The type and age of houses in the benefit area: this can be obtained from a more detailed field survey or Google Street View (supplementing the data from the NRD). Using the floor plans and photographic illustrations on MCM-Online (see MCM-Online Supporting Information for Chapter 4), the age of the houses in the benefit area can be determined.

Provided this information is available, the appraiser can then evaluate potential direct damages using the generalised standard residential depth/damage data on MCM-Online.

However, unlike the overview study, a more detailed analysis of intangible benefits is required at an initial level of analysis.. Rather than simply applying the weighted average figure of £225 per property per year, the intangible benefits need to be determined using Defra's risk reduction matrix (Defra, 2004b) (Table 4.34). In addition, it is recommended that a more detailed vulnerability analysis is conducted (see below).

Table 4.34 Intangible benefits associated with flood risk management improvements (2013 prices)

Standard of protection after – AFP (RP in years)										
Standard of protection before – AFP (RP in years)			0.007	0.008	0.01	0.013	0.02	0.033	0.05	0.1
			(150)	(125)	(100)	(75)	(50)	(30)	(20)	(10)
	1	–1	£353	£273	£254	£194	£93	£32	£15	£6
	0.1	–10	£272	£267	£248	£188	£86	£27	£10	£0
	0.05	–20	£261	£256	£239	£179	£76	£17	£0	–
	0.033	–30	£245	£240	£222	£162	£60	£0	–	–
	0.02	–50	£184	£180	£161	£102	£0	–	–	–
	0.013	–75	£83	£79	£60	£0	–	–	–	–
	0.01	–100	£23	£19	£0	–	–	–	–	–
	0.008	–125	£5	£0	–	–	–	–	–	–

Notes: AFP = Annual Flood Probability; RP = Return Period; Annual Benefits = Damages (before) – Damages (after).

Source: Defra, 2004b

Generalised standard residential depth/damage data

Identifying the variables used to classify dwellings should be a routine procedure in the field or from other sources. First, identifying the type of dwelling should present no real problem as these can be obtained using Google Street View (or any equivalent) and from direct field observation. By contrast, the age of dwelling may involve a small degree of subjectivity unless local authority planning departments can provide mapped information.

The photographs on MCM-Online (see MCM-Online Supporting Information for Chapter 4) should form a guide in the identification of the age and type of dwelling in any field survey. Once the number, type and age of the houses have been identified, the generalised standard residential depth/damage data provided there can be utilised. Building fabric and inventory costs for short- and long-duration floods at different depths can then be ascertained.

For example, for a detached, pre-1919 property, at a 0.1m flood depth, the damage figure for flooding of less than twelve hours is £25,591, whilst for floods of more than twelve hours' duration the corresponding figure is £50,858.

Intangible benefits

Research into the valuation of intangible health benefits concludes that the potential value of avoiding such impacts is, on average, £225 per household per year. In addition, this research concluded that the most important factor when calculating potential intangible impacts is the flood risk (Defra/Environment Agency, 2004). As a result, government guidance now requires appraisers to consider how the level of exposure to household flood risk varies with, and without, the proposed scheme. At the outset, this requires the appraiser to determine the level of risk, such that:

- For areas of uniform risk (such as housing on level ground behind a flood defence), damages are based on common standards of defence of an area.
- For areas of greatly varying risk (sloping ground away from a river), damages are based on individual levels of property flood risk (Defra, 2004b, 6).

Once the standard of protection before and after the proposed scheme intervention has been established, it is then necessary to apply the risk redistribution matrix to establish the values of the potential intangible benefits (Table 4.34).

- For areas of uniform risk, identify the standards of protection, use the matrix to obtain the intangible benefit per household and apply this to the total number of households in the benefit area.

- For areas of greatly varying risk, separate houses into different band areas depending on the standard of protection before and after the potential scheme intervention, evaluate the intangible benefits for each band and apply to all households in each band. Combine for the overall figure.

This is the recommended approach for including the intangible health impact benefit values in this Manual. However, these figures should not be subjected to a distributional impact analysis because the values in Table 4.34 already account for such impacts.

In the same vein, where coastal erosion appraisals are being conducted (see Chapter 7), current government guidance advises the following:

> The long-term equivalent of £200 per household per year capitalized over 50 years using current Treasury rates is some £5000. It is therefore proposed that where coastal erosion schemes are designed to significantly postpone (by more than 25 years) imminent loss of property (i.e. the property concerned has more than 80 per cent chance of loss in the next 5 years without the scheme), the damage reduction or benefit per property should be enhanced by this capital value.
>
> (Defra, 2004b, 8)

Vulnerability analysis

A vulnerability analysis for households comprises a method indicating the likely impact of floods of different severities on the households affected. Because of its experimental nature, we do not advise formalising the presentation of the results of such an analysis due to the possibility of making the analysis appear over-precise. Nevertheless, users are recommended to consider the variables used in the calculation of the Flood Hazard Research Centre Social Flood Vulnerability Index (SFVI) (Tapsell *et al.*, 2002), a composite index consisting of the variables described in Table 4.35. In addition, users are recommended to assess the following:

- the number of residents in the flood-prone area (these may be disaggregated by flood frequency, if required);
- the approximate proportions of households in each social grade;
- the proportion of residences which are bungalows, basement flats or ground floor fats (often occupied by the elderly and infirm);
- predicted flood depths (depths of over 0.6m can be life threatening);
- flood warning lead-times; and
- other flood characteristics, including the location of residences close to flood risk management measures which may be overtopped or breached.

In undertaking a vulnerability analysis, it is sensible to concentrate on estimating the number of households that will suffer the most severe conditions and that are the most vulnerable. The variables in the SFVI (Table 4.35), as well as those given above, offer this potential. Users should obtain figures for each of these variables for the population in the floodplain areas affected.

It would be wise to give higher priority to schemes that protect more vulnerable people, so that the gain to society is greatest from the expenditure of the necessary resources. Such an analysis should assist the user in determining the scale of the likely intangible impacts of flooding on householders.

In this regard it will be noted from Chapter 1 (Table 1.4) that the Defra Outcome Measures that count towards the allocation of central government Grant in Aid towards risk management schemes include a supplementary element for deprived areas, in relation to both flood risk reduction schemes and schemes to reduce the effect of erosion at the coast. In some respects this supersedes the analysis of social vulnerability described above, but it should be noted that economic deprivation is only one of the variables counted in our SFVI, the others being the health status of those affected (i.e. prior illness), the age profile of the population at risk (i.e. the elderly) and the presence of single-parent families.

These are the variables we found to have greatest explanatory power with regard to assessing vulnerability to flooding (Tapsell *et al.*, 2002). However, the Grant in Aid formula has perhaps a different purpose in prioritising areas where there is economic deprivation (and hence lower potential flood damage of the – more meagre – assets at risk) rather than areas where vulnerability to flooding is greatest. It is clear that both are important.

Table 4.35 FHRC Social Flood Vulnerability Index (SFVI)

Index	Variables	Rationale	Data needs
Elderly (over 75 years)	Residents aged 75 and over as a proportion of all residents	The age of 75 was chosen because epidemiological research has shown that after this age there is a sharp increase in the incidence and severity of arthritis (and other conditions) and this illness is sensitive to the damp, cold environmental conditions that would follow a flood event	Small Area Census data
Lone parents	Single parents as a proportion of all residents	Previous Middlesex University research has shown that lone parents (of either gender) are badly affected by floods because they tend to have less income and must cope single-handedly with both children and the flood impacts, with all the stress and trauma that this can bring	Small Area Census data
Pre-existing health problems	Residents suffering from limiting long-term illness as a proportion of all residents	Research by Middlesex University has shown that post-flood morbidity (and mortality) is significantly higher when the flood victims suffer from pre-existing health problems	Small Area Census data
Financial deprivation	Unemployment: unemployed residents aged 16 and over as a percentage of all economically active residents aged over 16	The financially deprived are less likely to have home contents insurance and would therefore have more difficulty (and take a longer time period) in replacing household items damaged by a flood event	Small Area Census data
	Overcrowding: households with more than one person per room as a percentage of all households		
	Non-car ownership: households with no car as a percentage of all households		
	No-home ownership: households not owning their own home as a percentage of all households		

4.9.3 Detailed project appraisals

As with an initial project appraisal, a fully detailed appraisal requires the analysis of intangible benefits using the risk reduction matrix and the analysis of the vulnerability of benefit area occupants. It also needs to allow for the damage effects of warnings and property-level protection measures. A full feasibility appraisal differs from an initial appraisal in the detail of the benefit assessment for direct damages.

In fully detailed appraisals, it is appropriate to differentiate houses in the benefit area by their type, age and the social grade of the occupants (as indicated in Table 4.31). This means that the detailed direct damage data provided on MCM-Online can be used. In order to reflect socio-economic equity considerations, this data should – where it is deemed to be 'necessary' and 'practical' (HM Treasury, 2003) – be subjected to a distributional impact analysis.

Detailed standard residential depth/damage data

To make full use of the detailed standard residential depth/damage datasets on MCM-Online, the social grade of the occupants of the houses in the benefit area should be established. Because the social grade variable derived from the 2011 Census data relates to the census output area as a whole (approximately 125 households), and not to individual households, the social grade of individual occupants is calculated on the basis of averages. For example, if 60 per cent of the dwellings in the census output area fall into the C2 category and 40 per cent fall into the DE category, the depth/damage data could be weighted as follows:

- The damage at 0.9m for a C2 inter-war semi-detached house at a flood duration of <12 hours is £27,688.
- The damage at 0.9m for a DE inter-war semi-detached house at a flood duration of <12 hours is £20,507.
- Thus, the damage for the average semi-detached inter-war house in this census output area is £(27,688 × 0.6) + £(20,507 × 0.4) = £24,815.

Therefore, once the social grade of the occupants has been established, the detailed direct damage figures on MCM-Online can be used in the appraisal. This is the most refined data available for estimating direct residential flood damages.

Identification of approximated social grades

The Office for National Statistics (ONS) provides approximated social grades based on the 2011 Census which use the same classification as this Manual. To identify the grade of households in the benefit area, appraisers should refer to the ONS datasets which are available for every census output area in England and Wales. The easiest way to download these is from the ONS's Labour Market Statistics website: http://www.nomisweb.co.uk.

Table QS611EW: Approximated Social Grade can be accessed and downloaded using the 'wizard query' and can be found under 'detailed statistics' then '2011 census – quick statistics'. This table provides 2011 Census estimates that classify 'household reference persons' (HRPs: the people responsible for owning or renting or who are otherwise responsible for the accommodation). In the case of joint householders, the person with the highest income takes precedence and becomes the HRP (ONS, 2013), aged sixteen to sixty-four, living in households by approximated social grade, for census output areas, regions, counties, parliamentary constituencies, etc.

The 2011 Census data are in the process of being updated and readers are advised to check for regular updates on the official 2011 Census website (http://www.ons.gov.uk/census).

It should be noted that Table QS611EW effectively excludes the over-sixty-fours, unless they are in employment, as well as the long-term unemployed. This problem can be resolved, or at least lessened, by evaluating the proportion of over-sixty-fours in a benefit area using ONS Table KS102EW: Age Structure (available from the same source as Table QS611EW). If the proportion of over-sixty-fours is found to be especially high, for instance above the national average of 15.9 per cent, data on social grade for this age group may be obtained from private sources, such as Experian's Mosaic database. The Experian data, like many commercial data products, are based on postal, rather than census, geography. A comparison between postcode units and census output areas is available from the ONS at: http://www.ons.gov.uk/census.

4.9.4 Distributional impact analysis

The Treasury's 'Green Book' (HM Treasury, 2003) recommends, where it is 'necessary' or 'practical', that to incorporate social equity considerations into flood risk management and coastal defence appraisals, potential benefits should account for distributional impacts. Determining if it is 'necessary' or 'practical' then depends on a number of circumstances, including:

- The likely robustness of any calculation of distributional impacts. Whether a community at flood risk can be identified with reliable data and categorised according to their prosperity or social grade.

- The type of project being assessed. Whether the assessment will contribute to an appraisal that demonstrates equity and fairness to people.
- The scale of the impact associated with a particular project or proposal. Whether the time and effort in undertaking the assessment is proportional to the scale of the overall appraisal, either at a strategic or scheme level.

(Defra, 2004b, 3)

If a distributional analysis is not required, the standard depth/damage curves that focus on the property type and age should be used, without accounting for social grade. If a distributional analysis is required, total weighted factors should be applied by approximated social grade (Table 4.36). However, these total weighted factors for C1 and C2 will generally have a negligible effect on the distributional impact assessment. Therefore, use of total weighted factors is recommended only where AB or DE approximated social grades are predominant. Total weighted factors may then be applied to adjust the standard depth/damage curves, provided on MCM-Online, to obtain potential damages avoided, taking account of distributional impacts (Defra, 2004b, 4).

A number of points are of importance in this government guidance:

- Weighted and non-weighted results should be presented, with the former complemented by a sensitivity analysis.
- Where property 'write-offs' are considered, average values should be based on average 'no risk' values of properties of similar type in the region (see Horsfield, 2012, for a map of the regions). For individual property potential damages, the appropriate weighting factor should be applied.
- Where averaged depth/damage data is used (i.e. for a pre-1918 detached house), as opposed to that for different social grades of the house's occupants, further weighting should *not* be undertaken, as these averages are already weighted for this effect.

Table 4.36 Total weighted factors by approximated social grade

AB	C1	C2	DE
0.74	1.12	1.22	1.64

Source: PAG (Defra, 2004b)

- In areas with a high proportion of rented accommodation the grade of the owner of the property should be taken into account for building damages and that of the occupier applied to content damages.

(Defra, 2004b, 4)

In this respect the research cited above developed a figure of £200 (at 2003 values) per household per year as a quantification of the adverse health effects of flooding (Defra, 2004b). This measure of 'intangible impact' is detailed in the FCERM-AG (Environment Agency, 2010a) and is widely used, although it tends not to contribute more than *c.*£6,000 per house (PVd). As such, it appears to be far too low.

The best estimate of these intangible effects is perhaps now linked to a value of £2,513 per household per event, based on future climate change metrics (Ramsbottom *et al.*, 2012). This value relates to a figure of £1,065 per person and the assumption that there are 2.36 persons per house from the 2001 Census (still applicable following the 2011 Census). The figure of £1,065 is made up of:

- the existing approximate £225 value (updated from the £200 referred to above) used by Defra as the willingness to pay, per household per year, to avoid stress and health effects caused by flooding; and
- £970 as a central value of medical and productivity costs for an average of four months; these are a blend of general practitioner care, cognitive behaviour therapy and non-direct counselling (JBA, 2012).

This figure of £2,513 is now used more commonly than £200/£225, perhaps simply because it is a larger figure! Clearly more research is needed here.

4.10 ASSESSMENT AND CONCLUSIONS: REMAINING ISSUES AND LESSONS FROM EXPERIENCE

Residential flood damage data may be used in as detailed or as generalised a way as required for the purpose of the benefit assessment. Dwellings and their occupants can be noted without reference to age or type or property or the social grade of resident. In this circumstance, WAADs and residential sector averages are then used to estimate damage potential for different flood stages. Alternatively, depth/damage data relating to all three variables can be applied or simply the type and age of the house noted, irrespective of the social grade of its occupants.

Likewise, an in-depth vulnerability analysis can be conducted, taking into account a detailed understanding of the number of vulnerable households. Alternatively, a generalised picture of the population may be used in a desktop/strategy assessment. Similarly, a generalised intangible health damage figure of £225 per household per year can be applied, or the more detailed risk reduction matrix can be applied.

Whatever level of aggregation is chosen, there will be errors traceable to the original data sources. It is almost impossible to quantify these errors but every attempt has been made to minimise them. However, certain 'health warnings' and additional issues remain:

- Damage susceptibility estimates: professional opinion varies on the precise effect of flood water on inventory and building fabric items. Susceptibility must be continually updated as more information becomes available.
- Inventory and building fabric data: standard checklists have been devised which are not exhaustive. Due to a lack of accessible data, the extent of variation in ownership of certain items is unknown without mounting extensive field surveys.
- Quality: error is involved when categorising the quality of inventory items.
- Average remaining values: ARVs are not empirically assessed. Items are generally assumed to be approximately halfway through their lives, which may distort potential damage estimates, especially in newly established households.
- Secondary data sources: applying nationally based data to small areas locally may lead to errors.
- Ground floor plans: individual properties will vary from these specifications to some degree.

5 Flood damage to non-residential properties

Flood risk areas often contain non-residential properties (NRPs) and in some cases they can be a significant factor when contemplating major expenditure on flood risk management measures. This chapter contains methods and data which allow direct and indirect flood loss potential for NRPs to be assessed. It also contains guidance and data by which the damage reducing effects of NRP-specific flood resistance and resilience measures and flood warnings may be gauged because these may significantly reduce flood damage potential.

NRPs include high street buildings (e.g. traditional retail outlets, banks, public houses); traditional-style buildings (e.g. offices, schools, libraries); industrial-style buildings (e.g. industrial units, warehouses, supermarkets, sports centres, manufacturing plants); modern framed buildings (e.g. office blocks, community centres); health and religious buildings (e.g. hospitals, churches); and others (e.g. car parks, electricity substations, playing fields).

Section 5.1 introduces methodologies used for NRP loss estimation and associated key principles. Section 5.2 explains the rationale for developing the new methodology for assessing potential flood damages to NRPs and outlines the evolution of the methodology and data collection for this Manual. Section 5.3 outlines the new NRP categorisation. Section 5.4 presents the main components of flood damage and Section 5.5 gives details of the susceptibility of these components to damage. Section 5.6 focuses on direct flood damages and the production of standard depth/damage data. It also explains depth/damage data for various flood scenarios (e.g. for circumstances where there is a flood warning or where flooding is of long duration). Section 5.7 discusses indirect flood losses, gives guidance on when these may be significant and indicates a general uplift factor for economic losses. Section 5.8 is a cautionary explanation – and provides three examples – of circumstances in which NRP losses (as well as other losses) might be expected to be unusually high. Section 5.9 addresses the potential damage reducing effects of property-level protection measures (i.e. resistance and resilience measures) and flood warnings. In this section, partly in order to reduce the reader's need for cross-referencing, we reproduce the principles and formulae introduced and applied in Section 4.6 of the residential damage chapter, but with slightly differently expressed formulae and different parameter values which are appropriate for NRPs. The focus in Section 5.10 is on how to apply the depth/damage data in initial studies and more detailed appraisals. This section also discusses the damage data now available to users on MCM-Online. Section 5.11 includes a note on updating the data, and some concluding comments are included in Section 5.12.

5.1 INTRODUCTION

The data in this chapter are the result of fresh research into NRPs and provide the basis for estimating flood loss potential in this sector. We have attempted to simplify the task of estimating these potential damages further and to address some of the problems with these data which were apparent in the 2005 'Multi-Coloured Manual' (MCM) (Penning-Rowsell *et al.*, 2005a) and its 'Blue Manual' and 'Red Manual' forerunners (Penning-Rowsell and Chatterton 1977; Parker *et al.*, 1987).

We have reduced the number of NRPs given a land use code from sixty-six in the MCM (2005) to forty-three. These forty-three NRP categories relate to the property codings in the Environment Agency's National Receptor Dataset (NRD) (Environment Agency, undated c) and are used here to 'represent' most NRPs. The MCM-Online will provide a match between the NRP categories and the MCM codes used in the NRD (see MCM-Online Supporting Information for Chapter 5). We provide ways of estimating potential losses designed for either simplified, 'high-level', initial analyses and/or meso-scale appraisal, and for appraisal using a higher level of resolution and detail sometimes more suitable for project appraisals and/or at the micro-scale (i.e. scheme scale).

When undertaking an appraisal for a scheme that might have a life of fifty-plus years, it is appropriate to consider general rather than detailed categories of NRP. For example, a traditional retail outlet (of the kind often found in high streets) may be a stationer's shop in year 1, a pawnbroker's shop by year 10, a shop selling mobile telephones by year 20, and so on. In recent years, the frequency of changes of this sort has been increasing. It is therefore more appropriate to consider these types of retail outlet as a single category – say, a 'traditional' retail outlet, an example of one of the categories of NRP which equates with 'High Street Shop' in the Environment Agency's NRD.

The most important change from the MCM (2005) is the method utilised to construct the NRP direct damage dataset. In 2005, the NRP damage dataset was derived from over eighty visits to sampled NRPs where lengthy interviews with managers or owners were undertaken with different degrees of success. Sometimes it proved very difficult for respondents to provide answers to the questions asked. The NRPs selected included properties which had been flooded as well as ones which had not been flooded but which were located in flood risk areas. Small sample sizes within some of the NRP categories proved to be the principal weakness of this approach, coupled with the time and expense required to repeat such an exercise when the data required updating.

We have now replaced this method of data acquisition and construction of potential loss data with one which directly parallels the now well-accepted 'synthetic' method (informed by empirical flood loss data) used to compile the residential potential damage data. With the exception of the Industry/Manufacturing sub-sector, the new method is based primarily upon the now-significant and increasing degree of building design standardisation and commonality which exists within the damageable characteristics of many types of NRP. It is also based on the compilation of detailed inventories and values for NRP building fabric, other contents and clean-up operations for different types of NRP to which damage susceptibility values are applied to yield depth/damage data. By using firms of experienced quantity and building surveyors – who have years of experience of assessing flood damage potential – and standard industry rates (e.g. for building structure and fabric replacement or refurbishment), as well as government secondary data sources on stock values, the new method combines empirical evidence with a synthetic approach to damage data construction directly comparable with that used for the residential damage data. The depth/damage data are based on *economic* values rather than the financial values used by the insurance industry.

A new feature of the NRP depth/damage data is that complete datasets are now provided for three flood durations (short, long and extra-long). This replaces the previous approach, in which uplift factors were utilised.

The data inputs to the new NRP damage datasets represent many hundreds of different items, a large number of £/m^2 standard rates routinely used by quantity surveyors and experience with replacement and refurbishment requirements as well as damage susceptibility values based on many years of experience.

Because of the high degree of variability and uniqueness among industrial and manufacturing properties, the synthetic approach taken for other NRPs has its limitations when used for this sub-sector, so a complementary method has been used to construct potential damage data. This method draws on both UK and Dutch empirical research into the damageability of manufacturing premises as well as recent experience of flood damage to modern manufacturing plants in Thailand (see CSIS, 2011).

Although it is difficult to make a meaningful comparison between the sample size of the dataset used in the MCM (2005) and the present method, the latter is based on a vastly greater number of data items and samples. It is also much easier and affordable to update these data than was the case with the previous method.

5.1.1 Different approaches to NRP flood loss estimation

Several different types of approach have been used to estimate flood damages to NRPs (Kreibich *et al.*, 2010; Admiraal, 2011). All relate a flood parameter to a corresponding value of loss to produce a loss function. Some flood damage models are based on absolute loss functions. These measure damage in terms of money, such as the directly priced damage. Others are based on relative loss functions – where the loss is shown as a fraction of the percentage of the estimated total replacement value of a property or area and its contents. Relative loss functions depend on total replacement values that often include elements of betterment, which makes them unsuitable for use in national economic efficiency benefit –cost analyses generally employed in appraising flood risk management projects in the UK. Therefore, in this chapter, and in our previous manuals (e.g. Penning-Rowsell *et al.*, 2005a; Parker *et al.*, 1987), we employ absolute loss functions, plotting probable monetary damage against flood depth and other parameters.

Empirical and synthetic approaches for flood damage development may also be distinguished (Kron, 2007). Empirical approaches collect NRP data after floods, for example, by interviewing property owners or managers. Examples include the Australian ANUFLOOD model (employing absolute loss functions) (Queensland Government, Department of Natural Resources and Mines, 2002) and the German MURL and Hydrotec models (using relative loss functions) (Emschergenossenschaft and Hydrotec, 2004). The German flood damage database HOWAS is another example of the use of the empirical approach (Merz *et al.*, 2004), and Kreibich *et al.* (2010) used the approach in developing their FLEMOcs model. This empirical approach was employed to collect the data for NRPs for the MCM (2005) but proved to be time-consuming and costly, given that for some categories of NRP it yielded some small sample sizes and fundamental updating is resource intensive.

Synthetic approaches are based on detailed considerations of 'what if?' (i.e. what is the probable damage to a particular building type having a particular function if flooded to, say, 1.0m depth?). In this approach, experts (e.g. building surveyors, quantity surveyors) are used to generate quantitative answers using standard cost information. This approach is the basis of the residential flood damage data in this Manual and is used for the first time for NRPs here. A key attraction of this approach is the comparative ease of updating the flood damage data over time. Combinations of both empirical and synthetic approaches are also used as in the USACE HAZUS-MH model (Scawthorn *et al.*, 2006a; 2006b). To some extent, the approaches used for both residential and NRP damage estimation in this Manual employ a mix of empirical and synthetic approaches (i.e. the approaches are predominantly synthetic but are also informed by experience of actual cases of flood damage and collection of empirical data).

5.1.2 Key principles influencing the NRP potential flood loss data

A number of key principles have influenced the research leading to the production of new methods for assessing flood damage data for NRPs. These principles have been followed in order to provide data that are simple in structure, relevant to flood risk situations and accompanied by a breakdown of flood damage components to allow users in restricted cases – should this be justifiable – to construct their own depth/damage curves from the data we present combined with data they collect themselves. The principles are as follows:

- In this Manual we have reduced the number of property or land use categories from sixty-six in the MCM (2005) to forty-three for which flood damage data are now available online (see MCM-Online). This is intended to simplify use of the data and to recognise the fact that meso- or micro-scale appraisal for flood risk management purposes involves appraising flood damages over a long time perspective (typically one-hundred years, or, say, twenty-five years for property-level protection measures), and that one particular type of premises in the flood risk area today (e.g. a warehouse) could become quite different in five or fifty years' time (say, a supermarket or a sports centre).

- We have obtained data for all NRPs at risk from coastal or fluvial flooding in England and Wales, and the selection of NRP categories is partly based on this information. We have focused upon the thirty or so most frequently occurring NRP types found in Flood Zones 2 and 3 in the NRD. Subsequently, we have added some further categories of NRP which we consider enhance the sub-sector mean damage data explained in this chapter. This represents at least 95 per cent of the number of NRPs in the NRD.
- We have attempted to include all new technological developments in NRPs over recent decades and other major societal changes that have markedly affected changes to flood damage potential. The growth of out-of-town superstores and modern framed buildings are examples.
- We have disaggregated data for six components of flood damage (building structure and fabric; building services; moveable equipment; fixtures and fittings; stock, raw materials and contents; and clean-up costs). The MCM (2005) NRP data did not include the sixth component of damage – clean-up costs – and this is a key difference between the two sets of damage data.
- We have provided depth/damage datasets for all NRPs for three different flood durations: (short – twelve hours or less; long – greater than twelve hours and up to three days; and extra-long – three–seven days).
- A brand-new method has been used to derive the NRP flood loss potential data. The 2005 data was derived entirely from over eighty visits to and interviews with the managers or owners of NRPs. This was a time-consuming and somewhat difficult exercise (as well as costly to repeat) that involved contacting representatives at head offices of companies with properties located in flood risk areas to obtain generic data about the range of property types that they owned or leased, and their susceptibility to flooding. Discussions were held with representatives from both organisations whose premises had been flooded and some that had not yet been flooded. The 2005 approach suffered from small sample sizes in some of the NRP categories. The new dataset presented in this chapter replaces these data based on data acquisition and construction of potential loss data using an approach which is very similar to that used to compile the residential potential damage data. We have compiled and built up detailed inventories for: (a) building structure and fabric, (b) building services and (c) fixtures and fittings and have added to them values for (d) moveable equipment, (e) stock and (f) clean-up costs – to which damage susceptibility values have been applied to yield depth/damage data. By using experienced quantity and building surveyors and by using standard industry rates (e.g. for building structure and fabric replacement or refurbishment) as well as government secondary data sources on stock values, the new method combines empirical evidence with a synthetic approach to damage data construction directly comparable with that used for the residential damage data. The data inputs to the new NRP damage datasets represent many hundreds of different items, a large number of £/m^2 standard rates routinely used by quantity surveyors and is based on many years of experience with replacement and refurbishment requirements and damage susceptibility values.

It is not possible to make a meaningful comparison between the sample size of the dataset used in the MCM (2005) and the present method. The latter is based on a vastly greater number of data items of different types and many samples. It is also much easier and affordable to update these data than was the case with the previous method.

5.1.3 Factors influencing NRP flood damage potential

The parameter most commonly used to determine NRP potential losses is flood water depth. Early NRP damage data (e.g. Penning-Rowsell and Chatterton, 1977; Smith and Greenaway, 1988) were based partly upon observation of actual flood damage and partly on the assumption that flood depth was the key variable explaining variance in flood losses. However, today there is fairly conclusive empirical research evidence about the most important factors influencing NRP flood losses (Table 5.1). Using principal components analysis (PCA), flood depth has been shown by Kreibich *et al.* (2010) to be the most important determinant of commercial flood losses and almost all studies now take this perspective (e.g. Admiraal,

2011). Kreibich *et al.* (2010) examined commercial flood losses in over 600 companies affected by floods in Germany in 2002, 2005 and 2006 and found that flood depth explained more variance in these damages than any other parameter. There is no reason to believe that German NRPs are sufficiently different to invalidate the transfer of these research findings to England and Wales. In the new NRP dataset introduced in this chapter, damage potential is presented using seventeen increments of flood depth.

The second most important factor determining commercial flood loss potential is building size, measured by Kreibich *et al.* (2010) by business volume or the number of employees, and measured in this Manual by the spatial dimension in square metres of the property (the footprint of the building(s) only, excluding grounds and car parks within the property boundary). In the dataset presented here, flood loss potential is given in terms of pounds sterling per square metre at January 2013 pound values, which takes account of this variable.

Research reveals that the third most important variable influencing commercial flood damage potential is 'precautionary measures' (Kreibich *et al.*, 2010). These measures may be in the form of property-specific warning-dependent or warning-independent resistance measures or resilience measures taken by the business, or they may take the form of 'community-based measures', such as de-mountable flood defences. In this Manual, the standard NRP flood loss datasets assume that no precautionary measures are taken. However, we also provide guidance and data designed to assess the potential damage reducing impact of flood warnings and related property-level protection measures (but not de-mountable flood defences deployed more remotely from a property). These are the subjects of Section 5.9 of this Chapter.

Table 5.1 Factors determining commercial flood losses ranked in importance

Factor	Rank order
Depth of flood water	1st
Property or building size	2nd
Precautionary measures	3rd
Flood warning lead-time	4th equal
Flood duration	4th equal
Commercial sub-sector	4th equal
Flood water contamination (sewage, chemicals and oil)	7th

Source: From empirical data analysed by principal components analysis by Kreibich *et al.*, 2010

Flood warning lead-time was found by Kreibich *et al.* (2010) to be the fourth most important variable affecting commercial flood damage potential, alongside flood duration and the commercial sub-sector of the activity. Beyond that, flood water contamination was the least important variable. In the data presented here, damage datasets are provided for three flood durations (short – twelve hours or less; long – greater than twelve hours and up to three days; and extra-long – three–seven days).

The NRP flood loss data presented here are based further on the following parameter assumptions:

- flood water velocity is low and not more than 0.5m per second;[1]
- the pollution/sediment load is Category 3 with some Category 2 (70:30 split) (see Chapter 4, Table 4.11); and
- the flood water is freshwater (i.e. not saltwater, although a general uplift factor for saltwater flooding is provided).

It should be noted that the commercial sub-sector (herein NRP category) was not found by Kreibich *et al.* (2010) to be as important as the first three determining factors in explaining variance in commercial flood damages, although sub-sector contributes to explaining at least some of this variance. Categorisation of NRPs using a large number of categories is therefore unlikely to be particularly helpful in explaining damage variance, which is one reason why we have progressively reduced the number of NRP categories over time and advocate using NRP sub-sector mean damage data.

5.1.4 Comparison of 2005 and 2013 NRP flood damage data

The 2005 NRP damage data can be adjusted to January 2013 pound sterling values using the Consumer Price Index by factoring the 2005 damage data by 1.244. Comparison of the 2005 and 2013 depth/damage values reveals that some property categories have higher damage values in 2013 compared with 2005 and some have lower damage values. Overall, taking the NRP sector as a whole, the 2013 depth/damage values are 3 per cent higher than the adjusted 2005 values. Among the NRP categories that have higher depth/damage values in 2013 are churches, warehouses, public houses, banks, non-specific offices, surgeries and car showrooms. The NRP categories that have lower depth/damage values are high street shops, supermarkets, schools, libraries, hotels, theatres, sports and leisure centres and fire or ambulance stations. These differences are mainly due to the application of a different flood damage estimation methodology (see Sections 5.6.1–5.6.4), which is based on a larger sample of data inputs of many types compared with the methodology used in 2005.

5.2 THE RATIONALE FOR DEVELOPING A NEW METHODOLOGY AND DATABASE FOR NRPS

5.2.1 The different approaches evaluated

Three basic approaches to creating a new methodology and flood damage database for NRPs were examined before the approach in this Manual was selected.

The first approach was to continue with the existing MCM (2005/2010) empirical methodology, database and updating procedure but to enhance this by: (a) conflating some property categories; (b) expanding the evidence base by undertaking more head office (HO) interviews; (c) increasing the number of cases contributing to mean damage values; and (d) using published and unpublished secondary sources for updating purposes (e.g. Office for National Statistics (ONS) data sources).

This approach is more labour intensive in the database construction phase than the existing approach because we would need to undertake significantly more HO interviews than previously and sample sizes within categories would still remain comparatively small (though greater than one in many cases). The MCM (2005) recommended using the Consumer Price Index to update flood loss values. Examination of the UK Government's ONS online data has revealed that the Resource Cost Index of Non-Housing Building could be used to improve the accuracy of the updating of the 'building and fabric' component of NRP flood loss data and this would represent a small step forward. However, the empirical approach was rejected because of the difficulties and high costs involved in repeating and significantly extending the eighty-plus NRP site HO interviews used for the 2005 data and subsequent further high costs of repeating this exercise at regular intervals to update the damage data.

A second approach considered was to develop an entirely new methodology and damage database principally by utilising secondary data sources, bolstered by very limited primary data collection. To develop this approach there would need to be sufficient appropriate secondary source data, of sufficient quality, to construct a new NRP categorisation and to develop a new damage database, with updating being feasible using these published secondary sources. On further investigation this second approach proved to be infeasible. There is now a wide range of datasets available online via the ONS. Although some of these sometimes appear to be close to some of the six components of NRP damage (i.e. building structure and fabric, and building services) detailed examination revealed that they are insufficient and in most cases not suited for use in constructing NRP flood loss data (these datasets are obviously not collected for this purpose).

The third approach is the main one adopted here and involves developing a series of 'cost models' to represent now-typical categories of NRP (i.e. typical in England and Wales's floodplains). The cost models are principally a synthetic approach and are developed by compiling detailed lists of inventories and inventory values and by applying damage susceptibility factors to these at a range of flood depths and durations (Table 5.2). Professional expertise (i.e. in chartered quantity surveying and chartered building surveying) has been used to construct these cost models using industry standard cost rates for 2012/13, stock data from secondary sources and the actual flood damage experience of the surveyors. Subsequent updating can then be undertaken by sub-contracting the updating of the cost models to appointed professionals capable of undertaking this exercise. In Table 5.2 each inventory item is taken in turn, multiplying the rate per unit (£) by the size in m^2 and then by the appropriate susceptibility factor for each flood depth to arrive at potential damage values (£) for that inventory item. This process is repeated for all inventories in each damage component and then summed to produce total depth/damage estimates for each property category.

Table 5.2 An extract from a cost model for six categories of high street buildings

Structure and fabric **Short duration, indicative case** **A. High street buildings**															
Building type/Flood depth			A	B	C	D	E	F	−1	0.75	0.5	0.25	0	0.25	0.5
			Traditional retail	Traditional retail	Bank/ building society	Café/ Restaurant	Public house	Showroom	Susceptibility						
	Unit	Rate/ unit	Small	Large											
Floor area (GF)(SQ M)			120	400	200	250	350	500							
Rake gravel, roll	Sq M	4.4	0	0	0	0	20	50	0	0	0	0	0	0.1	0.2
Replace contaminated gravel	Sq M	14.6	0	0	0	0	20	60	0	0	0	0	0	0.1	0.4
Paving and re-bed	Sq M	22.9	0	0	0	0	40	50	0	0	0	0	0	0.1	0.1
Repair asphalt/tarmac finish	Sq M	10.2	0	0	0	0	120	200	0	0	0	0	0	0	0.1
Ditto; basement level	Sq M	10.2	0	0	80	0	0	0	0.12	0.14	0.16	0.18	0.2	0.3	0.3
Asphalt base and fill	Sq M	26.4	0	0	0	0	0	0	0	0	0	0	0	0	0
Concrete finish replace	Sq M	35.75	30	80	0	0	40	50	0	0	0	0	0	0	0
Ditto; basement level	Sq M	35.75	0	0	100	180	150	0	0.1	0.15	0.15	0.2	0.2	0.2	0.2
Re-point brick paviour	Sq M	9	0	0	0	0	30	15	0	0	0	0	0	0.3	0.3

Replace brick paviour	Sq M	60.5	0	0	0	0	30	15	0	0	0	0	0	0.1	0.1
Single skin wall with piers	Sq M	96.8	8	32	0	10	20	0	0	0	0	0	0	0	0
Solid brick/concrete/stone wall repair	Sq M	220	36	96	225	310	380	115	0	0	0	0	0	0	0
Rendering to walls	Sq M	32.5	10	44	0	0	40	65	0	0	0	0	0	0	0.1
Ditto; basement level	Sq M	32.5	0	0	120	0	120	0	0.1	0.2	0.3	0.3	0.3	0.4	0.4
Concrete post and panel fence	Lin M	79.2	0	0	20	15	20	25	0	0	0	0	0	0	0.1
Chain link fence	Lin M	60.5	0	0	0	0	0	12	0	0	0	0	0	0	0
Service security barrier	No	140	0	0	0	0	0	0	0	0	0	0	0	0.5	0.6
Replace security barrier	No	5,000	0	0	0	0	0	0	0	0	0	0	0	0	0

Note: This tabulation focuses on a small part of the building structure and fabric inventory and the susceptibility factors applied to these values to produce depth damage estimates. The case shown is for a short-duration flood and the indicative suite of susceptibility factors.

This approach has the merits that it has already been accepted for the construction of residential property flood loss data in the MCM (2005) and is the basis of the residential property flood damage data herein. Indeed, the approach was first published in the 'Blue Manual' (Penning-Rowsell and Chatterton, 1977) and in England and Wales has remained the preferred approach ever since for residential flood damage estimation.

For the Industry/Manufacturing sub-sector we have drawn upon secondary source data from Dutch and other research which indicates maximum flood damage values and depth/damage relationships for different types of manufacturing, and upon an in-depth study of flood damage potential in a large manufacturing plant in the UK. The method is explained in more detail in Section 5.6.2.

5.2.2 NRPs located within flood risk zones in England and Wales

According to the Environment Agency's NRD (Environment Agency, undated c) there are currently approximately 201,962 NRPs in England and Wales within Flood Zones 2 and 3. Flood Zone 3 represents a high risk of flooding and an annual probability of 1 per cent or greater from river flooding and 0.5 per cent or greater from tidal flooding. Flood Zone 2 is the zone in which the risk of flooding is low or medium, with an annual probability of 0.1–1.0 per cent from rivers and 0.1–0.5 per cent from the sea. The twenty-five most frequent NRD NRP categories account for 184,627 (i.e. 91 per cent) of the NRPs located in Flood Zones 2 and 3 – clearly, a very high percentage. The highest proportion of these NRPs are high street shops (17.2 per cent), with offices making up another 11.2 per cent of the total (Table 5.3).

When five additional of the most highly damageable (£/m^2) NRD NRP categories (as indicated by 2005 NRP damage data) were added to the most frequently occurring twenty-five NRP categories, these thirty NRP categories account for 191,844 of the NRPs in Flood Zones 2 and 3 (i.e. 95 per cent of the total). These thirty NRP categories then formed an initial basis for developing the NRP flood damage data employing the new methodology explained above (see MCM-Online for NRP data). Subsequently, fifteen further categories were added, primarily in order to improve the quality of sub-sector mean damage data.

In England (for the purpose of local taxation) and Wales the Valuation Office Agency (VOA) assesses rateable value for NRPs which are then grouped according to rateable value into four categories: Retail, Office, Industrial (i.e. factories and warehouses) and Other (including 'non-bulk' NRPs, such as public buildings). Public houses are classed as other/non-bulk by the VOA but may be assigned the attributes of the retail bulk class. The new MCM sub-sector codes (Table 5.4) have been allocated to the appropriate bulk classes.

Table 5.4 Classification of NRP sectors used in the 'Multi-Coloured Manual' (2005)

NRP sub-sector number	Sub-sector
2	Retail
3	Office
4	Distribution/Logistics
5	Leisure/Sport
6	Public Buildings
8	Industry/Manufacturing
9	Miscellaneous

Table 5.3 Risk exposure of selected NRP categories: number of properties located in Flood Zones 2 and 3

High Street Shop	Electricity installation	Office	Factory Works Mill	Public House	Car Park	Warehouse	Hospital	Surgery/ Health Care Centre
31,683	22,833	20,758	7,104	5,075	3,392	2,535	72	2,571

Source: Enivornment Agency, undated c

5.2.3 Developments and trends in NRP land use affecting flood damage potential

The trends in the NRP sector largely relate to the development and spread of new technology, to the location, size and type of NRPs, and also to the largely irreversible restructuring of the economy following the financial crisis in 2008. The major trends are summarised in Table 5.5.

The spread of new technologies and methods, such as those which have enabled online shopping, is transforming not only high street shopping but logistical and delivery services. This has led to high street business closures and to the development of large distribution depots at key transportation hubs. The steady transfer of retail outlets to large retail parks has led to a greater proportion of retail products being sold in large, generally single-storey, steel-framed and concrete buildings, with a reducing

Table 5.5 Developments and trends in NRPs influencing flood damage potential

Developments and trends	Factors affecting developments and trends
Commercial/office sector	
Growth in demand for out-of-town business park space and flight from old office space.	Cost pressures, behavioural factors, the search for innovative office solutions, transport infrastructure development and public sector relocation policy are driving a trend towards new office space in lower-cost locations. This is increasing the amount of obsolete office stock so that many existing office buildings will require a change of use.[1]
Varying levels of growth and contraction of particular types of businesses in England and Wales.	Significant reduction in office space required and occupied by the public sector but an increase in office space predicted in the automobile, engineering and aerospace sectors.
Significant cost differences between locations and individual towns.	Greater demand for local hubs coupled with public sector shrinkage and relocation trends, new transportation developments and the high costs of office space in south-east England are boosting the prospects of regional markets to attract office occupiers.[1]
Higher-density office space (the 'shrinking office') coupled with increases in remote working, 'hot desking' and 'hotelling'.	Cost pressures, behavioural factors and trends in ICTs are combining to reduce office space requirements and to increase office occupation density.[1]
Overall there is likely to be a decrease in UK office space in the 2010–20 period.	Restructuring of the UK economy sparked off by the 2008–09 economic recession and poor economic performance thereafter.[1]
Development of serviced office space.	A growth in 'turnkey' flexible office space offered by companies such as Regus and Bizspace. The UK, and London in particular, is currently the world's largest and most developed serviced office market with a growth of 7.5% in the number of serviced office centres between 2008 and 2011. In 2011, London had 8% of the world's serviced office centres. Birmingham and Manchester are also primary centres with cities such as Bristol, Milton Keynes, Nottingham, Leeds, Sheffield and Belfast being secondary centres. Standardised office provision based on work station numbers. This allows smaller businesses to enjoy the amenities comparable to those of large companies, increasing office quality and standards.[2,3]

Table 5.5 continued

Developments and trends	Factors affecting developments and trends
Development of high-rise office towers.	Growth in office space going into high-rise towers with prestige addresses. High-rise towers maximise gross floor area while minimising the ground level spatial footprint. Towers usually have extensive basements. Found mainly in major metropolitan areas.[4]
Retail sector	
In-town retail	
High street shops constitute the largest proportion of NRPs in the floodplains of England and Wales, although this proportion has reduced significantly. They now comprise 17.2 % of the total number of NRPs in Flood Zones 2 and 3 compared with 23 % in the 2005 MCM.	High street shops have, on average, high flood damage potential owing to a combination of high density and high stock values although stock values have also been decreasing.
In the 2005 MCM, we reported that the high street was entering a period of a consolidation and change although not apparently about to reduce in proportion (see above). In 2013, the picture is very different with high streets suffering significant decline except in premium locations such as London's West End. Many thousands of traditional high street brands and outlets have disappeared.	The economic recession since the collapse of Lehman Brothers in 2008 has accelerated change and generated a major shake-up of high street retailing and the consolidation of many types of business. This trend has been accelerated by the rapid rise of online retail purchasing, increased confidence in the security of delivery and the increased pressure of the competitiveness of out-of-town retail parks and shopping malls.[6]
Between October 2008 and January 2011 the following changes occurred: Building society offices –28 % Employment agencies –13 % Estate agents' outlets –9 % Betting shops +5 %	
In contrast, online retail sales increased by 18 % in 2012 and are likely to more than double between 2010 and 2020.[5]	
Neighbourhood grocery stores have become more popular and numerous.	Medium-sized stores such as the Tesco Express grocery and food stores have become common and are often located either in high streets or more commonly in key locations in urban neighbourhoods or in villages. Combination outlets, such as petrol filling stations with a food and grocery

Table 5.5 continued

Developments and trends	Factors affecting developments and trends
	store, have also become more common, usually on key routes through suburban areas. Fuel and the large grocery/food retailers have been combining forces in this way to offer more convenient ways to shop and purchase fuel. These are also responses of the retail sector to restrictive planning policies seeking to control the domination of the retail sector by the leading brands building further very large stores.[7]
Changes in the retail sector show marked regional differences.	The north of England has lost most shops proportionately and even formerly buoyant major city centre high streets (e.g. Bristol's Bond Street) have seen significant closures (here 14% of outlets were reported to be vacant in February 2013).[8]
Consolidation of businesses leading to fewer, larger outlets and distribution depots.	Cost and efficiency pressures, magnified by the recessions, have forced retailers to consolidate. For example, between April 2010 and April 2011, the sales of convenience stores rose by 4.9% on the previous year but the number of convenience stores declined by 0.4%. Increased sales in convenience stores have been driven by structural changes such as smaller households, longer working hours and building value perception. The average size of food and grocery retail logistics depots has been increasing and by 2012, 94% of stock was supplied by large centralised retail warehouses rather than direct store delivery from suppliers.[7,9]
Out-of-town retail	
Out-of-town retail parks are now numerous and often more attractive to retail customers than high streets.	The emphasis in retail shopping has switched from the high street to retail parks (as well as to online sales). These large retail parks are now dominating the retail landscape with their vast floor spaces and highly flood damage susceptible products. Retail parks appear to be more robust in adverse weather conditions than high streets, where parking is often much more difficult. For example, in February 2013 Experian reported that the wet and windy weather conditions in 2012 contributed to a 7% year-on-year reduction in retail sales overall but that the inclement weather had not deterred visitors to retail parks, which witnessed a vibrant 3.9% year-on-year increase in sales.[10]
	Contributory factors include tightening disposable incomes and the related search for value amongst retail customers and planning pressures and policies.
Floorspace in retail parks has been growing, driven in part by discount or 'value' retailers (http://www.shopping-centre.co.uk/news/fullstory.php/aid/5211/Discount_stores_drive_growth_out_of_town.html)	
The sale of non-food products (e.g. clothing, electrical products and white goods) has increased significantly in diversifying supermarkets and other types of outlet in out-of-town retail parks.	Large grocery retailers have been prompted to diversify into non-food items partly in response to UK Government measures to control the expansion of the leading players by imposing planning restrictions and impeding the development of large out-of-town superstores.[7]

Table 5.5 continued

Developments and trends	Factors affecting developments and trends
Demand for online sales and the economies of scale of large, centralised logistical depots has led to a growth in large logistical and distribution depots close to major transportation routes in various parts of the country. Growth of business and innovation parks in out-of-town locations.	Driven by advances in personalised ICTs and a search for greater efficiency. Examples of innovation science parks, generated by research advances in science and technology, can be witnessed around Cambridge and Oxford.
Industrial sector	
Manufacturing accounts for between 11% and 12% of the UK's national output. However, although still a very important component of the UK's economy, accounting for a high proportion of its exports, manufacturing has been in decline in terms of its importance in the UK economy (yet the UK is still ranked 9th in the world for its manufacturing output). Absolute change in the manufacturing sector has been positive at +50% between 1982 and 2007.[11]	Manufacturing is capital intensive: (a) light industry employing significant labour and equipment and machinery, and (b) heavy industry which is more capital intensive and dependent upon weighty machinery and smaller workforces. Equipment and machinery assets have very variable susceptibilities to flood damage. Research and development spend in the top 100 UK manufacturing companies has held up well despite the recessions but there is a long tail of smaller companies with a very low spend of this nature.
Significant spatial variations exist in the importance of manufacturing.	The East and West Midlands (24% in total) are the regions with the highest proportion of employees in manufacturing. The next most important region is the north-west, whereas London (2.8%) has the lowest proportion.
The industrial warehouse sector has been growing in recent years.	Warehouses have blossomed into distribution centres. Industrial warehousing capacity has been growing rapidly in the Western Corridor: Slough, Heathrow and the Thames Valley to the west of London.[12]
Self-storage is one of the fastest-growing sectors in the service industry.	The self-storage industry has continued to grow despite the recession, particularly in London and the south-east.
Other sectors	
A trend towards fewer, much larger, themed, public houses that focus on providing a food experience as well as drink.	Driven by changing customer behaviours and desires, drink-drive laws and benefits of economies of scale and search for greater efficiencies.
A steady growth in the number and size of fitness centres with both private and public sector facilities.	Changing lifestyles.

Table 5.5 continued

Developments and trends	Factors affecting developments and trends
The total number of vehicle servicing and repair outlets fell by 21% between 2001 and 2013 yet there has been a growth in the after-sales servicing repair market in terms of turnover.[13]	There has been a decade-long fall in vehicle servicing volumes.[13]
Generic	
Growth in generic business continuity planning aimed at reducing risks to staff, plant and equipment and customers associated with physical damages and disruptive losses. Much more prevalent in large corporate business and some medium-sized business, but adopted much less by small businesses.[14]	Driven by shareholders, auditors, legal requirements in some cases, management training, experience and promotion of BCP in UK business. Potential to reduce both direct and indirect flood damages.[14]

Sources: 1 GVA (2011/12); 2 Instant Offices Ltd (2011); 3 Gibson (2003); 4 Aysin and Ozgen (2009); 5 Ordnance Survey (2011); 6 *Independent* (2013); 7 Agri-Food Trade Service (2011); 8 Food and Grocery Experts (2012a); 9 Food and Grocery Experts (2012b); 10 Experian (2013); 11 Institute of Mechanical Engineers (2012); 12 Jones Lang Lasalle (2012); 13 Trend Tracker (2013); 14 Parker *et al.* (2012)

proportion of these products being sold in traditional brick-built and/or smaller outlets.

Between 2008 and 2013 negative economic growth and the shrinking of the UK economy placed severe pressures on almost all NRP sub-sectors to seek efficiencies and reduce costs. This has led to the major restructuring of many businesses, the disappearance of many well-known high street brands and the consolidation of retail outlets, distribution facilities and production plants into a smaller number of larger sites. Grocery and food chains have continued to develop larger and larger supermarkets in out-of-town business parks which have attracted more and more retailers, including discounters, seeking the economies of scale and customer parking facilities which could drive their businesses through successive economic recessions. Seeking greater market penetration, major supermarket chains have also developed highly standardised, medium-sized supermarkets in key locations across urban areas.

Most NRP developments have been in the form of multiples of one type of similar unit (e.g. retail units, offices or industrial sheds), which has meant that the property market has become highly specialised and segmented. The service sector has been the fastest-growing and most rapidly changing sector of the economy in England and Wales. In this sector, as in the other commercial and industrial sectors, many companies now rely heavily on computing and information technology equipment and telecommunications that require extensive use of cabling and electrical services, although wireless connections have also become common. These services are sometimes located at ground floor levels or below ground floors, for example in under-floor ducting. They are highly susceptible to flood damage.

Overall, these trends in NRPs have led to complex transformations in the composition and spatial distribution of flood damage potential as well as to rising NRP flood damage potential. They highlight the need for an effective method for estimating NRP flood damage potential.

5.3 THE EVOLUTION OF OUR MCM NRP CLASSIFICATION SYSTEM

The first NRP flood damage data for the UK were produced under research sponsored by the Natural Environment Research Council and were published in the 'Blue Manual' (Penning-Rowsell and Chatterton, 1977).

5.3.1 The 'Blue Manual' land use classification system

For the 'Blue Manual', a land use classification was derived from a procedure adopted by Bristol City Corporation, City Engineer and Planning Officer's Department (1971), which was largely in accordance with recommendations made in a paper presented to the Ministry of Housing and Local Government (1968) emphasising that the land use classification should be 'activity based'. The Bristol City Corporation classification was extended to be user – rather than activity – oriented. The classification was comparable to the Standard Industrial Classification (SIC) Minimum List Heading for manufacturing industry and the Census of Distribution for retail shops, and it was hierarchical. NRPs were grouped into four sectors, each with a number of categories and sub-categories:

- Non-Residential Domestic: two categories, thirty-three sub-categories;
- Retail, Retail Services and Offices: six categories, seventy-seven sub-categories;
- Manufacturing and Extractive Industries and Public Utilities and Transport: nine categories, 148 sub-categories; and
- Public Utilities and Transportation: nineteen categories, forty-nine sub-categories.

Therefore, overall there were thirty-six categories and 307 sub-categories, for each of which depth/damage data were made available. The 'Blue Manual' (Penning-Rowsell and Chatterton, 1977, 135) maintained that within-group variance (of damage potential) was unlikely to be greater than between-group variance for any of the sectors, categories or sub-categories, although that was more conjecture than proven.

5.3.2 The 'Red Manual' land use classification system

The 'Red Manual' (Parker *et al.*, 1987) used a modified version of the land use classification system in the 'Blue Manual'. All Sector 8 (Manufacturing and Extractive Industry) properties were re-classified and re-coded according to the SIC Revised 1980 version of the Central Statistical Office. As a result, the nine categories of manufacturing property in the 'Blue Manual' were expanded to eighteen categories in the 'Red Manual', with 216 sub-categories. There was no change to the land use classification for the other NRP sectors (i.e. Retail and Related; Professional and Offices; Public Buildings and Community Services; and Public Utilities and Transportation).

5.3.3 The 'Yellow Manual' land use classification system

The 'Yellow Manual' (Penning-Rowsell *et al.*, 1992) utilised the land use classification system developed for the FLAIR report (N'Jai *et al.*, 1990) and was based on that used in the 'Blue' and 'Red' manuals. Under the Retail and Related sector, a number of shop categories were added to update the categorisation. These included video libraries, pet shops, florists and toy shops. Under the General Stores category, hypermarkets and garden centres/nurseries were also added. The Services, Design Studio category had one addition – antiques. No changes were made to other sectors.

5.3.4 The 'Multi-Coloured Manual' (MCM) land use classification system

The MCM (Penning-Rowsell *et al.*, 2005a, 82) recognised that the FLAIR NRP categorisation used in the 'Yellow Manual' was by then deficient in a number of ways which were important in flood loss assessment.

The categorisation and data no longer reflected current trends, particularly in the Retail, Office and Logistics sectors, which, it was reckoned, rendered much of the damage data redundant. The temptation had been to add more and more land use/property categories in an attempt to cover as many NRP types as possible. However, it was now felt that there were too many land use categories and sub-categories for

NRPs. The categorisation of NRPs was also felt to be confusing because of its complexity. Standard errors were now thought to be very large for much of the sub-sector and category data, reflecting the small sample sizes of individual datasets.

This led the MCM to address these deficiencies in NRP data by developing a radical new approach which meant simplifying the categorisation presented in FLAIR (N'Jai *et al.*, 1990). A 'top-down', inferential or judgemental approach was largely taken for NRPs. These were considered to be more heterogeneous than residential properties (RPs) and therefore there is a higher degree of uncertainty regarding NRP damages and a larger potential error margin. The NRP types selected reflected the distribution of NRPs within the Environment Agency's former indicative floodplain (IFP) map and also reflected many of the new trends in the development and expansion of NRP property markets. Some 93 per cent of all NRPs were found to fall within the top twenty categories by frequency within the IFP map, and in just five sub-sectors (i.e. Retail Shops and Stores; Public Houses; Offices; Warehouses; and Workshops).

The resultant NRP categorisation for the MCM (2005) excluded the Public Utilities and Transportation category used in previous manuals as these were addressed differently, and were further divided into ten categories and sixty sub-categories, each of which was allocated an MCM code. Various modifications were incorporated (e.g. the Retail Stores sub-category was deleted and such stores were put into either the High Street Shops or the Retail Warehouses sub-category). A Hi-tech Offices sub-category was added. Three additional warehouse sub-categories were created for electrical goods, non-frozen foods and frozen foods. Six other property types were added (e.g. Beach Hut; Hairdressing Salon).

Whereas the 'Red Manual' had thirty-eight categories and 279 sub-categories of NRP, the MCM (2005) reduced this to a total of sixty-six categories and sub-categories. However, at the same time, the MCM (2005) identified five main components of NRP flood damage for each category.

5.3.5 The current land use classification system as it applies to NRPs

In the current land use classification system, as it applies to NRPs, there are seven one-digit coded sub-sectors of the NRP sector, including Miscellaneous (Table 5.6). Earlier in this chapter (see Table 5.1), we explained that (a) flood depth, (b) size of NRP and (c) precautionary measures are likely to be the most important variables determining damageability of NRPs in floods. These are the variables most likely to explain most of the variance between NRP potential damage values, with flood duration and NRP sub-sector of lesser – but still some – significance. For this reason, we now consider it unnecessary – and indeed potentially misleading – to attempt to distinguish between as many categories of NRP (i.e sixty-six) as was the case in the MCM (2005). Fine distinctions between NRP categories are also considered inappropriate because, as the recent turbulent restructuring of the economy of England and Wales demonstrates, many NRP premises are likely to change ownership and use during the life of a flood risk management scheme, which is typically fifty years or more.

We have also taken into account the need for the new NRP flood damage data to be used in conjunction with the Environment Agency's National Receptor Dataset (NRD) so that the new damage data can be mapped on to NRD land use types. The seven sub-sectors of NRP in the new land use classification (Table 5.6) are therefore very similar and capable of being translated to those in the MCM (2005), the land use codes of which are utilised in the Environment Agency's NRD.

Our quantity and building surveyors advised that it still makes sense to relate potential flood damage to the basic functions of properties (e.g. retail, office, industry uses). However, in order to limit the sub-categories to a sensible and practical number, and one which is user-friendly, we have grouped similar types of NRP into a single category, generating a significantly reduced number – forty-three, rather than the MCM (2005)'s sixty-six. The best example of this may be within retail units, where a standard shop unit could be used alternatively as a bank with much higher joinery quality and 'fit-out' costs, or as a coffee bar or restaurant where flood damage could occur to commercial catering equipment, tables and chairs, etc. Supermarkets and hypermarkets are categorised here alongside high street shops mainly because they are generically very similar (e.g. their stock contents are similar).

The new NRP land use classification also takes into account type and method of construction, which is an important variable in determining variance in damage

Table 5.6 New MCM NRP land use categories, land use codes, sub-sector and category descriptions

NRP sub-sector	MCM land use code	Sub-sector and category description
High street buildings (retail, retail and vehicle services)	2	**Retail**
	211	Traditional retail (small)
	211	Traditional retail (large)
	213	Supermarket
	213	Hypermarket
	216	Kiosk
	223	Car showroom
	234	Public house
	236	Café/restaurant
Offices	3	**Offices**
	310	Small office
	310	Large office
	310	Office block
	310	Large office block
	320	Bank/building society
Distribution/Logistics	4	**Distribution/Logistics**
	410	Warehouse
	410	Large storage depot
	410	Warehouse (heavy engineering)
Leisure and Sport	5	**Leisure and Sport**
	511	Hotel
	511	Large hotel
	518	Theatre
	521	Playing field
	523	Sports centre
	525	Sports stadium
	526	Marina
Public Buildings	6	**Public Buildings**

Table 5.6 continued

NRP sub-sector	MCM land use code	Sub-sector and category description
	610	School
	610	University/college
	620	Surgery
	630	Community centre
	640	Library
	650	Fire/police station
	660	Hospital
	660	Large hospital
	690	Religious building
Industry	8	**Industry**
	810	Industrial unit/workshop
	820	Food factory
	820	Chemical factory
	820	Building materials factory
	820	Base metals factory
	820	Metal products factory
	830	Heavy industry: motor manufacturing
	840	Sewage works
Miscellaneous	9	**Miscellaneous**
	910	Car park
	910	Multi-storey car park
	960	Electricity sub-station

Note: Site-specific surveys are likely to be required for industrial premises to take account of the large variety of industry types and potentially high damage estimates.

to building structure and fabric. Flood damage potential has been assessed for the following basic functional/constructional types: (a) high street shops; (b) traditional-style buildings; (c) industrial-style buildings; (d) modern frame buildings; (e) health and religious buildings; and (f) non-standard buildings. Buildings of similar size and function can be built in a variety of ways. While we do not anticipate the need to consider the age of a building when assessing flood damage costs, the division of building types into different forms of construction almost automatically includes an acceptance that many 'traditional' methods have been replaced by more modern materials, with the resultant flood damage costs differing widely. For example, older buildings frequently have brick supporting walls with timber suspended floors, whereas newer buildings may have a steel or concrete frame and slab concrete floors.

A further factor taken into account in the latest NRP classification system is that the cost per square metre of constructing a new building varies substantially in relation to the size of the building. For example, although functionally very similar, a 'corner shop' supermarket will be very different in size to a modern hypermarket. However, the cost (in pounds per square metre) of the reconstruction and refurbishment of a supermarket will be less than for a corner shop (assuming the quality of reconstruction and refurbishment is the same) because of economies of scale. This has been allowed for within the NRP classification (Table 5.6) and our damage data by having typical small/standard and large variants of traditional retail outlets, supermarkets and hypermarkets, hotels, offices, office blocks and hospitals.

The seven NRP sub-sectors (Table 5.6) are broadly similar in number to those used by German and US flood damage researchers and practitioners. Kreibich *et al.* (2007) use NACE categories (*Nomenclature statistique des Activities economigues dans la Communaute Europenne*) for classifying companies damaged and surveyed after flooding. Companies are categorised into one of five sub-sectors which are amalgamations of NACE (Agriculture; Manufacturing; Commercial (including retail, wholesale, hotels, restaurants, etc.); Financial; and Service (including public administration)). The US Hydrologic Engineering Center (HEC) in Davis, California, describes its approach to property or land use categories in its user guide (HEC-FDA, 2008). HEC allows users of its flood damage reduction analysis software to enter whichever damage categories they wish. However, the software is strongly oriented to distinguishing between just five categories (Residential; Commercial; Industrial; Open Space; and Public Facilities) and HEC specifies that a maximum of twenty categories may be used. However, it advises using sub-categories based, for example, on 'occupancy types'.

5.3.6 OS AddressBase land use codes

At the time of writing, the Environment Agency was not using OS AddressBase, but in order to future-proof land use codes we recognise that users of the new NRP flood damage data may need to undertake a land use survey using AddressBase – a new address and location file from GeoPlace LLP and the Ordnance Survey.

AddressBase has been borne out of the requirement of public sector organisations to have one definitive source of accurate spatial address data. In order to drive the initiative forward the Ordnance Survey and the Local Government Group entered into a joint partnership called GeoPlace LLP from where the AddressBase products are managed.

AddressBase offers a definitive supply of spatial address data with greater emphasis than previously on non-postal locations (such as car parks), not-yet-built properties and property changes. It uses a unique land use classification and property-type classification. If, in future, the Environment Agency opts to use AddressBase, the new NRP sub-categories and MCM (2005) codes will be matched and will appear on MCM-Online.

5.4 THE SIX MAIN COMPONENTS OF FLOOD DAMAGE

In order to establish accurate flood loss values, there is a need to clarify what is included under the different components of damage for NRPs.

In the MCM (2005) NRP flood damage data, five components of flood damage were included, whereas in the latest damage data a further component – clean-up costs – has been added. A significant strength of the flood damage modelling approach used here is the careful consideration and assembly of this many damage components. The use of just one NRP damage figure (i.e. building and contents damage, undifferentiated) is common amongst those models referred to in Section 5.1.1, although several differentiate as many as three damage components (e.g. loss to buildings, equipment and inventory; Kreibich *et al.*, 2010).

The six main damage components employed here are:

- building structure and fabric;
- building services;
- fixtures and fittings;
- moveable equipment;
- stock (or raw materials/work-in-progress/finished goods, other contents, as appropriate); and
- clean-up costs.

MCM-Online (Supporting Information for Chapter 5) provides tables of components of flood damage for each of the new functional/constructional NRP sub-sectors. The tables there are intended for use as a guide to aid assessment of flood damage and losses. Examples of the damage items for high street shops are illustrated in Table 5.7.

These components will obviously vary to some extent depending upon the type and size of the properties concerned, and similarly will vary in their susceptibility to flood damage. Each of the components is discussed below.

Table 5.7 Flood damage components for high street shops (Code 21)

Building structure and fabric	Building services
Single skin wall with piers	Repair/replace security installations
Solid brick/concrete/stone wall repair	Test and certify fire/smoke alarms
Rendering to walls	Repair/replace fire/smoke alarms
Service personnel gate	Test and certify electrical external lighting
Renew personnel gate	Repair/replace external lighting
Brick/concrete garage/store	**Fixtures and fittings**
Repair refuse area/chamber	Replace tea point facility
Dry and treat sub-floor	Replace kitchen facility
Check dry cables sub-floor	Replace catering kitchen facility
Overhaul/repair tmber ground floor	Repair racking and shelving
Replace timber ground floor	Replace racking and shelving
Repair first floor area	Repair furniture and fittings
Replace first floor structure	Replace furniture and fittings – medium quality
Rake and repoint brickwork	Repair specialist joinery fittings
External render paint	Replace specialist joinery fittings
Replace shop front	Replace vanity units
Glaze and paint windows	**Clean-up**
Overhaul ext door and frame	Clean concrete paving/yard
Overhaul glazed entrance doors	Clean gullies rod and flush drains
Repair/paint ext door And frame	Clean personnel gate
Repair/paint glazed entrance doors	Clean refuse area/chamber
Replace ext door and frame	Clean sub-floor
Replace glazed entrance doors	Clean timber ground floor

Table 5.7 continued

Building structure and fabric	Clean-up
New aluminium/upvc windows	Clean first floor area
Overhaul internal door – first floor	Clean stone features
Replace internal door and frame – first floor	Clean ext door and frame
Overhaul veneered door and frame	Clean ext door and frame; basement level
Replace overhaul veneered door and frame	Clean ext door and frame; glazed entrance doors
Block plaster remove and redo	Clean; specialist/revolving door
Stud partitions redo	Clean; roller shutter
First floor partitions redo	Clean internal door
Take down and renew dry lining	Clean internal door – first floor
Repair plastered ceiling	Clean veneered door and frame
Replace plastered ceiling	Clean plastered ceiling
Repair suspended ceiling, bulkheads	Clean suspended ceiling, bulkheads
Replace suspended ceiling, bulkheads	Clean hardwood flooring
Replace vinyl tiling	Clean wall tiles
Replace tiled flooring	Clean concrete floor areas
Skirtings average	Clean concrete floor areas; basement level
Emulsion paint to plaster	Clean staircase/handrail
Repair wall tiles	Clean metal stair/balustrade
Replace wall tiles	Clean concrete upper floor/beams
Replace sanitary fittings	Clean concrete upper floor/beams; mezzanine floor deck
Building services	Clean facing brickwork
Check/inspect and repair plumbing	Clean internal wall faces
Service boiler	Clean curtain walling and framing
Thorough overhaul of boiler and/or insulation	Clean shop front
Overhaul major mechanical services plant	Clean walls; basement level
Replace damaged elements of mechanical plant	Clean windows, sills, etc.
Overhaul/repair/replace specialist equipment	Clean internal plaster surfaces
Test and certify electrical power installations	Clean vinyl/tiled flooring

Table 5.7 continued

Building services	Clean-up
Repair/replace power installations	Clean carpet tiles
Test and certify electrical lighting installations	Thorough clean kitchen facility
Repair/replace lighting installations	Thoroughly clean commercial/catering kitchen
Test and certify IT/Bt/tv installations	Clean and overhaul sanitary fittings
Repair/replace IT/Bt/tv installations	Clean/overhaul vanity units, etc.
Test and certify security installations	Clean/overhaul tea point facility

Notes:

High street shops are represented by traditional retail shops of many types: supermarkets; hypermarkets; showrooms; retail kiosks; public houses; and cafés and restaurants.

Moveable equipment includes non-fixed fittings; computers; photocopiers; tills; furniture; display tables; display equipment/gondalas; baskets and stands; pallet trucks; freezers; forklifts; ladders; packing machinery; delivery vehicles; tagging equipment; etc.

Stock includes raw materials, work-in-progress and finished goods and other contents, the value of all of which was assessed from secondary data sources.

5.4.1 Building structure and fabric

Building structure and fabric has been broadly divided into the seven functional/constructional types discussed above. Different structures and fabrics will have different levels of susceptibility to flood damage. For example, for many older buildings, the building structure will comprise load-bearing brick or timber. For modern buildings, the structure will mostly comprise a steel or concrete frame (or steel encased in concrete), with various types of infill (e.g. brick; blockwork; masonry; dry linings; glazing; wood; corrugated sheeting; and various other light-weight cladding materials). Some buildings may have a combination of framed and load-bearing structures. Additionally, some buildings (particularly older ones) may include a basement or cellar, which may also increase susceptibility to flood losses and damage.

Building structure and fabric are often likely to be similar across a range of non-residential properties, and damages are likely to be more related to the age, size and location of property and partly to function. For example, properties located on an out-of-town retail or business park are likely to comprise a steel-framed modern structure, while properties located in a high street are likely to comprise a pre-1960s brick building (some of these properties may be Grade 1 or 2 listed). For certain businesses (e.g. banks and public houses), these older, pre-1960s buildings may comprise the majority of all premises.

Many companies fit out the interior of their premises in the same way, regardless of the type of building structure and fabric (e.g. most 'fast-food' restaurants). In many cases, the building 'shell' is purchased or leased and the interior fitted out according to company specifications. However, for some organisations, their premises do not represent a 'typical' type of building structure as they own many different types of property, with the type often dependent upon the particular location. Many companies report that building structure and fabric are not key issues when purchasing or leasing premises. The main criterion is a good location near to relevant markets or distribution centres. The many companies that lease rather than own premises are not responsible for the building structure and fabric, which remains the responsibility of the landlord.

5.4.2 Building services

The vast majority of relatively large NRPs tend to have the following basic services:

- electricity (including electrical heating);
- gas (including gas-fired heating);
- water supply and plumbing;
- telephone/internet lines/routers;
- cabling/ducting, below floor and/or in ceilings;
- air conditioning and ventilation; and
- alarm/security systems and/or CCTV.

The required services will vary depending upon the size and function of the relevant organisation. For example, not all premises have air conditioning or CCTV. Very large premises may also have their own pump houses, sub-stations and transformers. Evidence from the severe floods which affected the Rojana Industrial Park in Thailand indicates that factories that have a main electricity distribution board which is flooded will need to replace this (*Bangkok Post*, 2011). Some businesses and organisations have a policy of not installing gas services on their premises. Companies may also define certain items (e.g. lifts) as either services or fixtures and fittings.

5.4.3 Moveable equipment

Moveable equipment varies greatly from sub-sector to sub-sector of NRP. Some broad indications as to what constitutes moveable equipment for different sub-categories include: in a typical office – furniture, computers and other office equipment; in a high street shop – shelving, tills and baskets; in a distribution centre – forklift trucks and vehicles.

In some cases there may be difficulties distinguishing moveable equipment from fixtures and fittings (e.g. different types of shelving). There is no hard and fast rule for this. The items suggested in this chapter are those indicated by representatives of various organisations for different types of NRP and merely serve as guides for what might or might not be classified under this component. Moreover, although something may be defined as moveable, during a flood situation it is highly likely that the item would not actually be moved (e.g. if it was very heavy or if there was insufficient time or person-power available).

5.4.4 Fixtures and fittings

Fixtures and fittings largely comprise those items integrated into the building structure, such as:

- shelving and racking (retail outlets);
- kitchen facilities and tea points;
- reception fittings and furniture;
- flooring;
- carpets;
- counters;
- lighting; and
- static equipment and machinery, such as programmable logical control and robotic systems.

Some fixtures and fittings (as well as moveable equipment) may be used in separate storage or staff areas rather than in trading/customer areas (e.g. some furniture and kitchen and bathroom fittings or appliances).

In the Industry/Manufacturing sub-sector the factories most likely to experience heavy flood damage are the capital-intensive ones with weighty, static equipment and machinery, particularly those with electronic and automated machines. Although mechanical components can often be repaired by stripping down machines and applying lubricants to parts, electronic machinery will usually need to be replaced in order to avoid further damage from electrical shorting.

5.4.5 Stock/raw materials, etc.

This component tends to be the most variable in value terms of the six components of flood damage between sub-categories of property. For retail outlets, stock will comprise the goods that they sell, while for manufacturing companies it will comprise raw materials, work-in-progress or finished goods.

For some organisations and companies (e.g. those in the public sector or service industries), there will be little or even no stock as such, although office or medical supplies would qualify under this heading for damage assessment purposes. Depth/damage figures for stock are therefore very low for some types of property.

Some NRPs stand to lose a significant amount of stock after the publicity of being flooded even if their products have been barely affected, or in some cases unaffected, by physical contact with flood water. This is because of the need to protect brands against any possible sources of market contamination. An example is the Honda car manufacturing plant in Thailand, which flooded in 2001, where 1,000 vehicles were destroyed in the aftermath of the flood to reassure customers that no damaged vehicles would ever be sold.

5.4.6 Clean-up costs

Cleaning up involves just about everything that has been in physical contact with flood water, including materials affected by capillary action. Entrances, doors, glazing, walls, floors, kitchens, sanitary equipment, paving and fixtures and fittings will all require thorough drying and cleaning following flooding. Some NRPs are particularly susceptible to high clean-up costs. Examples include the food and beverage processing and manufacturing industry, where hygiene and food safety is of paramount importance, including for reasons of brand protection.

5.5 SUSCEPTIBILITY OF NRPs TO FLOOD DAMAGE

Guidance on the susceptibility of components of damage is available on MCM-Online. The direct flood damage potential of NRPs is estimated using details of the effects that flood waters of varying depths and durations would have on each of the six components of damage. The susceptibility of the six components can vary enormously from component to component, and between and within categories and sub-categories of NRPs. Susceptibility variance is complex and so the primary differences have been taken into account rather than focusing on the myriad smaller differences which can be found in reality. Certain key elements that drive damage values have been identified in benefit assessments, and these need to be considered here.

The height of components of damage (i.e. the height at which an item of damageable components starts to be damaged and the typical maximum height range over which damage occurs) is the key determining factor. Thus a 1.5m-high book shelving unit loaded with books may start to be damaged at just 0.01m. Damage to the shelving unit and the books it contains will then accumulate as flooding gets deeper until total damage occurs at around 1.4m. 'Susceptibility factors' describing the proportion of total damage expected at 0.01m, 0.25m, 0.75m, 1.0m and 1.5m may then be used to produce a depth/susceptibility relationship which is then applied to the value of the damageable component to produce an estimate of damage potential.

Depth of flooding is obviously a key factor affecting susceptibility and levels of damage. A few centimetres will affect flooring and floor coverings. Up to 0.3m of flood water is likely to affect some building fabric and services as well as much fixed and moveable equipment and many inventory items. At 1.0m of flooding many items would be written off and this would increase incrementally with increased depth of flood water. In certain types of business, such as furniture manufacture, damage is especially depth dependent, as around 95 per cent of work-in-progress may be stored at ground floor level, along with 60 per cent of finished goods. Timber goods are particularly at risk of damage from high moisture levels, including condensation.

There are a number of other drivers of susceptibility:

- the functional and constructional type of the property;
- the size of the premises;
- storage arrangements and positioning of services, fixtures and fittings, equipment and stock;
- time of flooding;
- refurbishment cycles;
- high versus low technology;
- temporary and semi-permanent structures.

Certain assumptions regarding susceptibility can be made and the following sections give more detailed guidance on the above drivers of damage values and characteristics of properties that might explain why some NRPs have higher levels of susceptibility than others.

5.5.1 The functional and constructional type of the property

Much of the difference in susceptibility of the components of flood damage can be explained by the function of the property and its constructional type. For example, traditional-style buildings (represented here by offices, library/fire/police premises, theatres, hotels and schools) are likely to have similar itemised damage components in terms of building structure and fabric and therefore similar damage susceptibilities.

The damage to stock and/or raw materials is a function of the amount of that stock, its quality (i.e. more or less expensive) and its susceptibility to flood damage. In general, high street shops and other retailers will have a higher susceptibility to stock damages and losses than other NRP categories, but perhaps a lower susceptibility for, say, fixtures and fittings. Similarly, industrial premises may also have high susceptibility for raw materials and work-in-progress. It is not uncommon for certain types of retailer (e.g. video rental companies) to have zero insurance cover on their stock; therefore, moving such stock in a flood is a high priority.

Fresh food produce is highly susceptible to flooding and would be written off on the grounds of public health risk, even if not touched by flood waters. The tightening of public health regulations aims to ensure that contamination from *E. coli* and other organisms should not be minimised but completely eradicated. Public confidence is paramount and it is not thought to be worth the risk of losing this by selling contaminated goods or reusing contaminated fixtures and fittings. Some canned or bottled foodstuffs might be salvaged if stored at higher levels, above flood waters; similarly, frozen or chilled goods in freezers might be saved if not directly affected by flood waters. However, these goods would soon be lost if power supplies were interrupted.

Health and safety issues are also paramount for health centres or other environments where stringent hygiene conditions are needed when conducting research experiments. Even clothes shops have stringent regulations on post-flood disposal to avoid litigation if products, even slightly contaminated by flood water, are not disposed of in a regulated manner. Some items of clothing, such as outdoor garments, might be salvageable if they have not come into direct contact with flood water, although damage from high air moisture content is again possible.

Paper goods (e.g. books or stationery) and electrical goods are particularly susceptible to damage from damp conditions and high air moisture content, even when untouched by flood waters. Some of these items may be written off for possible public health or safety reasons, and because items such as these need to have pristine appearances in order to sell.

Services tend to be much more susceptible to flood damage than building structure, largely due to their location at ground or lower floor levels. One problem is that of separating flood susceptibility and damages for building structure and fabric from that for services. Serious damage to building structure is unlikely to occur except in very deep flood events, or from damage resulting from high-impact coastal or river flooding, although depths greater than 1.0m are likely to result in some damage to building structure and specifically to building fabric. There can be immediate damage (although often fairly superficial) resulting from immersion and the force of the flood waters, for example:

- consequential damage largely due to the effect of water remaining within the structure for any length of time and damaging plaster, mortar and paintwork, and the effects of cleaning and drying-out processes;
- subsequent deterioration of timber due to impeded ventilation and increased moisture content; and
- deterioration of building fabric due to dampness in brickwork, spalling and efflorescence, and subsequent settlement of damp and eroded foundations.

Steel and concrete frames are less susceptible to damage, but building shell linings, such as plasterboard, are more susceptible as these may absorb flood water and contaminants and will need replacing. Older (pre-1960s) buildings can be more susceptible to flood damage due to their building fabric. Timber structures, flooring and fittings may be highly susceptible to flood damage (e.g. in pre-1960s buildings and public houses), as is wooden flooring in school halls and leisure centres. However, certain types of infill, cladding and insulation on more modern buildings may also be susceptible. Sound-proofing insulation in cinemas is particularly susceptible to flood damage.

With a newer building, a decision may need to be made, after a flood, whether to take out the internal walls and/or infill and leave the main structure and rebuild internally. Occasionally, in the case of very severe flooding, a decision may be made to totally rebuild the premises. For pre-1960s buildings, it can often be more expensive to refurbish than to rebuild, with cost premiums around 10 per cent or more.

A more recent development in out-of-town retail parks has been the use of steel-framed pre-fabricated building structures that are transported in a number of large pieces and assembled on site. Although the frames may be resistant to flooding, the infill material is often highly susceptible, and most of the structure may need to be replaced after a serious flood.

5.5.2 Size of premises

The size of premises will proportionately affect levels of flood damage and losses, and it is one of the most significant factors in determining damage levels. Average floor size for various NRP sub-categories varies considerably, depending upon the function of the property. There are often large differences between NRPs in the same sub-sector, and between units or outlets owned by the same company (Table 5.8). The indicative floor sizes provided in Table 5.8 should be used where no floor space data is listed in the NRD.

5.5.3 Storage arrangements and positioning of services, fixtures and fittings, equipment and stock

Many site surveys conducted by the Flood Hazard Research Centre indicate that key elements of damage relate to the vertical distribution of stock, raw materials or finished goods. The height at which such goods are fixed, stored or displayed is therefore important in assessing flood losses – the higher the goods, the less chance there will be of damage. A number of points can be highlighted.

Sub-ground areas

Goods and equipment stored in sub-ground areas, such as basements and cellars, are particularly susceptible to flood damage, either from flood waters

Table 5.8 Indicative floor sizes for NRPs

Property type	Floor area (m^2)	Property type	Floor area (m^2)
Traditional retail (small)	120	Sports centre	5,400
Traditional retail (large)	400	Office block	1,250
Bank/building society	200	Large office block	3,000
Café/restaurant	250	Large hotel	2,700
Public house	350	University	2,400
Showroom	500	Community centre	1,600
Small office	300	Sports stadium	25,600
Large office	750	Hospital	3,600
Library	900	Large hospital	10,000
Fire/police station	900	Surgery	270
Theatre	3,000	Church	1910
Hotel	800	Kiosk	12
School	3,300	Marina	1,860
Brick warehouse	2,400	Electricity sub-station	48
Industrial unit/workshop	150	Playing field	21,850
Warehouse	1,500	Car park	3,500
Large storage depot	4,000	MS car park	2,700
Supermarket	4,800	Sewage works	4,950
Hypermarket	8,000	Industry/manufacturing	880

rising from below due to seepage or, more commonly, from water pouring down from above once threshold levels have been exceeded (e.g. in older public houses and high street premises). Even after flood waters have subsided, damage due to damp penetration and condensation on the walls is possible. More modern buildings do not often have sub-ground areas.

In the case of petrol stations, fuel supplies in underground tanks are at risk of contamination by flood waters if the pumps are flooded by deep water. The fuel tanks themselves are usually double-lined and should survive immersion.

Single-storey properties

Single-storey properties, such as some retail warehouses, are susceptible to flood losses as firms often have no alternative storage facility for stock or equipment should a flood warning be received. Most warehouses and large retailers keep stock rooms almost full, so there is little opportunity to stack goods out of the reach of flood waters.

Storage methods

Methods of storing stock items may also affect susceptibility to flood damage. One example is goods stored in cardboard boxes. These are vulnerable to flood damage as the cardboard soaks up flood water and the goods inside may become damp and damaged (e.g. shoes). Moreover, if boxes are piled one on top of the other (e.g. television sets), once the cardboard boxes become damp they tend to collapse. Only 0.05m of water may be enough to cause a large pile of goods to topple and sustain damage. Anything with damaged packaging is likely to be written off and is also likely to be affected by high air moisture content. Public or consumer confidence and brand protection are key factors here.

Positioning of items

The positioning of items such as cabling for electrical equipment and computers, motors (e.g. for tills, lifts, escalators or forklift trucks), pumps, gearboxes and programmable logic controls can also affect susceptibility to damage. Where these are located at ground floor level or below this obviously increases the risk of damage. Similarly, where motors are located near to ground level on equipment, this increases their susceptibility to flood damage. However, machinery may be stripped down, cleaned or power washed and repaired, and electric motors may be renewed. Although companies may position cabling in suspended ceilings above ground to save damage to computing equipment, etc., if services such as electricity are affected by flood waters, companies may experience an interruption in trade and operations before these can be repaired or reinstated, and are still therefore susceptible to losses. In some service properties (e.g. modern banks), services for customers are often located on first floors of buildings and are therefore less susceptible to flood damage. Customer interview rooms may also be located in these areas, although access may be restricted by flooding at ground level.

Particular services, such as lifts and escalators, can be highly susceptible to flood damage, and can in turn result in companies experiencing loss of trade if these are affected. In some properties the lift motors and electrical equipment are located in plant rooms on the roof, which makes them less susceptible to damage. However, where the machinery is located at ground or lower ground levels, susceptibility is obviously increased. Storage companies would suffer large restrictions in the use of upper floors within their premises without lifts being available to transport goods. Similarly, department stores would lose business if escalators and lifts were not operating, as many people would not visit the upper sales floors. Rubber handrails on escalators are susceptible to warping when wet and will need replacing. Additionally, the machinery driving the escalators is located at the lower end and is highly susceptible if located at ground or basement levels. A typical £250,000 escalator unit may cost in the region of £30,000 to £40,000 to clean and repair.

Self-storage units

With self-storage units there is a danger of condensation within the units. However, the contents are generally the customers' responsibility and they often hold the only key to their unit. Loss of business would be a major problem for these companies if lifts and electrical services were to be affected by flood damage.

5.5.4 Time of flooding

Most retailers (e.g. high street shops) typically report that sales and business are concentrated or peak at certain times of year, most generally in the run-up to and during the Christmas and New Year period, Easter and other bank holidays, and school holiday periods. The Christmas period is the most important time of year for retail sales and services, and also generally – but not always – coincides with the flood season. Shops are particularly susceptible to large financial losses from flooding at these periods, potentially from both lost stock and loss of trading.

Other categories of business may have busy periods at different times of the year. One example is self-storage premises, which tend to become busy from spring onwards as people begin to think about moving house and need space to store personal items. Meanwhile, for some companies, business is most brisk in the run-up to the Christmas period. For hotels that rely heavily on business clients, September–November tends to be busy for conferences, while the summer months during school holidays tend to be quieter. For other hotels (e.g. those located in holiday resorts), school holiday periods may be their busiest periods.

Susceptibility to losses may also differ depending upon the day of the week of any flooding. For example, supermarkets report that greater losses would be experienced from Thursday to Sunday as these are the busiest trading days.

It should be noted that the damage data included in this chapter have not been weighted to adjust for seasonality, as this will vary depending upon the function of premises. The stock data used in the flood damage data are also an annual mean value, taking into account values of stock at year beginning and end.

5.5.5 Refurbishment cycles

Timing and levels of refurbishment may affect companies' losses from flood damage, in that if premises have been recently refurbished, losses will be greater than for another company that was shortly due to refurbish and would have incurred the repair and renovation costs anyway.

Companies have varying policies regarding the regular refurbishment of their premises. For some, this is done on an 'as and when necessary' basis, while others (particularly larger companies) have regular rolling programmes of refurbishment for the different components of their premises. Refurbishment may also be partial or total each time for each component.

Refurbishment also varies depending upon the component in question. For example, building structure, fabric and services might be refurbished less frequently than fixtures and fittings. Certain fixtures and fittings might be salvageable, cleaned and recycled in other premises after flooding (e.g. metal shelving). However, in many cases these items would be scrapped and replaced with new items, as key imperatives are a speedy resumption of business and reliability of equipment. Wooden and chipboard fittings are particularly vulnerable due to warping.

Depth/damage values included in this chapter for fixtures and fittings and moveable equipment assume that they are halfway through their lives and have been assigned an average remaining value (ARV) of 50 per cent (see Section 4.3.2).

5.5.6 High versus low technology

The more a particular company or organisation relies on new technology, the more susceptible they become should that technology be affected by flooding.

Large companies, such as national supermarket chains and banks, rely heavily on computing equipment, with tills or computers from each store or branch linked to the head office or support centre. If these tills or computers were affected by flooding then the businesses would not be able to operate and would lose trade and, more importantly, customer confidence as a result.

High-precision, expensive electronic machinery (e.g. in printing companies) is also highly susceptible to flood damage. Computing and software companies may have hardware worth millions of pounds, particularly in head offices and support centres, although this may not all be located at ground floor or below ground floor level. Many large companies are increasingly adopting strategies of risk reduction by setting aside space in other buildings or different parts of town in case of natural or other types of disaster (e.g. terrorist attack), so that operations are able to be quickly relocated with minimum disruption to business or operations.

5.5.7 Regional variations

Evidence shows that building, fit-out and refurbishment costs are affected by a building's location. Localised variables, such as market factors, the physical characteristics of a particular site, accessibility and topography will combine to produce a unique flood damage cost. Building fabric, refurbishment and fit-out costs often tend to be higher in south-east England than in other parts of England and Wales. Users may need to take these regional variations into consideration when calculating potential flood losses. Useful data on how to factor in variations in these costs (by region and county) may be found in reports from the Royal Institution of Chartered Surveyors' regular special reports on 'Review of Rehabilitation Costs' (e.g. Building Maintenance Information (2001)). These values may be reported alongside, but not used in, an economic appraisal undertaken to seek national funding for particular schemes.

5.6 DIRECT FLOOD DAMAGE DATA

The data have been assembled to represent at least 95 per cent of typical NRPs found in England and Wales. Although not based on statistical means, the damage data are considered to be representative of the national average potential flood damage. They enable consistent data to be utilised in estimating the benefits of flood risk management projects and aid benefit comparability across England and Wales to enhance consistency in decision-making.

Uncertainty is inherent in loss estimation and it is frequently reported that similar buildings experience vastly different damage and losses during floods. To the extent that any particular NRP selected from a flood risk area in England and Wales deviates from the representative average, the damage data can be expected to under- or over-estimate actual damage potential. The data must be considered as average for groups of similar NRPs. This has always been the case with standard flood damage data previously produced and published.

5.6.1 Production of standard depth/damage data

Direct flood damage potential has been estimated by assembling data on damageable items within each of the six damage components for each of forty-three representative NRPs (Table 5.6). Standard ground floor sizes (m^2) have been used to base each cost model upon (e.g. traditional shops – $120m^2$; schools – $3,300m^2$; supermarkets – $4,800m^2$). Where economies of scale are most likely to affect costings, two sizes of property have been used (e.g. office block – $1,250m^2$ and $3,000m^2$). The size statistic in the cost models refers to the square metre size of NRPs (excluding grounds), although in some cases damage to external paving, asphalt and other surfaces, as well as to fencing, has been included. Groups of susceptibility factors have been applied to the value of the damageable items based on previous knowledge of the vertical distribution of damageable items and experience with damage estimation.

For each NRP category, a table has been developed with the value of damage for depths at thirteen flood intervals, including threshold level (0.00m; 0.25m; 0.50m; 0.75m; 1.00m; 1.25m; 1.50m; 1.75m; 2.00m; 2.25m; 2.50m; 2.75m; and 3.00m) and, where applicable, cellar damages (not related to depth) (–1.00m; –0.75m; –0.50m; –0.25m) broken down by the six damage components. Examples of the depth/damage curves for high street shops (retail) (previously MCM code 211) are shown in Figure 5.1. From these data, sets of depth/damage curves have been produced which are available on MCM-Online.

Damage data are provided for three flood durations (short – twelve hours or less; long – greater than twelve hours and up to three days; and extra-long – between three and seven days). Other assumptions regarding the characteristics of flooding are specified in Section 5.1.2. It should be noted that, unless specified as otherwise (e.g. in the case of depth /damage adjusted for the damage reducing effects of flood warnings), these data assume that no damage reducing actions are taken by property owners, tenants or managers (e.g. no implementation of resistance and resilience measures).

The data for fixtures and fittings and moveable equipment, as with residential data (see Section 4.2.1), are based on economic values as these components are assumed to be mid-way between 'as new' and refurbishment or replacement. Therefore, 50 per cent of 'as new' values, or average remaining values (ARVs), are taken. Stock damages are assumed at 100 per cent of pre-flood value and building costs are taken at full replacement costs.

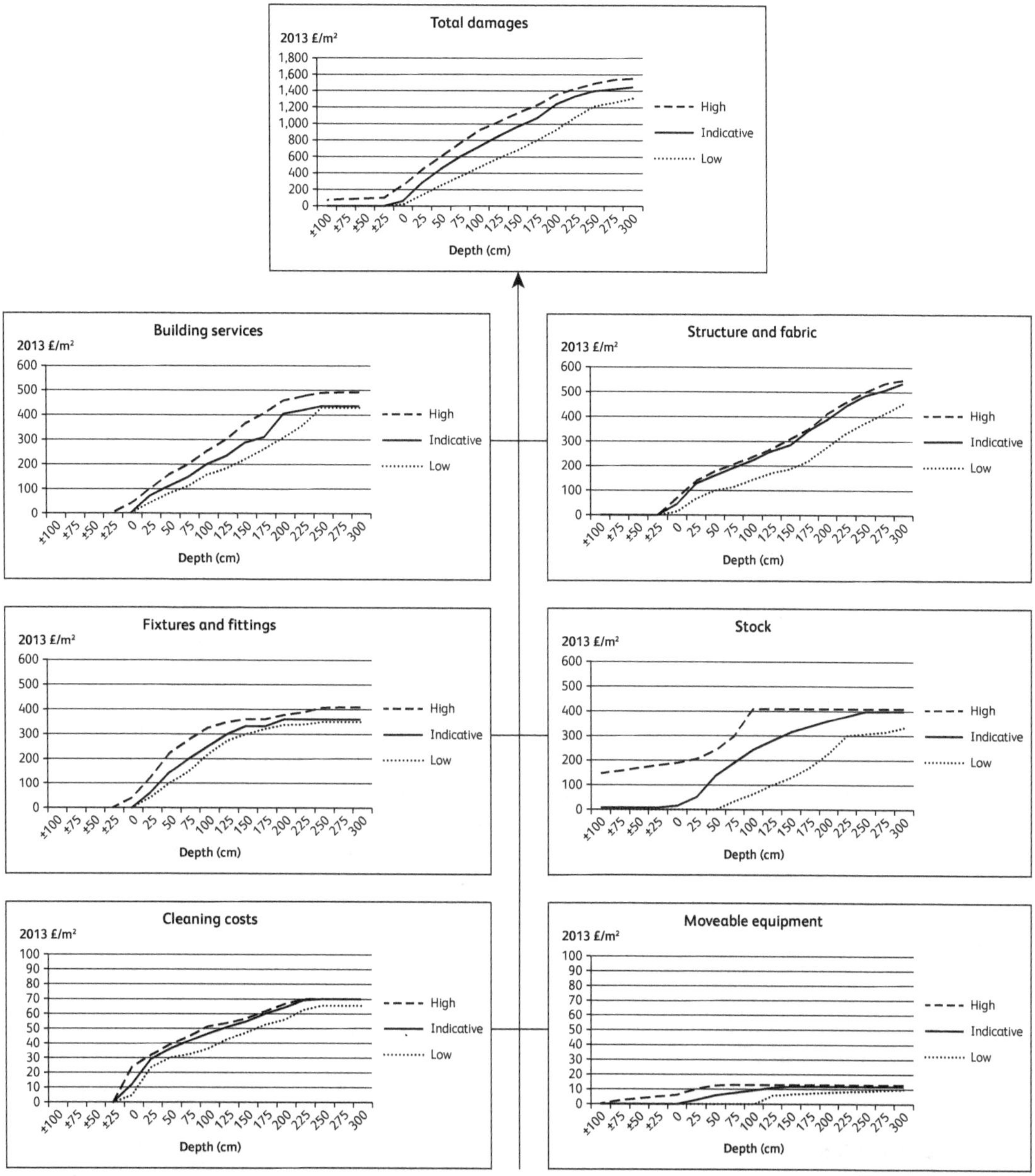

Figure 5.1 Depth/damage curve for high street shops (retail) (previously MCM code 211).

Notes: Scenario – no warning, flood duration less than twelve hours; 2013 prices; the £ damage scale used on the axes varies across the different components

5.6.2 Damage data for the Industry/Manufacturing sub-sector

Admiraal (2011) gives replacement values (per m²) for industrial properties located in ports based on the work of Klijn *et al.* (2007), Briene *et al.* (2010), Tebodin (1998; 2000) and FEMA (2010). Tebodin (1998) calculated the maximum possible damage values of port industries per square metre, and in a subsequent study (Tebodin, 2000) the maximum possible damage value per employee was also calculated. The maximum possible damage values are also presented as fractions of the replacement value (Tebodin, 1998).

These maximum possible direct damage values have been translated by us into depth-maximum damage values in £/m^2 at 2013 prices using the depth/damage factor curves provided by Admiraal (2011) to derive depth-maximum damage data for five categories of manufacturing plant: food, chemicals, building materials, base metal and metal products. The damage values include damage to building structure and fabric, equipment and stock/products only and for five flood depths only: 0.5m, 1.0m, 1.5m, 2.0m and 3.0m. Depth-maximum damage curves for the full seventeen flood depths used in the remainder of the NRP damage data herein and given on MCM-Online have been interpolated from these data to provide the standard damage data for these manufacturing categories. In addition, data from a detailed survey of a large car manufacturing plant in the UK have been drawn on to create depth/damage data for a further manufacturing category (i.e. motor manufacturing) (Chatterton, 2008).

The depth-maximum damage data for food, chemicals, building materials, base metals and metal products plants is generalised and of lower quality than the rest of the NRP data. This should be taken into account wherever manufacturing industry damages account for more than 10 per cent of total losses, and a site survey should then be undertaken to obtain more reliable data. The depth/damage data for motor manufacturing are of much higher quality although only three flood scenarios (i.e. three flood depths) were considered, from which damages for the other depths are interpolated.

5.6.3 Stock values

Stock values and damages are a potentially important component of NRP flood damage potential and were assessed using UK Government secondary data sources. A different approach was required here because the expertise of building and quantity surveyors did not extend to stock values, although they were able to assess stock susceptibility to flood damage.

The UK Office for National Statistics (ONS) regularly gathers and publishes an Annual Business Survey (see http://www.ons.gov.uk/ons/guide-method/surveys/list-of-surveys/survey.html?survey=Annual+Business+Survey+(ABS)). This provides annual business stock survey data for a detailed breakdown of commercial and industrial properties as well as for some other types. It provides statistical data on the total value of stock in each NRP category in England and Wales at year start and end, which can be averaged to find the mean stock level during a twelve-month period. For a number of years, the UK Department of Communities and Local Government published total floor space statistics for disaggregated classes of NRP, but this ended in 2008, when the UK Valuation Office Agency (VOA) took over and simplified this statistical source.

Unfortunately, the VOA's provision of total floor space statistics for NRPs appears to be experiencing difficulties and the data were unreliable when evaluated for use here. Instead, it was decided to use the total floor space data for NRP categories held in the Environment Agency's NRD. By matching and dividing mean total stock values by total floor space values for NRP types, mean stock values for each NRP sub-sector have been calculated. This approach to estimating mean stock values has a UK Government statistical source as its base and should be reliable. There are, however, judgemental decisions to be made using such data sources when it comes to matching these data with NRD NRP categories. Even so, the resulting data are the best quality available and should be reasonably reliable.

5.6.4 Damage susceptibility variance

From this information, upper and lower susceptibility bands and indicative susceptibility data have been defined for each component of damage for each NRP code. The 'low' and 'high' cases represent the susceptibility band, with the 'indicative' case also being provided. Further guidance on when to take the low, high or indicative susceptibility figures for the six components of damage is given in the form of notes for each susceptibility curve on MCM-Online. Figure 5.1 illustrates these bands for high street shops (category 21) for the base scenario. Data for the other NRP sub-categories are also to be found on MCM-Online. Depth/damage data are therefore derived from the average £/m^2 data for each of the six damage components in proportion to flood depth and percentage of damage.

Basement data misuse can inappropriately bolster estimates of damage potential and the present value of damage (PVd). The MCM-Online therefore does not provide data for properties with basements. However, the MCM-Online does provide data for selected properties that are likely to include a cellar, such as a public house or a restaurant. Here we assume 'cellar' to mean a room below ground with no functional use but limited storage. For functional basements that fill completely once ground floor threshold levels have been exceeded by flooding, it is likely that all contents and equipment would be written off or would need cleaning and repairing, and that refurbishment and redecorating would be necessary. It is therefore recommended that these data be ascertained only if field-based land use checks indicate that properties have basements in the appraisal area and when flood water ingress to these sub-floor areas is hydraulically determined; that is, when property threshold level will be exceeded. In such cases we recommend using the basement threshold level as the property threshold level in your calculation.

It is possible that some proportion of direct flood losses might be recoverable from salvaging fixtures and fittings, moveable equipment and stock (raw materials, etc.). One large high street retailer reported writing off £600,000 of stock following a flood, but recovered £50,000 from salvaging. However, due to health and safety fears and customer confidence, much less is now salvaged than in the past, and data included in this chapter do not include significant allowances for salvage values.

5.6.5 Standard errors

All flood damage functions (e.g. depth/damage, depth/duration/damage relationships and curves) are vulnerable to error because some hydrological variables are either omitted from the equation or simplified; because sampling errors of one sort or another occur; and because of measurement errors. The last might arise, for example, when the values assigned to either the components and contents of buildings are incorrectly estimated or the damage susceptibility values are over- or underestimated. We have made assumptions explicit regarding the characteristics of the floods to which the damage data apply to limit the first kind of error (see Section 5.1.2). In order to limit sampling error, the damage data have been constructed synthetically based upon the NRP types, examples and damageable items which our experienced building and quantity surveyors consider to be reasonably representative for England and Wales. Even so, judgements have had to be made in undertaking this research.

As far as values for damageable items of building fabric, etc., are concerned, wherever possible our surveyors have used industry standard cost rate data such as that available through the Royal Institution of Chartered Surveyors (e.g. the Building Cost Information Service). Damage susceptibility factors have been based on experience of flood damaged properties but are associated with judgement. For damage susceptibility, all of the depth/damage data are presented in terms of three susceptibility cases which are designed to describe the likely range of variance in susceptibilities – low, high and indicative. For the extra-long duration case, data are presented only for the indicative case. Under floods of extra-long duration, water will have infiltrated the fabric of the building which is susceptible to penetration, so the low and high cases will be the subject of very similar, if not identical, damage and are therefore represented by the indicative case.

Variance in the flood damage data arises from a number of sources, not least the very wide variety of NRPs represented in flood risk zones – ranging from high street shops to offices and warehouses to sports centres and hospitals – and the fact that some NRPs are unique in their main flood damage characteristics. We have, however, structured the NRP damage data in such a manner that its construction takes account of the leading determinants of flood damage variance found in recent research (see Section 5.1.2) and covers at least 95 per cent of the types of NRP found in flood risk areas, according to the Environment Agency's NRD.

NRP damageability will vary on two temporal scales: by season, because the value of stocks held are likely to be higher at some times of the year (e.g. prior to Christmas) than others (e.g. January); and cyclically, because the value of stocks held is usually drawn down during periods of recession and vice versa. It has not been possible to take into account these sources of variance, except by taking mean stock values for the year.

Variance is associated with each of the six principal components of NRP flood damage and it is not possible to produce simple error bands for the depth/damage data beyond those associated with damage susceptibility factors. The damage data for building stock and fabric and for building services are those in which there is the greatest confidence. Damage data for moveable equipment and fixtures and fittings may be considered less reliable, especially where substantial amounts of equipment and machinery may be present. The stock damage data should be reliable, as it is based on annual UK Government statistical surveys of businesses. It should be noted that damage values for the Industry/Manufacturing sub-sector should be regarded as indicative only and represent maximum damage values.

Overall, owing to the factors considered above, the data are associated with a range of uncertainty qualitatively estimated to be 50 per cent less or 100 per cent greater for the category damage values provided; and a factor of 25 per cent less or greater for the sub-sector damage. This is broadly in line with other flood loss models, such as the US HAZUS MH MR4 flood model (e.g. FEMA, 2010). Data Quality Scores (DQSs) are provided for the damage data on MCM-Online for these different damage components, and sensitivity analysis is advised based on these estimates, to determine whether decisions are sensitive to these variances.

5.6.6 Flood duration greater than twelve hours

Flooding of less than twelve hours is taken as short-duration flooding. For many NRPs, it is considered that longer durations of flooding would increase direct damage values. However, it is also widely felt that the majority of damage is done once flood waters have entered a property, and that depth of flooding is more important than duration.

However, longer flood duration may have more significant impacts upon insulation panels, which would need to be replaced, as well as certain types of building fabric, such as mortar. Wooden structures, boarding and flooring would also be likely to suffer greater damage. Longer recovery time might be needed to allow properties to dry out, and this may increase physical damage to the property, for which repairs may take longer. These factors are reflected in the depth/damage curves which are now available for short (less than twelve hours), long (more than twelve hours) and extra-long (three–seven days) flood durations on MCM-Online.

5.6.7 Coastal flooding

Coastal or tidal flooding may be more damaging than the equivalent depth of river flooding due to the salt and sand content of flood waters and wave impact. This can result in more damage and greater repair and cleaning costs. However, as flood waters from rivers are generally contaminated with sewage and/or chemicals, there is an argument that this would have the same general effect on damage as additional clean-up costs may be incurred. Many clean-up companies do not believe that damages would be substantially greater with coastal flooding.

Table 5.9 gives some indications of possible extra damage to building structure, fabric, services, equipment and inventory items that may result from coastal flooding. These data were calculated for the MCM (2005) and still apply. The average uplift factor across all flood depths is 1.22. It should be noted that these

Table 5.9 Possible extra damage from coastal (saltwater) flooding

Corrosion caused by saltwater (e.g. to steel piling, conduits, fencing and metal window frames)
Damage caused by waves, spray, sand and shingle including deposition of the latter
Potential scouring by water action around building structures
Damage to metal parts due to oxidation
Damage to building fabric and woodwork from penetration of salts
Damage to electrical wiring due to electrochemical action in the presence of saltwater
Pitting to plaster and other boards
Damage to paintwork due to salt being trapped between layers
Discolouration of external and internal decorations
Damage to metallic finishes
Damage to furniture and soft furnishings due to staining from salts
Clean-up is likely to be more costly
Potentially increased loss of production while plant is repaired or replaced (i.e. indirect losses)

Table 5.10 Uplift factors for damages from coastal flooding

Depth (metres)	Factor
3	1.15
2.75	1.16
2.5	1.16
2.25	1.17
2	1.17
1.75	1.19
1.5	1.19
1.25	1.21
1	1.23
0.75	1.29
0.5	1.33
0.25	1.41
0	1.17
−0.25	1.22
−0.5	1.18
−0.75	1.2
−1	1.29
Average from above	**1.22**

are only indications of uplift and should be treated with caution. Uplift factors are given in Table 5.10.

5.6.8 NRP sector and sub-sector weighted mean damage data and NRP category data

For initial/meso-scale appraisals, use the NRP sector weighted mean depth/damage data illustrated on MCM-Online. The mean is derived from all forty-three NRP categories (see Table 5.6) and weighted according to the frequency of properties in these categories within Flood Zone 3.

With the exception of properties in the Leisure and Sport sub-sector (5), for detailed project appraisals, the NRP sub-sector weighted mean depth/damage data is the preferred data (i.e. damage data for the seven NRP sub-sectors excluding sub-sector 5; see MCM-Online). Damage variance among property categories is too great within sub-sector 5 (Leisure and Sport) to use the sub-sector weighted mean. Instead, for this sub-sector, a weighted mean is provided for properties coded 51 (i.e. hotels and theatres) and the three-digit coded category level data should be used for properties coded 52 (i.e. playing fields, sports centres, marinas and sports stadiums). Alternatively, three-digit coded category data may be used throughout, but these are unlikely to increase the accuracy of the damage assessment (see MCM-Online Supporting Information for Chapter 5 for a match between the forty-three new NRP categories and the MCM codes used within the NRD). The forty-three NRP categories in Table 5.6 cover 95 per cent of NRPs encountered in the NRD.

The NRP overall sector weighted mean depth/damage data can be found on MCM-Online. This should be used for miscellaneous property and NRD properties coded '999' unless they are the three categories specifically listed in Table 5.6 (car parks, multi-storey car parks and electricity sub-stations), in which case the category data relating to these three should be used. Guidance on what to do if a land use/category is 'unclassifiable' according to information available is given in Section 5.10.

5.6.9 Capping flood damages at the market value of properties

In the UK it is government (i.e. Defra) policy to 'cap' the present value of estimated damages (PVd) at the market value of each building affected by flooding. This prevents flood damage estimates exceeding the market value of a property. Guidance is available in Section 5.10 on how this is undertaken.

5.7 INDIRECT FLOOD LOSSES AND DAMAGES

This section highlights some key issues relating to indirect flood losses. These are losses caused by the indirect effect of the physical damage caused by flood waters and are largely due to interruption of business. Table 5.11 indicates various indirect impacts that firms may experience.

The difficulty of obtaining accurate data on indirect NRP losses was highlighted in the earlier manuals, and was confirmed in our recent research. Where these

Table 5.11 Potential indirect impacts of floods on NRPs

Loss of income, trade and/or profit
Costs of disruption to business
Deferred trade and/or production
Inspection costs
Costs of transferring operations to another branch or site (e.g. transport costs, hire costs, etc.)
Rental of alternative premises
Loss of customers to a competitor
Supply chain is disrupted
Costs of additional working site(s)
Cold storage or refrigeration costs
Temporary staff costs, including costs of overtime, incentive payments, retainer fees, etc.
Costs of sub-contracting out work
Costs of responding to flood warnings (e.g. moving equipment and/or stock)
Temporary repair costs
Increased insurance premiums
Business re-launch costs including data recovery costs

losses are relevant, they are often difficult to estimate, and the figures are likely to include significant errors. Issues of confidentiality and security are also of concern to businesses, often resulting in refusal to release economic or financial data.

As a result of these factors, no specific indirect damage data are included in this Manual for NRPs, although for the first time a general uplift factor is advocated and guidelines on collecting indirect damage data are provided at the end of this section. For a more detailed discussion of indirect losses, the reader is referred to the 'Red Manual' (Parker *et al.*, 1987, chapters 1, 3 and 8).

5.7.1 Indirect economic losses

In a perfectly competitive (theoretical) market, most indirect flood losses would not exist when measured at the national economy level. This is because of transfers of business within the national economy: if one business property is closed as a result of a flood, all the sales or production lost at that site would be transferred at no cost to an alternative site. Although the flooded business would suffer financial losses, the costs to the nation would be zero because a flood-free competitor would have the production and/or sales transferred to them so that they made a compensating financial gain.

However, in the real world, economic indirect losses do occur: the greater the deviation from the perfectly competitive market of many small and identical producers, the greater the economic indirect losses will be. Thus, the greater the degree of specialisation and concentration within the economic sector concerned, the greater the indirect losses should be expected to be. Also, even when a flooded business loses sales to a competing business when sales are transferred, the flooded business is still likely to experience some extra costs (in addition to clean-up costs) which can be regarded as economic costs. If a flooded business is the sole producer or supplier of a service or product in the country, and flood disruption is such that trade becomes transferred to an overseas competitor, then this also amounts to a national economic loss.

To summarise, although businesses may suffer substantial financial losses, economic losses arise only when:

- the market is less than perfect and transferrable business in the nation does not get transferred;
- production, trade and sales are lost to an overseas competitor; and
- additional economic costs are identifiable.

The assumption has always been that additional economic costs of flooding falling upon businesses are close to zero. However, there is now reason to believe that these costs may be significantly more than that (although still nowhere near as great as indirect financial costs). Sources of additional economic cost could include the following:

- charges levied on flooded NRPs by customers for failing to meet supply agreements and not being able to fulfil level of service agreements (these charges may be substantial) (Chatterton, 2008);

- overtime or weekend payments made to staff for responding to flood warnings by moving equipment and stock (out of the reach of flood waters and then back to original locations);
- administration related to flood insurance assessors and insurance claims, and contracting flood insurance assessors;
- rescheduling and transferring operations to other branches, plants or sites; or making contingency arrangements for the hire of pumping, drying and other equipment, including vehicles (over and above clean-up costs);
- employing technical surveyors and contractors to inspect and assess utility services, equipment and machinery damage and to advise on replacements and repairs;
- keeping customers and suppliers informed; and
- miscellaneous other costs, such as those associated with finding replacement labour when workers have been laid off owing to prolonged business interruption and have subsequently found alternative employment.

These costs would need to be assessed net of transfer payments, such as VAT and excise duty on the cost of additional fuel.

NRP indirect losses can also be split into two further elements: those to the supplier and those to the customer or consumer. If the customer is effectively tied to the specific supplier, then the losses fall on the customer, whereas if the customer is free to choose, the losses fall on the supplier. It is, however, extremely difficult to quantify losses to the customer. These arise when the customer is forced to 'make do' with an inferior product, wait for the desired good to be available or travel further to reach the retail outlet. Economic losses to the supplier are incurred to the extent that it costs more to achieve the same sales; or sales are lost, primarily to overseas suppliers. In turn, these losses depend upon the length of time that the supplier's operation is disrupted by the flood: the duration of the flood itself *plus* the time taken to recover.

In the UK, indirect economic losses for NRPs are generally relatively small compared to direct losses, but they are still likely to be significant for the reasons stated above, because a flood may last for some time, and because its adverse impacts may be experienced for many weeks after the event. For example, in a survey of 415 companies affected by the floods of 2002 in Germany, Kreibich *et al.* (2007) found that the mean duration of business interruption was 43.1 days. Chatterton (2008) found that, for a major car manufacturing plant, significant flooding could lead to thirty days of maximum business interruption followed by another sixty days before disruption ended completely.

5.7.2 Uplift factors and related guidelines

In the majority of cases the proportion of indirect economic losses to direct losses for NRPs is likely to be relatively small – only a few percentage points above zero. Even so, this loss could be significant. This leads us to advocate the use of a general uplift factor for indirect economic losses and some simple guidelines that may be followed to indicate whether indirect economic flood losses should be investigated further to refine damage appraisals.

The costs of business interruption have been researched for both the autumn 2000 floods and the summer 2007 floods in England and Wales (Penning-Rowsell *et al.*, 2002; Chatterton *et al.*, 2010). In the floods of 2000, commercial business interruption insurance claims comprised 31 per cent of the total direct commercial damage claims. In the floods of 2007, business interruption costs were estimated to be 27.6 per cent of total direct business damages. Both of these figures represent indirect *financial* losses to the businesses affected, rather than *economic* costs. The economic costs of business interruption are therefore likely to lie somewhere between 0 per cent and approximately 27 per cent, but much closer to the lower end of this band than the upper. It is difficult to estimate this percentage, but it seems likely that the mean value lies somewhere between 2 per cent and 6 per cent of total NRP direct damages. Taking a cautious, conservative view, an economic loss uplift factor of 3 per cent (of total NRP direct damages) is advocated.

Indirect economic losses are most likely to be worth pursuing further in the following circumstances:

- where disruption of business is likely to lead to loss of exports or increased imports;
- where the economic sub-sector is highly concentrated so there is little capacity elsewhere to make up for the reduced sales at the flooded site;

Table 5.12 A guide to when to assess potential NRP indirect economic losses

NRP category, sub-category	Recommendations
NRP sector, sub-sector or category or individual properties	If NRPs contribute more than 30% of total direct damages, or an individual property contributes more than 10% of total direct damages, then a site survey will be necessary as indirect losses may be significant.
High street shops	Ignore indirect losses unless customers have to make a substantial journey to other similar shops and thereby incur significant travel costs.
Office	Ignore indirect costs in most cases. They are likely to be significant in only a few cases (e.g. in the case of call centres where business could be lost overseas or in the case of globalised business service sectors such as the financial service sector where the same may be the case). Many large firms now have business continuity plans to avoid these situations.
Industry/Manufacturing	Consider investigating indirect losses especially where information suggests that buildings are highly specialised in a concentrated sector and/or the company may face significant additional economic costs.
Distribution/Logistics	Ignore indirect losses unless the company may face significant additional economic costs.
Leisure and Sport	Ignore indirect losses.
Public Buildings	Ignore indirect losses.

- where the economic sub-sector is highly specialised so that there are few, if any, equivalent goods for that produced or sold at the site; and/or
- where the process that leads to the finished good is long (e.g. products from pharmaceutical research; built-to-order plant).

Where all or any of the above are relevant there is a need to investigate further indirect economic losses. If the above conditions do not apply, then any indirect costs are most likely to be the financial costs to the company rather than economic costs to the nation. Table 5.12 provides further guidance.

For those cases where indirect economic losses are likely to be significant, the following need to be determined:

- an estimated loss of 'value added' as a result of flooding, after taking into account any production which could be made up after the event (this largely applies to manufacturing companies);
- any additional costs suffered as a result of flooding (other than lost value added) less VAT and other tax elements;
- to where production would be lost and, crucially, whether the gainers are inside or outside England and Wales; and
- the proportion of production lost to the country.

With the above figures, it is possible to calculate an estimate of indirect economic losses by using the following formula.

Equation 5.1

$$\text{Indirect losses} = \text{Lva} + \text{A} + \text{F}$$

Where:

Lva = lost value added
A = additional costs
F = proportion of sales lost to foreign (or national) competitors

5.7.3 Indirect financial losses

Indirect financial losses are only relevant in local economic appraisals. They can be substantial for the businesses concerned and can vary greatly from company to company (even within the same NRP sub-category) depending on many different factors. For example, for car showrooms, indirect losses might range from £6,000 to £360,000 at under 0.5m of flooding, and from £100,000 to £2.2 million at 2.0m of flooding and above.

For NRPs such as banks and other businesses that rely heavily on the internet and/or telecommunications for their business, it is often very difficult to assess what might be their indirect losses from flooding. Financial loss of sales for a supermarket or superstore would be calculated on the basis of how long the store remained closed and whether car parks and supply warehouses are able to operate.

5.7.4 Loss of business confidence

A significant decline in flood risk management standards (when defences age and flooding probability increases), a major flood, or successive floods could have adverse impacts upon business confidence. However, this would be the case only if significantly declining flood defence standards and increasing flood risk are perceived as problems in the business community, perhaps through media attention. In the more extreme cases this can lead to a business district becoming 'blighted'.

In these circumstances it is likely that businesses will close and relocate to a flood-free location, in which case the cost to the nation would be close to zero, although local financial costs could be higher; for example, through loss of property rental income.

In the Environment Agency's Thames Estuary (TE2100) Options Appraisal, the indirect impacts on business confidence were high-level estimates of the potential disinvestment which might occur in London's financial services sector as a result of declining flood risk management standards in the future (Environment Agency, 2009b). In this virtually unique case it was expected that, because of the globalised nature of this business sector, some financial services sector businesses would relocate from London to alternative cities, such as Frankfurt, leading to a loss of value added and to a national economic loss (in addition to a financial loss to London's economy). 'Business leakage' was assumed to accelerate to a maximum of gross value added by the sector when the standard of flood risk management falls to below 1:5 years.

But London's financial services sector is very much the exception and it is therefore considered inappropriate to pursue any quantitative loss estimates in other scheme appraisals, although any accompanying benefit narrative should report business confidence concerns if they are present. There may be some negative impacts of floods or 'near misses' on some businesses, and a major flood is likely to heighten fears of business owners, particularly if infrastructure disruption is severe, but the evidence is that most firms recover fairly quickly. An exception is where there is a chronic flood risk leading to regular flooding of businesses and consequent 'blight'.

5.8 UNUSUAL AND COMPLEX CIRCUMSTANCES

There are circumstances in which the use of the standard NRP flood damage data presented here will radically underestimate flood losses. An example of these circumstances is where contamination of flood waters by chemicals or oils leads to substantial additional damage. Oil damage may lead to the damage to buildings being three times higher than in the case of largely uncontaminated waters (Kreibich *et al.*, 2010).

One such case occurred in New Orleans in 2005 in the flood that accompanied Hurricane Katrina. St Bernard Parish, in which the Murphy Oil Corporation's oil refinery is located, was flooded to a maximum depth of 4.3m when the storm surge from the hurricane caused the Mississippi River Gulf Outlet leveé to fail. The oil refinery, as well as a neighbouring one, is located in the 100-year floodplain. One of many oil storage tanks was dislodged during the flood, releasing over a million gallons of mixed crude oil, contaminating 1,700 homes in Chalmete and Meraux, causing enormous direct damage, including damage to the oil refinery. Most of the home sites were cleaned by Murphy and placed back into residential use. Following a major court action against Murphy in 2007, plaintiff residents were awarded US$330 million in financial compensation for negligence. In this case contamination caused additional damage to the oil refinery facility and had the knock-on effect of causing enormous damage to the residential community, but it could equally have affected other NRPs.

As the much higher than standard damage was caused to residential properties it was counted in that sector of damage, but had the oil contaminated other NRPs, rather than houses, the additional damage would be counted in this sector. However, indirect loss to the national economy might still be limited (because business would be transferred to flood-free competitors) unless an NRP exported its products.

Another example followed the severe floods in Cumbria in 2009. The port of Workington was closed owing to large volumes of material deposited by the River Derwent in the port's turning basin and access channel. Although some port functions, such as cargo handling, could continue, port commerce was adversely affected for several months until an emergency dredging operation removed large volumes of material.

A further example concerns flood damage to historic and listed NRP buildings, such as churches, old bank buildings, public houses, old mills and factories (English Heritage, undated). Historic buildings may not necessarily be listed and may, for example, be unlisted buildings in conservation areas. Although in some respects relatively resistant to flood damage, the flood damage to such buildings can be substantially more costly to repair compared to more modern buildings. Damages may therefore be significantly greater than indicated by the standard NRP depth/damage data. Historic building elements that are particularly vulnerable to flood damage include stone, brick and mortar walls, timber frames, wattle-and-daub panels, timber boarding and panelling, earthen walls and floors, lime plaster walls and ceilings and many decorative finishes as well as antique fixtures and fittings and contents. Organic materials, such as timber, swell and distort when wet, and suffer fungal and insect infestations if left damp. The hiring of specialist assistance to inspect, conserve, repair or restore historic property after floods, and the specialist treatments required, can add significantly to the costs of flood damage. Where such NRP buildings are present within a flood risk area for which a flood damage assessment is being made, standard NRP depth/damage data should be used in the first instance. These data may be modified subsequently if a site survey is undertaken which indicates the true level of likely damage costs.

These are just three examples of many potentially unusual, complex and unique cases in which the losses to the NRP sector might be grossly underestimated by simply applying standard depth/damage data. Such cases can be allowed for in initial studies or project appraisal reports (PARs) only if they can be reasonably anticipated, but appraisers should be aware of them.

5.9 THE DAMAGE REDUCING EFFECT OF PROPERTY-LEVEL PROTECTION (PLP) AND FLOOD WARNINGS

Measures taken by NRP property owners, tenants and managers to reduce flood damages are becoming more common as 'Kitemarked' flood resistance products become more common. However, in England and Wales they currently appear to be more common in the residential sector than among NRPs because greater attention has been paid to the former (e.g. through PLP grant schemes) and large NRPs are more difficult to protect using these techniques. For relatively small NRPs, there is a much lower capital investment required to implement these measures compared with large-scale engineering projects and they provide property owners and managers with the opportunity to manage the consequences of flooding where costly flood risk management schemes cannot be justified.

Another factor which may prompt businesses to consider installing PLP measures and to sign up for the delivery of flood warnings is the increased use of business continuity planning (BCP). Even so, recent research shows that BCP is much less commonly used by small and medium-sized enterprises (SMEs), compared with large businesses, and since SMEs are far more numerous than large enterprises only a small minority of NRPs can be expected to have installed flood resistance measures at this point in time. Recent research in the coastal flood risk zone in Devon has revealed that even businesses located in such zones sometimes fail to undertake BCP and often, where BCP is undertaken, physical PLP measures are less likely than other BCP responses, such as data back-up systems and insurance (Parker *et al.*, 2012).

Resistance measures, such as de-mountable door and window guards, airbrick covers and sump pumps, are designed to prevent the entry or build-up of flood water within a property. They may be manually or automatically operated, with operation sometimes coupled with a timely flood warning. Data from Saxony, Germany, in which a large number of NRPs were flooded in August 2002, reveal that 49 per cent of 415 companies had taken at least one flood precautionary measure (even though only 25 per cent had flood experience), although this included physical (i.e. building), financial and behavioural precautions (Kreibich *et al.*, 2007). The most common physical measure was flood-proofing of air-conditioning and exhaust air vents and flood-proofing of tanks, silos and storage containers. Kreibich *et al.* (2007) believe that the installation of these flood-proofing measures was probably often due to high standards and regulatory factors, rather than direct flood risk awareness and response. The least popular measure was adapting building structures to flood risk.

Resilience measures treat mainly the internal components of properties to make them less damage susceptible when flood water enters. Examples of these measures include replacing timber floors with concrete ones, raising electrical circuits and outlets, and replacing gypsum plaster with water-resistant material. These measures are permanent once fitted and do not require a flood warning to be implemented. However, a timely flood warning can allow property occupants to remove damageable contents to flood-free locations or to an upper floor so that flood damages are saved. (The potential benefits of flood warnings for reducing residential flood damages in this way are discussed in Chapter 4.)

5.9.1 Resistance and resilience measures

Just how common PLP measures are within NRPs in England and Wales is currently unknown, although, as in Germany, some commercial properties certainly have some physical elements of their building services flood-proofed because of normal product standards and regulatory reasons. The variability of NRPs is considerable and therefore the PLPs required to protect them can sometimes be significantly different from residential properties. Larger NRPs are more suited to – and more likely to be protected by – temporary, de-mountable flood defences that are remote from the property itself, rather than PLPs. Indeed, a study of the economic benefits of PLPs by the Committee on Climate Change (2012) considered only smaller NRPs, such as small high street shops and offices.

Research undertaken for the Environment Agency between 2010 and 2012 led to the development of a benefit assessment tool called the Appraisal of Non-Structural Responses (ANSR), which is based on research on flood warnings and the development of the Flood Warnings Response Benefit Pathways (FWRBP) model as part of the FLOODsite research project (Parker *et al.*, 2008; see also Chapter 4 herein). The ANSR tool specifies simple formulae and parameter values for estimating the benefits of PLP measures and flood warnings. These are discussed below.

Resistance measures

As with residential properties (Chapter 4), there are two possible approaches for estimating the benefits of resistance measures. The first involves increasing the ground floor property threshold level in the benefit area property database by 0.6m for those properties known to have resistance measures installed. This applies only to the number of properties which have a ground floor size of 320m^2 or less, with resistance measures at each return period considered in the appraisal. NRPs of more than 320m^2 are much less likely to be protected by resistance measures because doing so is usually much more complex, costly and therefore unattractive. However, if it is known – for example, through a site survey – that a large NRP has resistance measures installed, these may be taken into account as a special case. In some cases larger NRPs may be protected instead by de-mountable defences installed either very close to properties or more remotely from them. Again, the damage reducing impact of such de-mountable defences may be taken into account by raising the ground floor property threshold levels by the height of the de-mountable defences. (In this case some account should also be taken of the reliability of these defences, which current experience suggests is a little less than 100 per cent because of factors such as warning reliability and deployment issues.) Wicks *et al.* (2012) concluded that warning resistance measures are, on average, only approximately 75 per cent effective and so flood damage savings will be over-estimated if this is not taken into account. This means that the estimated flood damage savings at each flood return period need to be factored by 0.75 before they can be used.

With the first approach it is necessary to be able to identify which properties in the benefit area, up to the maximum ground floor size of 320m^2, have resistance measures installed. Records or mapping may exist which provides this information or, alternatively, a field survey may need to be undertaken. As with residential properties, if the appraisal is an initial one, and depending upon the scale of the appraisal, where the number of NRPs with resistance measures is very small (say 5 per cent of properties or less) it is probably not worth taking account of the effect of resistance measures in an appraisal. Otherwise, the estimated damage reducing effect of resistance measures must be included in the appraisal. This is because resistance measures are likely to be installed only in flood zones affected by relatively frequent events (say, up to 1:40 years), and these events contribute more heavily to average annual damages than longer return period events.

The second approach involves using formulae in which either a national average or local parameter values are used (e.g. for uptake of resistance measures). Where local parameter values are unavailable, a national average value may be used as a surrogate. An evaluation of residential take-up of flood resistance measures in England and Wales undertaken in 2011 and 2012 as part of the SC090039 research project for the Environment Agency (Wicks *et al.*, 2012) revealed that an estimated 8.1 per cent of properties (mainly residential) in the NaFRA 'significant risk' category (in which there are an estimated 192,000 NRPs) had installed flood resistance measures (Environment Agency, undated d). Here 'significant risk' is where the chance of flooding in any year is greater than 1.3 per cent (i.e. >1:75 years), which is the flood probability range in which the costs of installing resistance measures are most likely to be cost-effective. Here an assumption is made that where flooding probability is less than 1.3 per cent (i.e. <1:75 years) the likelihood of flood resistance measures being deliberately installed in NRPs is zero. This is a conservative estimate because May and Chatterton (2012) concluded that the evidence for resistance measures being cost-effective related to properties where the chance of flooding was greater than 2.5 per cent (i.e. >1:40 years). Of the 8.1 per cent uptake value (UP), the uptake of these measures was 4.8 per cent for warning-dependent resistance measures (WDRM) and 3.2 per cent for warning-independent resistance measures (WIRM), some of which are automatic.

These national data can be taken only as a rough guide to the prevalence of resistance measures because of the current paucity of data on NRP uptake and benefit. There are some indications that the uptake of WDRM and WIRM in NRPs may be less than for residential properties but available data relating to this are currently poor. When better-quality data are available these NRP uptake values will be refined and will appear on MCM-Online.

Other than UP (3.2 per cent), the effectiveness (EF) of these measures also influences the effect of WIRM on damage reduction (i.e. warning-independent resistance measures benefits (WIRMB)). Resistance measures require maintenance, may be misplaced, installed incorrectly and have a relatively short life (e.g. twenty years), and Wicks *et al.* (2012) give a value of 75 per cent for EF based on the work of Thurston *et al.* (2008). Resistance measures are also effective only in relatively shallow flooding; between 0.6m and 1.0m flood depths they may begin to leak and are inadvisable where flood depths are likely to exceed 1.0m. Because of these factors, we assume here that the maximum flood depth to which resistance measures are effective is 0.75m. The following formula provides a rough prediction of the current frequency of adoption and effectiveness of WIRM in NRPs (i.e. FREQ3).

Equation 5.2

$$FREQ3 = UP\ (0.032) \times EF\ (0.75) = 0.024\ (\text{i.e. } 2.4 \text{ per cent})$$

Where there is sound local knowledge of the degree of uptake, a local value may be used to replace the relevant value in the above equation.

For WDRM we also need to take into account that they depend upon property owners and managers receiving, and being available to receive, flood warnings – termed 'reliability and availability' (RA). RA depends in turn on: (a) coverage of the flood warning service; (b) uptake of this warning service; (c) effectiveness of the warning service; and (d) the availability of potential warning recipients. Wicks *et al.* (2012) analysed a range of evidence and concluded that RA was between 22.7 per cent and 30.0 per cent, although the higher of these two figures was taken as most likely (here we round this value to 30 per cent). Furthermore, these measures also depend on the proportion of property owners operating the measures (OP), which Wicks *et al.* (2012) estimate as 63 per cent. For WDRM, therefore, a prediction of the current frequency of adoption, operation and effectiveness (FREQ4) is provided by the following formula.

Equation 5.3

$$FREQ4 = RA\ (0.30) \times UP\ (0.048) \times OP\ (0.63) \times EF\ (0.75) = 0.068 \text{ (i.e. 0.068 per cent).}$$

Again, where there is sound local knowledge of any of these parameter values they may be used to replace the relevant value in the above equation.

May and Chatterton (2012) provide benefit (i.e. damage reduction) figures for a detached residential property of $62m^2$ ground floor size (see MCM-Online Supporting Information for Chapter 4) which, in the absence of benefit data for NRPs, may be used as surrogate damage reduction values. The values are for a flood exceedance probability of 2 per cent. For automatic resistance measures, damage reduction is £1,850 (building fabric), £450 (building inventory) and £800 (clean-up), totalling £3,100 or £50/m^2. For manual resistance measures, the equivalent figures are £1,000 (building fabric), £200 (building inventory) and £600 (clean-up), totalling £1,800 or £29/m^2.

The above data may be used to estimate an approximate value for the damage reducing effects of resistance measures on NRPs, assuming that a flood warning system is operational for manual measures:

Step 1: Calculate the number and then the total ground floor size (m^2) of NRPs up to and including $320m^2$ in size at risk at each return period in the benefit area up to the 1:75 years flood probability threshold where flood depth is not expected to be greater than 1.0m (TGA).

Step 2: The following equation may then be used to estimate the total damage reduction (in pounds) owing to NRP WIRM.

Equation 5.4

$$WIRMB£ = TGA \times DR \times UP \times EF$$

Where:

WIRMB£ = estimated damage reduction (i.e. benefit) by employing WIRM

TGA = total ground floor area of NRPs (of $\leq 320m^2$) located in benefit area within 1:75 years flood risk area and where flooding is not likely to be greater than 1.0m

DR = damage reduction: £50 per m^2 at 2012 £ values

UP = uptake of WIRM factor: 0.032

EF = effectiveness factor: 0.75

For NRP WDRM, the equivalent equation is given below.

Equation 5.5

$$WDRMB£ = TGA \times RA \times DR \times UP \times OP \times EF$$

Where:

WDRMB£ = estimated damage reduction (i.e. benefit) by employing WDRM

TGA = total ground floor area of NRPs (of $\leq 320m^2$) located in benefit area within 1:75 years flood risk area and where flooding is not likely to be greater than 1.0m

RA = reliability and availability: 0.30

DR = damage reduction: £29 per m^2 at 2012 £ values

UP = uptake of WDRM factor: 0.048

OP = operated: 0.63

EF = effectiveness factor: 0.75

To summarise, the above equations are only rough guides to the value of NRP damage reduction through use of resistance measures for the following principal reasons:

- the uptake data are based mainly on residential uptake, rather than NRP uptake;
- the uptake factor (UP) applies to the number of properties, although here we apply it to TGA, which is a surrogate measure for property numbers;
- DR in £ per m^2 is derived from research into detached houses; and
- only reduction in direct damage is included (i.e. indirect and other losses, such as health impacts, are not included).

The equations are, to a degree, progressively more reliable as sound local parameter values are substituted for the national average values.

Resilience measures

For England and Wales, Wicks *et al.* (2012) estimate that the uptake (UP) of resilience measures in residential properties in benefit areas up to the 1:75 years flood probability in England and Wales is 8 per cent, although this estimate is fairly crude owing to lack of research evidence. The effectiveness (EF) of these measures is considered to be less than for resistance measures and is estimated at 50 per cent.

May and Chatterton (2012) provide two benefit (i.e. damage reduction) figures for a detached residential property of $62m^2$ ground floor size, which may be used to derive a surrogate damage reduction value. These two values are for resilience measures with and without resilient flooring, the average for which is £2,525 (comprising building fabric, building inventory and clean-up costs) or £40.7 per m^2.

A modified form can be used to estimate an approximate value for the damage reducing effects of resilience measures on NRPs (i.e. RISMB):

Equation 5.6

$$\text{RISMB£} = \text{TGA} \times \text{DR} \times \text{UP} \times \text{EF}$$

Where:

RISMB£ = estimated damage reduction by employing resilience measures
TGA = total ground floor area of NRPs, up to and including $320m^2$ in size, located in benefit area within 1:75 years flood risk area and where flooding is not likely to be greater than 1.0m
DR = damage reduction: £40.7 per m^2 at 2012 £ values
UP = uptake factor: 0.08
EF = effectiveness factor: 0.50

Very similar simplifying assumptions to those applicable to the procedure for estimating the damage reducing effects of resistance measures are applicable to resilience measures. A further simplifying assumption which can affect the reliability of these estimates is that in some cases resistance and resilience measures may be used in combination. The UP may improve in the future and/or sound local evidence of uptake of resistance and resilience measures may be available, in which case the uptake value can be altered to match the evidence more closely.

The parameter values used in the above equations (i.e. for calculating WIRMB£, WDRMB£ and RISMB£) are derived from a national study of England and Wales and they are mean values. They can be substituted by local values where there is evidence to support a more customised local appraisal. For example, if it is known that, say, 10 per cent of properties have WIRM in a particular benefit area, then the UP parameter value may be altered from 0.032 to 0.10, and so on with other parameter values.

In practice, when undertaking a benefit assessment, a decision will have to be made about (a) whether or not to take into account the damage reducing effects of resistance and resilience measures (if they exist and ignored benefits may be exaggerated) or (b) making allowance for them, possibly by using the above equations.

5.9.2 Potential reduction in losses by moving moveable equipment, stock and other contents following receipt of flood warnings

Data reported from the discussion meetings with company representatives undertaken in 2004 for the MCM (2005) illustrated that, in theory, flood warnings could make a substantial difference to losses for certain types of property and under certain conditions. The majority of businesses perceive the need for flood warnings. The question remains as to how much damage and loss could be avoided on receipt of a flood warning if appropriate actions are taken to move contents out of the reach of flood waters.

Table 5.13, drawn from data supplied during discussion meetings in 2004, indicates the percentage of potential savings that may be used to calculate the reduction in damages (from the base flood scenario) for different flood depths for moveable equipment and stock following the receipt of flood warnings of four hours or more. The figures were produced by calculating the average reduction in losses on receipt of a warning by each depth across all NRP sub-categories where data were available. No data for indirect damage reduction are available.

The average potential loss reduction across all flood depths is 43 per cent. Note that this figure relates to only two components of damage (moveable equipment and stock); it is not 43 per cent of total damages. Although the increasing use of temporary barriers and defences could also result in future damage reduction to building structure and services and fixtures and fittings, these have not been considered for this Manual.

The figures in Table 5.13 should be taken only as a guide and represent potential rather than actual warning benefits, but they are applicable to NRPs of all sizes. Potential flood damages saved will depend upon: (a) coverage of the flood warning service; (b) uptake of this warning service; (c) effectiveness of the warning service; and (d) the availability of potential warning recipients.

Table 5.13 Potential direct flood damage reduction following receipt of a flood warning (lead-time >4 hours): moveable equipment and stock

Depth	Percentage potential direct damage reduction with a flood warning lead-time of >4 hours	
(metres)	Moveable equipment (%)	Stock (%)
3	40	38
2.75	40	39
2.5	41	39
2.25	42	40
2	43	40
1.75	43	41
1.5	43	42
1.25	44	42
1	45	43
0.75	45	44
0.5	45	42
0.25	44	36
0	43	38
–0.25	43	53
–0.5	43	54
–0.75	43	54
–1	43	53
Average	**43**	**43**

Availability of flood warning recipients to receive and respond to a flood warning is affected by many different factors. For example, Kreibich *et al.* (2007) found that in the River Elbe floods in Saxony in 2002, even when some business managers had received an apparently timely flood warning, they were unable to reach their premises in order to take damage reducing actions because of road flooding. These variables are summarised in the 'reliability and availability' parameter (RA) referred to above, which Wicks *et al.* (2012) concluded was between 22.7 per cent and 30.0 per cent, although the higher of these two figures was assumed to be most likely.

Damage reduction in NRPs by moving assets is also influenced by the EF variable (the effectiveness of warning responses). Kreibich *et al.* (2007) discovered that, on average, business damage mitigation was 52 per cent effective: the same percentage that was found to be applicable to residential properties by Parker *et al.* (2007).

The percentage damage reductions in Table 5.13 should therefore be factored by 0.30 and then by 0.52 to take account of these two variables. Flood damage data which take account of the potential percentage reductions in damage to moveable equipment and stock, factored by 30 per cent, are available on MCM-Online.

Even then this is only a guide to damage saving through movement of moveable equipment, stock and other contents. This is because, as Kreibich *et al.* (2007) discovered, some businesses which took actions to move moveable items out of the reach of flood waters did so without receiving an official flood warning. In fact, the managers of many companies (32 per cent of those affected by the floods) had recognised the flood danger through their own observations or those of their employees. The Elbe floods were of two types – slow rising and flash floods – and it was probably unlikely in the case of the slow rising floods that 'self-warning' would not have occurred. In the UK, most fluvial floods rise comparatively quickly, to some extent reducing the scope for self-warning compared with the Elbe case, but self-warning is still likely to be significant.

Only potential savings for warnings of four hours or more are included in Table 5.13. In the 2004 surveys, few companies reported substantial savings with less than a four-hour lead-time. For many companies, anything less than eight hours would be of limited use, and Kreibich *et al.* (2007) found that those companies receiving an average of fifteen hours of warning lead-time or more were those that were able to save all of their equipment and stock, whereas those that had not managed to save anything received an average of less than ten hours' warning lead-time.

5.9.3 Factors influencing the effectiveness of flood warnings

Table 5.14 indicates factors likely to have a positive impact upon the effectiveness of flood warnings. However, businesses often fail to respond to a warning, or respond ineffectively, due to a poor understanding of warnings, inadequate warning time and little flood experience or preparedness.

Table 5.14 Factors likely to increase the effectiveness of flood warnings for NRPs

Flood warnings are likely to be more effective in reducing NRP flood damages when:
There is a dedicated and customised flood warning service for companies which sign up to receiving such a service.
There is a warning lead-time of at least four hours and preferably eight–ten hours or more.
Specific information is provided in the warning message or supporting information on the timing and likely maximum height of flooding.
The business has a generic business continuity plan within which there are specific provisions for flood risk and this plan includes not only a risk assessment and a mitigation plan but procedures to be followed when a flood warning is received.
Staff are aware of their responsibilities and the actions they are expected to take when a flood warning is received.
There are sufficient able-bodied staff or contractors available to move equipment and goods and to take mitigating actions.
A business has made mutual support arrangements with other businesses to help it respond to a flood warning, including those in its supply chain.
Appropriate refrigeration is available for storing perishable foodstuffs, drinks, pharmaceuticals, etc. elsewhere.

Moreover, the first priority for most businesses on receipt of a flood warning is to ensure the safety of its staff and, if appropriate, its customers or clients, rather than undertake damage mitigating actions. For example, schools and residential homes for the elderly would hopefully always evacuate children and residents before taking any other actions. Also, some companies would not ask staff to help to move equipment and stock or to take mitigating actions on health and safety grounds. In addition, forklift trucks are subject to regulations concerning their operation, and only qualified people are allowed to operate them.

It is also unrealistic to expect that large amounts of stock and equipment would be saved in all premises if a warning were received. This would not be practical or even feasible in some cases; for example, if there is a need to organise staff from other locations or outside contractors to move items. In addition, there might be a need to arrange an additional location for storage of these items, if insufficient storage is available on site.

It may, therefore, be possible to move only smaller items of equipment – for example, computers – or the most valuable items in the time available. If nearby roads were flooded, moving items to other locations would not be an option.

5.9.4 Warning lead-times

Warning lead-time is important. There would generally be a big difference in the amount of damage/loss that could be mitigated with a two-hour warning compared with a warning of eight to ten hours. For many companies, particularly larger ones, anything less than eight hours' warning lead-time would be of limited use, except to ensure the safety of staff and customers/clients, to secure the premises and/or perhaps to save some of the most valuable items. Parker *et al.* (2012) found that a high proportion of businesses had secure arrangements for data back-up, usually off site. Most businesses could save some stock or equipment given a long lead-time, and for some the savings could be substantial.

5.9.5 Flood information

Currently, except for infrastructure operators, such as Network Rail in the UK, commercial and retail organisations generally receive the same flood warning information as household residents. This is often not thought to be accurate enough for many large-scale operations to take decisions on mitigating actions, due to the serious business implications of suspending trade, services or manufacturing. Evidence from our research suggests that smaller businesses would be more likely to take appropriate actions on receipt of a warning, and may be able to save a substantial amount of stock or equipment as a consequence.

Specific information on the time, level of impacts and duration of the flood is required by businesses and organisations before any mitigating actions are likely to be taken, and this information is often unavailable. The difficulty of producing a warning with sufficient credibility for people to act poses practical limitations, as the prediction of rare events carries an inevitably high risk of false alarms if real events are not to be missed.

5.10 HOW TO USE THE NRP DEPTH/ DAMAGE DATABASE

When carrying out a flood loss assessment, a decision needs to be taken on which level of data is proportionate to the study being undertaken. Guidelines are given here for three types of study: (a) initial and meso-scale; (b) detailed project appraisal, micro-scale and strategy reports; and (c) site surveys (most suitable at the detailed project appraisal and micro-scale levels). Here 'meso-scale' refers to a national, regional, sub-regional or catchment scale, whereas 'micro-scale' refers to project appraisal reports (PARs) and higher-resolution studies.

Whenever undertaking an appraisal, flood damage data need to be selected and applied thoughtfully rather than mechanically in order that the data match reality as closely as possible, notwithstanding that the potential for error is ever present.

5.10.1 Initial studies

Data requirements for NRPs are as follows (see MCM-Online Supporting Information for Chapter 5 for a working example):

- The number of properties in each sub-sector and category:
  - Retail (2);
  - Offices (3);
  - Warehouses (4);
  - Leisure (51), Sports categories (521, 523, 525, 526);
  - Public Buildings (6);
  - Industry (8);
  - Miscellaneous (910, 960).
- All depth/damage data for NRPs is in pounds per square metre, so the area of the ground floor space of the above sub-sectors and categories is required. This is available on MCM-Online Supporting Information for Chapter 5.
- The current standards of flood risk management protection provided for the benefit area.
- The weighted annual average damages (WAAD) (see Section 4.9.1) are then taken from MCM-Online for each sub-sector and category and multiplied by the appropriate ground floor area.

5.10.2 Detailed project appraisal, micro-scale and strategy reports

For project appraisal reports (PARs) and more detailed strategy reports, a step-wise approach to data assembly is suggested here. Note that a Data Quality Score (DQS) 1–4 should be allocated for the land use sub-sector or category of each NRP, ground floor area, the depth/damage data assigned to the sub-sector or category, and the property threshold (see Section 3.5 for an explanation of this experimental method).

Step 1: List the NRPs in the benefit area

- Determine the number by sub-sector or category of NRPs in the benefit area primarily by using the NRD (DQS 2) and further enhanced by other data sources, such as the OS Mastermap (due to be replaced by OS AddressBase) (DQS 2), Google Street View (DQS 2) and preferably a site survey (DQS 1) for selected properties (see Section 5.10.6).
- Selective field checks are always recommended to authenticate data quality.

Step 2: Determine each property's ground floor area (building's/buildings' footprint only)

Determine ground floor area using one of the following sources (each with differing DQS). Selection will depend upon available budget and timescale.

- Use the NRD to determine each property's footprint (DQS 1).

If unavailable:

- Determine area by field measurement (DQS 1); or
- Use GIS tools to measure the area from OS 'Mastermap' or equivalent (DQS 1); or
- For specific or unconventional properties, use http://www.royalmail.com to determine property postcode, then http://www.voa.gov.uk to determine the current valuation, which also gives the total ground floor area of each property valued (DQS 1); or
- Use the indicative floor sizes provided on MCM-Online (DQS 3).

Step 3: Linking the new MCM data to NRD codes

- Link the NRD (MCM) codes to the weighted sub-sector means on MCM-Online. For some categories (most notably Sport (52) and Miscellaneous (9)), it will be necessary to use the corresponding three-digit NRP category code instead.

Step 4: Allocate depth/damage data

- MCM-Online gives the preferred depth/damage data for each NRP MCM code with appropriate data quality. Normally, the six weighted sub-sector means should be used rather than the three-digit categories, *except* in the case of properties coded 51, for which the indicated aggregated weighted category mean should be used, and properties coded 52, for which three-digit category data should be used.

Step 5: Undertake present value of damages calculation

- Use proprietary software to calculate estimated property present value of damages (PVd).

Step 6: Determine market value as required for 'capping' analysis (see Section 3.5.4)

- As FCERM-AG (Environment Agency, 2010a) guidance requires the expected PVd not to exceed the property's market value, rateable value will need to be determined from one of three sources, with varying data quality. Selection, as for Step 2, will depend upon budget and timescale.
- Use http://www.royalmail.com to determine property postcode, then http://www.voa.gov.uk to determine current valuation, which gives the rateable valuation for the property concerned (DQS 1). Knowing the bulk class yield will allow conversion to a suitable market value (see Environment Agency, undated d).

Step 7: Filtering

- Rank each property in the benefit area by its PVd or capped value (see Chapter 3). Check the data quality of at least the top ten contributing NRPs as, from experience, these contribute a significant proportion of total PVd, representing the potential for significant uncertainty.
- If, after the filtering process and improvement of data quality, any NRP contributes more than 10 per cent of PVd or capped PVd then a site survey should be undertaken to confirm these damages.

5.10.3 Vacant NRPs

There is a question about whether or not allowance should be made for vacant NRPs. Because of the recession and the restructuring of the retail sector, up to 14.3 per cent of high street shops (48,000 in total) in the UK were estimated to be vacant at the time of writing (*Daily Telegraph*, 2013). Many of these will not therefore contain valuable, damageable equipment, fixtures and fittings and stock, reducing their damageability. Arguably, then, lower depth/damage data estimates – reflecting damage to building structure and fabric, building services and fixtures and fittings only (i.e. three of the six flood damage components) – should be used for the high street shops sub-sector.

However, project appraisals are usually for a fifty-year period or more and during this period, shop vacancies may fall to near pre-recession levels (although the structural changes affecting the high street appear to be driven by irreversible changes). Since the greatest contributions to net present values (i.e. discounted annual average damages) occur in the earlier years of the life of a scheme, we suggest that an allowance is made for vacant high street shop premises in detailed project appraisals but not initial studies. Depth/damage data for the high street shop sub-sector for building structure and fabric and building services (aggregated) are available on MCM-Online. Vacancy depth/damage data for other sub-sectors are unavailable.

5.10.4 Avoiding errors

Errors in applying MCM NRP data in appraisals include systematic errors which arise through properties being assigned the wrong MCM code and where Flood Zone 2 or 3 is taken as the area of direct benefit rather than using more accurate hydro-dynamic modelling techniques. Such systematic errors can both under- and overestimate damage potential. Measurement error may also occur through incorrect measurement of the crucial property ground floor building size variable or through using inaccurate property threshold levels.

5.10.5 Avoiding errors associated with NRD non-addressable property points

The NRD was updated in 2011 to include all non-addressable properties greater than 25m^2 ground floor footprint. These often make up a significant minority of the NRD complement. The potential for making gross errors with non-addressable NRP property points is large and can render an appraisal grossly inaccurate unless measures are taken to avoid this. Non-addressable property points are typically large commercial buildings on industrial estates with a single post house covering all properties. They could also be myriad unclassified properties attached to an addressable property, such as an annex to a hotel. They will appear in OS Mastermap topography but will not have a corresponding address point. Using the NRD, these properties are coded 999 and refer to properties with a floor space greater than 25m^2. Using the NRP sector mean damage data is one way of addressing these properties but this can grossly exaggerate damage estimates. To guard against this, the rule is to ground-truth the true nature and function of any 999 NRP which has a ground floor size of 500m^2 or more and allocate either the appropriate sub-sector mean or category mean damage data to it. Ground-truthing may be achieved through a site visit or by using Google Street View.

5.10.6 Site surveys

The variety of NRPs is considerable, and average/standard depth/damage data given in this Manual may not be appropriate, in which case a site survey of the property will be needed. A standard NRP site survey proforma may be downloaded from MCM-Online Supporting Information for Chapter 5.

Whether or not to carry out a site survey is ultimately a matter of judgement but due to the amount of time required, surveys should normally be reserved for large properties with high flood frequencies. This may include, for example, project appraisal reports where, after several present value of damages (PVd) iterations (e.g. stabilising footprint, threshold, land use type), an NRP is still contributing significantly to overall PVd (e.g. >10 per cent).

The need for a site survey should also depend on the sensitivity of the decisions to be made to the valuation obtained: will it affect the FRM decision? Moreover, in certain sectors where there is a rapid change in properties through time, it may be preferable to adopt average values than to assume that a particular high damage potential operation will necessarily stay in this location for the duration of the appraisal period. If a site survey is used, it would be prudent to test against average values at least as a sensitivity check.

If a site survey is deemed to be necessary, the following can be used as a guide when choosing whom to approach within the organisation concerned to help complete the questionnaire:

- small firms – the owner;
- medium-sized firms – the plant/company manager;
- large, complex firms – the risk manager, managing or financial director, accountant, insurance claims officer, estates, facilities or property manager.

In many cases it may be necessary to consult more than one person in order to obtain all of the necessary information. It is advisable to obtain the name of the most appropriately qualified person (or persons) within the relevant organisation to approach for the information required. A letter or e-mail explaining the nature of the research should be sent to this person and followed up with a telephone call one week later to arrange a suitable time for an interview.

The length of the interview is likely to be up to one hour for small to medium-sized firms. For larger firms, longer may be needed, and a plan of the property and its components might need to be obtained. If a site comprises several different buildings of varying floor heights, it is recommended that separate damage assessments are made for each principal building.

In place of carrying out a site survey using a lengthy questionnaire, as advocated in an earlier version of this Manual, it is suggested that a simplified approach be taken, mainly focusing on the following questions (see MCM-Online Supporting Information for Chapter 5 for a simplified site survey questionnaire):

1 What is the cost of rebuild (building structure and fabric)? (Note that this is for the building unit(s) and not the total footprint of the property, which includes surrounding grounds, car parks, etc.)
2 What is the value of the services installed?
3 What is the value of the moveable equipment?
4 What is the value of the fixed equipment?
5 What is the value of the stock, raw materials and work-in-progress?
6 Are sales losses to overseas competitors likely to be significant during a flood affecting your premises? If so, what are they likely to be?

During the site meetings it may be necessary to remind respondents on which component they should be focusing. Experience shows, for example, that people might tend to start talking about stock while they should be concentrating on fixtures and fittings. It is essential that damages are not double-counted and that respondents focus on one component of damage at a time. In addition, respondents may need to be reminded to include vehicles, equipment and any other goods kept in yards or car parks, as long as they are the property of the company or organisation concerned.

Further reminders may be necessary regarding the issue of basements or cellars, and how these might affect damage figures and depth/damage curves. The number of storeys in the building also needs to be taken into consideration with respect to the correct proportional levels of damage and flood losses being recorded. For example, for a building with two or more storeys, only damage components on the ground floor (and basement or cellar, if applicable) are likely to be affected by flood waters. Therefore, damage figures need to be proportionate to level of damage for the building as a whole (see the site survey questionnaire on MCM-Online Supporting Information for Chapter 5).

There is no need to obtain exact figures from informants for losses and damage, but realistic rounded estimates of damage and loss potential. For example, this could be to the nearest £1,000 for smaller firms, or the nearest £10,000 for larger organisations, where indicative values of building fabric, stock, etc. can run into millions of pounds.

Fixtures and fittings and moveable equipment are assumed to be mid-way between 'as new' and refurbishment or replacement. Therefore, 50 per cent of 'as new' values, or average remaining values (ARVs), should be taken.

Depth/damage data given by respondents should always be checked for feasibility before leaving the premises (particularly in relation to the points given above), as it is much easier to address potential queries while on site.

The values for each damage component are converted to values per square metre of the buildings in question and entered into an appropriate spreadsheet for the nearest MCM code of the property in question to obtain correct susceptibility levels. The susceptibility to damage for each component is assumed unchanged from the previous research and new depth/damage data are automatically generated based on the revised component values derived from the site survey. In short, valuation of component damages is revised with respect to a specific property and applied to existing susceptibility curves.

5.10.7 Customising depth/damage data: 'mixing and matching' the data

Normally the standard NRP depth/damage data should be employed in appraisals because the data represent national averages for different NRP categories. If a more customised approach is utilised it should be very clearly flagged up in the appraisal, and its impact compared with the use of the standard data should be reported as part of sensitivity testing.

A 'mix and match' approach to damage assessment can be employed. This enables the user to build their own depth/damage curves if they so desire (and where appropriate), from an understanding of data derived from the land use survey of the characteristics of the NRP properties at risk and the six constituent damage components, in tandem with the guidance in this chapter. The average per square metre damage values for components can be changed using MCM-Online, should the need arise in specific circumstances.

5.11 UPDATING

It is recommended that any updating by users of damage data for NRPs provided in this chapter should use the Consumer Price Index (CPI), and this will be done regularly for the data on MCM-Online.

National consumer price indices within Europe, such as the Retail Price Index (RPI), vary considerably in terms of their coverage of goods and services and in their construction. Therefore, Eurostat, in conjunction with the National Statistical Offices of EU member states, developed the CPI as a comparable measure of inflation.

This index is also preferred to the Producer Prices Index (PPI) as the vast majority of NRPs are non-producers. The base date for data in this chapter is January 2013.

5.12 OTHER ISSUES AND LESSONS FROM EXPERIENCE

This chapter has highlighted the complexity of providing accurate data on NRPs compared with residential properties. The range and diverse functions of NRP types, their size, and the varying degrees of susceptibility for each component of damage make it more difficult to construct these data.

Existing research indicates that the type and function of an NRP is not the most important determinant of potential NRP flood damage. Flood depth, property size and precautionary measures all come before the category of NRP in the influence they exert on flood losses. It can therefore be misleading to base too much emphasis on NRP category and to spend time trying to differentiate overall damage values by reference to NRP sub-categories. NRP sub-sector damage means are almost certainly just as – if not more – reliable than using higher-resolution category damage data.

There will inevitably be errors in the data supplied with this chapter. It is not possible to quantify all of these errors, although every attempt has been made to keep them to a minimum. Certain points to note and 'health warnings' remain, for example:

- Data on the extra-long duration and coastal flooding, and on the potential reduction in losses following receipt of a flood warning, need to be treated with some caution. Equally, data on damages below ground floor levels should be treated with care.
- Data relating to potential damage saving related to PLP are tentative and subject to many assumptions. The steps which need to be taken to estimate the damage reducing effects of PLP and flood warnings (related specifically to the movement of moveable equipment and stock) are set out in MCM-Online.
- ARVs on fixtures and fittings and moveable equipment have not been empirically assessed.
- As with residential properties, professional opinion varies as to the precise effect of flood waters on the various components of damage. Susceptibility must therefore be continually updated as more information becomes available.
- Values associated with moveable equipment and fixed equipment and machinery (the latter being included in fixtures and fittings) should be treated with caution.

Error is present in any flood damage data and therefore it is wise in any appraisal (at any scale) to subject these data to sensitivity testing by considering the effect of altering damage susceptibility values and by examining the robustness of the damage estimates to plus or minus 25 per cent assumptions. There is also the possibility of errors arising at the time of appraisal, so care should be taken to avoid these.

It should be apparent that a considerable amount of further research could, and arguably should, be done to improve NRP flood damage data. But any extra research effort should be proportionate to the added benefit it generates, and therefore some avenues are probably not worth exploring much further. The main areas of research which do need to be addressed in the future include:

- exploration of the comparison and 'calibration' of standard NRP damage data employing insurance company data on NRP flood damage claim data which, to date, insurance companies have been most reluctant to release in sufficiently detailed form;
- detailed evaluation of NRP flood damage in a select range of NRP categories in the aftermath of flood events, and the potential for calibration of standard damage data with these data; and
- further examination of the benefits of PLP measures in different types of business premises.

NOTE

1 Kelman (2002) has researched high-velocity floods on coastal residential properties but there is no definitive British work on high-velocity flooding of non-residential properties. In such cases where flood velocity is greater than 0.5m per second there is a potential for failure. However, the probability of failure depends on characteristics such as age and maintenance of structure and the degree of exposure to flooding, all of which are uncertain. Any failure of property would result in damages equivalent to the market value of the property concerned and total or partial loss of contents and equipment (see MCM-Online).

6 Other flood losses

Utility services, schools, hospitals, transportation networks and emergency costs

The previous two chapters have presented data on damage to property, largely in the form of direct damages caused by floods entering these properties and causing damage or loss. This chapter explores losses accrued through the flooding of infrastructure assets (such as electricity sub-stations, roads, schools and water treatment plants). The losses experienced due to the disruption of services and transportation networks will have impacts outside of the flooded area as well as within it, and in some cases the impacts may be geographically far reaching. In addition, this chapter builds on previous work as it considers for the first time appropriate procedures for informing the selection of resilience measures for infrastructure assets.

The approaches presented focus primarily on information – predominantly from case studies and previous floods – on flood losses caused as a consequence of floods disrupting utilities and transport networks as well as other indirect losses sustained. However, where information is available about the direct costs to infrastructure, this has also been provided: with some assets having depth/ damage curves. In no sense are these case studies necessarily typical of UK flood situations; but they do demonstrate the concepts underpinning project appraisals in this field.

Thus Section 6.4 examines the potential losses due to the disruption of public utility losses, and presents case studies from the summer 2007 floods where losses were substantial. Section 6.5.1 examines the disruption of road traffic during flood events and Section 6.5.2 provides an approach for estimating the costs of rail disruption. Section 6.6 presents research results for assessing the costs of disruption to both hospitals and schools and Section 6.8 discusses costs to the Environment Agency and emergency services. One of the key features of the methodologies presented in this chapter is that a proportional approach is necessary and that for many infrastructure assets a site-specific approach to project appraisal should be undertaken when assessing these losses.

6.1 INTRODUCTION

This chapter presents information and appraisal guidance on indirect flood losses (see Chapter 1) caused when floods disrupt utility services and communications, and impose extra costs on those involved in managing floods and in the recovery phases. In assessing these indirect flood losses it is always necessary to ensure that they represent national economic losses, to the nation as a whole, rather than financial losses to the organisations and authorities involved. In the case of transport networks and service disruption this may be quite difficult as in some circumstances service costs may be transferred between service providers. For utilities, therefore, we should consider the extra costs associated with disruption caused by the flooding in addition to any regular costs. Also, when assessing road traffic disruption losses, we should include only the marginal extra costs imposed by the flood event in creating the need for diversions and the consequential extra distance travelled, rather than count the costs of the total journey.

This chapter directly builds on the work presented in the MCM (2005) (Penning-Rowsell *et al.*, 2005a, chapter 6) and moves that research forward, building on lessons from recent flooding and developments since the recommendations of the Pitt Review (Pitt, 2008). Additionally, it develops previous research as recorded in the MCM (2005) chapter, the 'Red Manual' (Parker *et al.*, 1987), chapters 6, 7, 8 and 10, as summarised in the 'Yellow Manual' (Penning-Rowsell *et al.*, 1992, 103–04), as well as deriving some information and insight from the 'Blue Manual' (Penning-Rowsell and Chatterton, 1977, chapter 5). However, all the data herein and on MCM-Online replace any previously published data.

The impacts of the loss of key services such as water or electricity utilities highlights how flood impacts can radiate beyond the immediate vicinity of a flood area. Understanding and quantifying such potential losses can provide a more comprehensive account of flood damages for inclusion in project appraisals. The 'Red Manual' developed a methodology for the evaluation of indirect flood effects, including the damage and disruption to public utilities (electricity and gas supply, water distribution, sewerage systems/sewage treatment and telecommunications). This was revised for inclusion within the MCM (2005).

Evaluating utility damage and disruption has been advanced through Posford Duvivier undertaking research with the Met Office into the impacts of severe weather. This research was carried out between 2000 and 2002 as part of a benefit evaluation of the National Severe Weather Warning Service (Posford Duvivier *et al.*, 2002). These methods are 'pragmatic' techniques for the evaluation of loss and disruption and relate to fog, ice accretion and wind weather events as well as fluvial or tidal flooding. These methods are comparable to those needed to assess flood risk reduction benefits, and are therefore among the bases of the material presented here.

The summer 2007 floods caused widespread impacts to infrastructure in England and Wales and in particular there was an increased focus on securing continuity of service of utilities during flood events and the need to minimise disruption. For instance, the closure of the Neepsend electricity sub-station in Sheffield caused a loss of power to a population of around 40,000 people (Pitt, 2008). In addition, over 42,000 people lost power for twenty-four hours in Gloucester due to the flooding of Castle Meads electricity sub-station, and the closure of the Mythe water treatment plant left a population of 350,000 people without water for over two weeks (Pitt, 2008). These impacts, the subsequent recommendations of the Pitt Review (2008) and the incorporation of flooding into the climate change agenda have recognised the importance of improving the resilience and resistance of infrastructure which needs to be taken into account when assessing the potential for future losses.

The Adaptation Sub-Committee (ASC) (2012) has quantified the proportion of 'critical national infrastructure' (CNI) at risk from significant or moderate flooding ('significant' and 'moderate' here refer to the NaFRA categories of flood risk (Environment Agency, undated d)), suggesting that there are, for example, over 18,000 electricity assets (7 per cent of all national assets) and 1,400 sewage treatment-related assets (25 per cent of all national assets) in these categories in England and Wales. In addition, they recognised that many vulnerable groups are located in at-risk areas, including over 1,000 care homes (with 30,000 residents), and 1,600 schools (7 per cent of all schools nationally) with a total of 500,000 pupils. Therefore, the potential risks and impacts of flooding on these types of infrastructure may well be worthy of inclusion within a project appraisal.

Another significant development since MCM (2005) has been the introduction of partnership funding (see Chapter 1). In some ways this reinforces the need to undertake a more comprehensive assessment of the potential losses in the infrastructure sector (and therefore an appraisal of flood risk management (FRM) benefits) as the asset owners/managers/regulators are viewed as potential contributors to FRM investment funding.

Before detailing specific approaches to appraisal for different sectors it is necessary to identify the general types of losses that accrue when services are disrupted:

1 The physical susceptibility of a plant and/or its supporting networks. This relates directly to the physical damage potentially caused by flood waters and therefore to the performance of the asset.

Henceforth, this will be referred to as the 'direct damages' component of losses.
2. The wider economic impact. This will include the disruption caused to locations both inside and outside of the flood risk zone.
3. Wider, less tangible impacts that affect those living both inside and outside of the flood risk zone.

Each of these losses may impact services and infrastructure to different degrees, the severity of which may depend upon (after MCM (2005) and Cabinet Office, 2010):

- the *dependency* of properties/businesses/other infrastructure served by utility plants and networks;
- the ease and cost of *transferability* of production to sites not affected by flooding (e.g. the degree of redundancy in the system): if a service can be replaced easily by another service, it is said to have high redundancy/transferability; and
- the *duration* of any disruption.

The effort devoted to the assessment of any type of loss should be proportionate to its impact, and although it may be technically feasible to assess the potential losses of many assets, it may not be effective or necessary to do so. Consequently, the potential losses from some infrastructure assets are deemed to be less significant than those from others. The initial step within any project appraisal, therefore, is a prioritisation of the potential losses which should be included for quantification within an economic assessment.

6.2 PRIORITISATION OF LOSSES FOR INCLUSION IN PROJECT APPRAISAL

Undertaking an assessment of the potential losses from the disruption of services and transport networks during flooding can be time consuming and resource intensive. Therefore, careful consideration should be given to deciding whether it is worthwhile to undertake the specific and full appraisal methodologies presented in this chapter, using proportionality methods based on direct damages to include disruption effects as a proportion of total losses (see Table 6.4), or whether the simpler alternative of merely enumerating their relative status (e.g. rank of relative impacts in a qualitative statement in the project appraisal) (Section 6.3) should be followed instead.

Therefore, the 'filtering' process based on that developed by Defra for shoreline management plans (Defra (and Welsh Office), 2001) and presented in MCM (2005) remains broadly appropriate. It has been refined and updated to include increasing efforts by service providers and network owners to understand the susceptibility of their assets to flooding and increase resilience since the summer 2007 floods. Additional guidance about prioritisation in assessing potential economic losses, specific to different sectors, is provided within the following sections and should be viewed in conjunction with the appraisal summary tables (ASTs) within the Environment Agency's FCERM guidance (2010a; 2010b).

Through this filtering process, a shortlist of assets is prepared as candidates for detailed economic appraisal. Those that do not make the shortlist are merely enumerated (Table 6.1) to give qualitative weighting to the appraisal and provide details for any prospective multi-criteria analysis. This process of prioritisation should be applied as a precursor to any initial study or detailed project appraisal to determine the overall importance of utility and infrastructural damages and disruption in the wider context of potential flood damages (or indeed coastal erosion) to residential and non-residential property. For example, it is not sensible to evaluate in detail the monetary losses from the disruption by flooding of a minor roadway when lengths of motorway may be affected at a flood probability threshold of 1:25 years and many residential properties may also be impacted.

The prioritisation process is illustrated in the flow diagram in Figure 6.1 and consists of five steps:

- Step 1: Identify those assets at risk of flooding.
- Step 2: Determine the likelihood of flooding of assets.
- Step 3: Determine the criticality of the assets to flooding.
- Step 4: Utilise a risk matrix for prioritisation.
- Step 5: Assess the impact of resistance and resilience measures.
- Step 6: Quantify losses per high- and very high-risk assests.

For any qualitative analysis, enumeration of the assets at risk of flooding might be done through the use of a

Table 6.1 Enumeration, descriptors and valuation measures to gauge the scale of the infrastructural risk

Infrastructure type	Enumerator/descriptor	Valuation measures
Roads	Length (in km) of motorways, A, B, minor within the floodplain flood thresholds	User numbers (cars, HGV, LGV, PSV) Flood-free alternatives
Railways	Length (in km) of intercity, regional, local, commuters tracks flood thresholds	Number of passengers of different types (commuter, business, other), trains per day
Electricity transmission	KV, lengths, thresholds of flooding of plinth	Supply catchment, population served
Electricity distribution	Size of sub-stations threshold of flooding	Supply catchment, population served
Gas pressure, pumping stations*	Type and number	Supply catchment, population served
Water treatment works	Type and number (pumping station, booster station, etc.) thresholds of flooding	Supply catchment, population served
Sewage treatment works	Type and number (biological filter, activated sludge, pumping station, etc.) thresholds of flooding	Drainage catchment, population served
Telecommunications**	Exchanges, cabinets, pillars, threshold of flooding	Population served

Notes: *Water distribution and supply mains, trunk sewers and gas lines can all but be ignored unless likelihood of fracture is high (e.g. on exposed river crossing or where it might be threatened by the ground around it becoming saturated so that it floats and threats the pipe work joints); ** Redundancy is now high with universal application of mobile telephony. Telecommunication losses and disruption can all but be ignored unless physical damage is likely with high probability within an exchange.

number of different metrics. Examples of these are presented in Table 6.1.

One of the key elements of the prioritisation process is the use of a 'total risk' matrix. This is presented in Table 6.2 and combines Steps 2 and 3 of the process in Figure 6.1 to provide a classification of the likelihood of damage or disruption and the scale of this impact. Impact needs to account for both the scale of the flood loss or disruption (e.g. the number of people affected) and the joint effect of susceptibility, dependency and transferability on the infrastructure system. If, for instance, susceptibility of an asset is low, the need for transferability will also be low. However, if susceptibility is high then the

Table 6.2 Total risk matrix

<table>
<tr><td rowspan="3">IMPACT</td><td>Significant</td><td>Medium risk</td><td>High risk</td><td>Very high risk</td></tr>
<tr><td>Moderate</td><td>Low risk</td><td>Medium risk</td><td>High risk</td></tr>
<tr><td>Low</td><td>Negligible risk</td><td>Low risk</td><td>Medium risk</td></tr>
<tr><td colspan="2" rowspan="2"></td><td>Low</td><td>Moderate</td><td>Significant</td></tr>
<tr><td colspan="3">LIKELIHOOD</td></tr>
</table>

Note: The low, moderate and significant likelihood bands follow the NaFRA likelihood bands (Environment Agency, undated d) and the significant, moderate and low impact categories are defined for each receptor type.

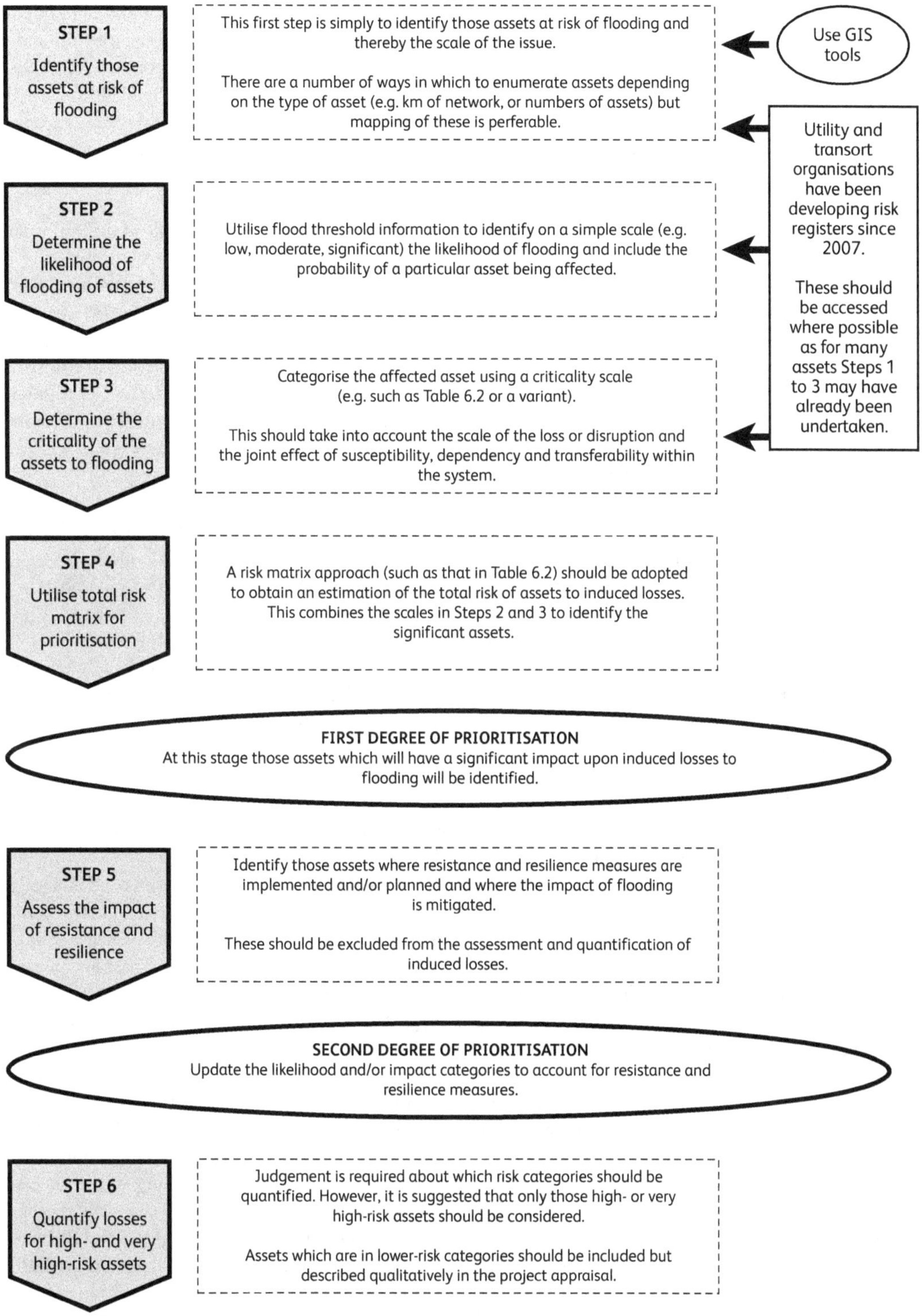

Figure 6.1 Prioritisation process for selecting those assets to quantify potential losses

direct damage potential might be high, but if redundancy in the system is also high (i.e. the transferability of a service is relatively easy) then the potential losses accrued will be low (see Parker *et al.*, 1987, chapter 2).

As shown in Figure 6.1, many utility and transportation organisations have begun a process of assessing the susceptibility of their assets to flooding and other risks and have developed appropriate risk registers. In part, this is a response to the Pitt Review (Pitt, 2008) recommendations, but it has also been prompted by the Climate Change Act 2008 (HM Government, 2008), whereby infrastructure providers are required to prepare reports about how they are assessing the risk and preparing for climate change impacts, including flooding. (These reports can be found at http://www.defra.gov.uk/environment/climate/sectors/reporting-authorities/reporting-authorities-reports/.) Although these provide some information about the general risks to different types of infrastructure, for security reasons risk registers for specific assets are considered to be sensitive information and therefore access is restricted.

However, where this information is accessible to project appraisers, the prioritisation process will already have been undertaken and can be adopted in place of Steps 1 to 3 in Figure 6.1. In most cases organisations have adopted a risk matrix approach similar to that presented in Table 6.2, the results of which can be adopted and utilised in a comparable way.

Deciding upon the specific categories to quantify is subjective, and judgement is required. In most cases any element of utility/infrastructure which is categorised as 'high risk' or 'very high risk' in the matrix should be considered and further investigations undertaken to refine and confirm the risk evaluation. These assets should then be quantified and included within the project appraisal's economic assessment. Other infrastructural elements which are in the lower risk bands should not be ignored, but could be described qualitatively within a project appraisal report.

For example, a 450kV electricity power line crossing a coastal floodplain with its lines located at spring tide level (under a 'do nothing' scenario) would be at 'very high' total risk. A water treatment plant providing water to a large town at risk from flood events with a thirty-year (3.3 per cent) threshold would be at 'high' total risk. Both of these scenarios would involve in-depth quantitative evaluation of losses. In contrast, an electricity sub-station (switchgear) serving a population of over 5,000 but flooding with a threshold of less than 1:200 years (0.5 per cent) would have 'low' total risk, so only a description of the scenario and enumeration of the assets would be necessary in the project appraisal report.

A more comprehensive example of the importance of context and prioritisation process (or 'filtering', as it was then termed) was provided for the Caldicot Levels in South Wales in MCM (2005) (Penning-Rowsell *et al.*, 2005a, 111–115). Table 6.14 in Section 6.4.4, provides an initial investigation of utility and infrastructure elements drawn from the following sections as well as a basic assessment for common utility assets utilising the risk matrix approach.

6.3 LESSONS FROM PREVIOUS FLOODS – PERCENTAGE UPLIFTS FOR ASSESSING POTENTIAL LOSSES

A less resource-intensive approach to assessing the potential losses due to the flooding of infrastructure which has been adopted in recent years by project appraisers has been to 'uplift' the potential direct damages by a percentage factor. These percentage values have been calculated based upon the actual losses estimated from previous national-scale floods in 2000 and 2007 (Penning-Rowsell *et al.*, 2002; Chatterton *et al.*, 2010) and are presented in Table 6.3. This table provides best-estimate damages (in pounds), and where possible presents these as percentages of both total household damages and total household and business damages. Table 6.4 provides the relative losses to each transport and infrastructure sector and the likely split between direct and indirect damages.

The appropriateness of use and transferability of the values to different flood situations will depend primarily on the context of the situation being examined. These values have been generated from some of the most severe flooding experienced in England and Wales in the last fifty years and therefore should not be considered typical of all flood situations. In some cases the range of potential percentage factors is quite wide; for example, road losses range from 2 per cent (in 2000) to 16 per cent (in 2007). Other estimates have a narrower range; for instance, rail

Table 6.3 Damages sustained during the 2000 and 2007 floods and those damages as a percentage of total household damages and total household and business damages

	Best estimate (£million)	% as a proportion of total household damages	% as a proportion of total household and business damages
2000 floods			
Property damages	570	–	–
Road traffic disruption	13	2.3	n/a
Railway network	34	6	n/a
2007 floods			
Households (buildings and contents)	1,200	–	–
Businesses (buildings, contents and disruption)	740	–	–
Electricity	138	11.5	7.1
Gas	<1	0.1	0.1
Water (and waste water)	186	15.5	9.6
Roads	191	15.9	9.8
Rail	36	3	1.9
Telecommunications	<1	0.1	0.1
Schools	49	4.1	2.5

Source: Costs taken from Penning-Rowsell *et al.*, 2002; Chatterton *et al.*, 2010

Table 6.4 Economic damages sustained to selected transport, utility and public services infrastructure during the 2007 floods

	Infrastructure impacts	Total damages (£million)	% of total cost associated to direct damages to property/ infrastructure	% of total cost associated with disruption to goods/ services	% of total losses within the sector*
Utilities	Water and waste water	186	65	35	28
	Electricity	138	6	94	20
	Gas	1	6	94	0
Transport and communications	Roads	191	45	55	29
	Railways	36	29	71	5
	Telecommunications	1	90	10	0
Services	Schools	49	76	24	7

Note: * Only selected infrastructure damages are presented here, therefore percentages do not add up to 100.

Source: Chatterton *et al.*, 2010

losses range from 3 per cent (in 2007) to 6 per cent (in 2000) of total household damages. However, further experience, such as the 2012 floods, suggests that using an uplift factor would not provide a good indication of total losses.

Therefore, these percentage values should not be used blindly as a 'fix' for assessing damages in these benefit categories. Where the likelihood of damages due to the disruption of services or damage to infrastructure is likely to be significant (based on assessment using the risk filter) a site survey is recommended. There are also likely to be situations (such as the flooding in 2012) whereby losses in this sector are unrelated to potential property and business damages (e.g. a flood affecting a major motorway running through a rural area), so these averaged values are not likely to be appropriate. Therefore, these percentage values need to be adopted cautiously when estimating potential losses and they should always accompany a full qualitative narrative and justification.

6.4 POTENTIAL FLOOD LOSSES CAUSED BY DISRUPTION TO PUBLIC UTILITIES

HR Wallingford (2012) has quantified the potential impact of flooding on utility infrastructure in research undertaken for the Climate Change Committee. For each of the utility assets listed below, the total number of service facilities at moderate and significant risk is provided, as is the percentage of the total facilities in England:

- electricity – 10,600 (6.6 per cent);
- gas – 250 (8.3 per cent);
- oil – 40 (11.1 per cent);
- sewerage – 970 (17.4 per cent);
- water – 490 (12.2 per cent);
- information and communication technology (ICT) – 940 (5.7 per cent).

Electricity has the most service facility assets at risk from flooding: a total of 18,500 at all levels of risk. However, these vary in size (and therefore also in the number of people who are affected by their flooding), and only 3 per cent of electricity assets are at significant flood risk. But 10 per cent of sewerage facilities and 7 per cent of water supply-related assets are at significant flood risk (i.e. greater than a 1.3 per cent chance of flooding in any one year) and therefore there is a high potential that these assets may need to be included within a project appraisal.

Utility disruption may in reality also have large interconnectivities with other areas of disruption. In a stakeholder workshop following the 2007 floods, Central Networks (owned by EON) examined some of the potential consequences of a high-impact/low-probability (HILP) event causing power disruptions to the Midlands regions. Some of the outcomes of this workshop (Central Networks, 2008) highlight the interconnectivities between utilities, networks and other services. Thus, some of the consequences of a power outage to this region might be as follows:

- Losing power supplies would affect the British Telecom towers, impacting upon local and national communications.
- Losing power supplies to electronic key pads could trap prisoners within local cells. Also, building alarms work through BT lines and therefore may be activated, meaning additional police resources may be required.
- Potential loss of CCTV to Birmingham city centre creating security and organisational issues.
- Eighty regional railway stations may be affected if power supply is compromised and this may also have an effect on rail signalling and the national train network. This may then have an impact far beyond the local flooding.

Although the methods presented in this sector (and others) present individual approaches to quantifying the potential impacts of flooding on utility services and other networks, these interconnectivities should also be considered; these and systemic impacts might be included within a narrative accompanying the project appraisal.

Since the knock-on impacts of electricity disruption affect most of the other services and networks, our approaches for appraising the impacts of a disruption to this utility are discussed first.

6.4.1 The impact of flooding on electricity power supplies

Due to the serious repercussions of severe power outages, both electricity and gas companies are under a legal obligation to ensure security of supply (Pitt, 2008).

The loss of electricity supply can have serious impacts on the daily lives of individuals and on the continuity of businesses. A short power outage may be considered as merely inconvenient for some; but for others who rely solely on electricity for all cooking and heating, any outage – and in particular an extended one – will significantly impact upon their ability to undertake daily activities. The loss of electricity can have widespread implications as it very quickly impacts on various other sectors, such as mobile communications and the ability of hospitals and other public services to run normally.

Additionally, the impacts of the flooding on electricity infrastructure can extend much more widely than the flooded area. If multiple or critical assets are flooded, electricity can be interrupted for whole cities or regions quite distant from the site of the flooding. A situation like this occurred in the 2007 floods whereby 130,000 households were affected by a disrupted supply in Yorkshire and Lincolnshire (Chatterton *et al.*, 2010), with some experiencing a power outage for five days. Box 6.1 provides a case study of the impacts of the summer 2007 floods on electricity services, when the estimated losses of £138 million accounted for 20 per cent of all infrastructure losses (over 4 per cent of all economic losses).

BOX 6.1 IMPACT OF POWER OUTAGE DURING THE 2007 SUMMER FLOODS

Costs to electricity and gas included the direct costs of curtailed supply, the costs of redirecting supplies to avoid curtailment and, where supplies were cut off, the loss of value to users. Information was obtained from supply company records and personal communication with utility supply managers. Total estimated costs to electricity and gas were estimated at £138 million, with over 90 per cent of this associated with the perceived cost (determined from willingness to pay values) of disrupted supplies to customers (Chatterton *et al.*, 2010). Generally, the electricity supply system remained intact during the flood events. Supplies were maintained by switching power distribution through different parts of the network where this was possible. This flexibility meant that major costs of disruption were averted. Estimated infrastructural damage was approximately £9 million.

Fortunately, a number of potentially very damaging incidents did not happen. The possible breaching of the Ulley Dam near Rotherham posed significant risk to the electricity sub-station at Brinsworth. The consequences of dam failure would have been very serious, placing the electricity and gas distribution networks of Sheffield and surrounding areas at risk.

The greatest estimated cost for power utilities was associated with the impact of deprived supply to customers and losses incurred due to curtailment in the value derived from using power, over and above the price normally paid for it (£0.07/kWh). Based on surveys of electricity consumers, the Department of Business, Enterprise and Regulatory Reform (2007) used an average willingness to pay estimate of £10 per kWh per customer to represent the value of avoiding electricity disruption, confirming the considerable 'consumer surplus' (benefit obtained above price paid) for security of electricity supply.

Two major power cuts occurred. Curtailment of supplies in Yorkshire and Lincolnshire in June affected 130,000 households, which experienced 1.9 million hours of interruptions to services in total spread over a five-day period. This involved a loss of 10,000 MWh and an equivalent (willingness to pay) cost of disruption of £100 million. The failure of the Castle Meads sub-station in Gloucestershire cut off electricity from 12,000 households for approximately twenty hours at an equivalent cost of *c.*£25 million. Additional rota cuts in some areas resulted in a further £5 million loss of value to users. This gives a total of *c.*£130 million for affected parties, mainly in Yorkshire, Lincolnshire and Gloucestershire. Thus the loss of perceived value to users accounted for more than 90 per cent of the total economic costs of flooding in the electricity sector.

Direct damages to electricity infrastructure

Chatterton *et al.* (2010) argue that more than 90 per cent of the total economic damages attributed to the electricity sector can be ascribed to the loss of value to users. However, direct economic damages to electricity infrastructure could also be included for assessment. In 2007, the direct damage to electricity infrastructure was estimated to be in the order of £9 million. This consisted of a number of significant losses, including Neepsend sub-station (£1.3 million) and Thorpe Marsh sub-station (over £1 million).

Depth/damage data are not available for distribution or grid sub-stations. In the latter, damage is potentially highly variable and depends on the configuration and siting of transformers, switchgear and other equipment. Depth/damage data are available, however, for primary sub-stations. Table 6.5 provides a summary of the depth/ damage data and presents estimated values per square metre at different depths of flooding for long-duration (>12 hours) and short-duration floods (<12 hours). Site surveys and further discussions with infrastructure owners would be required to assess the direct damages to grid and distribution sub-stations.

Disruption to electricity distribution

Electrical transmission in England and Wales is solely the responsibility of the National Grid and supply to customers is separate from distribution. Any contacts regarding transmission should be made via the National Grid. Any contacts regarding distribution should be made via the appropriate regional distribution network operators (DNOs) (ENA, 2013).

Assets associated with electricity transmission and distribution relate to stages in the voltage transformation process from super grid (first stage of transformation from the National Grid) to grid, primary and distribution sub-stations, each distributing electricity to fewer and fewer customers (see Table 6.6). The degree of redundancy (and resilience) in this distribution network is crucial to customer supply disruption.

Table 6.5 Average damages per square metre expected from different depths of flooding for a primary electricity sub-station

Depth (cm)		–50	0	25	50	100	150	200
Total damage (£/m²)	Short (<12 hours) duration floods (£/m²)	0	10	850	1,140	2,235	3,590	5,480
	Long (>12 hours) duration floods (£/m²)	0	25	1,410	2,225	3,345	4,695	5,530

Note: Values calculated using an average floor space for the asset of 48m²; see Chapter 5.

Table 6.6 Types of electricity sub-stations

Sub-station type	Typical voltage transformation levels	Approximate number in UK	Typical size	Typical numbers of customers supplied
Grid (super grid)	400kV to 132kV	377	250m × 250m	200,000 to 500,000
Grid (bulk supply point)	132kV to 33kV	1,000	75m × 75m	50,000 to 125,000
Primary	66/33kV to 11/6.6kV	4,800	25m × 25m	5,000 to 30,000
Distribution	6.6/kV11kV to 400/230V	230,000	4m × 5m	1 to 500

Source: ENA, 2009, 4

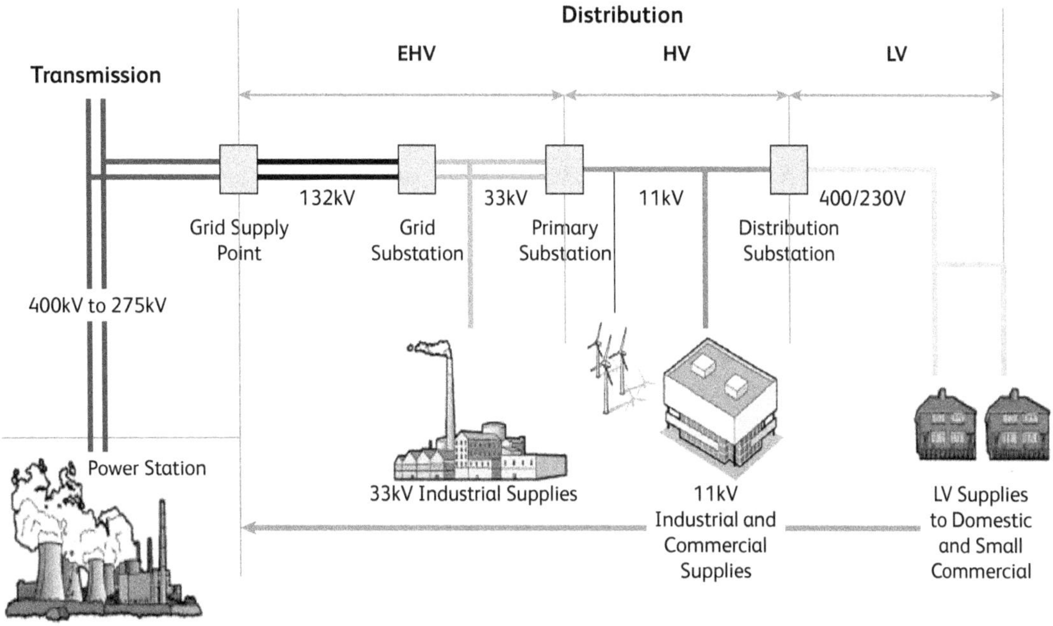

Figure 6.2 Typical electricity supply chain

Source: ENA, 2011, 12

As a rule, the higher up the distribution chain, the greater the degree of redundancy. For example, power stations are not single points of supply and power generation can be switched from other parts of the grid as loading demands. The transmission towers and cables supplying electricity to each stage of transmission are resilient to flood inundation with little, if any, effect on disrupting distribution, unless towers collapse under tidal surge conditions. However, the loss of circuitry through flooding does not usually equate to loss of distribution capacity. Super grid sub-stations have at least duplicate circuitry equipment and therefore have high built-in redundancy. From a physical damage perspective, the switchgear and control equipment are most susceptible to flooding, as transformers are generally raised.

Sub-stations with a single point of supply should be identified. Redundancy will usually decline with proximity to the point of customer supply. Particular attention should be given to grid and primary sub-stations when considering resilience.

If the site is a distribution sub-station and is not supplying critical infrastructure, societal losses should be considered in proportion to the severity of the consequences. However, it is likely that flooding of distribution sub-stations will create limited induced losses as customers supplied are also likely to be flooded and repairs to customer electrics/wiring are likely to be more protracted than repairs to the distribution sub-stations themselves (usually within twenty-four hours). In exceptional circumstances it may be appropriate to consider the induced losses to 'off floodplain' customers supplied by flooded distribution sub-stations.

As part of its periodic price review, Ofgem, the energy regulator, sets standards of service targets for companies which directly relate to the reliability of supply experienced by connected customers. These targets include guaranteed standards, covering:

- payments for customers without supply for more than eighteen hours; and
- payments to customers experiencing frequent interruptions.

However, the Electricity (Standards of Performance) Regulations 2010 (HM Government, 2010c) exclude the payment of compensation to customers as a result of severe weather conditions in some cases, thereby excusing compensation payments for many severe floods and storms.

To minimise downtime of electrical circuitry, network companies (DNOs) are all part of a mutual aid agreement – the North, East, West and South Area Consortium (NEWSAC) – to share resources during electricity emergencies. This agreement was invoked in summer 2007 when CE Electric (now Northern Powergrid) received assistance from Scottish Power and Central Electric. Electricity outage was significantly reduced as a result of these NEWSAC agreements.

The ENA's Engineering Report 1, '*Electricity Networks Climate Change Adaptation*' (2011), indicates a high impact from fluvial flooding and coastal defence breaching and coastal erosion to sub-stations, with an equally high risk to customer services. This risk is expected to grow with climate change into the future.

The loss of supply incidents in Yorkshire and Gloucestershire in 2007 all related to exceptionally high water levels reaching critical depths. These serious incidents – and that in Carlisle in 2005 – were a 'wake-up call' to the electricity distribution and supply industry. The *Electricity Safety, Quality and Continuity Regulations* (HM Government, 2002) – ESQCR Section 3(1) – had not been specific about how to keep sub-station equipment safe from flooding. The 2007 flooding precipitated ETR Report 138 (ENA, 2009) as a common approach to the assessment of flood risk and the development of targeted mitigation measures subject to benefit–cost appraisal. These are presented below and suggest an integrated approach to determining solutions that merge the aspirations of wider community-based schemes promoted by the Environment Agency and other flood risk management authorities and resilience aimed solely at electricity installations.

Following the publication of ETR Report 138 (ENA, 2009), the electricity industry has begun a ten-year programme to improve sub-station resilience. Under the *Electricity Distribution Price Control Review*, Ofgem (2009) set allowances of approximately £110 million for this priority work. As part of this latest price review DNOs have been asked to report on their activity in terms of flood mitigation schemes and flooding site surveys for each sub-station. The majority of DNOs did not complete any schemes in 2010/11 and the overall number of schemes completed was much lower than forecast. Of the 2010/11 forecast activity of forty-four sub-stations where flood mitigation was considered, only fourteen schemes were complete (eleven in just one DNO – Electricity NW Ltd – where one more scheme than forecast was completed). Risk assessments in the majority of the country are progressing very slowly.

Provision of an asset register of sub-station assets is unlikely as the Cabinet Office would classify such information as 'restricted' for public reference because of the possible menace of terrorist activities potentially associated with critical national infrastructure. Each project appraisal should identify assets at risk in dialogue with either the National Grid (Transmission) or the DNOs.

Table 6.7 Risk matrix for electricity sub-stations

<table>
<tr><td rowspan="3">IMPACT</td><td>*Significant: Grid sub-stations serving a population of >50,000*</td><td>Medium risk</td><td>**High risk**</td><td>**Very high risk**</td></tr>
<tr><td>*Moderate: Primary sub-stations with 5,000 to 30,000 population supplied*</td><td>Low risk</td><td>Medium risk</td><td>**High risk**</td></tr>
<tr><td>*Low: Distribution sub-stations with fewer than 500 people supplied*</td><td>Negligible risk</td><td>Low risk</td><td>Medium risk</td></tr>
<tr><td colspan="2" rowspan="2"></td><td>*Low: Flood Zone 1*</td><td>*Moderate: Flood Zone 2*</td><td>*Significant: Flood Zone 3*</td></tr>
<tr><td colspan="3">LIKELIHOOD</td></tr>
</table>

Prioritisation for appraising disruption to electricity distribution

The process of evaluation of the proportionality of an electricity sub-station to the total flood risk of a community is summarised below. This is adapted from ETR Report 138 (ENA, 2009):

- Step 1: Identify with typology all sub-stations in the floodplain under consideration and for which the National Grid or DNO is responsible. Identify the risk based on the likelihood and impact of flooding (adapting the total risk matrix in Table 6.2, for example as in Table 6.7).
- Step 2: Estimate the population served based on length of perimeter (Table 6.8) and the presence of 'V' list critical infrastructure (from DNO) provided in Box 6.2. This is a broad estimate. The results from discussions with National Grid or the appropriate DNO will, of course, be more accurate.
- Step 3: Establish whether the site is within an existing flood defended area and determine the condition of the defences and their actual standard of protection. Where defences are below the Environment Agency's set target condition grade and/or the standard of protection is below the resilience levels set by ETR Report 138 (ENA, 2009), establish the flooding threshold for key parts of the sub-station that will trigger disruption of supply to customers and critical infrastructure (Table 6.9).
- Step 4: If not in an existing flood defended area, establish whether the site has been made resilient against flooding with either permanent or temporary locally installed measures. If the measures are temporary, establish whether the site is in receipt of a flood warning (provided by organisations such as the Environment Agency, Natural Resources Wales and SEPA) and that the erection of temporary measures is practical within the lead-time of warnings provided. If the site is either not

Table 6.8 Estimations of population served based on the perimeter fence length

Sub-station type	Average perimeter fence	Ratio customers to metres of perimeter
Grid (super grid)	900m	250:1
Grid (bulk supply point)	300m	183:1
Primary	110m	136:1

Source: After ENA, 2009

BOX 6.2 'V' LIST CONSUMERS WHERE VITAL SERVICES ARE PROVIDED ACCORDING TO THE REVISED ELECTRICITY SUPPLY EMERGENCY CODE (REVISED JANUARY 2005)

- coal mines;
- major airports and associated control facilities;
- railway operations;
- gas reception terminals; gas storage installations including gas boosting and compression equipment; gas compressor stations and principal development and control sites which provide facilities for the control of gas supply systems and emergency procedures
- licensed electricity generators, including nuclear power stations so as to ensure the safety requirements of their nuclear licences continue to be met; electricity transmitters and distributors;
- essential water and sewerage installations; land drainage pumps;
- hospitals as agreed in consultation with health authorities and NHS trusts;
- ports and docks;
- postal, telecommunications and broadcasting services;
- oil refineries and vital oil pipeline pumping stations.

(DECC, 2006)

Table 6.9 Resilience levels for electricity sub-stations

Resilience levels without reliance or temporary flood protection measures		
Level 1	Flood protection not exceeding 0.1 % AEP	Grid sub-stations unless cost benefit determines Level 2 is appropriate
Level 2	Flood protection not exceeding 0.5 % AEP coastal and 1 % fluvial	Primary sub-stations unless cost-benefit determines Level 3 is appropriate unless where for limited additional marginal costs Level 1 can be justified
Level 3	Other protection levels including provision of limited alternative supplies	Limited to distribution sub-stations as appropriate

Source: Adapted from ENA, 2009

in receipt of flood warnings or these are inadequate to secure the site, consider the flooding thresholds for key parts of the sub-station and the potential for transferring other supply to customers and critical infrastructure.

- Step 5: If no flood intervention measures are in place or planned imminently by the DNO, establish the flooding threshold for key parts of the sub-station likely to disrupt supply to customers and critical infrastructure.
- Step 6: Establish the degree of network interconnection to minimise loss of supply to customers and critical infrastructure. Where transferability of supply is 'seamless', losses associated with flooding are only direct damages to the sub-station.
- Step 7: If the project appraisal is specific to the sub-station, establish the most appropriate flood risk management system, in conjunction with the DNO, to protect the sub-station (see Table 6.10).
- Step 8: Conduct a benefit–cost analysis methodology of preferred solution(s) including an assessment of societal risks. This includes the evaluation of damages by flood depth for critical plant and equipment and the cost of customer supply losses.

'Customer/minutes' loss as a result of flooding during the accounting period including the 2007 floods were only 4.2 per cent of total (for comparison, lightning and wind and gales contributed to over 20 per cent). However, the widespread losses of electrical power extend well beyond the obvious consequences, and the following should be included where possible as part of the assessment of societal losses:

- Loss of traffic lights can lead to traffic gridlock with knock-on effects on the ability of emergency services to respond.
- Mobile telephony will overload and fail within six hours.
- Domestic central heating (even gas fired) will fail and hypothermia is a real threat during winter flooding.
- Disruption of water supplies and sewage treatment and disposal could pose a serious health hazard.
- Petrol pumps, cash tills and cash machines will fail.
- Radio and TV broadcasts will fail to reach the affected population.
- Use of candles and alternative cooking practices could pose potentially serious fire hazard and dangers of asphyxiation.

The appraiser should create a template as to when each of the above benefits is worthy of further analysis. Thus, the ratio of property within the floodplain to those outside the floodplain serviced by a distribution sub-station subject to flooding (within Flood Zone 3) may determine whether induced losses should be assessed. This is probably worthwhile only if more than 50 per cent of the properties served by a flooded distribution sub-station are largely flood free (i.e. in Flood Zones 1 and 2).

- Step 9: Quantify the potential costs due to the disruption of services (Equation 6.1). The MCM (2005) provides an example of the flooding of electricity infrastructure and subsequent disruption in the Caldicot Levels and the same principle holds that the key values to quantify are: likely number of properties impacted and the duration of service disruption.

Table 6.10 Potential intervention measures for electricity infrastructure with their advantages and disadvantages

Intervention measure		Advantages	Disadvantages
Permanent	EA intervention measure (wall or embankment)	Removes flood risk to design flood level	High-cost solution and long 'solution' lead-time
Permanent	Buildings and critical assets protected 365 days per year	Access maintained and all apertures sealed with site not requiring to be manned during flood	Protection generally only effective to a height of 1m above ground level medium-cost solution
Permanent	Barriers and gates at critical openings in perimeter	Access to critical plant maintained	Site needs to be manned during flood incident medium-cost solution
Permanent	Sub-station critical assets raised	Removes risk of flooding to new design threshold	High-cost solution with long construction lead-time
Permanent	Sub-station relocation outside floodplain	Wholly removes flood risk	Very high-cost solution and disruptive to customers during construction
De-mountable	Buildings and critical assets where supports are permanent and panels etc. stored on site	Removes flood risk to design flood level	Medium- to high-cost solution and resource intensive during flooding with potential for operational failure
De-mountable	Site protection where supports are permanent and panels etc. stored on site	Removes flood risk to design flood level	Medium- to high-cost solution and resource intensive during flooding with potential for operational failure
Temporary	Site protection measures installed following flood warning	Low-cost solution	High deployment and training costs for erection, etc.

Source: Adapted from ENA, 2009

Equation 6.1

Estimated costs of disruption = number of households affected by power outage (i.e. total number of properties served by the sub-station or infrastructure affected) × hourly electricity consumption × willingness to pay value (per hour) to avoid power outage × estimated length of disruption to supply.

Ofgem (2011) provides an annual energy electricity consumption value of 3,300 kWh per household (equating to 9 kWh per household per day). The Department of Business, Enterprise and Regulatory Reform (2007) suggests a willingness to pay value to avoid the disconnection of power of £10 per kWh for domestic users. It also suggests that different consumers place a different value on the loss of supply depending upon their use and the time of day affected. Industrial and commercial usage may therefore vary considerably and might be ascertained via a site survey; however, an indicative willingness to pay for industrial users is £35 per kWh (Department of Business, Enterprise and Regulatory Reform, 2007).

In general, most repairs to distribution sub-stations would be effected within a twenty-four-hour period and therefore power would be restored to properties relatively quickly. However, those properties and businesses which are themselves flooded will suffer electricity outages for longer, because the property-level electrical fittings will also need repair. The specific impacts of these outages will depend upon whether residents are in temporary accommodation (and therefore may be less impacted by the lack of supply) or whether they have remained in the affected property. Therefore, in some situations it may be appropriate to estimate the number of households that might be flooded within the area served by a distribution sub-station and remove these from the total number of properties affected by the power outage.

In addition to the cost, although not governed by the regulations, to maintain reputation and as a gesture of goodwill, electricity companies may choose to pay some compensation to customers for the loss of supply, particularly if they are without electricity for a considerable length of time.

6.4.2 Gas infrastructure

As with electricity companies, under the terms of their operating licences, gas companies have a legal obligation to maintain an adequate security of supply (Pitt, 2008). This is enforced through the Gas Safety (Management) Regulations 1996 (HM Government, 1996), and although these are primarily safety related, they do include provisions for ensuring the continuity of supply. Gas supplies are principally piped from North Sea terminals (e.g. Easington, Yorkshire) or from Milford Haven. Gas transmission in England and Wales is solely the responsibility of the National Grid Gas.

Losses due to the damage of gas infrastructure in summer 2007 were limited and estimated to be less than £1 million (Chatterton *et al.*, 2010). As touched upon in Box 6.1, if the Ulley Dam had failed, the gas installation at Guilthwaite – one of three installations supplying gas to Sheffield – would have been flooded. However, the advanced warning of the threat of the dam's breach permitted an engineering solution to be employed in order proactively to mitigate the risk of disruption to the Sheffield supplies. A near miss occurred at Toll Bar, Doncaster, whereby water threatened to flood the primary gas main. This was ultimately prevented, but if it had occurred, the supply to 12,000 homes would have been threatened and millions of pounds' worth of damages would have been caused (Chatterton *et al.*, 2010). Indeed, mitigating potential impacts and maintaining supplies were key features of National Grid's response to the flooding in 2007. Distribution teams worked well in identifying issues of concern (such as at Kilham, Yorkshire, and flooded areas around Sheffield) and identifying ways in which supply could be maintained.

Since the 2007 floods, as part of its adaptation to climate change, National Grid Gas has been working to increase the resilience of its assets to flooding, including activities such as reinforcing river banks and further research about what the impacts of flooding are on pipelines and other equipment (National Grid Gas, 2010). Flooding is a risk considered and categorised within its adaptation risk assessment process, which has examined potential climate change impacts on each type of key gas infrastructure asset. Risks are categorised (on a four-point scale) according to the degree of material risk they pose to different assets and how robust business process and/or action plans are to deal with these risks.

For flooding, in the majority of cases it is considered that continuity of supply either would not be threatened or the level of risk is low in terms of the damage likely to be sustained. A gas compressor station was considered to be at risk of flooding, but supply was not thought to be threatened even if it were inundated. One other element of the network which scored an 'amber risk' (i.e. a score of three out of four) was the National Transmission Pipework (gauge pressure of approximately 70 barg). This was thought to be at risk as there is potential for these pipes to float if the ground around and above them is flooded. However, the main concern is that there is insufficient information about these risks, so further research is required to be able to quantify their susceptibility to flood water fully. Therefore, these assets should be considered when appraising the potential impacts of flooding on gas infrastructure.

Overall, the pressurised gas network is far more resilient than electricity distribution. The main concern remains the pipework and its pressure gauges where the ingress of flood water would necessitate a mass purge of the affected pipeline, similar to the situation that caused concern at Toll Bar in 2007. Chatterton *et al.* (2010) indicate that if water did enter pipes, the damage and disruption could be catastrophic.

Despite these extreme risks, following 2007, National Grid Gas was reported to be generally more concerned with the physical direct damages to infrastructure because, except in the case of multiple assets being flooded, there is generally a high degree of transferability within the system (Chatterton *et al.*, 2010). National Gird Gas (2010, 29) reports that 'if a site is flooded it is highly unlikely that in the short to medium term it will impact on the delivery of gas'. A greater risk to consumer supply is posed by equipment or communication failure – i.e. through an interruption in electricity or telecommunications – rather than through direct flooding of the gas network. This reconfirms that electricity is a priority infrastructure service for consideration in appraisal.

Should a gas installation be located in a floodplain under investigation then discussions with National Grid Gas or other distributors may be appropriate on the lines of the step-by-step guide for electricity (Section 6.4.1). In those situations where further analysis of a loss of gas supply is required, the calculation provided for electricity may be adopted (Equation 6.1). The Department of Energy and Climate Change (2012) estimates the annual gas energy consumption for the average UK home to be 18,000 kWh which can be used within the relevant disruption cost estimation.

6.4.3 The impact of flooding on water and waste water assets

The floods in summer 2007 highlighted the susceptibility of the water network to 'single points of failure' that can have widespread impacts on homes, businesses and industry in both the areas flooded and those considerable distances from the floods (Pitt, 2008, 237). It was estimated that up to 350,000 customers lost their drinking water supply for up to seventeen days (Pitt, 2008) and some institutions (e.g. some hospital buildings in Gloucestershire) were still reliant on bottled water some five weeks after the flooding (Whiteley, 2008).

It is useful to establish the geographical boundaries of the water supply and sewerage operating companies and the water supply-only companies that operate separately within these boundaries to facilitate contact (maps can be found at Water UK, 2007).

Potential direct damages to water infrastructure

Sewage treatment works (STW) are included as a non-residential property (NRP) asset within Chapter 5. Table 6.11 provides a summary of these depth/damage data and presents estimated values per square metre of the STW's assets at different flooding depths for long-duration (>12 hours) and short-duration floods (<12 hours). These provide indicative values only, however, and therefore site surveys or discussions with the infrastructure owner are recommended to verify these estimations and to appraise the potential damages to water supply infrastructure which are not included as depth/damage curves in Chapter 5.

The five-year price reviews and establishing resilience

Regulators have a key role in supporting the UK's resilience agenda, and the Pitt Review (Pitt, 2008, 265) recommended that this was recognised by placing a specific duty 'on economic regulators to build resilience'. One of the ways Ofwat (the water regulator) regulates is to set the price, investment and service 'package' that customers receive. This includes setting limits on the prices the companies can charge their customers. When Ofwat does this it must

Table 6.11 Average damages per square metre expected from different depths of flooding for a sewage treatment works

Depth (cm)		–50	0	25	50	100	150	200
Total damage (£/m^2)	Short (<12 hours) duration floods (£/m^2)	0	5	30	40	70	100	145
	Long (>12 hours) duration floods (£/m^2)	0	10	45	70	110	145	180

Note: Values calculated using an average floor space for the asset of 4,950m^2; see Chapter 5.

balance the interests of consumers with the need to make sure the sector can finance the delivery of water and sewerage services. Ofwat also needs to ensure the water companies are able to meet their other legal obligations, including their environmental and social duties. The Pitt Review's recommendations following the 2007 floods in England formed an essential part of these obligations and duties (Pitt, 2008). Box 6.3 provides an example of Severn Trent Water and its activities to increase asset resilience to flooding.

Ofwat sets a price review every five years. The last price review was in 2009. Some companies proposed investment to increase the resilience of their service to natural hazards (largely flooding). In Ofwat's Final Determinations (Ofwat, 2009) it allowed £414 million across the industry for network and asset resilience schemes (to be invested between 2010 and 2015). This will mean that by 2016, 9.6 million people will benefit in this respect from increased resilience to extreme hazards, such as flooding.

While the companies' business plans proposed resilience schemes at specific sites, the companies may decide, via these plans, to prioritise activity differently in order to achieve the service output in a more efficient manner. Therefore, although business plans will have a rigorous process for delivery, with rewards for early achievement and penalties for failure, the resilience schemes companies provide in reality can change from those set out in their business plans. Accountability for implementation of resilience measures lies firmly with the company. Therefore, we advise FRM appraisers to engage directly with the company whose assets are deemed to be at risk. It would be counter-productive to engage in community-based mitigation schemes where a water utility would gain benefit if individual plans were in train for resilience by water companies, and vice versa. Appreciation of preferred options for flood mitigation, alignment of funding and a full appreciation of community benefits are fundamental to appraiser/Environment Agency/water company dialogue.

BOX 6.3 INCREASING RESILIENCE: AN EXAMPLE FROM SEVERN TRENT WATER

Severn Trent Water supplies water services to around 8 million people, largely in the English Midlands. The flooding in Gloucestershire in 2007 highlighted inherent risks to its network which are not acceptable to the company or its customers. Setting out a twenty-five-year vision, in its Strategic Direction Statement (SDS) (Ofwat, 2013) the company identified the need to improve the resilience of its strategic network to reduce the risk of customers losing their water supply from all potential causes, including flooding.

The improvements proposed in this plan were for the provision of an alternative piped source of water to all communities larger than 20,000 people, as part of a prioritised programme of improvements over the next ten years. Experience during the Gloucestershire flooding showed that 20,000 people is the maximum number that can be reliably supplied with bottled water and bowsers in the event of an interruption to supply.

- 1.4 million people who are currently dependent on a single source (nearly 20 per cent of Severn Trent's customers) will benefit from an alternative source if their normal source of water fails.
- A further 0.6 million people (8 per cent of customers) who are currently dependent on a single pipe will be provided with an alternative piped supply.

This programme is designed to comply with the Security and Emergency Measures Direction (SEMD; DETR, 1998), which requires all water companies to maintain supplies in the event of a serious emergency. Severn Trent will therefore reduce the likelihood of failure by:

- protecting ten treatment works at risk of flooding from a 1:200-year flood event;
- removing a single point of failure from one critical site; and
- providing resilient power supplies at twenty sites.

For the 2014 price review, Ofwat is focused on setting outcomes and giving companies greater ownership. To help companies when considering their resilience, it has worked with the industry to set some best-practice resilience planning principles. Appendix B in *Resilience: Outcomes Focused Regulation, Principles for Resilience Planning* (Ofwat, 2012, 67) covers themes which emerged out of a survey Ofwat conducted with water companies in England and Wales, including how the companies can ensure continuity of service.

This was conducted to improve understanding of different company views on, and approaches to, resilience planning. Out of the twenty-one companies contacted, seventeen responded. Several of these companies focus efforts initially on identifying critical assets before undertaking more detailed analysis of risks for the more critical assets. One respondent noted the need for the assessment of resilience in different parts of the system to be more 'joined up'. For a number of companies, the focus for resilience has initially been on water assets, but several respondents stated their intention to look more closely at the resilience of waste water assets in future. One of these respondents considered that, for the waste water service, resilience is in most cases likely to be most effectively delivered through better preparedness for events, rather than implementation of costly capital schemes.

Some companies have developed indicators for defining and measuring resilience, as follows:

- the number of customers affected by a service outage;
- the number of customers served by only one treatment works;
- the number of customers that cannot be served by another source;
- the number of single-feed district meter areas (DMAs);
- the risk of property hours lost (combining likelihood, duration and population affected);
- the supply (DG3) interruptions of longer than twelve hours' duration; and
- the risk of flooding.

One company primarily investigates issues where a population greater than 10,000 is at risk of loss of supply for a significant period of time and how services can be maintained or restored in such cases. It does not measure resilience *per se* but suggested an approach linked to risk or the likelihood of failure; however, it was also considered that this would be far from straightforward.

Cost of severance of water supply and the Security and Emergency Measures Direction (SEMD)

Chatterton *et al.* (2010) indicated that flooding of Severn Trent's Mythe water treatment works led directly to the loss of piped potable water supply to *c.*350,000 consumers (138,194 properties) in the Gloucester area from 22 July 2007 for up to seventeen days (Pitt, 2008).

The overall costs to Severn Trent Water as a result of flooding at Mythe were estimated by the company at £29.6 million. This included the operational distribution of tankers and bowsers and quality testing some 23,000 samples after connection was resumed. The cost of bottled water supply and distribution to 350,000 consumers was estimated at £25 million – approximately £4.50 per person. The deployment of 1,400 bowsers to 1,100 locations was the largest of any incident of its kind in the UK. This illustrates the scale of damage and disruption when a major single source of supply utility serving a wide hinterland of users is flooded.

However, with respect to compensation to water users, in these exceptional circumstances payments under Ofwat regulations were not made. Using Ofwat (2008) guidance on the cost imposed on households when water is cut off, at £10/household/day, this would equate to a total of £23.5 million. However, this compensation agreement is waived in extreme weather conditions. In lieu, payments of *c.*£3.5 million were made by the water industry (Severn Trent) to a number of local organisations associated with managing the flood event, in recognition of the burdens imposed by disrupted supplies.

Water UK (2011) gives an overview of the *Security and Emergency Measures Direction* 1998 (DETR, 1998, as revised by Defra/Welsh Assembly Government, 2008), which requires all water companies to maintain supplies in the event of a serious emergency. This minimum requirement may be supplied through bowsers or tankers (typical costs of bowsers are available from the water companies; e.g. Bristol Water (2012)).

In 2004, the Water UK Council established a mutual aid protocol for all members to ensure delivery of water by companies during an emergency. This protocol (amended after 2007) includes agreements to share emergency equipment and to support affected member companies during incidents. This enhances the resilience and contingency options available to the industry as a whole. However, loss of supply for long periods without access to toilets or showers suggests that valuing alternative supplies is only 'the tip of the iceberg' of the real value of maintaining alternative supplies. Ofwat (2012, 38– 39) covers willingness to pay and includes a reference to the UKWIR (2011) report which should be followed where lack of resilience may present failings in the water supply chain. Possible water groups for consultation should include:

- all customers;
- water sensitive customers (e.g. hospitals);
- vulnerable groups;
- business customers;
- Customer Challenge groups/ Wales PR14 Forum; and
- representative bodies (CC Water as the statutory water consumer representative, the Environment Agency, business customer representatives, local authorities and minority groups).

Companies should engage customers on wider resilience issues that would help to improve understanding and identify priorities. Areas for engagement could include:

- exploring what customers understand by resilience and resilience planning;
- understanding their preferences for different levels of resilience and willingness to pay;
- understanding the extent to which customers value security and safety of supply (service benefits), value for money and affordability when it comes to resilience;
- understanding customer preferences for external benefits (environmental and social) whilst recognising any overlaps with other qualitative and quantitative methods when assessing such benefits;
- identifying customers' long-term aspirations for resilience and the appropriate pace for introducing investment;
- communicating any environmental and other requirements met through resilience planning and ways that have been considered for minimising costs of intervention options; and
- identifying which communication channels and mechanisms customers prefer for discussing and learning about resilience planning.

Severance of sewage treatment

Generally, sewage treatment and pumping facilities are not as susceptible to flooding as water supply facilities. Although there were failures of sewage treatment plant and pumping stations in 2007 (300 suffered some degree of flooding), most of the repairs were not large enough to show up as significant costs in the water companies' accounts.

Yorkshire Water suffered worst from the flooding of waste water treatment plant and associated infrastructure. Some £50 million of the £70 million estimated costs incurred by the company were within Hull, where much of the plant is located below high tide level. As a result of flooding, water treatment for services for 2,500,000 people (just over a million homes) was affected for an average of two days.

Box 6.4 provides an extreme example of the potential impact of flooding on waste water services undertaken as part of the TE2100 project (Chatterton, 2008).

Water related assets at risk: appraisal suggestions

The Cabinet Office (2011, 28) suggests a benchmark that 'as a minimum essential services provided by Critical National Infrastructure (CNI) in the UK should not be disrupted by a flood event with an annual likelihood of 1 in 200 (0.5 per cent)'. The guide goes on to indicate that the costs and benefits of individual projects should be considered when deciding which projects to fund and whether the benchmark can be achieved. The benchmark does not apply to other infrastructure that is not designated as critical national infrastructure.

There is a fundamental difficulty in creating a definitive list of water supply and sewerage infrastructure at risk from flooding (or any critical infrastructure, e.g. electricity sub-stations, for that matter). Any reference to sites/assets being critical infrastructure indicates that the asset is important and

BOX 6.4 BECTON SEWAGE TREATMENT WORKS, LONDON: AN EXTREME EXAMPLE OF THE EFFECTS OF TIDAL FLOODING

As part of the TE2100 project, extreme flood scenario testing was undertaken on the capital's major infrastructure and high-profile public buildings. The study at Becton illustrates the grave consequences of a tidal surge incident in the event of failure of London's tidal defences.

The waste water treatment works

A tidal surge inundating Becton STW would be an unprecedented environmental disaster. The works consist of sixteen primary treatment tanks, sixteen aeration lanes (with 32,000 aeration cones), final settlement and sludge incineration. Most of the tanks and treatment processes would be secure until major flooding, as the works have built-in resistance and resilience against minor inundation. However, under extreme tidal conditions inducing one or more breaches of flood defences, the saltwater contamination from the tidal surge would totally destroy the activated sludge and aeration process on such a massive scale that re-seeding would not be a practical treatment and the processes would be disrupted for up to five years.

Although low-lying City sewers would be inundated anyway, much of the sewage would come from areas beyond the flood risk zone and the aeration treatment would be bypassed, depositing 180 tonnes BOD (untreated sewage) per day into the Thames (along with 106 tonnes BOD from Crossness, south of the river, serving 2 million people south of the Thames, and similarly affected).

The Thames would quite simply 'die' and become similar to the pre-Bazelgette Thames of the nineteenth century, but on a far greater scale. Untreated sewage would accumulate and be moved up and down the Thames with every tide. The aeration tanks would become settling tanks but would be unable to cope with effective removal of BOD and ammonia. The conditions, especially in summer, would be very serious.

The corrosive salt would mean replacement of air blowers at a cost of £3,200,000 but these would require specialist manufacture with delivery taking many months. Sludge incineration and energy generation would cease but the effect of these activities is minor in relation to the bypass of raw sewage to the Thames.

The de-salination plant

North-east London will not be able to sustain a continuous water supply (presently from west London) in the near future without the new de-salination plant to be commissioned. A tidal surge would knock out the plant with serious damage to the mechanical and electrical equipment and HV transformers, with up to a two-year recovery period anticipated. High-quality stainless steel, which has a fourteen-month delivery and installation cycle, would need replacing and the whole plant cleaned and scoured to counter contamination and threats such as *E. coli*. Once flooded, the plant will be directly exposed to contamination from raw sewage from the STW plant. The cost of this is put at £50 million to £60 million following an extreme tidal surge, with likely similar re-commissioning costs for less serious inundation as much of the mechanical and engineering plant is below ground level.

(Chatterton, 2008)

could provide useful targeting information for those with a terrorist intent. Such information may require a protective marking (e.g. 'restricted'). Consequently, an appraiser must rely on the often incomplete data provided by the Environment Agency's National Receptor Dataset (NRD) as a starting point and follow up the results with contact with the water supply and sewage treatment providers themselves.

The process of evaluating the contribution of a water supply or water treatment works to the total flood risk of a community is similar to the step-by-step procedure outlined for electricity installations (see Section 6.4.1) but with a filter as used by NaFRA to account for proportionality.

- Step 1: Identify the risk based on likelihood and impact of flooding (adapting the total risk matrix in Table 6.2). Table 6.12 provides an adapted total risk matrix for sewage treatment and Table 6.13 does the same for water supply works.
- Decision Filter: Consider the next steps only for 'high' and 'very high' risk cells.
- Step 2: Establish whether the site is within an existing flood defended area and determine the condition of the defences and their actual standard of protection. Where defences are below the Environment Agency's set target condition grade and/or the standard of protection is below the optimum design standard proposed by the Environment Agency, establish the flooding threshold for key parts of the works likely to disrupt supply to customers and critical infrastructure (see 'V' list established for electricity and provided in Box 6.2).
- Step 3: If not in an area already benefiting from flood risk management measures, establish whether the site has been made resilient against flooding by the water company through either permanent or temporary locally installed measures. If the measures are temporary, establish whether the site is in receipt of flood warnings and that erection of temporary measures is practical within the lead-time of warnings offered.
- Step 4: If the site is not in receipt of flood warnings or these are inadequate to secure the site, consider the flooding thresholds for key parts of the works and the potential for transferring other supply/treatment capacity to customers and critical infrastructure.
- Step 5: If no flood intervention measures are in place or planned imminently by the water company, establish the flooding threshold for key parts of works likely to disrupt supply to customers and critical infrastructure.
- Step 6: Establish the degree of network interconnection to minimise loss of supply/treatment to customers and critical infrastructure. Where transferability of supply is 'seamless', losses associated with flooding are only direct damages to the works.
- Step 7: Establish the most appropriate flood risk management system in conjunction with the water company (see Table 6.10 for examples established for electricity, which provides a starting point for these).
- Step 8: Apply a conventional benefit–cost analysis of preferred solution(s), including societal and

Table 6.12 Risk matrix for sewage treatment

IMPACT	*Significant: >30,000 cumecs effluent dry weather flow*	Medium risk	**High risk**	**Very high risk**
	Moderate: 5,000 to 30,000 cumecs effluent dry weather flow	Low risk	Medium risk	**High risk**
	Low: <5,000 cumecs effluent dry weather flow	Negligible risk	Low risk	Medium risk
		Low: Flood Zone 1	*Moderate: Flood Zone 2*	*Significant: Flood Zone 3*
		LIKELIHOOD		

Table 6.13 Risk matrix for water supply

IMPACT	*Significant: >20,000 population supplied or 'V' list customers (see Box 6.2)*	Medium risk	**High risk**	**Very high risk**
	Moderate: 5,000 to 20,000 population supplied	Low risk	Medium risk	**High risk**
	Low: <5,000 population supplied	Negligible risk	Low risk	Medium risk
		Low: Flood Zone 1	*Moderate: Flood Zone 2*	*Significant: Flood Zone 3*
		LIKELIHOOD		

environmental risks. This includes the evaluation of damages by flood depth for critical plant and equipment and the cost of customer supply losses using cost of water under SEMD provision as a minimum cost, supplemented with willingness to pay data/surveys, as appropriate (similar to Equation 6.1). MCM (2005) (Penning-Rowsell *et al.*, 2005a) provides an example of appraisal for the Newport Waste Water Improvement Scheme, which highlights the process that could be applied; however, the risk matrices suggested in Tables 6.12 and 6.13 should replace the approach adopted in the 2005 example.

6.4.4 Summary of utility disruption costs

Table 6.14 provides a summary of the relative importance of utility and infrastructure elements, and utilising the risk matrix approach provides a basic examination of the susceptibility, dependency and transferability of these different infrastructure assets. Within this table a further variable – scale – has also been introduced to distinguish between whether few people/properties are affected and whether the effects of flooding would impact large numbers. This information is collated from information and methodologies in the previous section and provides a very broad overview of those assets whose disruption is potentially worth quantifying.

Although the information in Table 6.14 and in the following assessment provides a good indication of those infrastructure assets that are likely to require further investigation, appraisers should not automatically dismiss those which score lower. The table provides only a qualitative indicator of the proportionality of including the investigation of an infrastructure asset within an appraisal. There may be particular circumstances whereby an asset assumes greater significance; for instance, when it is likely to be flooded frequently or when a disproportionate number of people may be impacted by flooding. Therefore, appraisers should undertake their own filtering approach using the risk matrix in Table 6.2, and if in doubt should speak to the infrastructure owners/providers to determine their criticality.

In summary, we suggest that full monetary quantification of utility damages/losses is mainly required (i.e. proportionate) and likely to contribute significantly to the present value of benefits in the following situations:

- Tidal inundation of electricity transmission lines greater than 132kV unless flooding thresholds are less frequent than 1:75 years (1.3 per cent).
- Tidal inundation of electricity transmission lines of less than 132kV but only if flooding is more frequent than 1:25 years (4 per cent).
- Flooding of sewage treatment works when the risk of flooding is more frequent than 1:75 years (1.3 per cent) and the effluent dry weather flow is greater than 5,000 cumecs.
- Flooding of sewage treatment works when the risk of flooding is moderate (i.e. more frequent than

Table 6.14 Summary of impacts for utility/infrastructure components assuming that there are no flood resilience measures or actions taken to increase redundancy

Utility/infrastructure	Susceptibility	Dependency	Redundancy/ transferability	Scale 1 = few 2 = many 3 = very many	Total likely impact
Electricity transmission and distribution					
> 132 kV (fluvial)	Low	High	Low	3	Low
>132 kV (tidal) [1]	High	High	Low	3	High
<132 kV (fluvial)	Low	High	Low	2	Low
<132 kV (tidal)	High	High	Low	2	Medium
Grid (super grid) sub-station	High	High	High	3	Medium [2]
Grid (bulk supply point) sub-station	High	High	Medium	3	Medium [2]
Primary sub-station	High	High	Medium	2	Medium [2]
Distribution sub-station	High	High	Low	1	Medium/ low [3]
Gas transmission					
Gas pressure stations	Medium	Medium	Low	1	Low
Gas pressure stations	Medium	Medium	Low	2	Medium
Water and waste water treatment					
Sewage treatment works	Medium	High [4]	Low [5]	1	Medium
Sewage treatment works	Medium	High [4]	Low	2	Medium
Water treatment works	High	High	Medium [6]	1	Medium
Water treatment works	High	High	Medium [6]	2	High
Water pump stations	High	High	Low	1 and 2	Medium
Telecommunication systems					
Connection points – cabinet	Low	Medium	High	2	Low
Telecommunication connection points – pillars	Low	Medium	High	1	Low [7]

Notes: [1] Transmission lines across a coastal floodplain are likely to collapse during a severe tidal inundation. Also, if a transmission line is within an area flooded for any considerable period of time, maintenance of that structure will be difficult and the integrity of the asset threatened. [2] The absolute impact will depend upon the specific site plan and the location of equipment within it; in particular the positioning and height of the switching gear and transformers. [3] This is 'low' in the situations whereby the properties the sub-station is servicing are also flooded as the sub-station will be repaired before the houses. It is 'medium' in situations where the sub-station is servicing properties which remain dry (i.e. 'unflooded' properties). [4] Environmental damage through treatment bypass might be as important as physical damage. [5] A reminder that in this circumstance the redundancy remains low – unless measures have been taken post-Pitt to increase the transferability of the service. [6] Depends upon locality. [7] Redundancy of landline facilities is extremely high with saturation coverage of mobile telephones.

1:200 years (0.5 per cent)) and the effluent dry weather flow is greater than 30,000 cumecs.

- Flooding of water treatment works when the risk of flooding is more frequent than 1:75 years (1.3 per cent) and the population affected is greater than 5,000.
- Flooding of water treatment works when the risk of flooding is moderate (i.e. more frequent than 1:200 years (0.5 per cent)) and where the dependent population is significantly large (i.e. >20,000).
- Flooding of electricity grid sub-stations (including super grid and bulk supply point installations) when the risk of flooding is moderate (i.e. more frequent than 1:200 years (0.5 per cent)) as these serve more than 50,000 (and up to 500,000) customers.
- Flooding of primary and grid sub-stations when the risk of flooding is more frequent than 1:75 years (1.3 per cent), thereby serving a dependent population of greater than 5,000 people.

6.5 POTENTIAL FLOOD LOSSES CAUSED BY DISRUPTION TO TRANSPORTATION

The flooding of roads and railways can cause significant damage and disruption for travellers. The Environment Agency (2009a) estimates that 10 per cent of main roads and 21 per cent of railways are at risk from flooding; with two-thirds of both these roads and railways being at significant or moderate risk of flooding. The floods in both 2007 and 2012 saw substantial impacts on transport networks. Indeed, south-west England in November and December 2012 experienced significant transportation problems due to flooding which virtually cut off some areas from the rest of the country for several days. This section provides case studies of flooding to both road and rail and provides approaches to quantifying these losses in relation to flood risk management scheme appraisals.

6.5.1 Losses due to the flooding of roads

Damages to roads and resulting road traffic disruption can result in one of the highest non-property losses from flooding. Since the publication of the MCM (2005) (Penning-Rowsell *et al.*, 2005a), lessons have been learned from many instances of road traffic disruption due to flooding, and Box 6.5 provides an example of the road losses sustained due to flooding in 2007. Responsibility for roads falls between the Highways Agency, for major roads and motorways, and local authorities, for local roads, so it may be necessary to consult with these organisations when considering quantifying the potential losses from roads, depending on the types of roads affected.

A number of approaches to the estimation of the impacts of road traffic disruption are presented. These vary in their level of complexity and therefore the appraisal resources that they require. The selection of which method to utilise depends upon the scale of the likely disruption; where losses are likely to be significant, the more in-depth and detailed approaches are recommended. Before further details are provided, appraisers should consider the direct damages to infrastructure, discussed below.

Direct damages to road infrastructure

Road reconstruction costs following flooding will vary depending upon the type and scale of damage, the type of road impacted and the location of the required repair. Unit reconstruction costs for resurfacing a local road range between *c.*£15/m^2 for a quiet road to *c.*£50/m^2 for a very busy road (which might require a thicker surface layer and roadworks to be carried out at night or during off-peak times, incurring overtime costs) (Hertfordshire County Council, undated; Conway County Borough Council, 2013). If severe damage occurs or other road structures, such as bridges, are affected, costs may be considerably higher and will need to be evaluated on a case-by-case basis. The Highways Agency should be contacted separately for roads under their management as they will have different unit costs for repair and reconstruction.

Losses due to road traffic disruption

Estimating traffic disruption based on previous flood events is inadvisable as the severity of disruption can vary dramatically. Traffic disruption cost estimates for the summer 2007 floods, for example, highlighted a large range of between £22 million and £174 million (see Box 6.5; Chatterton *et al.*, 2010). As a percentage

BOX 6.5 LOSSES DUE TO THE FLOODING OF ROADS IN THE SUMMER 2007 FLOODS

The 2007 floods in England and Wales caused damage to roads and related infrastructure, requiring capital expenditure and rescheduled extra maintenance. The closure of roads resulted in extra costs due to congestion, and increased travel time and distances. Information in our research was obtained from a variety of sources: mainly insurance companies, the Highways Agency and local government authorities (LGAs). Estimated costs were in the order £191 million, with approximately half owing to physical damage and half to traffic delays. However, there is much uncertainty about the estimate of traffic delay and redirections.

Sixteen local authorities accounted for 86 per cent of the total road damage costs to all LGAs (*c*.£75 million out of a total of *c*.£85 million). Road damage costs in Sheffield alone were *c*.£15 million. Gloucestershire's roads were worst affected, with repairs of some sort necessary at 420 locations. These included thirty bridge repairs, fifteen slope stabilisation works and 135 major surfacing repairs. An early and detailed estimate made at 303 locations in the county suggested an estimated total repair bill of £34.2 million.

Approximately 50 per cent of emergency capital maintenance work in Gloucestershire related to unclassified rural roads, of which 42 per cent of this total related to grouped small schemes. It is noted that motorways suffered proportionately less damage due to higher standards of design and maintenance when compared with minor roads. Thus, virtually all costs associated with damage to roads and related infrastructure, such as bridges and culverts, were incurred by LGAs and identified from LGA accounts. The Highways Agency reported extra costs due to the flooding of only £33,000.

With respect to disruption, six motorways (M1, M4, M5, M18, M40 and M50) were closed at some time during the flooding. The M1 (Junction 31 to 34) was closed for forty hours because of the danger of dam breach at the Ulley reservoir (Rotherham).

Disruption costs are difficult to assign but have been estimated by the Highways Agency at £2.3 million for the M1 incident alone. Data on the type and magnitude of traffic flows were used to determine the cost of the extra time and distances travelled due to congestion at given 'nodes' on the road network. Interpretation of flood maps suggested obstructions at 200 flood/transport nodes. This gave a mean figure of direct traffic disruption of £98 million, but the range of possible values in this estimate is very large – between £22 million and £174 million, depending on accounting assumptions – leading to considerable uncertainty.

(Chatterton *et al.*, 2010)

of direct damages, traffic disruption for the 2007 floods was approximately 10 per cent of property damages (using the highest estimates for both), whereas for the autumn 2000 floods, this figure was nearer 2 per cent (Penning-Rowsell *et al.*, 2002).

The chief determining factors for traffic disruption costs include flood duration, the scale of the area affected (and therefore the number of roads) and specifically which roads are impacted (whether a flood causes significant knock-on effects to other parts of the network). Therefore, estimates based on previous events could lead to drastically over- or underestimated figures as losses are highly location specific.

Consequently, the three situations when the calculation of traffic disruption costs are most likely to be justified are when any of the following (or a combination of the following) is present:

- when the annual probability of the flood event that causes traffic disruption is greater than 20 per cent;
- when a significant part of the local network is affected; and/or
- when the duration of the flooding is several days or even weeks/months, as happened on the A361 in Somerset in 2012.

Four approaches for appraising road traffic disruption costs will be described and the suitability of each will depend on available resources and the likely severity of the road disruption:

1 The delayed-hour method: an averaged cost per hour for a delay on an average Highways Agency road.
2 The diversion-value method: the value of time based solely on the length of diversion (assuming that there is no reduction in traffic speed).
3 The speed–time method: reduced speeds are considered and a value of time applied for each diverted vehicle.
4 The origin–destination matrix method: using sophisticated transport appraisal and modelling tools (e.g. SATURN/PARAMICS).

The benefits of flood risk management include reducing the disruption of road traffic that occurs when roads are inundated. A road system is a network of links connected at nodes which make up the road junctions. The extent of the disruption then depends upon the local density of the links and the proportion of those links that are closed by the flood. For each annual probability of flooding, the losses from disruption are given by the additional costs to each vehicle using the road network where these additional costs are incurred by taking longer times to make the journey. Those extra costs are made up of the increased time costs and also the additional vehicle resource costs (vehicle operating costs – VOC). Resource costs exclude all indirect taxes (such as fuel duty and VAT) and are calculated as a function of a vehicle's speed (Figure 6.3). They are collated for a range of speeds and vehicle types in Table 6.15.

THE DELAYED-HOUR METHOD

A very crude disruption cost could be ascertained using averages of Highways Agency data and DfT values of time.

Assuming an average speed of 100kph (approximately 60mph), a single car delay of one hour on a motorway or trunk road will cost the UK £11.90. According to the Highways Agency's National Operations Group, the average vehicle flow per hour on the strategic road network is 1,794 vehicles (see Chatterton *et al.*, 2010). Based on this, we can estimate that the average delay of one hour will cost the UK a total of approximately £21,350. This figure can be refined if specific data about the hourly flow rate of the road are available and weighted accordingly if other vehicles (e.g. LGVs and HGVs) are included.

The averaged figure should be used only on Highways Agency roads (i.e. motorways and major trunk roads in England) as it will vary considerably at lower average speeds and on other roads. Table 6.15 can be used to refine the hourly cost per vehicle based on the average speed for the road(s) in question and this can then be multiplied by the calculated traffic flow for each particular road. This hourly figure will then need to be multiplied by the estimated duration of disruption. Table 6.17 provides likely delay durations for various flood return periods. Although we acknowledge this method here, and consider it superior to a percentage estimate based on property damages, more refined modelling should be undertaken where possible and if an appraiser thinks it is proportionate to do so.

The other approaches each adopt the same basic principle: if a road is closed, traffic will be diverted around this disruption point in the network.

Table 6.15 Total resource costs of travel as a function of speed (pence/km) (updated to 2013 prices)

Total resource costs (pence/km)								
Speed (kph)	5	10	20	40	50	80	100	120
Car average p/km	256	131	69	38	32	22	20	18
LGV average p/km	300	156	83	47	40	30	28	27
OGV1 p/km	331	175	96	55	47	36	–	–
OGV2 p/km	423	229	130	79	68	54	–	–
PSV p/km	1,901	974	510	276	229	–	–	–

Source: Data supplied by the Department for Transport

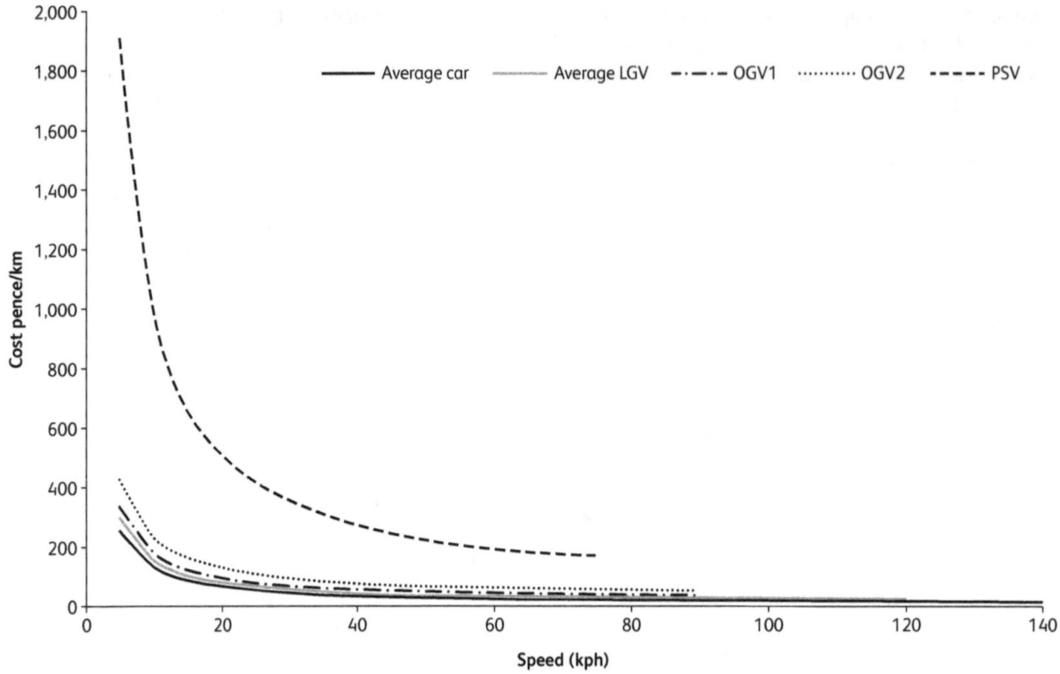

Figure 6.3 Cost–speed relations for resource costs

Source: Data supplied by the Department for Transport

Essentially, the second and third approaches are extensions of the first approach, but the level of detail increases, including how costs are attributed.

Thus, the additional costs incurred due to a flood can be estimated using Equation 6.2.

Equation 6.2

> Estimated costs incurred during disruption = number of vehicles delayed × additional cost per vehicle × flood duration (in hours)

When using this equation it is the total number of vehicles that take longer to make journeys that is important, including the traffic on those roads on to which traffic is diverted to avoid the flooded roads. However, excluded from the equation are those vehicles that are travelling to or from an address that is also flooded.

When considering the traffic disruption caused by flooding, the first question is whether it is worth calculating these benefits at all. Equation 6.2 should be used to derive an initial crude estimate of the likely benefits of alleviating traffic disruption, since otherwise the costs of calculating these benefits can exceed the present value of the traffic disruption benefits.

THE DIVERSION-VALUE METHOD

The simplest way to apply Equation 6.2 is to assume that cars will be diverted on to neighbouring roads and therefore the distance that they travel will increase; however, their speed will be unaffected. For example, suppose that 15,000 vehicles travel through the local network each hour and will have to travel, on average, 2km further, but their average speed (40kph) will not decline. In this scenario, the cost of that flood event will be equal to 15,000 × 0.38 (see Table 6.15) × 2 for each hour of the disruption due to flooding. If the flood lasts six hours, the costs of traffic disruption amounts to £68,400. In this instance, the figure is small and therefore it is disproportionate to refine this value through further modelling.

It is possible, although time consuming, to estimate the turning movements of vehicles at each road junction. This helps to improve the accuracy of disruption cost estimates by eliminating vehicles which are unlikely to travel through the sections of road at risk of flooding. The MCM (2005) (Penning-Rowsell *et al.*, 2005a, 118) provides a method for this procedure.

THE SPEED–TIME METHOD

The MCM (2005) provides a more in-depth method for calculating traffic disruption, and where possible this should still be used in conjunction with the updated figures provided here. In line with experience since 2005, we have attempted to produce a simpler and less time-consuming method which will still give an adequate estimate of traffic disruption costs. (More detailed modelling of local traffic conditions and driver behaviour may be the preferred option where the likelihood of road traffic disruption due to flooding is significant; for example, the Somerset floods of 2012.)

- Step 1: Determine which roads will be cut by floods of different annual probabilities and the durations of closure in each case. As an approximation, a road should be assumed to be closed when the middle of the lane is inundated and certainly when the crown of that road is flooded. Although this may be considered quite cautious, it is consistent with Environment Agency advice which attempts to prevent the public driving through flood waters.
- Step 2: Estimate the volume of traffic using each road in the local network (e.g. including those roads on to which traffic is likely to be diverted in a flood). Annual average daily traffic flows for all major and minor roads in Great Britain are available from the Data.Gov website (HM Government, 2013). The data are split by category of vehicle (car, LGV, HGV, etc.), which is relevant for calculating the different costs of travel for each vehicle type (covered in Step 5 below). An appraiser might also consult the Highways Agency and/or local authority, which retain a large amount of data about traffic flows.
- Step 3: Calculate the costs to traffic of using the local network under normal conditions. The Department for Transport provides free flow speeds for all built-up and rural road types at: http://www.dft.gov.uk/statistics/series/speeds/. Free flow vehicle speeds provide information on the speeds at which drivers *choose* to travel and their compliance with speed limits, but these should not be taken as estimates of *actual* average speed across the road network. Alternatively, the 'Free flow speed' column in Table 6.16 can be used for this purpose.
- Step 4: For each flood event, determine the routes that diverted traffic will take. For diversion distances, and where local expertise is absent, an online tool can be used; for example, the 'Get directions' feature on Google Maps provides distance information on the length of a selected stretch of road. As different traffic flow values are applicable to different types of road (single-carriageway urban roads, dual-carriageway rural roads, motorways, etc.) it is necessary to ensure that the diversion route is calculated using a separate distance value for each particular road type used (see Step 5). Many assumptions need to be made when establishing the likely routes for diversion and there is a need to concentrate primarily on diversions using major routes rather than minor roads.
- Step 5: Calculate the costs to traffic of using the network under flood conditions. The most difficult aspect here is to calculate how the non-flooded network will cope when the diverted traffic is added to it. Each road type has a free flow limit (see the second column in Table 6.16) and a capacity limit (see the third column in Table 6.16) and when this is reached speed flows will be reduced linearly. Equation 6.3 can be used to calculate the reduced speed of vehicles on the diversion routes.

Equation 6.3

$$\text{Speed} = VM/(1 + (VM/(8 \times DIS)) \times ((F/QM) - 1))$$

Where:

- DIS is the length of the road between junctions;
- F is the traffic volume in pcu (per car unit) equivalents; and
- VM and QM are as defined in Table 6.16.

When the reduced speed has been calculated for each diversion route, a cost per vehicle type must then be

Table 6.16 Speed–flow relations

Road type	Free flow speed (kph)	Free flow limit (pcu/h/lane)	Limiting capacity (pcu/h/lane)	Speed at limiting capacity (kph)
	VC	QC	QM	VM
	Free flow speed	Speed falls linearly over this range		
Rural motorway	90	1800	2600	76
Rural dual carriageway	79	1600	2400	70
Rural all purpose road	70	400	1800	57
Rural all purpose road – poorly aligned	50		600	50
Urban motorway	80	1700	1400	66
Urban dual carriageway				
With limited access and 80 kph limit	65	1400	220	56
65 kph speed limit	50	600	1100	30
Urban single carriageway road				
outer area	45	500	1000	25
intermediate area	35	350	600	25
central business area	25	250	500	15
Suburban – major radial or outer ringroads				
No major intersections	Speed limit	2000	47	–
< 1 major intersection per km		1700	27	–
1–2 major intersections per km		1200	20	–

Note: Department for Transport has confirmed that these 1981 values are still applicable.

Source: DfT, 1981

assigned using Table 6.15. This includes the value of time and the cost of running a vehicle (excluding indirect taxes, fuel duty, etc.) and is based on TAG Unit 3.5.6 (DfT, 2012b). It is now possible to calculate the total traffic disruption for the flood event using Equation 6.4.

Equation 6.4

Estimated potential costs of road traffic disruption = number of vehicles affected (for each vehicle type) × length of diversion (km) × total cost of travel per km (for each vehicle type) × flood duration (hours)

An estimate of flood duration (in hours) for each return period is provided in Table 6.17. This should not be used as a substitute for detailed flood modelling but should be applied cautiously where no site-specific probabilities and durations are available.

The origin–destination matrix method

The most sophisticated method of assessing road traffic disruption costs employs an origin–destination matrix and complex traffic modelling results. This is the most accurate of all the approaches. If an origin–destination traffic matrix is available for the area, then the flows on the different roads can be

Table 6.17 Indicative delay durations at different return periods

Likelihood of flooding	Delay duration (hours)
Up to and including the 5-year return period	6
Up to and including the 10-year return period	6
Up to and including the 25-year return period	12
Up to and including the 50-year return period	24
Up to and including the 100-year return period	48
Up to and including the 200-year return period	96

calculated using a transport model such as SATURN (which is a general traffic assignment model) or PARAMICS (which is a more commonly used micro-simulation model, like the old DRACULA model). Given the complexity of transport modelling, we recommend the support of a specialised transport modeller. More details can be obtained at: https://www.gov.uk/transport-appraisal-and-modelling-tools. Unless the network being modelled is small (e.g. fewer than twenty roads between junctions) then it is tedious to carry out an analysis without an available origin–destination matrix.

High-frequency events and floods which affect a significant part of the network

Roads are often the first points to flood in a floodplain, either because they run along the riverbank or because they cross the floodplain. Consequently, they may be flooded in very frequent events and perhaps several times a year. This needs to be accounted for within any calculations. However, in many cases the costs of raising the road and therefore solving flooding problems via an engineering solution will be lower than a flood risk management option. In these circumstances it may be appropriate to cap the PVd due to road traffic disruption at the least-cost solution of raising the road above the flooded level.

When considering floods which affect a significant part of the network, there are two sub-categories:

- disruption to sparse rural networks where diversion routes are long (e.g. 10km); and
- dense, heavily trafficked urban networks.

The former can be handled through the methods just described. However, the latter often involve dozens of links and, to be feasible, analysis of such networks requires the existence of an origin–destination matrix and a traffic model.

Long-duration floods

Since traffic disruption losses are calculated on an hourly basis, the total losses from floods lasting several weeks can obviously be very significant, as experienced in Somerset during winter 2012/13. This example and the lessons it provides are illustrated in Box 6.6.

During longer flooding events, awareness of road closures will increase and drivers will begin to find alternative routes, may travel during less busy times of day, may select alternative transport options or may choose not to travel at all, and therefore traffic on the diverted routes may begin to ease. Additionally, not all diverted routes will be full to capacity at all times of day, so traffic speeds may vary. There is a need to ensure that the costs of traffic disruption are not severely overestimated in these circumstances.

Summary: flood losses (roads)

Only in special circumstances are traffic disruption costs likely to contribute significantly to FRM scheme benefits. More commonly, direct damages to vehicles are likely to be more important – especially where underground car parks are flooded (see Chapter 4 for direct damage to vehicles). An additional complication is that some roads may form part of the flood risk management system – the crest of a road may protect the area behind it or, in urban drainage systems, the roads may contribute to flood conveyance and storage.

However, where major roads are inundated by floods with annual probabilities of less than 20 per cent, traffic disruption costs may constitute a significant proportion of scheme benefits.

BOX 6.6 SOMERSET, WINTER 2012/13 FLOODS

Flooding in Somerset in November and December 2012 and January 2013 caused the A361 to be closed for a total of sixty-nine days. The road was affected in three main periods, being closed for twenty-six days between 23 November and 19 December, for twenty-three days between 22 December and 14 January and for eleven days between 28 January and 7 February. The stretch of the road between Burrowbridge and East Lyng was particularly badly affected. In addition, the A39 was closed on numerous occasions over this period, most frequently at Bilbrook and Carhampton.

For the purpose of this analysis, it was assumed that all A361 traffic would divert via the A38 and A372 through Bridgwater. More complex assumptions regarding route choice would have required a full traffic model of the area, which was unavailable. Whilst it is recognised that this is only one of several routes available, most are of fairly similar length and would require approximately the same additional travel time. For example, travel via the A39, M5 and Toneway into Taunton is likely to be comparable. The results are not, therefore, considered to be significantly affected by this assumption.

TUBA (transport users benefit appraisal) has been used to calculate the extra costs associated with a diversion route. All the values of parameters used are taken from the Department for Transport's Transport Analysis Guidance: these include vehicle mix, occupancy, operating costs, values of time for different trip purposes, etc.

Other assumptions were as follows:

- daily diverted flow (average day): 5,266 (from Somerset County Council (SCC) data);
- daily diverted flow (average weekday): 5,608 (from SCC data);
- increase in travel time (using Google Maps): sixteen minutes (eastbound), twelve minutes (westbound);
- increase in travel distance: 9.4km (eastbound), 9.0km (westbound);
- length of closure (financial year 2012/13): sixty-nine days (ten weeks);
- vehicle mix is the same during all time periods;
- no allowance is made for any additional congestion (e.g. in Bridgwater).

The table below gives the TUBA's results. Note that the operating costs include fuel duty for businesses, and both duty and VAT for consumers.

Travel time	Consumers	£0.542 million
	Businesses	£0.526 million
	Sub-total	£1.068 million
Operating costs	Consumers	£0.220 million
	Businesses	£0.189 million
	Sub-total	£0.409 million
Additional greenhouse gas		£0.032 million
Total cost of ten-week closure		£1.510 million
Equivalent costs per day		£21,884

(Somerset County Council, 2013a)

Table 6.18 Summary of the data requirements and available sources of information

Data	Source
Traffic modelling tools	DfT: https://www.gov.uk/transport-appraisal-and-modelling-tools
Step 1: Road levels	GPS-based drive-through survey
Step 1: Flood durations	Hydraulic model
Step 2: Traffic flows	DfT: http://data.gov.uk/dataset/gb-road-traffic-counts
Step 3: Free flow speeds	DfT: http://www.dft.gov.uk/statistics/series/speeds/
Step 4: Diversion routes	Google Maps: http://maps.google.co.uk/
Step 5: Resource costs and values of time	DfT: http://www.dft.gov.uk/webtag/documents/expert/unit3.5.6.php

Again, we must stress that the four approaches detailed above should be used to obtain an informed (if somewhat rudimentary) disruption cost which can then be used to ascertain whether a more detailed analysis is required using sophisticated traffic modelling and specific, local data. Traffic disruption costs can easily be miscalculated due to the somewhat sporadic nature of vehicle flows and the assumptions which must be made relating to traffic nodes. For example, a recent appraisal by Somerset County Council estimated local disruption costs in a range from £1.6 million to £5.3 million using sophisticated modelling, whereas a cruder estimate of the same disruption put the figure at £24 million (Chatterton, undated). This overestimate was thought to be attributed mainly to the application of significantly reduced travel speeds over the whole period of the floods, whereas in reality traffic flows on the diverted roads were significantly reduced only during rush-hour or equivalent conditions.

6.5.2 Potential losses owing to rail disruption

In the past it had been assumed that railways will not be significantly affected by flooding because cautious Victorian railway engineers sited their track and facilities well above the floodplain. However, significant flood events in 2000, 2007 and 2012 in England and Wales have shown that damages and delays can and do occur. Between 2004 and 2010 flooding caused a total of 520,000 delay minutes with an associated cost of £38.2 million (Network Rail, 2011; here the calculation was based on a crude estimate of National Audit Office delay minute costs per service of £73.47, from Burr (2008)). This is an estimate of the cost of disruption relating to the economic impact to passengers of an average train on the UK railway network.

Indeed, the Environment Agency has estimated that 21 per cent of railways in England are at risk of flooding (Environment Agency, 2009a). However, despite the high potential for damages, much of the railway infrastructure in the 2007 floods remained resilient to flooding, with only forty-two sites out of the 265 affected sites sustaining severe damage (Dora, 2007). However, the impact of flooding on the railways may worsen as a consequence of climate change (Network Rail, 2011).

Losses due to flooding can arise in the following areas:

1. Damage to assets: Network Rail incurs direct damages to infrastructure assets, including track and circuits, embankments, structures and stations. There is also the potential for train operating company (TOC) rolling stock to be damaged, although generally these losses are likely to be minor in comparison.
2. Performance delay/cancellation costs: these occur as a result of delays and/or cancellations in the train service and will involve costs to Network Rail for compensation to TOCs and to TOCs themselves for loss of revenue and compensation to reimburse inconvenienced passengers.
3. Costs of alternative travel arrangements: when trains are cancelled owing to flooding the TOCs are

under an obligation to enable passengers to continue their journey and provide alternative transport (such as replacement bus services). Although these are noted here as a separate category of loss, Network Rail's compensation to TOCs will include an element to reimburse for these losses.

When appraising the benefits of schemes to reduce rail disruption owing to flooding, care is needed to calculate costs which can be prevented by those schemes: damage and performance costs may be caused simply by heavy rainfall which cannot be prevented by flood risk management investment. Indeed, only 30 per cent of the total delay minutes attributed to flooding between 2004 and 2010 were caused by fluvial flooding, with 65 per cent being due to localised heavy rainfall (Network Rail, 2011). However, delay minutes and cancellation of trains are a very good proxy to use for estimating the additional costs caused by the disruption of rail services by flooding.

Any delay in service or performance as a result of severe weather (which causes a section of track to be out of action) results in a compensation payment from Network Rail to a TOC under Schedule 4 (cancelled services) and Schedule 8 (delayed services) of the Track Access Agreement (see ORR, 2012 and the Railway Act 1993 (HM Government, 1993)). The compensation costs applied to the delays are based on a formula derived by the former Strategic Rail Authority and comprise two components: a *societal* element and a *marginal revenue* effect. Societal elements relate to issues such as pollution and road congestion and are quantified by passenger/minute of delay. The marginal revenue effect estimates the impact of good or poor performance on ticket revenue. Standard costs set out in these agreements are assigned to the delay/cancellation depending on the type of route affected, the operator affected and the location of the incident; with the busiest routes allocated the highest weighting. Thus, a delay close to London in peak rush hour will be assigned a higher delay cost than, say, a mid-afternoon delay in rural Wales. In short, assigned costs generally relate to fare value. A minimum threshold of two to three minutes is set for delays, beyond which an attribution code is allocated and a delay incident is logged.

Box 6.7 provides an example of the damages and disruption to rail travel during the 2007 floods, and Box 6.8 a case study from Somerset during 2012. Performance data have also been provided by Network Rail for disruption which occurred throughout 2012 and into the beginning of 2013. Network Rail records delays in four-week periods (i.e. thirteen periods (P) per year), with period 1 starting on 1 April. The data analysed for the 2012 flooding events include delay minutes, cancellations and delay costs between the beginning of February 2012 (2011/12_P12) and the end of January 2013 (2012/13_P11). Three attribute delay codes are used for flooding: flooding not due to exceptional weather; severe flooding that could not be mitigated; and station flooding. All three types of flooding have been included within the analysis here.

BOX 6.7 CASE STUDY: DAMAGE AND LOSSES TO THE RAILWAY NETWORK IN THE SUMMER 2007 FLOODS

The summer floods in 2007 caused estimated losses of £36 million in the rail sector, which equated to 5 per cent of the economic losses to infrastructure (see Table 6.4). Of these losses, close to £10.5 million are considered to be direct damages to the track or other infrastructure, with the remaining £25.6 million attributed to disruption costs. Some 265 locations in total were affected, but despite their widespread distribution only forty-two sites suffered significant infrastructure damage, with a further 116 sustaining some damage; 107 sites experienced only performance-related losses (Dora, 2007).

Dora (2007) presents an in-depth examination of Network Rail's Western territory (which includes areas in south-west England, through Bristol into the Thames Valley and into Herefordshire and Worcestershire). It highlights that during 2007 a number of lines were closed and some repeatedly closed. However, most lines were opened relatively quickly after flooding. The exception to this was a line near Ledbury, Herefordshire, which was closed for two weeks owing to potential damages to a nearby bridge.

Part of the Cotswold line which needed repairs was also closed for two weeks, and a line to the south of Oxford was closed for ten days. Another significant closure occurred due to a landslide on the Manchester Airport to Cleethorpes line, which took six weeks to repair and return to normal (Dora, 2007; Chatterton *et al.*, 2010).

Network Rail had well-developed plans in place to mitigate the impacts of the floods, with the establishment of Extreme Weather Action Teams (EWATs), rapid response teams on standby and procedures in place to receive up-to-date rainfall forecasts from the Met Office. Response activities were effectively prioritised to enable lines to open as quickly as possible with the minimum disruption to TOCs and passengers. Much work has also been undertaken since 2007 to understand the impact of flooding on the network and improve the resilience of assets to flood waters.

(Information from Chatterton *et al.*, 2010; Dora, 2007)

BOX 6.8 CASE STUDY: DAMAGE AND LOSSES TO THE RAILWAY NETWORK IN THE 2012 FLOODS IN SOMERSET

Damage and disruption owing to flooding in November and December 2012 were significant, with all except one railway line running through Somerset impacted. The mainlines to both Bristol (Severnside) and London (Great Western) were affected by flooding, which meant that for a period of time Taunton became the rail terminus in south-west England.

Significant closures affected three locations: Cowley Bridge (fifteen days), where significant sections of the track were undermined and washed away; Dawlish/Teignmouth sea wall (six days); and Honiton on the Exeter to Waterloo line (five days). It was estimated that close to 1,900 trains were cancelled or delayed during the flooding in November and December 2012, and the significant disruption also impacted local roads, which meant that rail replacement buses were unable to transport passengers affected by the rail disruption.

Initial calculations of the costs of the damage and disruption estimate that the costs are in the order of £7 million (which includes the restoration and repair of the lines, signalling equipment and ballast, costs of the many replacement bus services, and revenue lost and compensation payments). Of these, SCC (2013b) estimates the cost of the compensation payments made by Network Rail for just five of the most serious incidents to be £1.2 million (multiplying the number of delay minutes by the National Audit Office industry value for delay minutes).

As well as the costs detailed above, SCC (2013b) suggests that the severe transport disruption that occurred in late 2012 may have other additional losses to the regional economy. It argues that the disruption has caused a loss in reputation for the region and will cause a loss in confidence in the transport network that will potentially affect future business investment and tourism in the area.

(Information from SCC, 2013b)

Figure 6.4 illustrates the total costs attributed to the delay minutes experienced by services between the beginning of February 2012 and the end of January 2013, with the inset graph portraying the number of delay minutes and the number of services cancelled. The total delay minutes cost due to flooding during this period was £16.9 million. Flooding affected trains throughout the period of interest, although four periods dominated the losses due to flooding (P4, P7, P9 and P10), accounting for 87 per cent of all delay costs. The severest impact on rail services occurred in an eight-week period from the last two weeks in October to the end of December 2012. This period accounted for 59 per cent of delay costs, 51 per cent

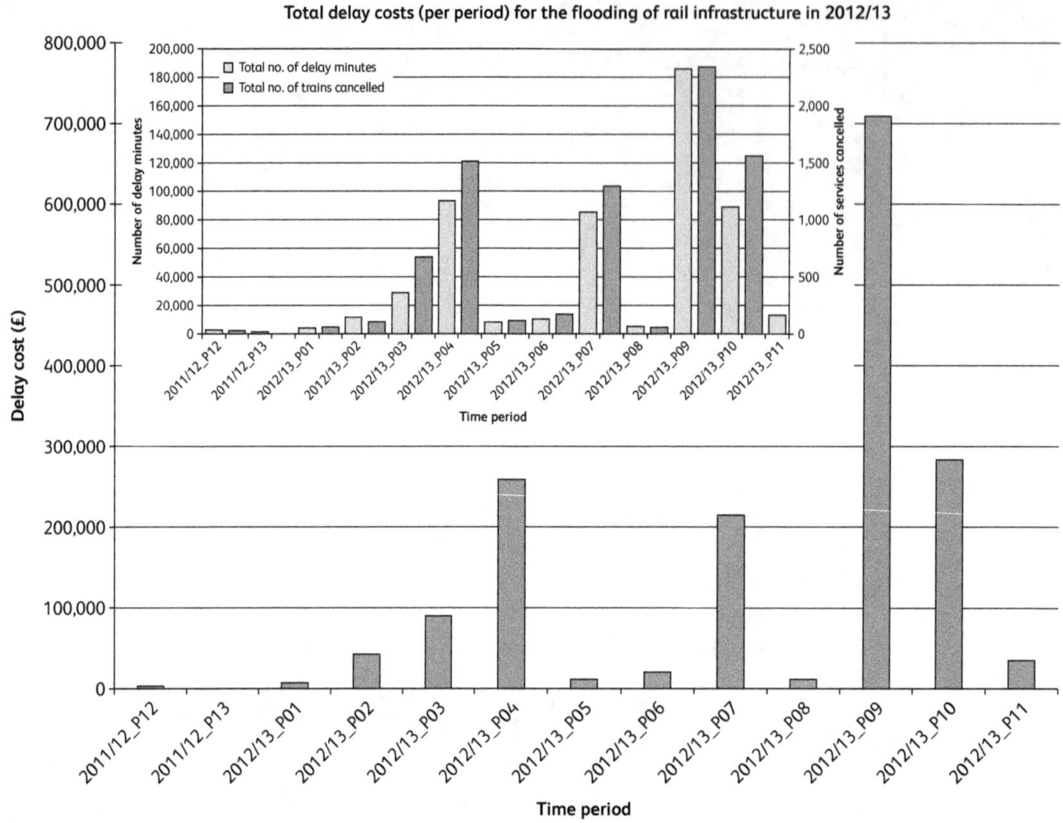

Figure 6.4 Total delay costs (per period) due to the flooding of rail infrastructure in 2012/13 (inset graph shows the total delay minutes and the number of trains cancelled)

Source: Data supplied by Network Rail

of delay minutes (273,982 minutes) and 48 per cent of all trains cancelled (3,897 trains).

The losses were also geographically concentrated. Figure 6.5 provides the breakdown of delay costs per route. This highlights a concentration of delay costs in the Western, London North Western and London North Eastern route areas. The high cost amongst the Western train routes (£5.2 million) was primarily accrued from mid-November to the end of December 2012 when flooding severely disrupted both road and rail transport in south-west England. The costs for the other two lines are more spread out across the year. However, both suffered peaks of losses in late June/early July, mid-September/early October and mid-November/early December 2012. These routes contain the main eastern and western intercity lines in England and therefore any disruption to these lines will have had substantial disruption and delay costs.

Figure 6.4 also highlights the number of services which were cancelled due to flooding, as these are not reflected in the Schedule 8 cost data presented here. These cancellations mean that actual losses will be significantly higher. In addition, the delay costs might represent only performance delay costs.

Using these data, the average performance cost to Network Rail is *c.*£28 per minute of service delay (i.e. not including the cancellation of services) for passenger TOCs and *c.*£17.50 for freight operating companies (FOCs). However, for TOCs, there is a wide range of average delay costs. This range is due to the different times of day and different services impacted and how these services are weighted within the compensation payments.

It is standard within the industry to multiply the delay cost by a factor of three for unexpected delays to provide a better estimate of the true cost of delays (DfT,

2009). If this factor is applied to the average delay cost in 2012, the average cost per delay minute is £84 per service. Adopting the average number of passengers per train (182) means that the average delay loss per passenger/hour is £27.70.

Since the 2007 floods, Network Rail has increased its focus on the current and future flood resilience of the rail network. Work has begun to improve the drainage of the network, and climate change factors have been included in drainage and structures design manuals (Network Rail, 2011). Other work to improve the resilience of assets to flooding and other weather-related risks will be implemented during planned maintenance and renewal of assets.

Mechanisms were already in place prior to 2007 to monitor bridges which may be subject to scour during

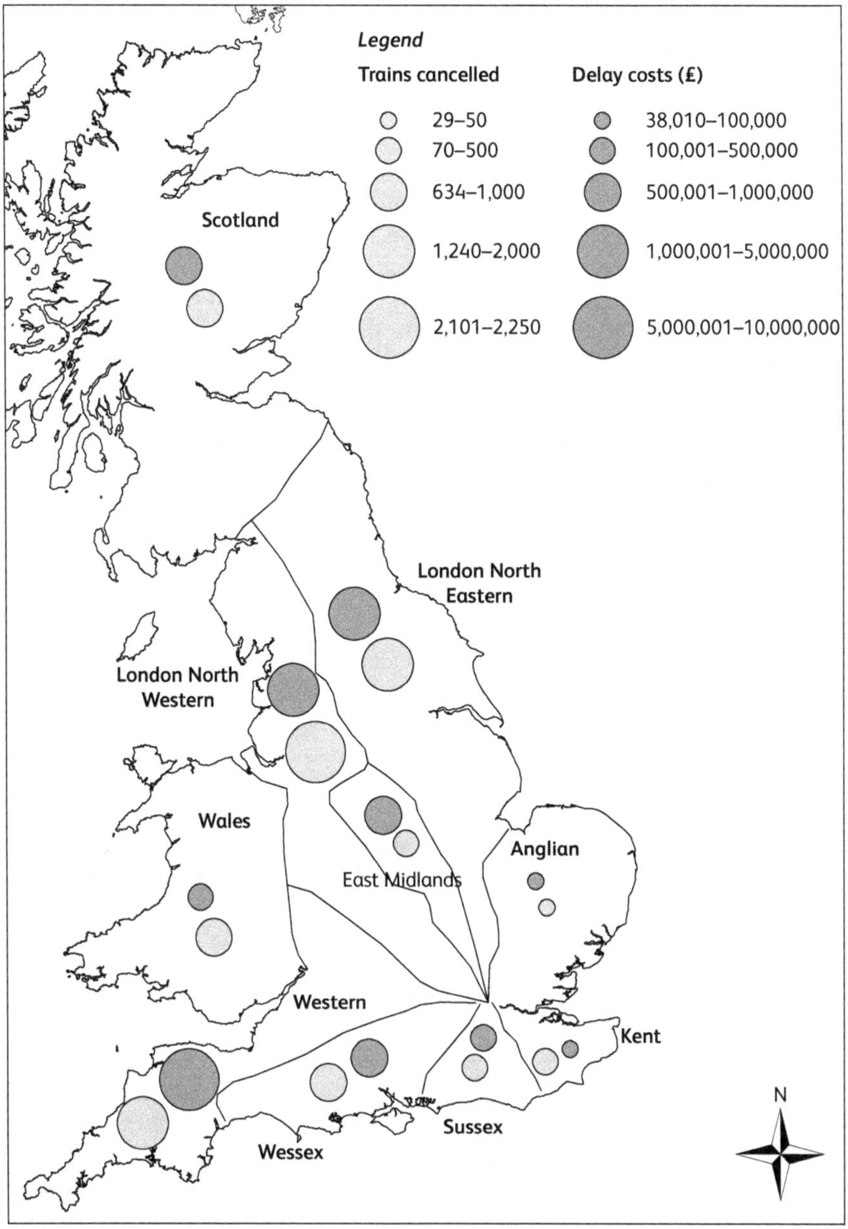

Figure 6.5 Geographical distribution of delay costs and number of trains cancelled

Source: Data supplied by Network Rail

a flood event and enable appropriate action to be taken. However, the mapping of assets at risk has been improved by Network Rail's asset management with the development of a GIS system which is used to identify and categorise risks to assets. This system evaluates many risks, but it includes flooding and utilises flood events of different return periods. These developments mean that Network Rail has significantly improved its understanding and categorisation of both the location and potential impacts of flooding on its network in recent years. It is strongly advised that FRM project appraisers make use of this knowledge and that any significant assessment of rail damage and disruption should include a site visit and discussions with the appropriate Network Rail route engineer.

Direct damage to rail assets is difficult to quantify as it varies considerably depending upon the circumstances of the flooding and the particular element of infrastructure affected; for example, embankment, track, signalling, etc. Therefore, a site survey to estimate potential direct damages is highly recommended, along with discussions with local Network Rail engineers. However, to inform estimates, the following indicative unit reconstruction values for Network Rail assets might be used: £4,000 per metre for embankments; £3,000 per metre for soil cuttings; and £4,000 per metre for rock cuttings.

A method of benefit assessment for disruption to the rail network

As discussed above, flooding of rail track and associated infrastructure will cause some services to be cancelled and others to be delayed. The more severe the flooding, the more severe will be the disruption and the larger the number of services cancelled.

However, an approach is needed to assess both the severity of the disruption and how to quantify these losses. The level of detail with which this is undertaken can be varied, whether average data or a more comprehensive analysis is used. However, railway damage and disruption are generally very context specific and therefore assigning a proportional (uplift) value to this figure is quite a 'stab in the dark'. Where damages and/or disruption appear to be significant (i.e. assets are highly susceptible to flooding, or a major rail line or key gateway route is impacted) it may be worthwhile undertaking a full assessment of the assets, including a site survey.

In research for the Met Office (Posford Duvivier *et al.*, 2002), one of the algorithms developed was designed to establish the losses associated with disruption of rail services through severe weather. The method established losses at postcode level for five weather events and for the five severities used in the Met Office's National Severe Weather Warning Service. The approach recommended here is an adaptation of this method for analysing the economic losses (to rail operators and customers) associated with the disruption of rail services owing to flood events.

Understanding, and quantifying, the likely number of cancellations and the duration of any delays is critical to the approach. This can be combined with statistics detailing the number of journeys made each day on each franchise service to estimate the scale of flooding impacts on the regular service. Posford Duvivier *et al.*'s (2002) severe weather (heavy rain) for delay and cancellation statistics can be used to represent the degree of disruption experienced during flood events. This gives the benchmark data for delay and cancellation in Table 6.19 and suggests the use of a 40/60 split for passenger trains between delay and cancellation in this simplified algorithm.

The following steps should be followed to calculate the costs of disruption:

- Step 1: Obtain or create a map of the rail network running through the area at risk of flooding or the potential flood risk management benefit area. This should include the specific TOCs which run services on the affected part of the network. Information on the routes that each of the TOCs operates can be found on the National Rail website (http://www.nationalrail.co.uk/stations_destinations/maps.aspx).
- Step 2: Determine the number of services impacted and/or the passenger journeys for the rail line at risk from flooding per twenty-four-hour period. This information can be difficult to identify

Table 6.19 Percentage delay/cancellation due to flooding

Rail service	Delay %	Cancellation %
Passenger service	40	60
Freight service	45	55

Source: Posford Duvivier *et al.*, 2002

as the Office of Rail Regulation (ORR) presents only global annual passenger numbers for each of the rail companies for detailed assessment in its National Rail Trends Portal (ORR, 2013). (These averaged data can be found on MCM-Online Supporting Information for Chapter 6.) Therefore, it is necessary to approach each TOC operating on the rail lines within the appraisal area to refine the global passenger numbers and identify the relative significance of the line.

- Step 3: Apply Table 6.19 (delay/cancellation percentage split from the effects of flooding) to the number of services/passenger journeys per twenty-four hours which would be affected by flooding. An estimate of the likely length of a delay to a service is also required. This can be quite complex and we recommend discussions with Network Rail engineers to identify an average number of delay minutes that a service might suffer if a route is affected.
- Step 4: Quantify the losses. Two methods are described below for quantifying these losses. The first relies on estimating how much compensation will be paid by Network Rail to TOCs and FOCs and therefore represents the additional costs due to flooding. The second uses the value of time (VOT) approach adopted by the National Audit Office (Burr, 2008) in its investigation of rail delays.
- Step 5: Convert the costs calculated per hour to annual average disruption. This requires an assessment of the depth and extent of flooding likely at different probabilities. Owing to the complexity and context-specific nature of the rail network it is preferable to undertake a site survey to fully understand the likely impacts of flooding and the likely length of a delay. Where this is undertaken, site-specific annual flooding probabilities should be applied. Dora (2007, 3) gives a diagram from the *Railway Rule Book* (RSSB, 2012) which provides depths at which rail infrastructure is impacted by flooding. In summary:

  - Up to 50mm below the top of the rail – normal working conditions.
  - Up to 100mm above the top of the rail – speed restrictions (maximum speed 5mph).
  - More than 100mm above the top of the rail – authority to allow trains to run must be given by the local manager (and therefore cancellations are likely).

Although this provides some information about the depths of flooding that affect the rail itself, in reality the track is only one element of railway infrastructure that may be affected by flooding. The inundation of areas around the track may also critically affect other assets, such as signal infrastructure, and affect the stability of embankments. It is also necessary to account for the impact of moving water, which may close rail lines at lower depths. Therefore, potential losses should be considered as soon as any Network Rail property is affected by flooding (including property assets, embankments, drainage, culverts, bridges, other crossings). These estimates can then be refined through discussion with local Network Rail route engineers.

In the absence of site-specific annual flooding probabilities (which are preferable), use the road traffic return period disruption durations provided in Table 6.17.

As with road traffic disruption, adopting this model will be complex for Fenland-type situations or during widespread flooding, when delays and cancellations will have significant 'knock-on' effects on each other. Disruption will escalate significantly as flooding becomes regional or if a key junction or station in the network is affected. An example of such aggregated effects is the disruption which occurred in the south-west of England in late 2012. On 26 November ATOC reported that 20 per cent of all national rail services were being impacted by flooding (ATOC, 2012). Situations where widespread disruption occurs – and therefore where both rail response and repair teams and rail replacement infrastructure are stretched – are likely to have losses disproportionate to the sum of the aggregated flooding incidents. Therefore, disruption figures calculated using the approach presented here represent a minimum economic cost of disruption, relating to the separate flooding of individual floodplain areas and rail links, rather than all-region impacts.

Quantifying disruption costs due to delay and disruption

Quantifying the costs owing to rail disruption can be undertaken utilising two approaches. The first method uses the compensation payments made to TOCs/FOCs by Network Rail to recompense for delayed or cancelled services, whereas the second method uses a value of time approach accounting for how much time travellers would pay to avoid a delay. The first approach uses the total number of likely services impacted by flooding; whereas the second utilises the number of passenger journeys impacted by flooding. If no information on the number of passenger journeys per twenty-four hours is available then the average number of passenger journeys per train, 182 (Burr, 2008), might be used to provide an estimated value based on the likely number of services affected.

The compensation payment method provides an estimate of disruption due to the delay of a service and is broadly related to the value of the fare being paid by a passenger. The second approach provides an estimate based on the value of time and attempts to quantify the inconvenience or lost work time caused by a delay or cancellation. Combining these estimates provides an upper estimate on the value of a disruption. This arguably includes some degree of double counting as the compensation value – if reclaimed by affected passengers – does provide some recompense for their inconvenience, however it may provide a closer estimate of the true costs of the disruption of rail services.

Utilising compensation payments

This utilises the average compensation costs that Network Rail pays to the TOCs under Schedules 4 and 8 of the Track Access Agreement. Table 6.20 provides indicative compensation values per delay minute and per cancelled service. Actual compensation values for each of the TOCs and FOCs, as agreed in the Track Access Agreements, is restricted information and therefore the indicative values provided are based on data of the delay costs and cancelled services between 2011 and 2013. Low, medium and high values are provided for passenger service performance delays (per minute) and cancellations (per service) to account for the wide variation between TOCs and the lines impacted. These values could be used to provide a range of the potential losses due to rail disruption. Alternatively, it may be appropriate to select one of the values depending on the significance of the rail line within the assessment area. For instance, if a busy rail line (such as a main commuter route or east/west coast mainline) will be impacted, the higher value should be applied. Conversely, if a less busy, rural route is within the assessment area, the low value may be more appropriate.

A single indicative value for delay and cancellation for freight is provided as, unlike passenger services, there is little variation in the compensation values. These values should be multiplied by the likely minutes of delay and estimated number of services impacted to provide an approximation of the potential losses due to flooding.

A value of time (VOT) approach to quantifying the losses

Similar to the compensation approach, this method also calculates loss based on the number of delay minutes experienced. However, this approach utilises willingness to pay approaches presented by the *New Approach to Transport Appraisal* (NATA; DfT, 2011b). Guidance is provided for assessing the benefits of developing new transport schemes (e.g. new roads) and a monetary value provided – based on willingness to pay survey techniques – that each type of transport user (e.g.

Table 6.20 Indicative compensation values for delayed and cancelled services

	Delay compensation value £ per minute per service *			Cancellation compensation value £ per service cancelled**		
	Low value (£)	Medium value (£)	High value (£)	Low value (£)	Medium value (£)	High value (£)
Passenger services	40	71	97	673	2,034	2,591
Freight services	–	18	–	–	1900	0

Notes: * Including a delay multiplier of 3; ** Including a cancellation multiplier of 3.

Source: Data supplied by Network Rail

Table 6.21 Values of time, based on the willingness to pay of each type of passenger per hour (2013 values)

	Value of time (VOT) £ per hour		
	Business passenger	Commuter	Other passenger
Original values per hour	£42.34	£5.83	£5.13
Uplifted to account for an unexpected delay	£127.02	£17.49	£15.39

Note: The resource cost estimate has been utilised in this instance as these values are net of indirect taxation. DfT (2012b) values have been updated utilising HM Treasury (2012a) GDP Deflator.

commuter, business user and others) would pay, per minute, to avoid a travel delay. These should be uplifted by applying the 'delay multiplier' factor of three (DfT, 2009) which Burr (2008, 46) argues is 'used by the rail industry to recognise that unexpected delays are more costly to passengers'. Table 6.21 provides a summary of these costs, which are provided for 2013 and have been updated from the 2010 values provided by the Department for Transport (DfT, 2012b).

These values can be used to approximate the costs of a delay to different passenger types and therefore calculate the costs of travel disruption when multiplied by the length of any delay. If a local analysis is to be undertaken, data need to be gathered on the proportions of the different types of passengers travelling per train, which will vary depending upon the specific line as well as the average number of passenger journeys per train (see Step 2, above).

Table 6.22 provides averaged data on these proportions for each TOC (a national and regional breakdown by journey purpose is available on MCM-Online Supporting Information for Chapter 6). Although these data can be used for a general assessment, specific lines and services may vary considerably in their composition and so a per railway line analysis should be undertaken where possible.

It is possible to use average data for an approximation of the costs of rail disruption. Burr (2008) utilised data from the Department for Transport, Network Rail and the ORR to calculate the average number of passengers per train and an average of the type of passenger split for all journeys throughout a week. This identified that an average train contained fourteen business travellers, ninety-five commuters and seventy-three other passengers. Burr then applied these to the VOT values for 2007 and calculated an average value of £73.47 for every minute a train is delayed. Applying the updated VOT values in Table 6.21 provides an updated value of £76.06 per delayed minute of a train.

Although the methodology of valuation has been refined since MCM (2005), the breaching scenario of

Table 6.22 Percentage breakdown of the journey purpose of rail travellers by TOC in 2012

Train company	Commute	Business	Leisure
Arriva Trains Wales	30	9	61
c2c	67	5	29
Chiltern Railways	37	23	41
CrossCountry	15	28	57
East Coast	12	30	59
East Midlands Trains	23	28	49
First Capital Connect	45	26	29
First Great Western	30	20	50
First Transpennine Express	24	14	62
Greater Anglia	57	18	26
London Midland	45	14	41
London Overground	64	3	33
Merseyrail	37	8	55
Northern Rail	38	9	53
ScotRail	39	13	47
South West Trains	53	15	32
Southeastern	61	12	27
Southern	50	16	34
Virgin	9	30	62

Source: Passenger Focus, 2012a; 2012b

the Caldicot Levels presented there (Penning-Rowsell *et al.*, 2005a, 124–125) follows a similar prioritisation process and therefore remains broadly relevant.

6.5.3 Summary: road and rail disruption owing to flooding

Both road and rail disruption will continue to be impacted by flooding in the future and effects may increase due to climate change. However, both the Highways Agency and Network Rail have made great progress in identifying those assets that are at risk. Appraisers need to ensure that they communicate with the necessary regional or local engineers to utilise this knowledge. It is essential also to take account of ongoing and planned activities to make road and rail assets more resilient to flooding to ensure that the benefits of these activities are not double counted within the benefits of project appraisals for flood risk management schemes.

6.6 POTENTIAL LOSSES TO THE PROVISION OF OTHER PUBLIC SERVICES

Flooding can also have significant impacts upon the provision of other public services. This includes the flooding of schools, hospitals and other public buildings, such as council offices, libraries, museums and community centres. For a number of these types of building (such as libraries and museums), their closure due to flooding may be considered a nuisance and non-direct losses are considered to be secondary to direct damages. A number of these types of properties have been assessed when considering non-residential properties (see Chapter 5) and therefore appraisals can be undertaken for direct damages.

However, for some public buildings, there is considerable disruption potential if they are flooded, causing significant health, social and economic impacts. Both hospitals and schools fall into this category and therefore both of these need to be considered within a project appraisal.

6.6.1 Potential losses from flooding of schools

In 2007, it was reported that 858 schools suffered flood damage (Audit Commission, 2007), of which some were impacted for a considerable period of time. Box 6.9 provides an example of the impact of floods in June 2007 in Kingston-upon-Hull, the East Riding and Yorkshire. Schools elsewhere in the country were also severely impacted by flooding in July 2007. In total, in Hull, 390 schools were affected, with 305 of them suffering only minor damage and minor disruption to pupils and eighty-two needing more significant repair (but they still opened normally in autumn 2007). The remaining three schools suffered such severe damage that pupils had to be educated in temporary classrooms or alternative accommodation (Chatterton *et al.*, 2010).

HR Wallingford (2012) reported that over 1,600 schools are in areas of flood risk, with 740 of them at significant or moderate flood risk, with a total of over 200,000 pupils enrolled. Therefore, there is a moderate to high potential for losses from the flooding and closure of schools in some areas.

Direct damage to schools

The flooding in 2007 provides some indication of the losses that can be experienced owing to the direct flooding of a school. Twenty-six local authorities with schools impacted by the flooding in June 2007 received £9 million in flood recovery grants, with almost half of this sum (£4.69 million) assigned just to Kingston-upon-Hull and the East Riding of Yorkshire. A further £3.6 million in grants was then allocated to twenty-one local authorities for repairs, etc. caused by flooding to schools in July 2007. However, these are only initial grants and may not cover the total losses suffered; therefore, these values provide only an indication of the scale of losses experienced.

Chatterton *et al.* (2010) report that Kingston-upon-Hull City Council suggests that the total costs of the repair and replacement of contents of school buildings was £12.2 million. However, it is perhaps inappropriate to give an average value per school as there was a great degree of variability in the level of damage different schools sustained. Direct damages to schools in Gloucestershire ranged from £1,000 to £90,000, with one further school (St David's Primary

BOX 6.9 THE IMPACT OF THE 2007 FLOODS ON SCHOOLS IN KINGSTON-UPON-HULL, THE EAST RIDING AND YORKSHIRE

Schools in Kingston-upon-Hull were particularly badly impacted, with only eight out of ninety-nine schools in the city unaffected by the flooding (Coulthard *et al.*, 2007). Researchers reported that ninety-five primary schools and thirteen secondary schools suffered temporary closure in June 2007. Not all of the disruption was caused by direct flooding to school buildings; many pupil days were lost because the flood waters prevented access (Chatterton *et al.*, 2010).

Estimates of the total number of pupil days lost in Hull range from 107,000 by Kingston-upon-Hull Council (Chatterton *et al.*, 2010) to 114,400 (Coulthard *et al.*, 2007). However, in either case the social impact was considerable and had significant knock-on impacts in the wider economy. Coulthard *et al.* (2007) use the total number of pupil days lost and minimum wage values to estimate a loss of £2.4 million to the wider economy through parents having to miss work to care for children. Chatterton *et al.* (2010) provide a minimum estimate of the value of school days lost at a total of £9 million, but state that this is probably an underestimate of the value of school days lost. Therefore, an estimate of total disruption is in the order of £11.5 million.

The Cabinet Office (2008) reports that 290 schools were damaged across Hull, the East Riding, Rotherham and Doncaster, affecting a total of 170,000 children. This equated to almost 400,000 pupil learning days lost. The situation could have been much worse, however. The flooding occurred in the summer term, after the exam period, minimising the direct impact upon pupils' education. The timing meant that some schools could close early for the holidays and the summer break allowed many of them to undertake repairs without further disruption to pupils.

Although Coulthard *et al.* (2007) report that in November 2007 (five months after the flooding) 400 pupils in Hull were still being taught in alternative locations, within a week of the floods over half of Hull's schools had reopened (Chatterton *et al.*, 2010) and all but two reopened for the start of the new school year.

(Information taken from the Independent Review Body into the 2007 Hull floods: Coulthard *et al.*, 2007; Chatterton *et al.*, 2010; and Pitt, 2008)

School, Moreton-in-Marsh) suffering £1.9 million in damages (Chatterton *et al.*, 2010).

Schools are included within non-residential property assets in Chapter 5. Table 6.23 provides a summary of these results and presents estimated values per square metre of floor space for flooding at different depths for long-duration (more than twelve hours) and short-duration floods (less than twelve hours).

Other direct costs which may be calculated are:

- costs of additional temporary classrooms and/or other accommodation costs; and
- costs to the school/council of added support services to pupils affected by flooding (e.g. the cost of additional counselling services).

These costs were reported by Kingston-upon-Hull City Council to be in the order of £700,000 and £514,000, respectively (Chatterton *et al.*, 2010), but elsewhere they may vary considerably from these figures, depending on the scale and duration of the flooding.

The disparity in the severity of flooding experienced between schools means it is difficult to suggest an average value for the cost of additional classrooms. However, if a conservative approach is taken and those schools which were reopened within a week are excluded from the calculation then the £700,000 divided by the forty-six severely impacted schools creates an average of around £15,000 per school. In all likelihood, however, the most severely affected three schools should be assigned a much higher proportion of these temporary accommodation costs.

Coulthard *et al.* (2007) indicate that in Hull, 3,000 pupils were directly affected by the flooding, and these pupils would have needed the most support from the

Table 6.23 Average damages per square metre expected from different depths of flooding for a school

Depth (cm)		–50	0	25	50	100	150	200
Total damage (£/m^2)	Short (<12 hours) duration floods (£/m^2)	0	15	135	200	315	405	555
	Long (>12 hours) duration floods (£/m^2)	0	45	175	265	395	490	640

Notes: Values calculated using an average floor space for the asset of 3,300m^2; see Chapter 5

counselling and other support services offered. This equates to a value of *c*.£150 per pupil.

Estimation of the costs due to school disruption

In addition to the direct damages associated with the flooding of schools, there are other potentially significant impacts associated with the closure, or part closure, of a school. The significance of these losses will vary depending upon the severity and duration of the flooding experienced and the type of school affected.

Losses from a school closure may include:

- The loss of parents' earnings (or number of staff days lost) due to the need to take time off to care for dependent children.
- The value of the loss of a pupil's education.
- Additional travel costs to alternative schools or temporary school locations. (This is very difficult to establish as it would require knowing the additional journeys of all pupils and staff.)

Each of these losses will be a function of the duration of closure and disruption to the school and the number of pupils affected (in particular younger pupils). Some interdependencies may also be noted. For instance, if a number of schools within an area or city are affected, the potential for sending displaced pupils to alternative locations may be reduced.

The following steps should be followed to assess the disruption losses due to the flooding of schools:

- Step 1: Identify the location of schools within and close to the flood risk area. Schools directly adjacent to the flood risk area should also be considered, as many schools in 2007 were temporarily closed due to pupils and staff being unable to access their schools. These schools may be impacted by flood warnings, any emergency actions and transport problems. Assess whether there are alterative flood-free routes to access the schools. Use this information to create a shortlist of schools for consideration for appraisal.
- Step 2: Identify the type of each of the schools on the shortlist and the number of pupils enrolled. Information about every school (including mapped schools, school type and number of pupils) is available from the Department of Education's performance tables (DoE, 2012a). Alternatively, the average number of pupils in a primary school in England is 300 and the average number in a secondary school is 1,000 (DoE, 2012b). These figures could be used, but actual values should be collected and used where possible.
- Step 3: Assess the likely impact of flooding on each school under different flooding likelihoods. This assessment needs to focus on the severity of the impact and the duration, and it should address the following questions:
  - Will the school be directly flooded or is it likely to be impacted only by access issues?
  - Is the entire school site impacted by flooding or would it be possible to continue to educate children in other school buildings once flood waters have receded and while buildings are being repaired?
  - How long is the school likely to be closed or partially closed? Note that half of all schools affected in Hull were able to open after one week.

It may be appropriate and proportionate to enumerate the losses for those schools which are likely to be closed for more than two days.

However, if several schools suffer only minor flooding or disruption, the cumulative losses at these multiple sites may be significant. The maximum disruption time for a school to be closed should be assessed at five days. After this period it would be expected that a school would be reopened or alternative accommodation secured (in either another location or temporary classrooms). Calculate the number of pupil days lost due to flooding for each school. This can be achieved by multiplying the number of pupils in the school by the numbers that are likely to be impacted (i.e. this will be all pupils if the whole school is closed or a proportion thereof if a school is partially closed) and then by the number of days the school is likely to be impacted.

- Step 4: Quantify the value of the loss of education. The first loss to calculate is a value attributed to the loss of a day of education (Equation 6.5). This step should apply to all of the pupil days lost. This is based on an approach adopted by both the National Audit Office (2005) to enumerate the cost of school absence and by Chatterton *et al.* (2010) to provide a minimum estimate of the value of school days lost. It is based on the equivalent daily costs to the local authority to educate a pupil. Annual pupil expenditure for every school in England and Wales can also be found in the Department of Education's performance tables (DoE, 2012a). Alternatively, national averages can be used, which for 2011/12 are calculated at £22.14 for primary schools and £23.67 for secondary schools per pupil per day, based on median values for all national (England) schools and 190 school days a year. Equation 6.5 should be applied to all pupils, irrespective of age.

Equation 6.5

Estimate of the value of school days lost = number of pupil days lost due to flooding × average daily expenditure per pupil

- Step 5: Quantify the paid productivity loss from parental absenteeism from work during the period of school closure. This second loss is based upon a method presented in Coulthard *et al.* (2007) and developed by Sadique *et al.* (2008) when considering closure of a school due to an influenza outbreak. Not all school-age children will require supervision if a school is closed. Therefore, secondary schools should be excluded from an analysis of the parent work days lost as it would be expected that the majority of children at these schools would be able to remain at home without supervision. Schools which educate a mixed range of children should be included but only an appropriate percentage of the pupil affected days should be taken to represent primary-aged children. There is also the need to account for: the presence of siblings within a school population (i.e. to avoid double counting the work days missed); that one parent may already be at home looking after younger siblings; that one parent may be unemployed; that some parents may choose to take annual leave (and therefore will not cause an economic loss); and that some parents may be able to make alternative childcare arrangements (i.e. grandparents or childminders). Indeed, the longer a school remains closed, the greater is the likelihood that many parents will be able to find alternative arrangements and a reduction in the wider impact on the economy through work days lost. Coulthard *et al.* (2007) divide the number of pupil days by two to account for the presence of siblings at the same school. However, we suggest that this is insufficient to account for all of the conditions described above, so we recommend dividing by three. Sadique *et al.* (2008) provide a more in-depth and complex way of analysing the percentage of the workforce that would be impacted by a school closure. This approach may be adopted if a more comprehensive analysis is required. Various values could be attributed to the loss of a working day. A minimum estimate might use the value of a work day lost at minimum wage (£43.46, based on the £6.31 per hour minimum wage for an adult and a 7.6-hour working day) or perhaps the cost of a day's work at an average wage (£76.77, based on an average hourly wage for a full-time person of £12.76 multiplied by a 7.6-hour working day) is more appropriate. Economic values (i.e. net of income tax and National Insurance contributions) have been provided, with gross pay values adjusted using the HMRC's Tax Calculator (HMRC, 2013) and a median hourly wage taken from 2012 values (ONS, 2012a). Equation 6.6 can be used to calculate the potential loss of work days due to the closure of a school.

Equation 6.6

Value of the loss of work days = (total number of primary-age pupil days lost/3) × average wage per day

6.6.2 Impacts of flooding on hospital services

Flooding to hospitals can result in major direct losses, but the associated disruption to hospital services can also have significant social and health impacts and ultimately, unless carefully managed, may result in fatalities or contribute to the early deaths of patients. Therefore, if hospitals are located within, or close to, a flood area, they need to be included within any project appraisal.

The wide variation in layout between hospitals and the presence of expensive equipment mean that a site survey is strongly recommended when appraising any hospital sites.

HR Wallingford (2012) reports that there are 590 care homes at moderate or significant flood risk in England and Wales, with over 15,000 client places, and that 7 per cent of medical services are located at flood risk (with 3 per cent of those at moderate or significant risk). Nevertheless, data about the impacts of flooding on hospitals are quite sparse. However, the summer 2007 flood event did provide some examples of hospitals being impacted by flooding in England and Wales. Pitt (2008) indicates that approximately ninety patients were evacuated from care homes and hospitals during these floods. The Department of Health (Whiteley, 2008) focuses on three main concentrations of flooding impacting hospitals: Gloucestershire; East Yorkshire (Hull); and South Yorkshire (Sheffield, Rotherham and Doncaster).

Gloucestershire Hospitals NHS Trust was primarily affected by an interruption of water supply, which was disrupted to some sites for up to eleven weeks (Gloucestershire Hospitals NHS Trust, 2008) and some hospital sites still had to rely on bottled water five weeks after the flooding (Whiteley, 2008). The lack of available fresh water and transport difficulties getting water to sites led to the extant Gold Command 'instructing the Trust to cease all hospital activity' (Whiteley, 2008, 1), leading ultimately to the cancellation of over 1,200 surgical operations and around 8,000 outpatient appointments (Gloucestershire Hospitals NHS Trust, 2008).

In Sheffield, there was no direct flooding of hospital facilities. The main issues related to continuity of electricity supply; however, in the most part hospital buildings with inpatients remained resilient to power outages due to prior investment in emergency generators. Whiteley (2008) discusses major problems due to transport grid-lock in the surrounding areas, meaning that many patients were unable to be discharged as they were unable to get home. However, traffic eased after a few hours and services within the hospital were minimally affected.

There were two main impacts in East Yorkshire. The Royal Hull Infirmary had to close all operating theatres for six hours following flooding of the main steam supply, and at Castle Hill Hospital in Cottingham the occupational health service was closed following the flooding of its building (Whiteley, 2008). A further risk was that electricity would be disrupted at the Princess Royal Hospital in Hull, but flood incident management activities by the Fire and Rescue Service and the Army ultimately prevented this.

In addition to the direct damages, concerns were raised about staff shortages due to a number of variables: their own homes had flooded; transport problems meant accessing hospital sites was difficult; and school closures meant that some staff needed to stay at home to provide childcare. Therefore, all outpatient appointments for one day were cancelled to ensure sufficient nursing staff were available for critical care (Whiteley, 2008).

Direct damages to hospitals

There are a few examples from National Health Service (NHS) hospitals in England and Wales which provide us with some indications of the costs of direct damages from flooding. Worthing Hospital has experienced flooding a number of times, most recently in summer 2012, when flood depths reached a maximum of 30cm and affected basements, the doctors' mess, the library and postgraduate medical centre, with a reported £350,000 damages. However, the Western Sussex Hospitals NHS Trust was required to pay only the insurance excess of £40,000 (NHS, 2012). No inpatients were affected by the flooding, although emergencies were diverted to other hospitals (Western Sussex Hospitals NHS Trust, 2012).

Table 6.24 Average damages per square metre expected from different depths of flooding for a hospital and a large hospital

Depth (cm)		–50	0	25	50	100	150	200
Hospital								
Total damage (£/m²)	Short (<12 hours) duration floods (£/m²)	15	55	340	535	925	1,220	1,655
	Long (>12 hours) duration floods (£/m²)	25	70	495	775	1,230	1,515	1,825
Large hospital								
Total damage (£/m²)	Short (<12 hours) duration floods (£/m²)	15	40	290	460	815	1,080	1,460
	Long (>12 hours) duration floods (£/m²)	22	55	420	670	1,070	1,320	1,590

Notes: Values calculated using an average floor space for smaller hospitals of 3,600m^2 and large hospitals of 10,000m^2; see Chapter 5

As stated above, we recommend a site survey to attribute both direct damages and disruption losses. However, hospitals were included within the analysis undertaken of non-residential property (NRP) assets in Chapter 5. Table 6.24 provides a summary of these results and presents estimated values of damage for a hospital (*c.*3,600m^2) and a large hospital (*c.*10,000m^2) when flooded at different depths for long-duration (more than twelve hours) and short-duration floods (less than twelve hours).

An assessment of the potential damages and losses to St Thomas' Hospital, London, was undertaken as part of the TE2100 study (Chatterton, 2008), which is summarised in Box 6.10. Additional information about flood risk to the same hospital trust can be found within a flood risk assessment undertaken in accordance with PPG25 sequential test for a proposed cancer centre at the hospital (Arup, 2012). This highlights many of the elements that may be affected in a complex of this size and may impact upon the continuity of the services on the site (e.g. sewer networks, water mains, etc.).

BOX 6.10 POTENTIAL DAMAGES TO ST THOMAS' HOSPITAL, LONDON, UNDER A TIDAL SURGE SCENARIO: AN EXAMPLE FROM TE2100 ASSESSMENT OF FLOODING IMPACTS IN THE THAMES ESTUARY

St Thomas' Hospital has a ground floor footprint of over 38,000m^2 and is part of the Guy's and St Thomas' NHS Trust, which in total has over 750,000 patient contacts each year. As a major teaching hospital for London, it provides general hospital services to three local London boroughs as well as additional specialist services to a wider catchment area. The potential threat of flooding has been recognised by the NHS trust, whose Emergency Planning Working Group investigated the impacts of a flood on the site and the services that they provide in 2007.

The TE2100 research concluded that flooding was likely to compromise a range of critical services and functions which are located in the basement or on the ground levels of the hospital. Flooding was therefore reported to have potentially major widespread impacts and 'lead to the *de facto* loss of routine hospital services throughout the Trust' (Chatterton, 2008, 7–8). Potential damages assessed to be caused by flooding by examining the hospital footprint under three different scenarios included:

- Medical equipment and stores:
  - £2.15 million of oncology equipment;
  - £1.5 million in pharmacy stores (equivalent to four or five days' supply);
  - £500,000 to £1 million of central store equipment;
  - £3 million of sterile services and decontamination units (two-thirds of operation kits); and
  - £5 million damage to PET radiation centre.
- Essential services:
  - major switch rooms;
  - water plant room affecting heating, drinking water and washing;
  - sewerage pumping station.

Flood depth	Impacts
Up to 0.1m above external ground level to account for sub-floor/ basement damages	Shut down of hospital (no services available for A&E, outpatients or wards)
	Direct damage: £10.5 million
	Clean-up and repair cost: £5 million
Up to 0.6m above external ground level	Shut down of hospital (no services available for A&E, outpatients or wards) with additional damages and repair costs included below
	Direct damages: £13.65 million
	Clean-up and repair costs: £10 million
Up to 2.0m above external ground level	Similar direct damages to the second scenario above but with the evacuation of the whole trust and an extended recovery period

Knock-on impacts of the flooding and closure of St Thomas' Hospital are likely to be widespread and severe. There is little spare capacity in London's busy hospitals (it is reported that they run at 95 per cent capacity). Additionally, under the flooding scenarios presented above, other local hospitals are also likely to be flooded. The impact on patient care is likely to be of far greater significance than the monetary damages caused by the flooding.

(Chatterton, 2008)

Losses due to the disruption of hospital services

Hospitals are complex facilities to investigate. As discussed above, they often comprise a number of buildings and/or sites and can have unique footprints.

As well as being dependent upon the usual range of services and networks (i.e. electricity, water, sewerage, communications and transport) hospitals depend on a range of other services which are critical to their running and the continuity of care. These include: catering; waste disposal services; clinical waste disposal; laundry; and many different stores services (including pharmaceutical, general stores, equipment, sterile). The availability and continuity of these services during a flooding event are further complicated by the fact that in some instances they are performed on site by the NHS trust itself, while in other contexts external private sector organisations carry out the services either on or off site. The flood events in 2007 highlight the vulnerability of hospital sites to disruption in services and utilities even if there is little direct flooding to a hospital's facilities.

In theory, there are a number of different ways in which the services of a hospital may be impacted, each of which may have economic cost implications. Some of these losses are direct in nature (such as the direct flooding of hospital infrastructure) whilst others are indirect (such as flooded roads leading to staff shortages and stopping patients attending their appointments). The total losses will also depend upon the scale of a closure (i.e. is only part of the hospital site affected?) and the duration of the impact.

Potential direct losses include:

- Direct damage to the building and fabric of the hospital.
- Damage to equipment and the closure or wards and other facilities (operating theatres, scanners, etc.) due to the direct flooding of hospital buildings.
- Costs of evacuation/transfer of patients to other hospitals or, in the case of some long-term-care elderly patients, to temporary alternative accommodation.

Potential indirect losses include:

- Losses attributed to the redundancy of hospital infrastructure (e.g. scanners and other equipment which cannot be used).
- Any increased costs due to the transfer of services to other health care providers. (If care is transferred *within* the NHS then there may be no net increase in costs, or only a marginal increase. However, there will be a greater increase in costs to the NHS if care is provided by a private hospital.)
- Increased staff and outpatient travel costs to alternative sites.
- Increased staff costs. If flooding prevents some staff from getting to work, it is likely that agency staff may be required to cover positions, entailing an increased cost.
- Closure of wards/equipment and the cancellation of appointments due to staff shortages.

Assessing the losses to a hospital is very complex in itself. Although there will be social costs to individuals through the cancellation of services or operations and a general reduction in the total available hospital resources, quantifying these impacts is complicated. For instance, strictly speaking, in the short term the closure of a ward may lead to an overall reduction in the direct costs to a hospital trust as running costs will be avoided.

In many situations, for example where acute care is closed (such as Accident and Emergency (A&E) departments or emergency operations), other health care facilities will need to cover these activities and so the direct costs will be transferred to these providers. In the most part these costs remain within the NHS and so there may be little increase to the UK as a whole. In addition, although hospitals may save money through a reduction in running costs, there may be many capital assets (such as scanners, theatres, etc.) lying idle, so there will be an overall loss to the country due to the non-operation of this equipment. Similarly, the terms of contract agreements with other service providers (e.g. laundry, catering and waste) may dictate that trusts are required to honour payment for these services which they are unable to use for a period of time.

However, the cancellation of services for any significant period may have longer-term implications for an NHS trust as it might mean that it is unable to reach its performance targets, thereby affecting its government funding for the next period. This occurred in the case of Gloucestershire Hospitals NHS Trust, which reported that missing patient targets meant that it did not 'receive the significant levels of related income that otherwise would have followed delivery of this activity' (Gloucestershire Hospitals NHS Trust, 2008, 33). However, this loss of income was mitigated through additional funding of £3.8 million provided by the South-west Strategic Health Authority and Gloucestershire Primary Care Trust. This money was said to cover the 'lost contributions to unavoidable expenditure (fixed/semi fixed costs)' and 'the net change in variable costs i.e. additional costs incurred to recover from the floods crisis less any costs not incurred as a result of reduced activity' (ibid). This net value of £3.8 million might therefore be taken as a proxy value for the net costs of damage and disruption due to flooding in this incident.

Understanding the dependency (and redundancy in the system) of the hospital services is critical to a decision about whether a hospital needs to be investigated further and whether the potential flood-related losses are significant. For instance, if a large inner-city teaching hospital is flooded then the number of people and services affected may be significant. However, there are likely to be other hospitals in the local area to which some of the patients and services may be transferred. A small rural cottage hospital may have fewer patients and therefore arguably may be less impacted by a flood. However, the likelihood of an alternative service provider within close proximity is also likely to be lower, so the costs of transferring services (e.g. through increased journeys) in these situations will be higher. One of the key elements is whether flooding will impact the NHS providing a service for emergencies (e.g. A&E) and other critical care, such as high dependency, critical operations, etc.) as these services need to be maintained and transferred to alternative providers. Impacts are therefore proportional to those key services which are being provided as well as to the transferability of those services.

If a hospital is within an area being investigated for an FRM project appraisal, a site investigation should be undertaken (including the location of the hospital's critical services) in order to appraise the impacts. Although there are some commonalities in the ways in which hospital services are organised within buildings, generally each hospital is unique and it is therefore necessary to understand the layout of a particular hospital (especially on the ground floor and in basements) and its location within the wider environment in order to appraise potential flood impacts and consequent scheme benefits. Therefore, we strongly advise assessors to engage directly with the hospital trust whose assets are deemed to be at risk.

Due to the inherent complexity, it is difficult to enumerate all of the potential losses here. However, this complexity should be captured at the very least within a qualitative narrative. Below are the basic steps that might be followed in order to appraise losses to hospitals:

- Step 1: Identify the location of the hospitals within the floodplain and close to the periphery of the floodplain (as these may also be impacted if major transport routes are cut).
- Step 2: For quantifying direct damages, identify the footprint of the hospital and those services at ground and basement level that may be impacted directly by flooding (i.e. number of wards and potential numbers of beds impacted that may be closed at different flood return periods). Utilise the NRP data in Chapter 5 to provide a basic assessment of loss that can then be refined through discussion with hospital authorities, especially if the hospital PVd constitutes a significant proportion (e.g. greater than 10 per cent) of the total potential losses.
- Step 3: Assess the likelihood that wards will need to be closed and the patients that might need to be transferred to alternative hospital/nursing home/hospice providers (e.g. if there is some redundancy on site, some patients may be cared for in other parts of the hospital site).
- Step 4: Where possible, enumerate the loss in value through the closure of a ward (i.e. the loss of bed space) (Equation 6.7) and the transport costs of the transfer of patients who would need to be evacuated from the hospital to another location (Equation 6.8). The first method of enumeration utilises the average cost of that asset per day to the NHS. Although not strictly an economic loss, it can provide a proxy value for the loss of use of a particular asset. For ward closures, identify the number of beds of a particular type that might be affected. For instance, different types of bed space have different day and night costs. Using the day costs of these will provide a maximum value. Table 6.25 provides data about the costs of beds. Where possible, it is also preferable to distinguish between the numbers of beds that cannot be used for general care, surgery and critical care, as the costs for these three types of care are quite different.

Equation 6.7

Costs due to the redundancy of beds = number of beds not available to be used × average cost per bed (distinguished, where possible, by the type of bed)

The same equation can be applied if medical procedures (such as operations) or other outpatient procedures are likely to be cancelled. NHS reference cost information is provided annually and gives

Table 6.25 Average costs of a hospital bed

Type of hospital bed	Average bed cost in the NHS*	Average bed day cost for elective and non-elective admissions** (surgery and general medicine)	Average bed day cost for critical care
Average cost of a bed per day (2011–12)	£261	£530	£983

Notes: * This value includes only the cost of NHS beds and excludes contracted-out services; ** The figures for critical care and general and surgery beds include the cost of treatment.

Source: Data supplied by the NHS

average costs of the various procedures in England and Wales. These can be found for 2011/12 on the government's website (Department of Health, 2012) and, if appropriate, used to enumerate the cost of cancelled services. For instance, the average unit cost of treatment in A&E is approximately £100, calculated by averaging all of the unit costs of admitted and non-admitted treatments at A&E (including minor injury services), taking account of the numbers attending A&E for each level of treatment in 2011/12. A value such as this might be used to give a very approximate value to the closure of this emergency service if multiplied by the number of people usually treated over the period a service is likely to be closed.

Equation 6.8

Costs of transporting patients = (number of patients who might need to be evacuated – the likely redundancy in the same hospital) × the average cost of a patient transfer

- Step 5: Qualitatively investigate other impacts. Discuss with the hospital trust the likely effects of flooding on the continuity of services. The continuity of the following essential services needs to be included within the narrative:

  - electricity supply;
  - water supply;
  - sewerage services;
  - laundry services;
  - catering services;
  - waste disposal services;
  - clinical waste services; and
  - sterile services.

  In addition, investigate other aspects, such as whether sufficient staff are able to access hospital buildings. Identify what the hospital is likely to do under different scenarios and how long disruption is likely to last. Learning from 2007, the likely impacts include:

  - the closure of wards which are threatened with flooding;
  - the closure of A&E services;
  - the cancellation of operations; and
  - the pre-emptive cancellation of outpatients and cancellation of other non-urgent treatment.

Broad annual statistics for hospitals in England and Wales can be found at the NHS Information Centre (http://www.ic.nhs.uk/home). This includes information such as the annual number of inpatients/outpatients, the numbers of hospital admissions and the total number of annual patient contacts. These can be used to calculate average daily contacts and to contextualise the level of disruption that might occur during floods. Target and performance statistics (such as how close to capacity a hospital runs) can also be used to understand broadly the level of redundancy in the system and the ability of the services to be transferred.

6.7 OTHER LOSSES IN RESPECT OF 'CRITICAL' INFRASTRUCTURE

There are many other assets which might be considered for inclusion within an appraisal for flood risk management measures. For instance, HR Wallingford (2012) assesses that there are eighty ambulance stations, 120 fire stations, forty police stations and fifty coastguard stations at moderate or significant risk of flooding. Flooding of any of these may impact on the ability of services to respond and fulfil their emergency response roles. However, as Category One responders under the Civil Contingencies Act 2004 (HM Government, 2004), each of these should have resilience plans in place and should have undertaken assessment of their own flood exposure. Therefore, if assets of this type are identified within an appraisal area then discussions with the appropriate emergency planners for the organisations and/or the local resilience forum should be able to provide useful appraisal information and begin to quantify the extent of the expected damage and disruption.

In addition, there are likely to be more universal facilities which would be at risk from flooding and may cause disruption to the public, businesses or industry following flooding. These assets may include: nuclear power stations; ports; airports; defence bases; and postal facilities. Some of these may not be directly flooded themselves but may instead be impacted by disruption to local transport networks.

Each of these more unusual facilities would need to be assessed on a case-by-case basis through a site survey and discussion with managers or owners. The following questions (amongst others) may assist this appraisal process:

- How susceptible is the asset to flooding?
- Is it likely to be flooded directly or impacted by the disruption of supply in other services?
- How long is the asset likely to be impacted?
- How many households/businesses/other services are affected by the disruption to this asset? What is the level of dependency on this resource?
- What is the redundancy of the service/resources? Is it a service that can easily be transferred to an alternative supplier? (For instance, in the case of the postal service, not only might mail transport be re-routed but there are alternative postal suppliers and alternative mechanisms for electronically delivering documents. Therefore, dependency will usually be low and redundancy high.)
- What do the asset owners/managers understand by resistance and resilience planning and what activities have they already undertaken?

The scale of the losses here, however, needs to be carefully assessed as the knock-on costs of disruption will be higher for some than for others. For instance, as described in Section 6.4.1, the impact of flood-affected power stations on disruption is relatively low as they are not single points of supply, and power generation can be switched to other parts of the National Grid as necessary. Therefore, in this case, losses will be entirely due to direct damages. For others, such as postal services, there may be a high degree of redundancy within the service, whereby any road or rail link might be used to transport mail, albeit potentially at an increased cost. Of course, if flooding is widespread then the postal network would be more vulnerable to disruption and would be impacted by electricity supply disruption similar to many other modern services. Locally, in flooded areas, services will be severely impacted, which may have knock-on impacts for both individual households and businesses, although with the increased uptake of more technology-based communication, such as e-mail, the significance of any disruption may have decreased.

It will be easier to assess the impacts on some establishments than others. Appraisers are unlikely to be able to gather more than approximate, generalised data for some of the facilities, as site-specific data may be unavailable. For instance, for security reasons, specific data are unlikely to be forthcoming from the Ministry of Defence, and similar issues may also be associated with penal facilities. Additionally, these large, valuable assets are more likely to have invested in resilience activities (or are likely to do so) and these activities/measures must be accounted for within appraisals to avoid the double counting of these benefits.

Despite the problems associated with including these more unusual facilities within appraisals, efforts should be made to do so. Some of these assets may make a significant contribution to the total benefits within an area and under the conditions of partnership funding the operators or owners of these facilities may be identified as potential beneficiaries from a flood risk management scheme and therefore potential contributors to funding.

6.8 EMERGENCY SERVICES AND RELATED COSTS, AND ENVIRONMENT AGENCY COSTS

The benefits of flood risk management include reducing the costs incurred by a number of organisations in tackling flood incidents and in the recovery process. This section covers research into both the autumn 2000 floods (in England and Wales) and the 2007 floods (predominantly in England). Our investigations were more detailed for the former event, and the results remain highly relevant, and the later research followed the same methodology and generally used the same data sources. The contrast in the results is important (see Section 6.8.5).

Depending upon the severity of the flood event, several emergency services may be involved in both emergency works and clean-up operations, during and after the flood event (for more details see Parker *et al.*, 1987, chapter 10). Extra staff time and materials may be required, and additional administrative costs may be involved. Authorities and bodies providing emergency services include the following:

- local authorities;
- police authorities;
- fire services;
- ambulance operations;
- the Environment Agency/Natural Resources Wales;
- voluntary services; and
- the armed forces.

It would appear from floods in the past that the majority of the costs are incurred by local authorities, police authorities and fire services, and that the major expense is often sandbagging. Within local district, county or unitary authorities, the departments involved can include social services, emergency planning, highways and engineering. When appraising a particular scheme, these departments, and the police and fire authorities, should be contacted for any appropriate information they may hold on their costs incurred during past flood events.

Care should be taken in this exercise to separate fixed costs from marginal costs. Local authorities and the Environment Agency have staff who are employed specifically to deal with emergencies, and a reduction in flooding will not necessarily lead to a reduction in these costs. Similarly, both the police and the fire authorities are themselves emergency services, and the reduction in flooding or coastal erosion would not necessarily reduce the costs to the nation of these services. Therefore, their fixed costs cannot legitimately be included within the benefits of flood risk management. Nevertheless, all these emergency services may incur extra costs as a result of particular flood events (marginal costs), which may be counted in the benefits of flood risk management.

Previous research has investigated these costs in detail – Parker *et al.*, 1987, chapter 10; Penning-Rowsell *et al.*, 2005a, chapter 6 – with the former using a scenario approach to develop synthetic emergency services costs, in the absence of a major recent flood. Parker *et al.* (1987) provided standard emergency services costs per property flooded in urban areas for a range of flood scenarios and flood durations. The maximum figure reported was £511 (January 1985 prices; equating to £1,265 in January 2013 prices). This figure serves as a context to the information contained below.

6.8.1 Emergency services and related losses: the Bellwin claims procedures

Our research for this section used information obtained in 2004/05 from the Department of Transport, Local Government and the Regions (DTLR; now the Department for Communities and Local Government (DCLG)) on emergency costs incurred by local authorities and others in the autumn 2000 flood events. Organisations active in the flood management and recovery phases are allowed to recover a proportion of their costs from central government under what is termed the Bellwin process.

Analysis of the Bellwin submissions in year 2000 from local and police authorities to the Department of the Environment, Transport and the Regions (DETR) indicates that 'Red Manual' techniques (Parker *et al.*, 1987) for the evaluation of emergency response costs during and following a flood event had undervalued these costs. This was partly because no representative data were available with respect to emergency costs prior to the widespread floods of autumn 2000. The costs of sandbagging operations and the emergency repair costs to roads, bridges and drainage infrastructure had previously been largely unavailable.

This section looks at both the Bellwin process and the national data for the year 2000. We also analyse (Section 6.8.3) the details of a specific claim in 2000/01 from North Yorkshire County Council (NYCC) and its seven district councils, one unitary authority (York) and the police authority within the area of NYCC's jurisdiction. Finally, we also review the same types of data sources for the local authority and emergency costs incurred during the 2007 floods.

The DTLR triggered the Bellwin process in 2000/01 to make available financial assistance to local authorities in England (as defined in Section 155(4) of the Local Government and Housing Act 1989 (HM Government, 1989)) in respect of storms, floods and other adverse weather conditions occurring between 1 October and 31 December 2000 and involving destruction of – or danger to – life or property. A similar trigger was initiated for Wales by the National Assembly for Wales. Claims above a pre-determined threshold were audited and eligible expenditure reimbursed through the DETR at 100 per cent of costs (as against the 85 per cent set out in the normal Bellwin procedures; see below) owing to the severity of the event.

The Bellwin system insists that eligible expenditure must be made 'on or in connection with the immediate action to protect life or property' (HM Government 1989, Section 155). The system of thresholds is based on the judgement that prudent authorities should budget to cover a proportion of the costs of emergencies from their own reserves and resources. Ministers have taken the view that it is reasonable to expect authorities to make provision for contingencies at a level of approximately 0.2 per cent of their total budget requirements. In general, the Bellwin rules of eligibility are designed to identify the marginal costs of the flood or other disaster event. This closely matches the national economic losses/costs, although the reimbursed cost of fuel, for example, is inclusive of taxes and therefore exaggerates economic costs. Unfortunately, however, no overall breakdown of Bellwin data is available to guide a more detailed analysis.

The following are eligible under the Bellwin rules:

- the cost of taking immediate action to safeguard life and property or prevent suffering or severe inconvenience; and
- emergency works usually within a deadline of two months from the date of the incident.

The following are not eligible under the Bellwin rules:

- insured costs, whether by the authority or a third party;
- costs recoverable from businesses; and
- capital expenditure.

In this respect, items of a capital nature or items insured by third parties are not eligible (for more details, see Penning-Rowsell *et al.*, 2002). Staff costs relate only to additional resource costs over and above normal salaries. It is recognised that the summary of local, fire and police authority costs will marginally underestimate actual costs because of omission of ineligible costs eliminated during the audit process.

The Bellwin scheme procedures were revised following the 2000 floods and a review and consultation exercise was held (DTLR, 2001b) to address the shortfall in funding which occurred in 2000. Annual guidance provided by the DCLG describes the current procedures and rules to which local authorities have to adhere when claiming Bellwin assistance (DCLG, 2012c). Usually, when the Bellwin scheme is activated, 85 per cent of emergency expenditure is funded; however, this was raised to 100 per cent for the autumn 2000, June 2007, November 2009 and June–July 2012 flooding incidents (Sear, 2012).

Due to the inadequacy of the Bellwin scheme arrangements in 2000, supplementary payments were made in England and Wales to local highways authorities for flood damage caused to roads and bridges in the autumn and winter of 2000/01. The payments ranged from £97,960 for Torfaen in Wales to £7.06 million for Kent, with totals of £41.9 million for England and £17 million for Wales, and a mean payment per authority of £1.9 million for England and £780,000 for Wales.

Local authorities in England and Wales are also eligible to apply for financial aid from the European Union Solidarity Fund (EUSF) in the event of major natural disasters, including floods (Council of the European Union, 2002). The summer 2007 floods qualified as one of the forty-nine EUSF interventions since 2002, whereby €162.3 million was granted in aid to the UK (European Commission, 2013). These funds provide another source of data relating to the recovery costs from the summer 2007 floods.

6.8.2 The Bellwin and severe weather capital grants scheme data for autumn 2000

MCM-Online (Supporting Information for Chapter 6) details the Bellwin thresholds and claims made by

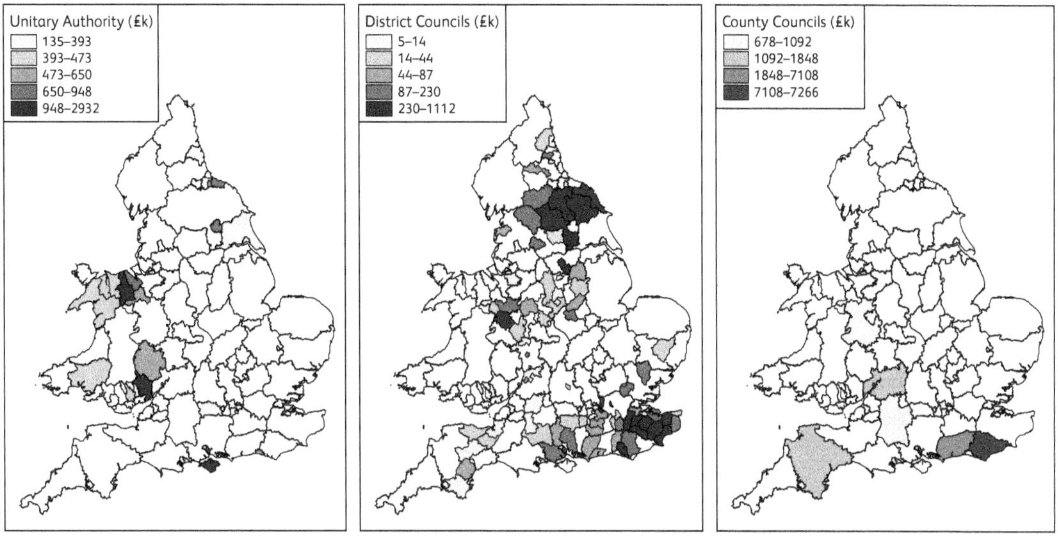

Figure 6.6 Bellwin claims by unitary authorities, county councils and district councils (£thousand)

Source: Penning-Rowsell *et al.*, 2002

each authority following the autumn 2000 floods. In summary, claims were made for £21.3 million, with total thresholds of claimants equal to £7.5 million, equivalent to emergency costs of a baseline £28.8 million. Claims for Welsh authorities amounted to £5.5 million, with total threshold values of £2.1 million. Most of these claims were from district councils, county councils (including, under the Bellwin regulations, fire services) and unitary authorities, as well as three police authorities (North Yorkshire, South Devon and Cumbria). Figure 6.6 shows the geographical distribution of Bellwin claims from local authorities (see also Penning-Rowsell and Wilson, 2006).

Most of the claims here were for overtime for emergency services, including filling and distributing sandbags. However, details are provided on Bellwin claims only when these are below £10,000, because in such circumstances they are not audited locally but are available for audit by the DCLG. (The scheme was administered at the time of the 2000 floods by the DTLR.) However, in the case of the autumn 2000 floods, only seven were below £10,000 (out of the 114 claims received for England and Wales). To fill the gap in the level of detail provided, the North Yorkshire case study (Section 6.8.3) sheds valuable light on the breakdown of emergency costs.

A number of important issues are raised:

- In terms of overall context, councils have a responsibility to co-operate in disaster situations. If this effort is for the benefit of one council rather than for all of them then it is legitimate for all Bellwin claims to go through that council. Thus, for example, Tendridge suffered a serious problem in 2000, and the London Borough of Croydon (downstream) undertook much of the work, constructing a temporary dam. Therefore, Croydon charged Tendridge District Council for its costs, and Tendridge then sent the claim on to the DETR (having deducted the threshold value).
- Some thirty-one local authorities did not claim as their claims were below the Bellwin threshold. Assuming an average emergency cost for each of these councils of half the threshold, emergency costs may be assumed to amount here to £3.1 million.
- In the case of fire authorities, these were not allowed to hold reserves and did not have Standard Spending Assessment allocations. Section 155 of the Local Government and Housing Act 1989 does not permit fire authorities to make direct Bellwin applications. Metropolitan fire authorities can claim from the Bellwin fund, but combined fire authorities claim through their constituent authorities (under the Fire Services Act 1947 (HM Government, 1947) local authorities can combine

to provide fire services). Bellwin claims are submitted to their constituent authorities by these fire authorities, or a bill is sent to the authorities in the areas where action is taken. County councils do have reserves, and therefore do have to pay these bills, but their Bellwin thresholds are high.

6.8.3 The situation in Yorkshire in autumn 2000: a case study

North Yorkshire's County Council (NYCC), its fire service and its police authority were all affected in 1999 to 2001 by a number of disasters, each of which required revenue expenditure over and above that allocated as a general reserve for emergencies. The Norton and Malton floods of 1999 were quickly followed by the widespread flood events of autumn 2000, affecting all district councils within the shire county's jurisdiction and the York Unitary Authority. The fuel delivery crisis of autumn 2000, the foot and mouth epidemic and the Great Heck train disaster soon followed.

However, the 1999 Malton/Norton flood event set a precedent for Bellwin claims to the DTLR (DETR, as was) which differs from the claims made by other county authorities (e.g. Gloucestershire) in that the DTLR allowed the majority of the county expenditure to be claimed via districts.

Table 6.26 Principal costs for NYCC

A	Gross revenue costs	£1,845,000
B	Disqualified costs (by auditor)	£6,000
C	Combined district Bellwin threshold	£129,000
D	Revenue costs incurred but not eligible through Bellwin	£119,000
E	Expected Bellwin payment (i.e. A–B–C–D)	£1,591,000

The councils' costs

Headline costs to NYCC are summarised in Table 6.26 (see also Figure 6.7). The £119,000 'disqualified' costs relate to expenditure within departments of NYCC for which the district councils have no responsibility (e.g. Social Services; Education). Thus, extra costs for home carers as a result of delayed journeys because of flooding, and sandbagging and pumping out flood water from schools were excluded from the claim. However, where two schools in Selby were used as evacuation centres, the costs were eligible as the schools were being used for a community-wide service. In short, where the district council and community issues were paramount, the NYCC costs were allowable in the district council claim.

The details of the district council claims were not available to us at the time of that research, but they

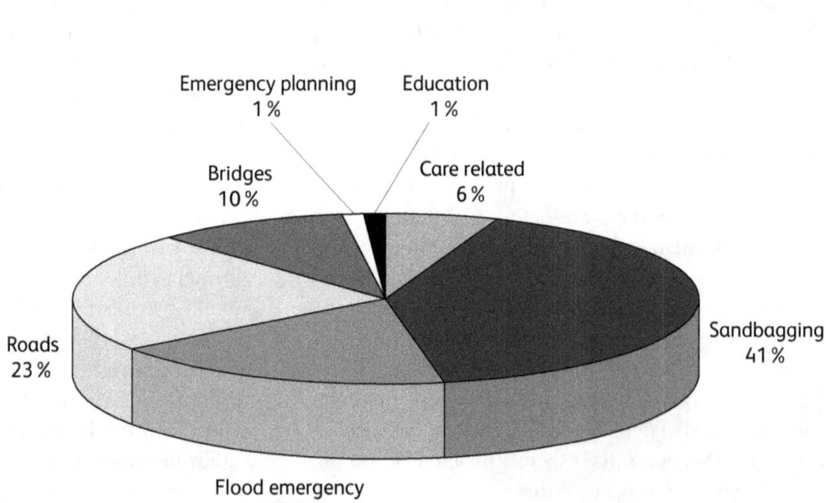

Figure 6.7 NYCC cost summaries

Table 6.27 Summary of NYCC, police and fire service emergency costs (North Yorkshire Police Authority)

Authority	Cost	% of total
North Yorkshire County Council	£1.845 million	39.34
(NYCC Bellwin claim)	(£1.733 million)	−36.95
Fire Service	£0.414 million	8.83
North Yorkshire Police Authority	£0.681 million	14.52
District Council*	£0.750 million	15.99
York City Unitary Authority*	£1.000 million	21.32
Total	**£4.690 million**	

Note: * District Council and York City costs remain estimated, with Selby accounting for half the District Council costs. Most of the District Council costs relate to staff overtime, particularly filling sandbags.

were generally less than £50,000 each (some did not even reach their Bellwin threshold until NYCC added to the claims). Most of the costs related to overtime for individuals staffing emergency centres and providing general assistance.

Tables 6.27 and 6.28 give details of all NYCC costs claimed through Bellwin. Figure 6.7 shows that 92 per cent of the total NYCC cost related to emergency repair of bridges, highways, flood 'fighting' and sandbagging. Indeed, sandbagging operations alone constituted 41 per cent of all NYCC costs. In addition to the £1.845 million headline figure for emergency works, an estimated £3.25 million was budgeted for replacing roads and bridges and £250,000 for capital repairs to forty flooded schools (five schools required major capital works).

Table 6.27 gives details of the Bellwin claim for fire services made within the NYCC claim as allowed by the Bellwin rules. Fire services in North Yorkshire are funded 77 per cent by NYCC and 23 per cent by York Unitary Authority; fire service cost claims were therefore made by NYCC and York separately. Overtime represents nearly three-quarters of the fire service costs, with supplies and transport making up the other quarter.

The North Yorkshire Police Authority

Under Section 25 of the 1996 Police Act, police authorities could opt for a direct claim to Bellwin. Having incurred substantial foot and mouth and Great Heck train disaster costs, in addition to the flood emergency costs, the police authority did not want to delay reimbursement decisions by making a combined claim with the local authority.

The fire service

The auditors refused very little of the claim as ineligible prior to submission. Between 80 per cent and 90 per cent of the costs relate to officer overtime (see Table 6.28), but overtime does not include 'consequential' overtime (i.e. overtime worked on non-flood duties because officers have been called out to flood duty). Most of the overtime involves setting up roadblocks; assisting local authorities, the Environment Agency and the military; facilitating evacuations; and attending flood-related road traffic accidents.

6.8.4 The Environment Agency's costs in 2000/01

The regional and area (where available) 'lessons learned' reports detail the total emergency response and emergency repair costs to the Agency in the autumn 2000 events (Environment Agency, 2001). Only the summaries of these are presented here.

The available data and results

Table 6.29 details the emergency repair costs of £10.292 million and emergency response costs of £8.442 million for England in the 2000/01 floods. The combined emergency repair and response costs for Wales totalled £780,000. Table 6.30 details the additional data archiving, flood reporting and Flood Line costs (both internal and external to the Agency). (See also Figures 6.9 and 6.10.)

The total of this extra statutory expenditure in 2000/01 was £2,368,800. Total Environment Agency costs directly associated with the autumn 2000 floods were therefore £21,102,800 (excluding Wales's emergency repair and response costs).

Table 6.28 NYCC, fire service and police authority costs associated with autumn 2000 floods

Category	North Yorkshire County Council, fire and police service spend in each						
	Selby		Ryedale		Harrogate		Hambleton
	£	No. of cases	£	No. of cases	£	No. of cases	£
Care-related services							
Nursing home evacuation	101,790	5					2,657
Total	**101,790**	**5**					**2,657**
Flood alleviation							
Sandbagging	471,379		15,917		53,508		48,715
Emergency costs	28,072		21,604		174,000		50,617
Total	**499,451**		**37,521**		**227,508**		**99,332**
Highways and bridges							
Emergency repairs – roads	1,877	1	58,959	19	111,381	9	37,922
Emergency repairs – bridges	5,809	2	2,287	1	62,761	7	20,268
Total	**7,686**	**3**	**61,246**	**20**	**174,142**	**16**	**58,190**
Emergency planning							
Evacuation	12,700						
Total	**12,700**						
Education related							
Evacuation – Sherburn school	10,015	1					
Total	**10,015**	**1**					
Total North Yorkshire CC	**631,642**	**9**	**98,767**	**20**	**401,650**	**16**	**160,179**
	£	%	£	%	£	%	£
Fire services							
Overtime/additional staff	102,479	72	23,164	72	24,355	72	27,808
Travel/subsistence	1,194	1	270	1	284	1	324
Supplies	15,818	11	3,580	11	3,764	11	4,298
Premises/telecommunications	1,310	1	296	1	311	1	356
Transport	22,324	16	5,046	16	5,305	16	6,057
Fire Service Total	**143,125**	**34.6**	**32,356**	**7.8**	**34,019**	**8.2**	**38,843**
North Yorkshire Police							
Overtime police	263,844	80.4	100,648	90	1,224	68.5	15,317
Overtime support	2,077	0.6	1,193	1.1		0	
Mutual aid (other forces)	34,073	10.4		0		0	
Travel, subsistence, accommodation	0	0	271	0.2		0	
Vehicle costs	750	0.2	7	0	400	22.4	46
Other costs	1,835	0.6	979	0.9	25	1.4	
Command/authority wide	25,523	7.8	8,696	7.8	139	7.8	1,296
Police Authority Total	**328,102**	**48.2**	**111,794**	**16.4**	**1,788**	**0.3**	**16,659**
Grand Total	**1,102,869**	**39**	**242,917**	**8.6**	**437,457**	**15.5**	**215,681**

Notes: Where number of cases, i.e. incidents, is quoted these are the most significant separately identified incidents; York is a unitary authority and North Yorkshire's costs do not apply here; all the North Yorkshire CC costs reflect the local assistance to the District Councils; District Council actual costs are not included; the North Yorkshire costs are the allowable Bellwin cost; a further

District council										
	Craven		Richmondshire		Scarborough		York		Total	
No. of cases	£	No. of cases	£	No. of cases	£	No. of cases	£	No. of cases	£	No. of cases
2							not available		104,447	7
2									**104,447**	**7**
	58,537		35,045		30,465		not available		713,566	
	5,866		20,344		19,140		not available		319,643	
	64,403		**55,389**		**49,605**				**1,033,209**	
18	45,626	12	18,434	12	126,891	17	not available		401,090	87
4	57,849	3	21,157	6	940	1	not available		171,071	22
22	**103,475**	**15**	**39,591**	**18**	**127,831**	**18**			**572,161**	**109**
									12,700	
									10,015	**1**
24	**167,878**	**15**	**94,980**	**18**	**177,436**	**18**	**0**	**0**	**1,732,532**	**117**
%	£	%	£	%	£	%	£	%	£	%
72	13,512	72	11,169	72	16,457	72	77,510	72	296,454	72
1	157	1	130	1	192	1	903	1	3,454	1
11	2,088	11	1,726	11	2,544	11	11,980	11	45,798	11
1	173	1	143	1	211	1	990	1	3,790	1
16	2,943	16	2,433	16	3,585	16	16,885	16	64,578	16
9.4	**18,873**	**4.6**	**15,601**	**3.8**	**22,989**	**5.6**	**108,268**	**26.1**	**414,074**	**100**
91.9	500	92.3	892	91.2	4,713	91.8	179,723	83.1	566,861	83.2
0		0		0		0	5,059	2.3	8,329	1.2
0		0		0		0	631	0.3	34,704	5.1
0		0	10	1		0	1,039	0.5	1,320	0.2
0.3		0		0	20	0.4	1,576	0.7	2,799	0.4
0		0		0		0	11,541	5.3	14,380	2.1
7.8	42	7.7	76	7.8	399	7.8	16,834	7.8	53,005	7.8
2.4	**542**	**0.1**	**978**	**0.1**	**5,132**	**0.8**	**216,403**	**31.8**	**681,398**	**100**
7.6	**187,293**	**6.6**	**111,559**	**3.9**	**205,557**	**7.3**	**324,671**	**11.5**	**2,828,004**	**100**

£119,000 was not eligible through Bellwin and the auditors dismissed a further £6,000. The percentage values reflect the monetary spend as a percentage of the grand total in the last column. Due to rounding, some percentages will not sum to 100.

Table 6.29 Environment Agency emergency repair and response costs

Flood defence committee/region £000s	Original budget 2000/01	1 % ceiling	Total emergency repairs*	Total emergency response*	Downpayment allocation**	Balance due (repayable)
North West	26,576	266	380	110	0	224
Northumbria	4,158	42	195	131	406	–122
Yorkshire	25,248	252	3,942	2,432	4,009	2,113
Midlands	30,658	307	1,750	786	968	1,261
Southern Region						
Hants & Isle of Wight	3,472	35	159	106	181	49
Kent	13,096	131	1,106	741	869	847
Sussex	11,214	112	401	838	851	276
Southwest	8,612	86	374	134	280	142
Avon & Dorset	5,043	50	276	187	163	250
Bristol Avon	3,323	33	163	55	78	107
Somerset	5,700	57	340	552	317	518
Thames	70,300	703	1,167	2,095	878	1,681
Head Office	0	0	–	10	0	10
SUB TOTAL	**207,400**	**2,074**	**10,253**	**8,177**	**9,000**	**7,356**
Anglian Region						
Essex	10,590	106	2	60	0	0
Great Ouse	11,232	112	25	75	0	0
Lincoln	13,588	136	–	90	0	0
Norfolk/Suffolk	16,063	161	–	10	0	0
Well & Nene	7,479	75	12	30	0	0
TOTAL ENGLAND	**266,352**	**2,664**	**10,292**	**8,442**	**9,000**	**7,356**

Notes: * Includes emergency repair and response costs for 2000/01 and 2001/02; ** Provisional allocation by Head Office Finance pending confirmation of final out-turn.

North-east and southern regions represent well over half of all costs. The disproportionately high emergency response costs in relation to emergency repair costs in the Thames and Anglian regions reflect the cross-regional assistance provided. Total emergency repair costs were 49 per cent of total Environment Agency costs whilst emergency response costs were 40 per cent. Costs of the preparation of post-event reports and collection of data, condition surveys and mapping by the Environment Agency and its consultants represented a significant 11 per cent of total costs.

Issues relating to the Environment Agency costs

EVENT COSTS VERSUS TOTAL RESOURCE COSTS

Given that the Environment Agency, like the police, has a role in emergencies, a key issue relates to 'additionality' (additional resource costs) versus what is spent year on year as a provision for an emergency, and establishing whether each region, in reporting the

Table 6.30 Environment Agency additional costs

Region £000s	External						Sub total	Internal	Total costs	Out-turn 31.2.2001	Estimated carry over
	A	B	C	D	E	F					
North East	0	444	148	138	23	36.3	789.3	267	1,056.3	0	300
Midlands	0	33	45	5	0	40.8	123.8	80	203.8	150	0
Southern	0	24	56	0	129	28.2	237.2	73	310.2	0	0
Anglian	0	0	5	0	0	46.5	51.5	0	51.5	52	0
Thames	0	79	37	40	57	74.4	287.4	102	389.4	731	0
South West	39	95	10	0	0	24.3	168.3	24	192.3	41	100
North West	0	3	0	2	9	33.6	47.6	55	102.6	103	0
Wales					46.8	15.9	62.7		62.7	0	0
Total	**39**	**678**	**301**	**185**	**264.8**	**300**	**1,767.80**	**601**	**2,368.80**	**1,077**	**400**

Notes:

A = Photogrammetry and LIDAR
B = Ground surveys
C = Aerial photographs
D = Condition surveys
E = Flood report lessons learned (estimated internal costs allocation to flood report £100,000)
F = Extra Flood Line call costs (memo dated 23.11.2000 from Manoch Kerman)

All the above attract extra statutory contribution of 100% subject to a ceiling of £2.5m. Costs which could be classed as pre-feasibility should be included in this schedule.

results reproduced in Table 6.29, applied similar rules in respect to this.

As with the Bellwin philosophy, the Department for Environment, Food and Rural Affairs (Defra) will grant the Agency any repair and response costs over 1 per cent of its annual budget provision (England), and this happened in 2000/01 for sums in excess of 1 per cent of the Agency's regional budgets. Claims were allowable only if the costs were above normal operating costs and therefore, in principle, their magnitude should usefully reflect the true marginal costs of the emergency response.

In short, base-level operating costs are not reflected in the claim and therefore, in principle, their use in project appraisals, suitably adjusted as marginal costs, would appear to be appropriate.

Health, stress and effects on personnel during and after the 2000 flood events

The autumn 2000 flood events (15 September to 15 November) saw enormous pressures on Environment Agency staff and it is likely that the length of the events had significant impacts on staff health, stress levels and general morale. The cost of absenteeism after 15 November as against a normal November could be a useful measure of the 'human' cost or toll of the flood events, but this 'knock-on' effect has not been included in the data given above.

The costs of Floodline and automatic voice messaging (AVM)

A vital part of the Agency's flood risk management armoury is the AVM/Floodline to assist the flood-prone public to mitigate their personal injury/stress and allow a reduction in property damage. The

Table 6.31 Flood-related expenditure for the sixteen most affected local authorities in the 2007 floods

Category of expenditure	Hull CC		Sheffield CC		East Riding DC		Oxfordshire CC
	(£)	(%)	(£)	(%)	(£)	(%)	(£)
General expenditure/costs [1]	8,943,503	21.5	3,330,100	14.8	4,202,845	14.5	479,763
Highways [2]	38,001	0.1	13,538,994	60.2	10,080,903	34.8	702,639
Recovery [3]	2,718,200	6.5	4,047,170	18.0	960,551	3.3	0
Schools [4]	13,448,000	32.4	469,646	2.1	9,973,116	34.4	1,464,710
Social services [5]	3,017,618	7.3	1,497	0.0	0	0.0	41,502
Housing [6]	13,109,972	31.6	1,079,850	4.8	3,539,760	12.2	508,451
Waste [7]	236,000	0.6	4,811	0.0	201,922	0.7	0
Total	**41,511,294**		**22,472,068**		**28,959,097**		**3,197,065**
	West Berkshire UA		Worcestershire CC		Doncaster MBC		Nottinghamshire CC
	(£)	(%)	(£)	(%)	(£)	(%)	(£)
General expenditure/costs [1]	742,352	16.3	10,839	0.2	3,456,693	24.9	–
Highways [2]	1,410,872	31.0	4,340,553	73.1	853,552	6.2	2,669,479
Recovery [3]	1,033,875	22.7	–	0.0	568,051	4.1	–
Schools [4]	1,103,742	24.2	1,261,956	21.3	2,564,359	18.5	1,399,036
Social services [5]	136,938	3.0	216,219	3.6	40,000	0.3	–
Housing [6]	706	0.0	12,572	0.2	6,393,674	46.1	–
Waste [7]	127,000	2.8	92,500	1.6	–	0.0	–
Total	**4,555,485**		**5,934,639**		**13,876,329**		**4,068,515**

Notes:

The shading highlights the highest percentage costs/expenditure by each local authority. Due to rounding, some percentages will not sum to 100.

[1] Significant costs/expenditure in this category include: overtime/temporary employment, lost council tax income, lost income from parking and sports centres, repair and restoration of parks and sports facilities, the hire of additional vehicles/machinery, repairs to street furniture, safeguarding dangerous structures.
[2] Includes work to clear debris, repairs to highways, pavements, etc.
[3] Includes grants to flood victims, advice centres, grants to charitable organisations, laundry facilities, disposal and safety of domestic electrical items.
[4] Repairs to schools, cost of temporary classrooms, support services for pupils.
[5] Reinstatement of properties, cost of alternative accommodation, lost income, evacuation of elderly social care patients, etc.
[6] Repairs to council houses, alternative accommodation to flood displaced, lost rental income, security costs of empty buildings.
[7] Additional landfill costs, penalties for not recycling, lost income at civic amenity sites.

Source: Chatterton *et al.*, 2010

	Cotswold DC		Herefordshire CC		Barnsley MBC		Cheltenham BC		North Lincolnshire DC	
(%)	(£)	(%)	(£)	(%)	(£)	(%)	(£)	(%)	(£)	(%)
15.0	2,040,467	92.2	165,055	4.0	154,621	2.3	8,248,900	96.5	949,821	18.0
22.0	0	0.0	3,114,663	74.7	5,141,080	77.7	–	0.0	4,227,345	80.0
0.0	171,917	7.8	325,592	7.8	433,549	6.5	13,200	0.2	99,286	1.9
45.8	0	0.0	563,695	13.5	127,589	1.9	–	0.0	8,627	0.2
1.3	0	0.0	0	0.0	18,634	0.3	–	0.0	–	0.0
15.9	0	0.0	0	0.0	731,317	11	282,000	3.3	–	0.0
0.0	0	0.0	0	0.0	13,513	0.2	1,000	0.0	–	0.0
	2,212,384		**4,169,005**		**6,620,303**		**8,545,100**		**5,285,079**	

	Rotherham MBC		Gloucestershire CC		Shropshire CC		Total
(%)	(£)	(%)	(£)	(%)	(£)	(%)	(£)
0.0	6,987,588	63.9	1,285,965	4.8	818,371	16.7	41,816,883
65.6	556,242	5.1	21,614,871	79.9	4,086,581	83.3	72,375,776
0.0	402,279	3.7	216,519	0.8	–	0.0	10,990,189
34.4	1,633,747	14.9	3,530,032	13.0	1,108	0.0	37,549,364
0.0	32,641	0.3	160,468	0.6	1,692	0.0	3,667,209
0.0	1,318,000	12.1	–	0.0	–	0.0	26,976,302
0.0	–	0.0	253,628	0.9	–	0.0	930,373
	10,930,498		**27,061,483**		**4,907,752**		**194,306,096**

Agency's National Flood Warning Centre has a significant budget to facilitate this process and, if flooding were reduced, these costs to the nation might also be somewhat reduced. Hence they could be taken, with care, as benefits of flood risk management.

The additional (i.e. over planned annual level) costs of the AVM/Floodline were estimated by the Environment Agency for the autumn 2000 events. The 467,200 calls to Floodline cost the EA some £300,000 (see the regional 'lessons learned' reports (Environment Agency, 2001)). However, it is widely accepted that some proportion of AVM capital and maintenance costs should also be included in these flood event costs, so they underestimate full costs in this respect.

6.8.5 Local government and other emergency costs in the summer 2007 floods

The summer 2007 floods also resulted in additional costs for local government authorities associated with damage to property, disruption to services and post-flood recovery. Emergency services were provided by police, fire and rescue and ambulance organisations as well as the Environment Agency. Research presented here is reported in more detail by Chatterton *et al.* (2010).

Distinctions are again made here between the impacts of the 2007 floods on:

- local government infrastructure and non-emergency services;
- expenditure by the emergency services; and
- expenditure by the Environment Agency.

Total costs in this three-part category are *c.*£223 million, including *c.*£85 million road costs, allocated to 'communication/roads' in Chatterton *et al.* (2010, Table 2.1), and £4 million expenditure on emergency services, allocated to 'emergency services' in Chatterton *et al.* (2010, Table 2.1).

Local government infrastructure and non-emergency services costs following the 2007 floods

Chatterton *et al.* (2010) undertook a detailed analysis of the audited accounts of the sixteen local authorities most affected by flooding in 2007. Total expenditure due to flooding for these local authorities was estimated to be approximately £194.3 million. Table 6.31 provides details of flood-related expenditure in 2007 grouped into seven categories: general expenditure, highways, recovery activities, schools, social services, housing and waste. Some of these costs relate to additional expenditure that was required (e.g. costs of temporary accommodation, additional staff costs or repairs to council housing) and others relate to lost income (e.g. due to the closure of car parks or sports facilities or council tax revenue).

As Table 6.31 highlights, the most significant category of flood-related expenditure varies between local authorities, depending on those assets which were affected by flooding. Overall, the most significant category of flood-related expenditure in 2007 was the cost of repair and reconstruction of infrastructure assets, which accounted for close to three-quarters of all local authority costs (Figure 6.8). These expenses included the considerable costs necessary to repair assets such as highways, schools and other council-owned property.

Of the one hundred-plus local authorities reported to be affected by the 2007 floods, eighty-two registered for the Bellwin scheme. Ultimately, forty-two of these received funds, with total grants totalling almost £19 million distributed.

Summary of emergency services costs in the 2007 floods (police, fire and ambulance)

The 2007 floods also placed an extra burden on police, fire and ambulance services for staff overtime, emergency supplies and provisions. Six police authorities reported costs above their normal operating costs, with two (Gloucestershire and South Yorkshire) claiming again under the Bellwin reimbursement arrangements and from the EU Restoration Fund. The total audited extra spend was £2.57 million, of which £1.3 million (40 per cent) was ultimately refunded (see Table 6.32).

Nine fire and rescue authorities reported above-normal operating costs, totalling £1.17 million, with South Yorkshire alone accounting for nearly 50 per cent of this. South Yorkshire, Humberside and Worcestershire between them claimed £770,000 under Bellwin (65 per cent of the total as shown in Table 6.32). Information was not available on

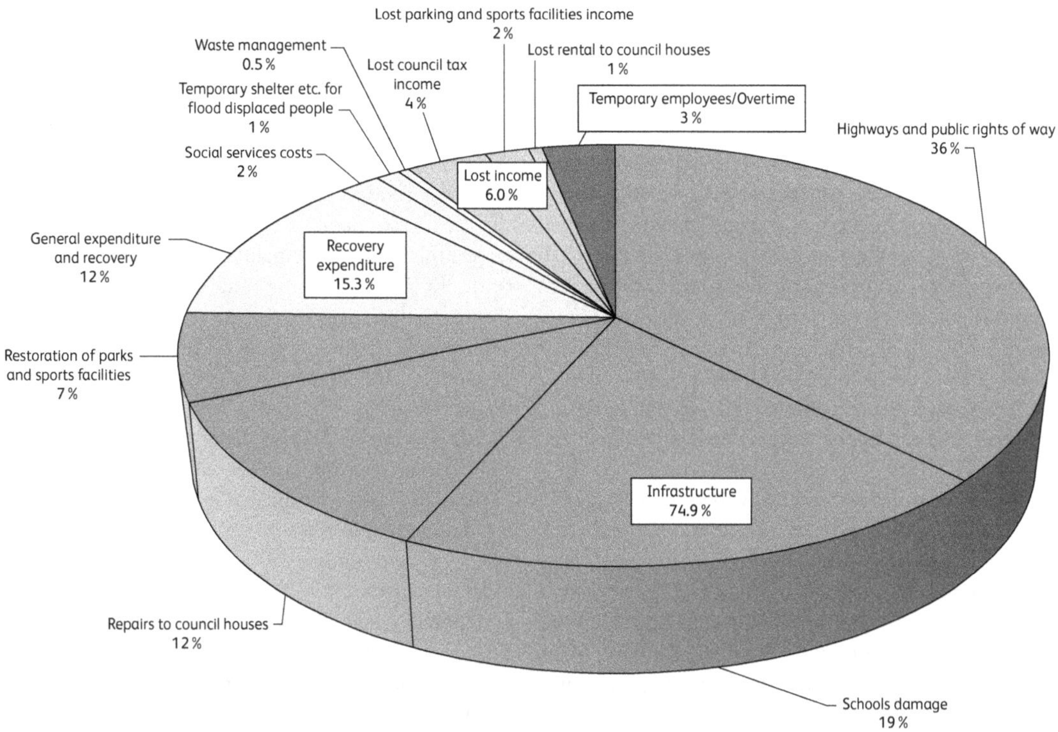

Figure 6.8 Flood-related expenditure and/or costs during the 2007 floods

Source: Chatterton *et al.*, 2010

Table 6.32 Grants and support provided to English local authorities and emergency services as compensation towards the costs of summer 2007 floods

	EU Restoration Fund (£million)	Payments under Bellwin (£million)
Local authorities	30.2	17
Police	0.3	1
Fire and rescue	0.1	0.8
Total	**30.6**	**18.8**

Source: After Chatterton *et al.*, 2010, Table 3.5; taken from audited accounts

additional ambulance service and related paramedic costs. Taken together, the identified additional costs for emergency services amount to £3.74 million, equivalent to less than 2 per cent of that spent by local authorities on infrastructure losses and non-emergency services.

Approximately 1,000 personnel from the Royal Navy, Royal Marines, Army and Royal Air Force assisted during the floods and immediate post-flood recovery, mostly in Gloucestershire. Approximately 350 personnel were used at any one time, as requested by the Environment Agency, for search and rescue, construction of flood defences and distribution of bulk and bottled water. Although their costs could be theoretically recharged under Military Assistance to the Civil Community Scheme, no such costs were separately reported by the sixteen local authorities incurring the highest flood-related costs in 2007. The use of military resources in this way was said to have provided training opportunities and did not impact on either pre-deployment preparation or support to current military operations.

Environment Agency costs in the 2007 floods

The summer 2007 flood incidents placed an additional burden on the Environment Agency in its role as the lead organisation for flood risk management in England and Wales. Approximately 500 'flood watch' and 'severe flood' warnings were issued. Floodline, a telephone information service, handled some 206,000 requests for recorded messages. The National Customer Contact Centre handled some 55,000 calls. Some 700 media interviews were given, and there were 2,600 media enquiries. Approximately 1,200 flood risk management staff and 1,000 other Environment Agency staff were involved in managing the events, and additional support was drafted in when required.

Accounts provided by the Environment Agency show additional emergency response costs of £4.43 million, mainly for extra manpower (65 per cent) and hire of equipment and contractors (24 per cent). There was also considerable damage to Environment Agency flood risk management infrastructure, estimated at *c.*£15 million, with one-third of this involving immediate emergency repairs and the balance scheduled for urgent repairs to permanent structures. Thus, total costs were £19.3 million, of which 22 per cent was for emergency, non-structural items.

6.8.6 Emergency services and the Environment Agency: data for project appraisals

Standard data

It is clear from the sections above that the emergency costs during flood events such as those of autumn 2000 and summer 2007 are substantial, but not all of these costs are allowable as contributing to the benefits of flood risk management schemes. This can be because some of these costs are covered elsewhere in the benefit assessment, but also because items such as sandbagging prevent damage in themselves, and project appraisals can assume that this damage does not therefore occur (i.e. freeboard provided by sandbagging should be allowed for in project appraisal flood modelling, although generally it is not). In project appraisals one should therefore *either* count the costs of sandbagging (and other emergency measures) and reduce the potential damage to those properties or roads affected *or* take the full damage potential there and not count the costs of sandbagging, etc.

In addition, some of the repair operations on roads and bridges can result in betterment, such that the bridge or road after the flood is in better condition than previously, as a result of the repair operations. In effect, this is deferred maintenance brought forward to the time just after the flood has occurred, rather than damage caused by the flood alone. Some of the damage occurred previous to the flood, and is repaired along with the damage that occurred during the flood.

Another problem in applying these data to project appraisals arises from how to scale the figures applied in particular project appraisal feasibility studies from the data as researched and reported above. Parker *et al.* (1987), as indicated above, used data giving emergency services cost per property flooded, in line with much other data in that (and indeed this) manual. There are clear relations between flood emergency costs and the numbers of properties flooded (Figures 6.9 and 6.10) which seem to indicate some form of 'economy of scale' of 'flood fighting': the bigger the event, the lower the unit costs. However, while this is interesting (and logical), it is not sensible to use this approach for scaling purposes, not least because many of the emergency costs are spent in *preventing* property being flooded, such that it is perfectly possible for there to be substantial emergency services costs without any property being flooded at all. This means that scaling by the number of properties flooded (Figures 6.10, 6.11 and 6.12), which at first sight seems sensible, can be inappropriate.

Therefore, an alternative approach has been adopted, as indicated in Table 6.33. This takes the total emergency costs under Bellwin (and below Bellwin thresholds), the severe weather payments and the Environment Agency costs for England and Wales, and allows only those costs appropriate to project appraisals (i.e. deducting 50 per cent of the roads and bridges costs, in lieu of betterment, and factoring the total Bellwin claims by 42.5 per cent, being the sum of those percentages that can be allowed as taken from the NYCC case study).

The result is an allowance of £60.8 million as total emergency costs in the 2000 floods applicable to project appraisals. The details within Table 6.33 show how this has been calculated. Expressing this amount

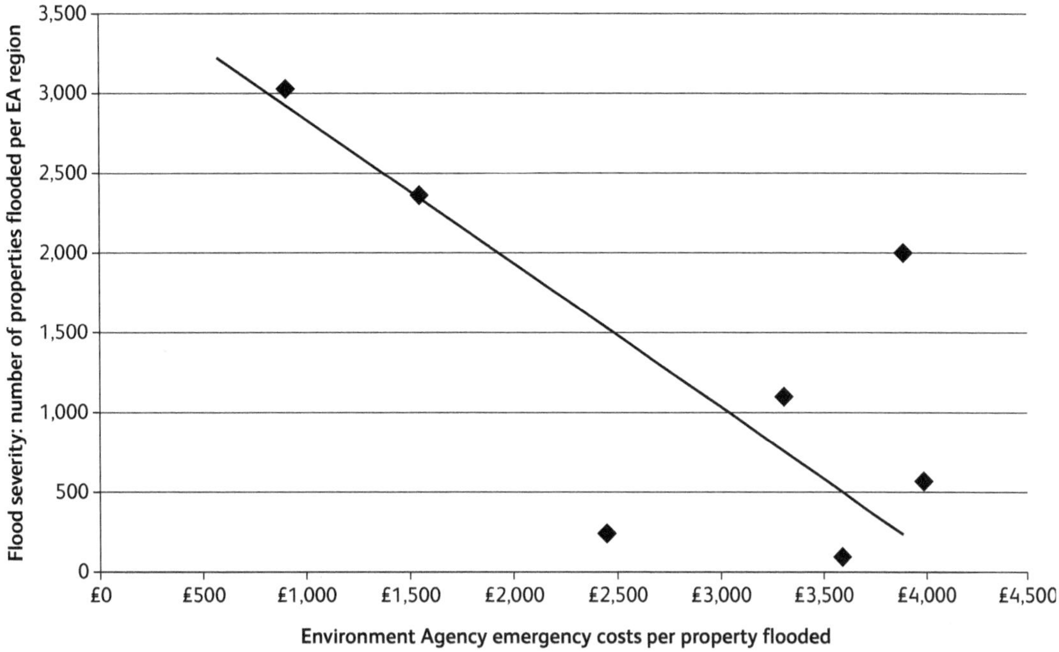

Figure 6.9 The Environment Agency's emergency costs: the declining unit costs per property flooded with flood severity

Note: Trend line fitted by eye

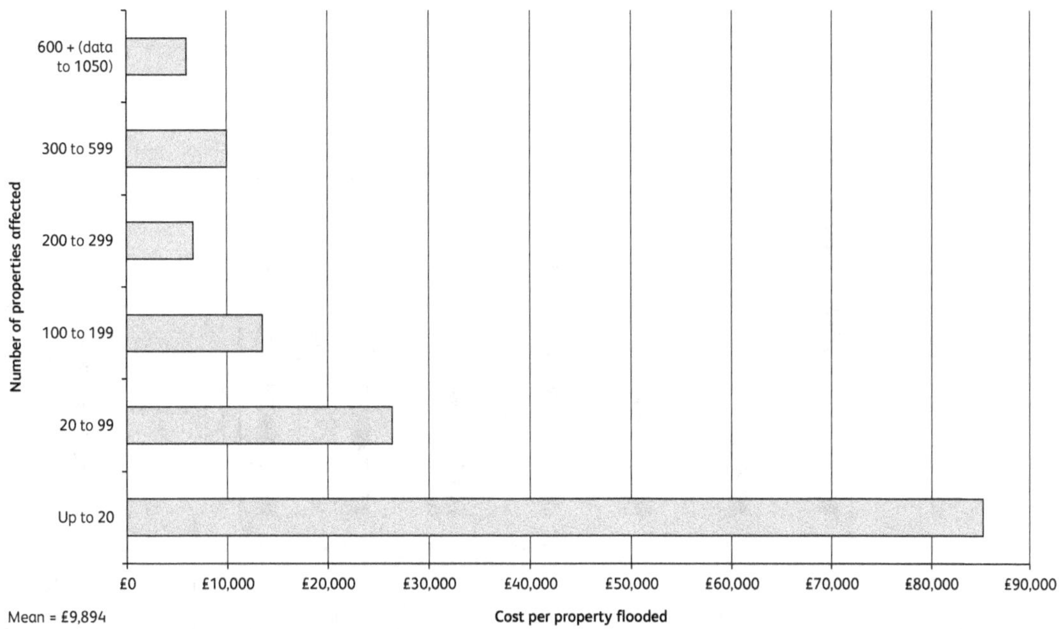

Figure 6.10 Emergency services costs per property flooded (per property band)

Notes: Excludes authorities with no properties flooded; properties include both residential and commercial properties

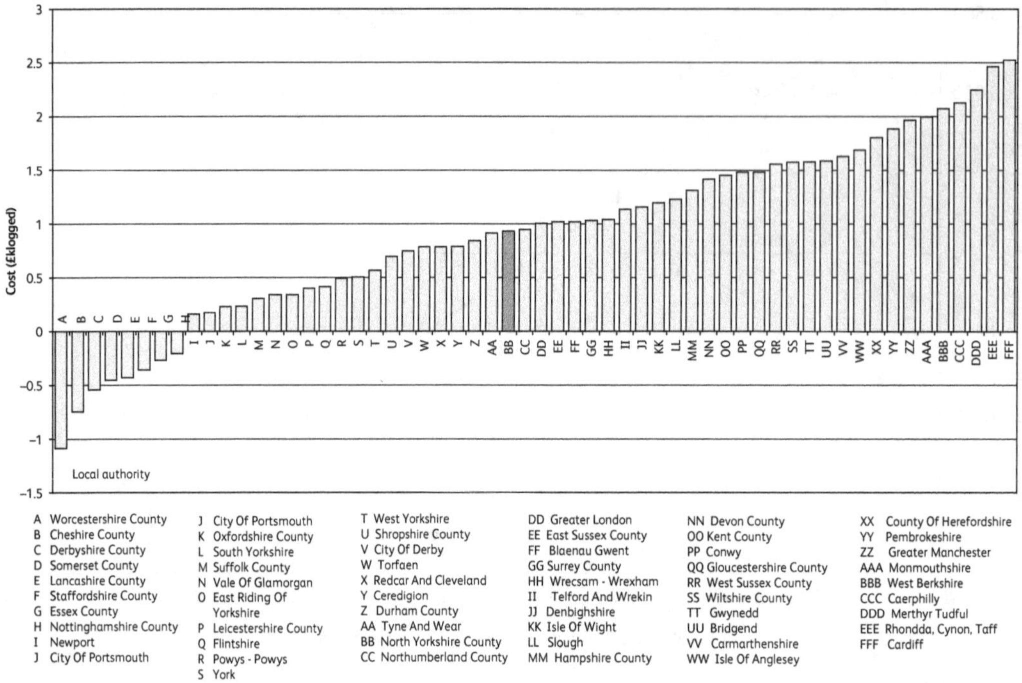

Figure 6.11 Emergency services costs per property flooded in the autumn 2000 floods

Notes: Excludes authorities with no properties flooded; properties include both residential and commercial properties; North Yorkshire County Council highlighted

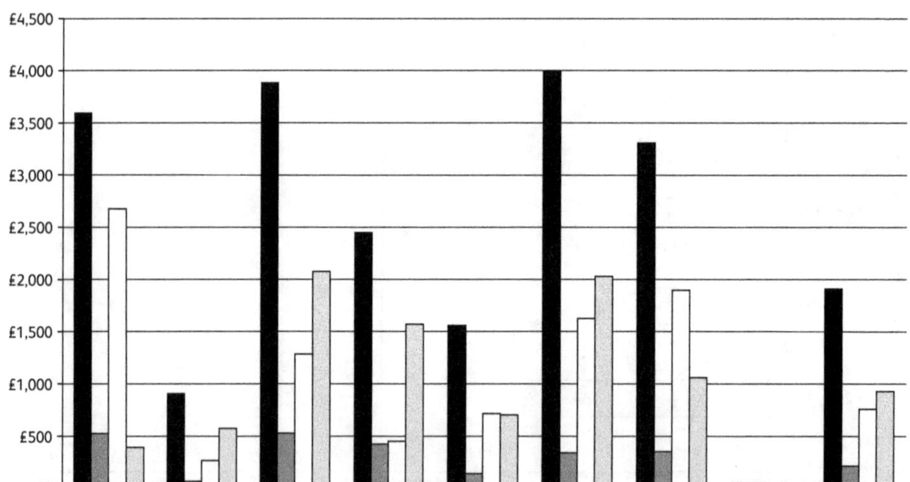

	Anglian	Midlands	North East	North West	Southern	South West	Thames	Wales-incomplete data	Means
■ Total costs	£3,591	£906	£3,884	£2,449	£1,551	£3,988	£3,307	£37	£1,902
■ Extra statutory costs	£520	£67	£529	£424	£131	£337	£353	£37	£214
□ Total emergency response	£2,677	£260	£1,283	£455	£714	£1,628	£1,898		£761
□ Total emergency repairs	£394	£579	£2,072	£1,570	£706	£2,023	£1,057		£928

Figure 6.12 The Environment Agency's emergency costs: regional data and overall means

Note: All data per property flooded in that EA region

Table 6.33 Overall emergency costs as applicable to project appraisals (autumn 2000 floods)

A. Autumn 2000 total losses [1]			
		Financial £ millions	Economic £ millions
Property losses		946	570
Professional partners (PP)		39	39
Capital grants to PP		58	58
Environment Agency		21	21
Road traffic disruption		73	13
Railway network		51	51
Agriculture		195	35
TOTAL (£ millions)		**1,383**	**787**
B. North Yorks CC emergency cost (Bellwin) breakdown			
Cost item	**% of total**	**Allowed [2] amount (%)**	**% of total allowed**
Sandbagging	41	0	0
Roads/bridges	33	50	16.5
Education	1	100	1
Emergency planning	1	100	1
Care related	6	100	6
Flood emergency	18	100	18
TOTAL	**100**		**42.5**
C. Emergency costs applicable to project appraisals			
Cost item	**Amount**	**Allowed [2] amount (%)**	**Allowed amount**
Total Bellwin:			
England	£28.30	42.5	£12.00
Wales	£7.60	42.5	£3.20
Costs below thresholds [3]	£3.10	42.5	£1.30
Severe weather payments:			
England	£41.90	50	£21.00
Wales	£17.10	50	£8.60
Environment Agency costs [4]:			
Emergency repairs [5]	£11.10	50	£5.50
Emergency response	£9.20	100	£9.20
TOTAL	**£118.30**		**£60.80**
As % of economic property losses of £570m			10.70%

Notes:

[1] From Penning-Rowsell *et al.*, 2002. [2] Judged to be proper economic costs, not counted elsewhere in benefit–cost analyses. The figure for roads recognises some betterment after repair (hence the 50% taken). [3] Taken as 50% of thresholds. [4] England and Wales. [5] As for roads, above, some element of betterment here, hence 50% taken.

as a percentage of the total economic property losses in autumn 2000 gives 10.7 per cent. This, therefore, represents a multiplier on top of total property damages that accounts adequately and appropriately for those 2000/01 emergency costs.

The same approach was adopted for assessing the total emergency and recovery costs during the summer 2007 floods in England. The total emergency costs (Table 6.34) are £110 million, that is 5.6 per cent of the total economic property losses (Chatterton *et al.* 2010, Section 3.8). The difference in terms of percentage between the 2000 and 2007 floods may be explained by another effect of the economies of scale. The 2007 summer floods affected a higher number of properties (up to 73,000 versus approximately 10,000 properties in 2000), but a lower number of local authorities claimed under the Bellwin scheme (thirty-eight versus eighty-seven). In other words, the figure obtained from autumn 2000 reflects widely dispersed flood-affected communities (Penning-Rowsell and Wilson, 2006), whereas the figure obtained from the summer 2007 floods reflects a smaller number of more densely populated communities in a more substantial event (Figure 6.13).

Site-specific assessments

There will be circumstances in project appraisals when the use of the standard data as given above is not appropriate, or the data are not considered accurate enough for project appraisal purposes. In this case, it will be necessary to collect data from the authorities relevant to the area in question. This is not easy, particularly in the absence of a recent flood, and care needs to be taken to ensure that fixed and marginal costs are separated, in order to identify solely the latter for counting within project appraisals. Notwithstanding the above comments, a standard checklist is provided on MCM-Online (see Supporting Information for Chapter 6) as a guide to obtaining these data.

6.9 LESSONS FROM EXPERIENCE

The flood situations experienced in 2007 and 2012 have both highlighted the often widespread and lengthy disruptive impacts of flooding on utilities and transport networks. From an overall event perspective, disruption to roads, electricity supply and waste water and water supplies remained the most significant risks in this sector in 2007, which reinforces their prioritisation for inclusion within project

Table 6.34 Overall emergency costs as applicable to project appraisals (summer 2007 floods)

Cost item	Amount	Allowed [1] amount (%)	Allowed amount
Total Bellwin and roads:			
Bellwin	£30.20	42.5	£12.84
Roads infrastructure	£175.00	50	£87.50
Environment Agency costs [2]:			
Emergency repairs [3]	£14.80	50	£7.40
Emergency response	£2.20	100	£2.20
TOTAL	**£222.20**		**£109.94**
As % of economic property losses of £1,942m			5.57%

Notes:

[1] Judged to be proper economic costs, not counted elsewhere in benefit–cost analyses. The figure for roads recognises some betterment after repair (hence the 50% taken).

[2] England and Wales.

[3] As for roads, some element of betterment here, hence 50% taken.

Source: Chatterton *et al.*, 2010

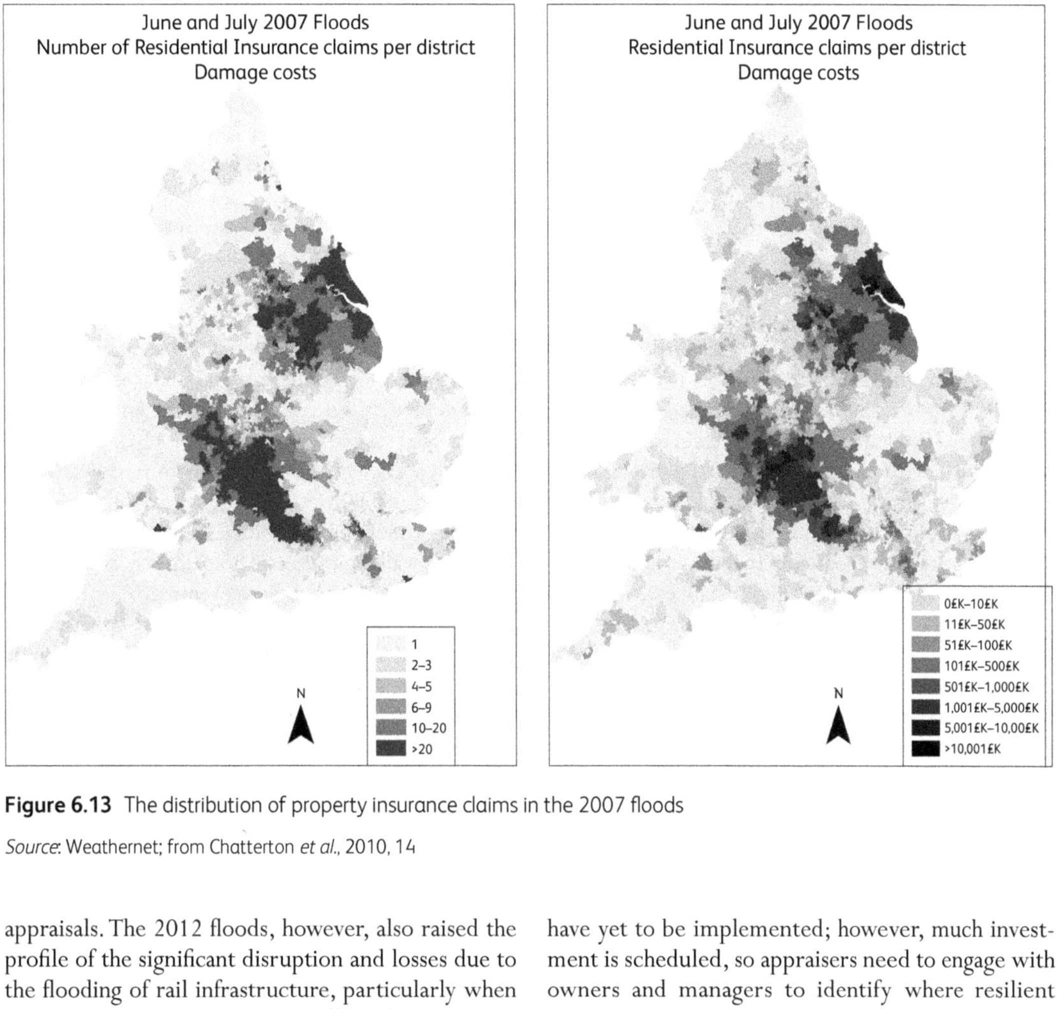

Figure 6.13 The distribution of property insurance claims in the 2007 floods

Source: Weathernet; from Chatterton *et al.*, 2010, 14

appraisals. The 2012 floods, however, also raised the profile of the significant disruption and losses due to the flooding of rail infrastructure, particularly when key junctions or stations are affected, as routing around flood problems in some areas is impossible.

A key development from the Pitt Review recommendations, and also due to the reporting requirements of the Climate Change Act 2008, has been much effort by service and transport providers to identify the risks of their assets to flooding, both now and in the future. This work is also likely to improve in the future. Although, due to security restrictions, this information is not always easy to access, where possible appraisers should utilise it to increase understanding of the likelihood and impact of flooding on infrastructure. There has also been a big effort – and much pressure – since the 2007 floods to make infrastructure more resilient to flooding. It is still too soon to tell how significant the move towards increased resilience has been and, in reality, few plans have yet to be implemented; however, much investment is scheduled, so appraisers need to engage with owners and managers to identify where resilient measures are imminent or have already been implemented.

Resilience measures need to be considered by appraisers in the broadest terms and not in isolation from other aspects of flood risk management because damage and disruption consequences are related not only to the efficacy of resistance/resilience but also to the introduction of non-structural FRM measures, such as: hazard identification, mapping and avoiding siting new installations in at-risk areas; substitution and provision of reserve capacity; the presence and effectiveness of flood forecasting and warning; flood incident management and business continuity planning; and the inclusion of infrastructure providers within emergency exercises.

In terms of priority within project appraisals, with regard to the damage and costs reported in this chapter, we would emphasise the importance of properly and fully assessing estimated road traffic disruption costs when high-volume road networks are affected by flooding at low return period events (i.e. with an annual probability of greater than 5 per cent). We would not, however, advocate assessing road disruption costs to minor parts of the road network (e.g. less than B-designated roads). The effects of floods on utilities appear to be very site specific, so we would recommend making local investigations to determine whether there would be impacts and associated costs in flood events, rather than attempting to develop and use standard data.

The costs to the nation of emergency efforts during flood events can be considerable, but appear to vary according to the type of flood (severe and very severe) and the nature of the area affected (urban and rural). The costs for minor floods may be proportionately less than reported here, but we have no data to guide decisions in those circumstances.

In summary, the key lessons for appraisers to consider are:

- Losses due to the damage of infrastructure and the disruption of services and networks can be significant.
- Proportionality of evaluation is crucial (where to concentrate, on what to concentrate and in what detail) to optimise appraisal resources.
- Interdependency and collaboration with all utility and commercial sectors is fundamental to understanding the true impacts and consequences of flooding.
- The degree of resilience and resistance measures introduced will determine the scale of consequences.
- The number and type of non-structural FRM measures will also moderate consequences.
- Improved infrastructure risk databases are now widely adopted by asset owners and should be accessed where possible.
- Many detailed case studies from past events are now available for a wide range of infrastructure types which is facilitating a better understanding of the scale of impacts. These do, however, tend to focus on the more severe events.
- Good professional contacts exist with sound technical knowledge and these should be utilised within an appraisal process.

7 Coastal erosion risk management

Potential losses and benefits

This chapter explains the procedures and techniques for assessing the potential benefits of investment in coastal erosion risk management. These benefits principally arise from delaying the processes of erosion, and thereby delaying the loss of land and property for the duration of the life of the proposed risk management works.

Key points to understand are:

- Erosion is effectively permanent and irreversible.
- This means that future uses of that land or property are lost.
- Decisions about investment versus no investment must start from a realistic evaluation of the 'do nothing' option.
- Coastal risk management works, which are designed to arrest this process of erosion, normally have a finite life.
- Hence the benefit from a particular coastal risk management project should be seen as a temporary – but usually lengthy – extension to the useful life of the land and property protected.
- The most reasonable assumption thereafter is that the original long-term erosion rates as before will resume.
- Coastal risk management projects are compared with a 'do nothing' option. This 'do nothing' option may involve 'walk-away' and hence the prospect of substantial erosion of coastal property (see the Environment Agency's (2010a) guidance on 'do nothing').

7.1 SUMMARY OF INFORMATION

This chapter discusses the procedures and techniques for assessing the potential benefits of investment in coastal risk management. These benefits principally arise from delaying the processes of erosion, and thereby delaying the loss of land and property for the duration of the life of the proposed risk management works.

- The main procedure (Section 7.4) presents the recommended methods for assessing the value to society of the delay of these losses, measured as the value of the increased length of time for which the assets are available for use.
- The inputs to these calculations require an assessment of the extent of erosion in the future, based on historic records or a probabilistic assessment.
- Complementary inputs are the value of the assets at risk, the anticipated duration of the life of the project and the current discount rate.
- Assessing the losses to agricultural land can be treated in a similar way, but infrastructure losses are more complex since properties may become unusable if their infrastructure is lost but they themselves are not.
- Flow charts are presented in this chapter, summarising the overall process and the assessment of infrastructure losses (Figures 7.1 and 7.5).
- Some data on the value of assets at risk is presented, but this chapter concentrates on methods and techniques, since much property value data will be site specific.

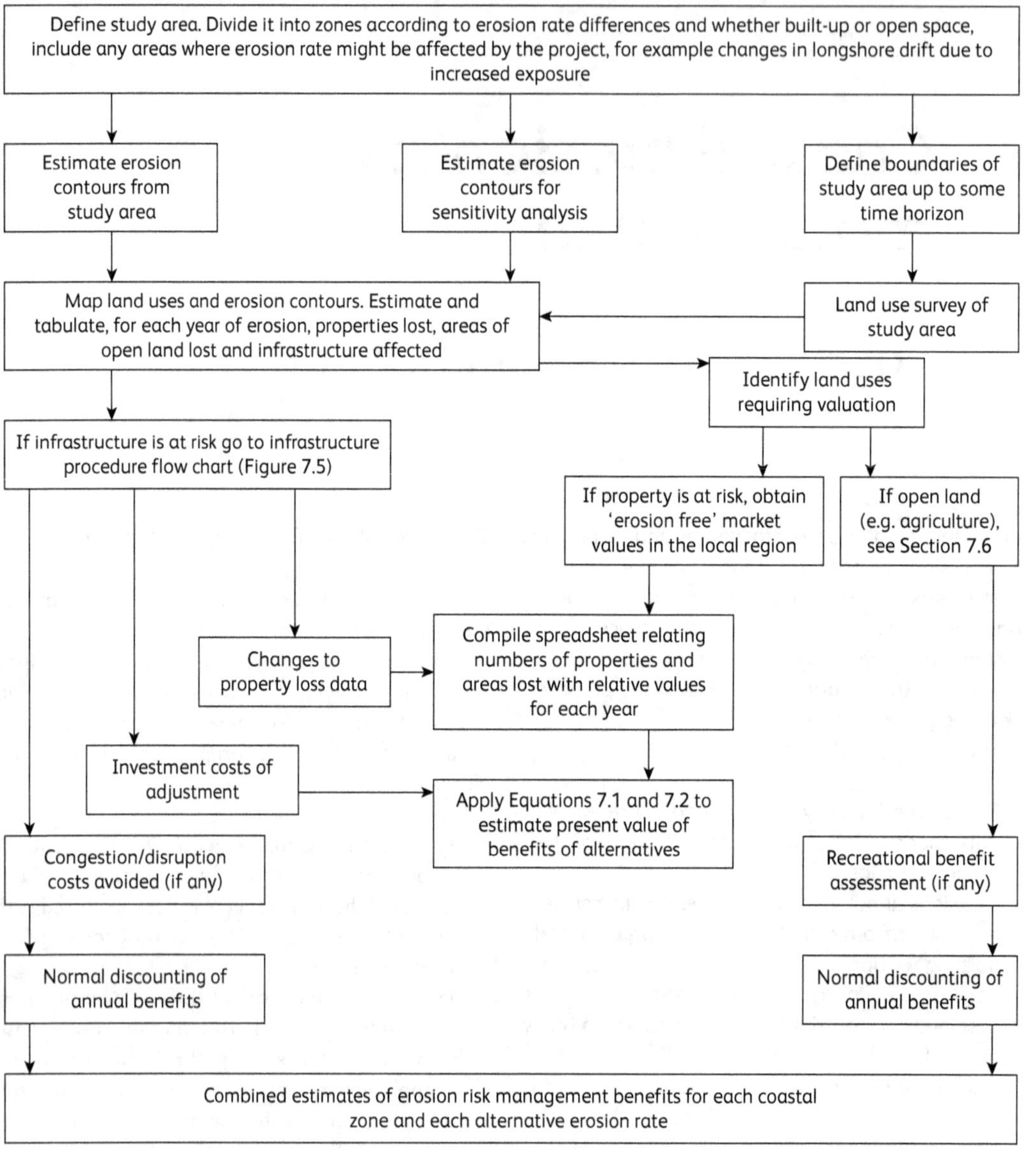

Figure 7.1 Flow chart of assessment process

7.2 PROBLEM DEFINITION

We need to be clear about terminology (Environment Agency, 2010d, 6), and thereby differentiate between:

- Coast protection: Coastal erosion risk management involving physical protection. This may include measures for stabilising landslides.
- Coastal erosion: The detachment of material, transport through the cliff system, deposition on the foreshore and its removal by wave action. The definition encompasses any one, or all, of these elements, and does not require direct wave action on the base of the cliff.
- Landslide: The failure of a mass of rock or soil under gravity, often an important part of coastal erosion processes.
- Cliff recession: The landward retreat of the cliff

profile (from cliff base to cliff top) in response to the coastal erosion processes (often including landsliding).

Crucially, only erosion effects that are actually related to *the sea* are eligible for Grant in Aid under the Coast Protection Act 1949 (HM Government, 1949), not landsliding *per se*. Landslips caused by groundwater, rainfall, etc. would not be eligible in this way.

7.2.1 Erosion at the coast

At specific locations around the British coastline, erosion can be a serious problem. Buildings, land, infrastructure and non-built upland may all be lost to the sea, either directly through erosion or from landslips and landfalls following erosion of the coastline.

- The most important difference here from losses owing to flooding is that erosion is effectively irreversible: once land has been eroded, or the cliff-top edge is deemed too close to a property for that property to be used safely, the land or property cannot be regained.
- This irreversibility means that future options for the use of that land or property are also lost. Hence there is a permanent loss at one point in time of a capital asset (i.e. land and buildings) which had offered a future stream of uses and therefore values.
- By comparison, flooding results in the loss through flood damage of an annual income flow (for example, in agriculture), and also the loss of a part of the capital stock which is reparable or replaceable (i.e. contents damage or building fabric loss or repair costs). Flood damage can be characterised as an annual average damage or loss for ever (without risk management works), whereas loss of land or property through erosion is a one-off loss.

Although the losses from coastal erosion are permanent and irreversible, coastal risk management works, which are designed to arrest this process, normally have a finite life.

- Hence the benefit from a particular coastal risk management project should be seen as a temporary – but usually lengthy – extension to the useful life of the land and property.
- This is in contrast to the permanent loss of the property from coastal erosion.
- The effective life of coastal risk management works will, therefore, be important, in that this is the length of the extension of the life of the property and land protected.
- There will undoubtedly be uncertainty over the expected life of coastal risk management works but an estimate of their useful and effective life is needed.

For gauging the effective life of any works, understanding and quantifying deterioration rates are important for estimating and planning programmes of maintenance that contribute to an asset's whole-life costs (WLCs), and for the day-to-day maintenance and renewal intervention activities (HR Wallingford *et al.*, 2009).

The most reasonable assumption after the end of the effective life of any works is that the original erosion rates will resume as before, although here there is often a high level of uncertainty with regard to how the coast will respond following the removal or failure of coastal protection. For example, the coast will often 'catch up' and rapidly erode to a new equilibrium profile or dramatically 'overshoot', as happened in Happisburgh in north Norfolk. Nevertheless, taking a long-term perspective, there are areas where rates of erosion will result in significant losses over a fifty-year time period. If coastal risk management is provided to a fifty-year design life, over the next fifty to one hundred years a large number and a varied mixture of buildings, infrastructure and open space are all at risk from erosion by the sea. And, remember, only erosion effects that are actually related to *the sea* are eligible for grants under the Coast Protection Act 1949 (HM Government, 1949); different funding streams may be available in relation to other causes of coastal risks.

The recommended approach for assessing the benefits of coastal risk management is summarised in Figure 7.1. The key points embodied in this approach are as follows:

1 Estimates are needed of erosion rates and cliff-top edges projected for fifty or even one hundred years into the future.
2 Alternatively, a probabilistic approach to erosion can be taken, resulting in a range of probabilities that a particular parcel of land or property will be eroded and therefore lose its usefulness.

3 A procedure is provided for evaluating the losses due to erosion, or the extension to the expected life and use of the property and land due to a delay in the erosion process resulting from investment in coastal risk management.
4 Techniques are provided for finding the appropriate values for properties whose market prices are likely to be affected by perceived erosion risk.
5 A procedure is provided for valuing potential infrastructure losses where these may be of several types.
6 Techniques are given for finding the appropriate economic values for agricultural land threatened by erosion.
7 Some guidance is given for finding the appropriate economic values for other open space threatened by erosion.
8 Some guidance is given for addressing real price and value changes through time where these could be significant.

The underlying assumption here is that coastal risk management projects are compared with a 'do nothing' option. This 'do nothing' option may involve 'walk-away' and hence substantial loss of coastal property. This is relatively easy to envisage for unprotected frontages where there is experience of ongoing erosion. It is much more difficult to envisage where existing defences have prevented recent erosion, but it should be borne in mind that these could fail at some time in the future without further action (Green and Penning-Rowsell, 1999).

Much of the research leading to the development of these techniques emphasises standard data and a simplified approach. Nevertheless, erosion problems are highly site specific, and the exact nature of the property affected by erosion is crucial to the magnitude of the losses and therefore the benefits of its risk management. It is inevitable, therefore, that the techniques recommended here comprise a balance between local factors and data and a standard national approach.

7.2.2 Policy options

In assessing the worthwhileness of investing in coastal risk management schemes, a number of policy options need to be considered:

- projects in different locations (i.e. where is the greatest priority from the national perspective?);
- different schemes which provide the same level of risk management for an area but use different technical solutions at different costs;
- schemes at a given site which offer different levels of risk management or have different risks of failure;
- different standards or levels of upgrading or renovation of existing schemes;
- different dates when a project against erosion should be implemented;
- different dates when an existing scheme could be renovated or upgraded;
- implementing no engineering strategy, but improving information on erosion risks and warning systems, and changing land use planning policy; and
- doing nothing (see Environment Agency, 2010a).

As with all benefit–cost analyses, all options should be compared with the last item listed above: doing nothing. As always, when faced with property falling off cliffs or in some other way being threatened by erosion, this appears to be a callous option. However, it should rather be seen as the starting point for an analysis whereby the nation allocates its scarce resources in an efficient way, and does not repeat mistakes of the past where in a few locations large sums of money were spent on unsustainable projects where the long-term costs of risk management far exceeded the value of property and assets retained.

7.3 ASSESSING THE EXTENT OF THE EROSION PROBLEM

7.3.1 Erosion rates and erosion 'contours'

Without some detailed knowledge of erosion rates in the future, it is not possible to calculate the benefits of delaying this process.

- The technical assessment of the erosion problem should result in a set of predicted erosion 'contours' for the coastline in question for at least the projected life of the proposed works.
- More distant future delays of erosion, for example from year 150 from the present to year 200 from the present, can also yield significant benefits, but these are so heavily discounted using normal

discount rates as usually to be insignificant. The only exceptions will occur where the value of property beyond the erosion 'contour' representing, say, project life plus 25 per cent, is so massive that its value, when discounted back to present values, is still likely to be significant (e.g. the major urban and hotel developments at some coastal resorts). Also relevant here, however, is property/land which represents a large risk to the environment if eroded (e.g. highly contaminated land, nuclear sites, landfill, etc.), which will persist long after the life of the project itself. Exceptions may also exist for heritage, archaeological and other highly designated sites.

- Erosion predictions will not be certain, and will have to be based on averages from the likely effects of storms of different magnitudes. The best that are likely to be achieved are predicted future erosion rates based on past erosion rates, informed by knowledge of the underlying geology and any anticipated changes that might affect the erosion process, such as rising sea levels or updrift coastal risk management works. Estimates of erosion rates should be made in this way, and sensitivity analysis used to gauge the significance of benefit totals to the assumptions made.
- It has to be recognised that predictions based on historical erosion rates are likely to become increasingly less useful for predicting future erosion as we start to depart from previously seen rates of sea level rise (i.e. with accelerating sea level rise over the next hundred years). Increasingly, it may be necessary to look towards modelled rather than historical data. The latest advice on sea level rise is given by the Environment Agency (undated b). This advice replaces Defra's Supplementary Note to Operating Authorities – Climate Change Impacts (Defra, 2006), and it is provided as supplementary information to Defra's policy statement on appraisal (Defra, 2009) and the Environment Agency's Appraisal Guidance (Environment Agency, 2010a), as well as supporting the FCERM National Strategy for England.
- Whilst recognising sea level rises owing to climate change, we must not forget the effect of isostasy which, in the long run, will also be of major significance for the evolution of our coastlines.

The result of this analysis should be erosion 'contours' showing the predicted location of the edge of the cliff or sea after erosion at the average rates predicted, using five-year intervals (or shorter if erosion is rapid). The five-year erosion contours are used to identify the band of years when a property would most likely become uninhabitable (i.e. either lost or located so near the edge of the cliff or sea as to be abandoned as unsafe).

These bands will show up on the Environment Agency's National Coastal Erosion Risk Maps (http://www.environment-agency.gov.uk/homeandleisure/134808.aspx). Instead of showing erosion rates, these maps show the erosion zone expected over three timescales – from 2010 (when this data was developed) up to approximately 2030; up to approximately 2060; and up to approximately 2110. It is shown as a range (e.g. '0.5–3.2 metres') that represents the upper and lower limits of the erosion we should reasonably expect up to that time. However, predicting coastal erosion is an uncertain science, and providing an exact position of the shoreline in the future would give a misleading impression of precision and certainty. Coastal planning documents, such as shoreline management plans (SMPs), also show this information. Long-term predictions are required for planning coastal management policies into the future, which requires evidence upon which to base planning decisions.

Therefore, for properties at risk from erosion, there will be some minimum acceptable safety margin ('sm' in Figure 7.2) between the cliff top and the building. Figure 7.2 illustrates a set of hypothetical erosion contours, and a safety margin (e.g. two years' erosion). For businesses, it may be more useful to estimate the floor areas of each property type involved, since it may be easier to obtain property values on this basis. For agriculture, the area of land of each grade and cropping pattern or land use lost in each year (or band of years) are the data required.

In addition, public utilities and infrastructure will be affected by the projected erosion, and some infrastructure elements will be affected before the property themselves would be unusable due to the direct risk of erosion (see Section 7.5). Here, it will be necessary to establish the approximate location of service lines to these properties. Again, safety margins on access roads may be enforced, or utility lines – or sub-stations on which they depend – may be closer to the coast somewhere else other than at the location of the erosion-prone properties.

This may mean that a property could be regarded as 'written off' before the property itself is touched

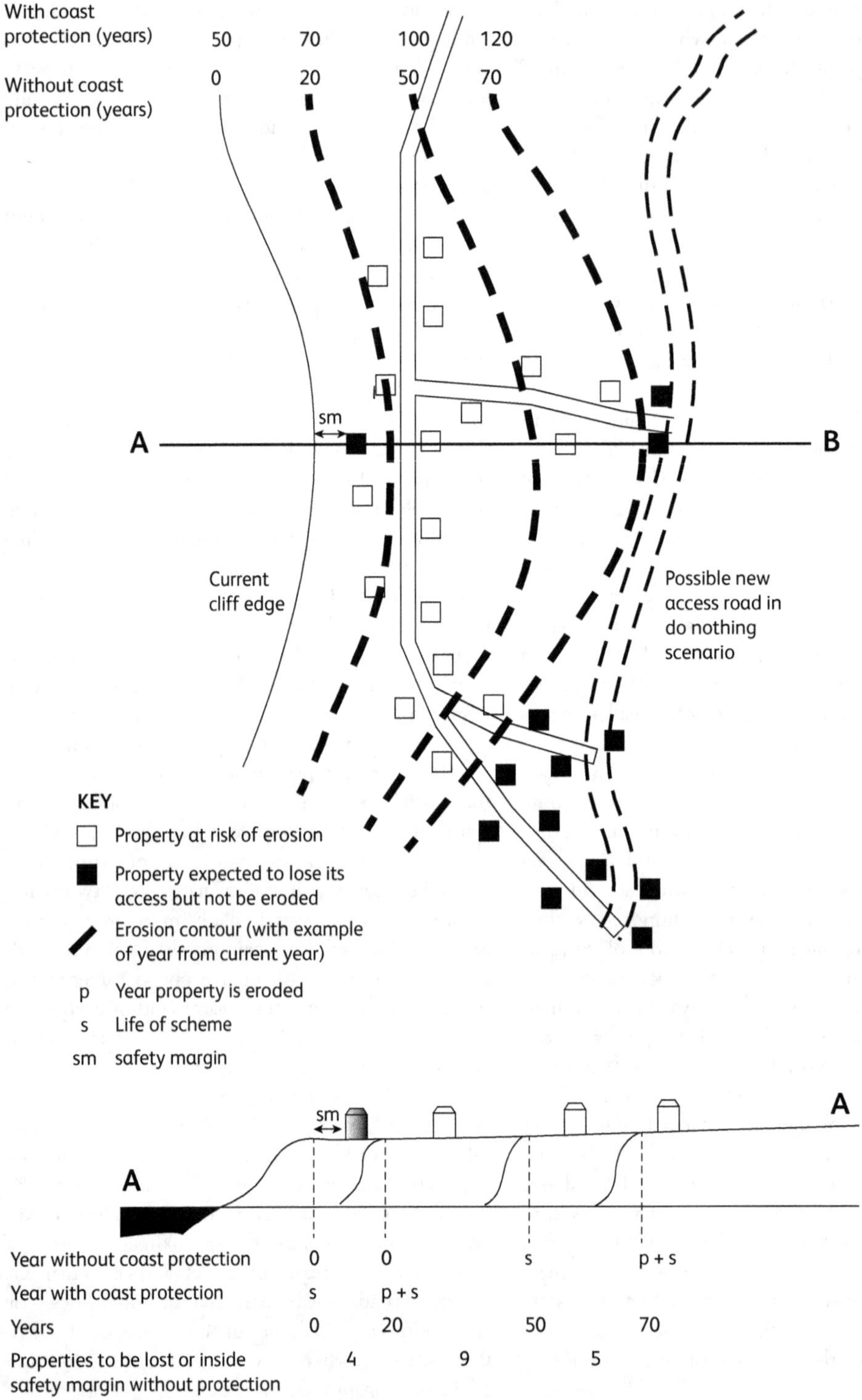

Figure 7.2 Hypothetical coastal erosion contours

by the erosion contour plus the safety margin. Procedures detailed in Section 7.5 ensure that losses associated with infrastructure erosion are estimated in the most appropriate way.

7.3.2 The probabilities of erosion: principles

The management of impacts of erosion at the coast is all about changing the probabilities of erosion events or the probabilities of the failure of past coastal risk management schemes. For example, the purpose of strengthening a promenade system is to reduce the probability that the risk management thereby provided will be lost when the structure fails.

However, it can be necessary to take account of complex chains of events. In coastal situations, it will frequently be necessary to calculate the joint probability of the simultaneous occurrence of a tidal surge, high tide and on-shore wind. In such chains, the probabilities of the different events may be independent or result from a common cause; these probabilities may also change over time.

For example, Figure 7.3 shows a fairly common situation on the coast where a soft cliff is protected by a wall, which in turn is protected by a beach. As the beach is eroded the probability that the wall will fail increases and when the wall has failed the cliff will start to erode. Similarly, a promenade typically forms a coastal risk management or sea defence function as well as having a recreational value; only once the promenade has completely failed will the land behind it start to be eroded.

In consequence, some understanding of probability theory and of reliability engineering is necessary when undertaking benefit–cost analyses of coastal risk management and sea defence schemes. FCERM-AG (Environment Agency, 2010a) provides an introduction to the relevant conditions. However, in general, there is very little statistical data upon which to base estimates of the probabilities of failure of existing schemes, or cliff-top erosion rates, and it will often be necessary to resort to engineering judgement to estimate the probabilities of property and land loss 'with' and 'without' a scheme.

Moreover, these probabilities in turn change over time as coastal risk management structures decay, or as a result of climate change, as this can affect either runoff or sea levels, and for other reasons. Changes in the probability of failure are also critical when considering a 'walk-away' baseline option: once maintenance and repairs of the existing structure have ceased, the probability of failure should be expected to increase as time passes.

Where there is an existing flood risk management structure or when erosion is actively occurring, the key decision is when to undertake works: for example, whether to replace or strengthen the existing structure before it fails or to wait until this risk has

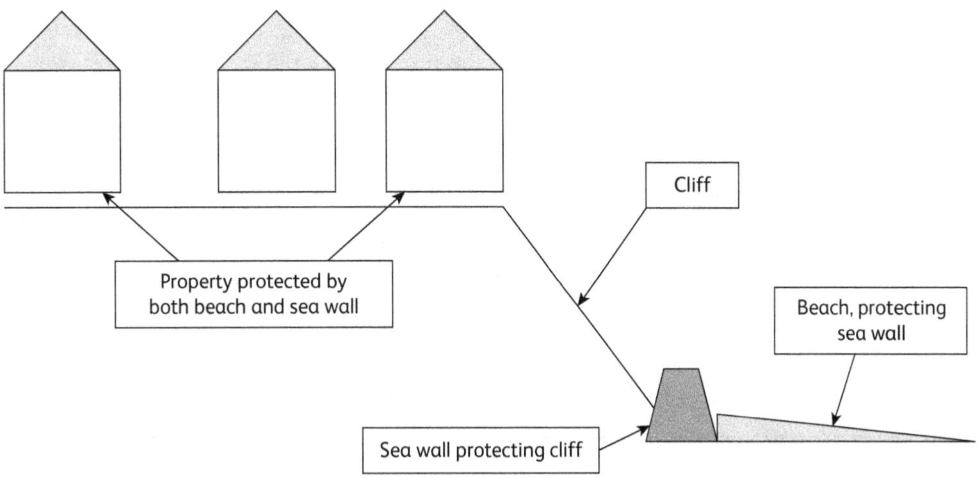

Figure 7.3 The nature of complex systems at the coast

increased, or whether to allow low-valued land to be lost through erosion where, in addition, the sediment released through erosion may have an economic value in beach replenishment and hence coastal risk management down-drift.

7.3.3 Calculating benefits by assessing the probabilities of erosion: practice

Since erosion is often episodic, with sudden losses of land and slides of cliffs, the use of erosion lines (Section 7.3.1), whereby it is assumed with certainty that erosion will reach a certain point inland in a given year, can be misleading.

Figure 7.4 (I) illustrates a cliff where, in any given year, there is a probability of different magnitudes of slides. Figure 7.4 (II and III) shows two possible states after two years have passed: in the first instance, only the house nearest the cliff has been lost, whilst in the second case both houses have been lost.

- Therefore, the use of a probabilistic approach should be considered, although the extent to which it is important to do so depends upon the distribution of probabilities of cliff falls and hence losses over time. It will be appropriate to undertake a sensitivity analysis to determine whether or not it is essential to carry out a probabilistic approach.

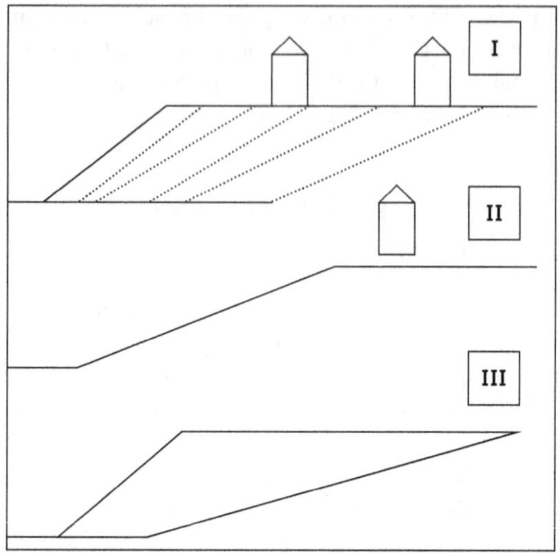

Figure 7.4 A cliff where, in any given year, there is a probability of different magnitudes of slides I; II and III show two possible states after two years have passed with different probabilities or rates of erosion

Table 7.1 Basic data for a hypothetical project to delay coastal erosion

Property	Value (£)	Mean year lost
House A	80,000	4
House B	60,000	7
Three mobile homes	3,000	10
Public house	240,000	13
House C	120,000	16
House D	90,000	17

- For example, Table 7.1 gives some basic data for a hypothetical project to delay coastal erosion and Table 7.2 gives a best estimate of the probability that house A will be lost in any given year where the same probability function also applies to all of the other properties. If it is assumed that the scheme has an engineering life of twenty years (at which point it fails), then the present value of erosion benefits is £215,758 (i.e. approximately £216,000).
- If, instead, we assume that each property is lost in the year at which the probability of loss is maximum (i.e. year 4 for house A), then the present value of erosion benefits is £205,000.

So, in this case, adopting a probabilistic approach to erosion makes very little difference to the estimate of the benefits. However, where the distribution of probabilities (as in Table 7.2) is very asymmetric, there can be much larger differences in calculated benefits.

Table 7.2 A best estimate of the probability that House A will be lost in any given year

Year	0	1	2	3	4	5	6
Probability	0.05	0.1	0.15	0.2	0.35	0.1	0.05

Economic appraisal spreadsheet GEHO0310BSDE-E-X.xls (Environment Agency, undated a, 2010a) enables the probabilistic approach to be applied. If the probability of loss for a given property is set to 1.00 in a given year then the method can be used deterministically.

In recent research, certain tools have been developed for probabilistic assessment of erosion of the coast over long timescales and large geographical areas, such as the SCAPE (Soft Cliff And Platform Erosion) model, which has seen successful application in the work completed by the Tyndall Centre's 'coastal simulator' (Dawson *et al.*, 2009), and has been widely applied by North Norfolk District Council (Walkden and Hall, 2005; 2011).

This meso-scale model of the North Norfolk coast has the ability to simulate an eroding shore system over scales that are large in coastal modelling terms (i.e. decades and tens of kilometres, rather than months and hundreds of metres). The development of the model shows that it is possible to simulate the long-term behaviour of a soft rock/beach coast subject to extensive human intervention. Importantly, the longshore effects of local management decisions can be represented, so the regional as well as local costs and benefits of coastal management interventions can be evaluated.

So far, this SCAPE model has been developed specific to the locality in question, but it is now receiving increasing practical attention and application. Such models could revolutionise the appraisal of coastal erosion and coastal flooding risks by allowing us to move away from historical rates and incidences of both, and model likely long-term trends into the future more accurately. It is an area of research that needs to be watched closely for the implications it will have for many aspects of coastal risk management.

7.4 VALUING DELAYS IN LOSSES, AND THE EXTENSION OF THE LIVES OF PROPERTY AND LAND

7.4.1 The concept of benefit as a delayed loss

As indicated above, the benefit of coastal risk management works is an extension to the life of, or the delay in the loss of, erosion-prone property and land for a period of time equal to the life of the risk management works, assuming that erosion after the end of the works' life would proceed at the same rate as it would have done without the project.

Thus, a property that is predicted to be lost by erosion in twenty years' time without risk management would be expected to be lost in seventy years' time with effective coastal risk management works having a life of fifty years. Therefore, the value of the benefits of coastal risk management is profoundly affected by the timing of the extension of the life of the property.

The possible approaches to the problem of finding an appropriate economic value of these benefits are as follows:

1. To find a figure for the annual value of the property or land, and from this calculate a present value for the additional time that the property can be used, given the extended life of that property based on known erosion rates and the intended life of the coastal risk management scheme.
2. To find an economic (market) value for the property or land as if it were not at risk, and to take the difference in discounted values of the loss with and without risk management as the benefit of coastal risk management.

In adopting these methods, which are similar, we are rejecting the possibility of finding economic (market) values for property with and without erosion risk, and calculating the benefits of coastal risk management as gauged by the difference between the two values.

This is because such an approach (the 'hedonic' approach) is fraught with difficulties and there are probably insufficient properties at risk from erosion in Britain to enable realistic values to be obtained for such situations. Also, markets appear to ignore risks such as erosion when determining valuations, because such elements seem to be minor components of the long list of variables affecting house and property prices/values.

7.4.2 The procedure for valuing property life extension I: using freehold property prices

The procedure here to overcome the problem of finding realistic market values for erosion-prone properties involves the following stages:

1 Finding a market-based property price which is not affected by erosion risk but otherwise reflects the attributes of the erosion-prone property concerned.
2 Using a formula to determine the present value of the use of that property until the time when it is lost through erosion at current erosion rates.
3 Using a formula to determine the present value of the use of the property with the extended life provided by the coastal risk management scheme (i.e. the life as above plus the anticipated lifetime of the scheme).

The two formulae identified above are as follows:

Equation 7.1

$$\text{PV (without scheme)} = \text{MV}\ (1 - 1\ /\ (1 + r)^{p})$$

Equation 7.2

$$\text{PV (with scheme)} = \text{MV}\ (1 - 1\ /\ (1 + r)^{p+s})$$

Where:

- p = the expected life of the property with no coast protection project
- s = the expected effective life of the coast protection project
- PV = present value
- where r = discount rate

$$\text{PV asset value} = \text{MV} \times (1 - [1\ /\ (1 + r)^{\text{year of loss}}])$$

$$\text{PV asset loss} = \text{MV} - \text{PV asset value} = \text{MV} \times [1\ /\ (1 + r)^{\text{year of loss}}]$$

PV benefit = PV asset value (with scheme) – PV asset value (without scheme); or

PV benefit = PV asset losses (without scheme) – PV asset losses (with scheme)

Both calculations of PV benefit produce the same answer.

The benefit of carrying out the scheme is the difference between the two values of present value which represent the gain from s years of equivalent annual benefit. The procedure, very simply, involves the calculation of the discounted value of the property with coastal risk management minus the discounted value of the same property without any proposed erosion-delaying works. The greater the life of the scheme, the larger the benefit, but not proportionately, because property use further into the future is discounted more heavily than use made in the medium or short term.

In the 'Yellow Manual' (Penning-Rowsell *et al.*, 1992), a slightly different – and now generally outmoded – approach was adopted, providing extension of life factors (ELFs) for different lengths of scheme life and different discount rates. This gives the same result as provided by the two equations, but does not explicitly compare the 'do something' and 'do nothing' values (because the ELFs incorporated both the anticipated erosion in the future and the anticipated life of any coastal risk management works).

A number of assumptions underlie this method. These are as follows:

1 The freehold market value, without erosion risk, reflects the current value to society of the property, including the land on which it is built. This value should reflect the discounted flow of services provided to society by the property and land. If there is reason to believe that the market prices are distorted from economic or social prices then some adjustments should be made.
2 The future life prospects for the property are the same (ignoring erosion risk) whenever the viewpoint. Hence no deterioration in the property is assumed to take place through time, and the property lifetime is assumed to be infinite. This seems appropriate since the land would, without erosion, have an infinite life and similar uses could continue to be made of it.
3 Likewise, the social value of the property, ignoring its future lifespan, is assumed not to change over time. Thus, as our society becomes richer, more constrained, or poorer over time, the social value of the resources at risk could change. Since this is unpredictable, as with other relative price changes, the assumption of unchanging relative social value appears justified.

7.4.3 Erosion-free property prices

The property and land prices required for the discounted erosion delay factor method presented in the previous section are market freehold values, not adjusted for erosion risk. That is, they are based on the normal expected life of that property type and an infinite life for the land on which the property is located. Therefore, we do not use the possibly degraded value of properties about to be lost through erosion; nor do we use the 'artificially' enhanced values of property adjacent to cliff-top edges reflecting the amenity attractions of the sea view. Hence, Tables 7.3 and 7.4 provide highly aggregated datasets for values of the main types of dwelling found in this country.

These values can be used in the above equations, but greater reliability may be achieved by obtaining values locally for the specific types of property to be affected by the project, since property prices in Britain vary considerably by location. The values used for residential property should reflect its location type – such as being near the sea – but it should be safe (i.e. based on properties which do not have an erosion risk).

Locally appropriate property prices can be obtained through:

- the coast protection authority's own evaluation department, if it has one; and/or
- local estate agents. It should be relatively easy for agents to estimate typical or average values for this type of property which ignore the risk of the properties being lost through erosion without a coastal risk management scheme as well as such factors as a sea view.

On average, it is less time consuming and loses little accuracy to take 'average' or 'typical' values for a property of a given type, or value per square metre in the case of businesses, rather than making individual evaluations of specific properties.

Table 7.3 House prices and average annual rental values in the UK by region

Region	New dwellings price (£)	All dwellings price (£)	Average annual rent at March 2013 (£)
North East	171,684	142,240	7,080
North West	177,750	160,760	7,524
Yorks & Humber	181,427	162,841	6,948
East Midlands	191,053	171,689	6,792
West Midlands	195,398	180,808	7,200
East Anglia	255,538	249,032	8,388
Greater London	328,023	397,963	14,724
South East	290,749	292,459	9,528
South West	230,264	224,919	8,808
England	232,131	242,138	8,555
Wales	173,636	158,143	6,528
Scotland	223,611	179,275	6,804
N. Ireland	148,896	129,610	6,876

Notes: Original rental figures have been annualised by multiplying by twelve months; please note some variation to government regions – see Horsfield, 2012.

Sources: ONS, 2012b; Homelet, 2013

Table 7.4 Residential property prices and annual rent by dwelling type (England and Wales)

Average prices by property type		
	Property price (£)*	Annual rent (£)**
Detached	253,352	13,372
Semi-detached	154,072	8,132
Terraced	122,939	6,489
Flat/maisonette	152,971	8,074
All	162,080	8,555

Notes: * Prices from Land Registry, 2013; ** Annual rent for each property type has been calculated as a proportion of the average annual rent in England; see Table 7.3.

7.4.4 The procedure for valuing property life extension II: using property rental values

An equally viable way of valuing property is to determine the rental it commands in the marketplace: the value of having that property to occupy/use for a given period of time (say, a month or a year). Recently, a change in the process of valuation has been developed, using property rental values in this way, rather than freehold values (as above; other assumptions listed there also apply).

This has the effect of simplifying the benefit calculation, because rental values are already easily expressed as an annual sum, thereby obviating the need to translate capital values into annual amounts. Therefore, now we are faced simply with the discounting of a stream of annual payments into a present value, allowing for the points in time when the erosion management scheme starts and ends its effective life.

Rental values also have some advantages over freehold capital values, in that the latter can be distorted by a number of factors, such as the imposition of stamp duty, which should have the tendency to depress freehold capital values on the basis that the purchasers of these properties have to pay both the stamp duty and the price of the house. On the other hand, rental values averaged over a region or nationally may not be typical of the kind of rental incomes obtainable for coastal properties by their landlords. In addition, the number of rented properties may be few and far between in some locations, making valuation difficult. Countering this, there will be plenty of holiday rental properties at the coast, but their weekly rental values will not be a good indication of the total *annual* rents obtainable from these properties, allowing for vacant periods. Therefore, we need to proceed with care.

We need to determine the rental value of risk-free properties (as reasoned above for freehold house prices). The government, through Defra, has suggested an annual average rental value of £6,000. However, this appears to be rather a low figure in that the average value of freehold property in England is *c.*£242,000 (Table 7.3), implying a very low rate of return on capital in terms of the rental income obtained (i.e. less than 2.5 per cent), although that average freehold value is affected by high prices in London, where there is no erosion risk. Nevertheless, this value of £6,000 appears to be out of line with the average market rate of return on rented properties, which approximates 5 per cent to 6 per cent. Our market-based rental values are given in Tables 7.3 and 7.4 from the sources cited there.

The equations for calculating the benefits of protecting property using rental values (using data as in Tables 7.3 and 7.4) are as follows.

Equation 7.3

$$\text{PV (without scheme)} = a/b \times (1 - 1 / (1 + r)^{P})$$

Equation 7.4

$$\text{PV (with scheme)} = a/b \times (1 - 1 / (1 + r)^{P+s})$$

Where:

- a = the annual rental for that property
- b = the rate of return from renting that property (i.e. the 5 per cent or 6 per cent, as indicated above)
- P = the expected life of property with no coast protection project
- s = the expected effective life of the coast protection project
- PV = present value

Again, the benefit of carrying out the scheme is the difference between the two PVs. Using this approach, it remains necessary to use values for erosion risk-free properties, and to determine the point in time in the future when the at-risk property would no longer be habitable, given no erosion risk management scheme. We also, again, need to know the life of the proposed scheme(s), and therefore the length of time over which erosion is delayed for each property concerned. Discounting the annual rental values between the first and second points in time, above, gives the present value of protecting that property from erosion to set against the costs of implementing the erosion risk management scheme.

The move to use rental values is part of the government's scheme of partnership funding (OM3), with its Grant in Aid (GiA) funding formula, as used in the flood risk management area (see Chapter 1). It appears to be having interesting consequences. A report from East Anglia in this respect indicates that the new formula has been applied to the Cromer Coast Protection Scheme on the north Norfolk coast. Cromer has had a scheme in review since the late 1990s but it continually failed to achieve the priority scores that were in place at the time and investment was therefore not forthcoming. The scheme is a £9 million sea wall and groyne refurbishment project which protects *c*.750 properties. Under the new system, it scored significantly higher than the funding threshold, to the point where the scheme was able to proceed without contributions from third parties, with work starting in 2013. A significant proportion of the protected properties are within one of the 20 per cent most deprived wards in the country, which helped to boost the GiA formula score.

The Environment Agency in Essex is involved in a project at Clacton, another long-considered scheme. This proposal does not score so highly in the formula owing to a physically wide public amenity area between the properties and the cliffs, meaning residential property erosion damages are not realised in any assessment until well into the future. However, this project is still likely to proceed with contributions from third parties supplementing the available GiA (Environment Agency, 2013c).

7.4.5 House price and rental value trends

Coastal risk management works – like many other civil engineering projects – are appraised for an expected project life of perhaps fifty or even one hundred years. Hence the benefit assessment is based on a prediction of the economic values of the losses which may be prevented over many years. Whilst general inflation is ignored in social benefit–cost analysis, since real values are unaltered by it, potential changes in relative real prices are relevant (HM Treasury, 1991).

- It is not normal in assessing the benefits of flood risk management to incorporate real changes in the values of property or inventories, despite evidence that real inventory values do change through time as populations become more affluent. This is because a predictive model of changes in prices would be needed in order to justify a different set of future relative prices, and such a model is not available and is likely to be impossible to calibrate over more than a few years into the future.
- The same arguments could apply in the case of residential prices and coastal risk management works, in that the real value of houses has changed in the recent past. The 'Yellow Manual' (Penning-Rowsell *et al.*, 1992) presents some data showing real house price changes through time, indicating a differential growth rate in house prices over other retail prices of approximately 2.5 per cent between 1946 and 1998.
- However, many factors influence the price of property, ultimately reflecting supply and demand. In addition, though, the attributes of housing change through time, such that houses are now superior in many characteristics to those that were valued in 1946. It might be possible that houses in coastal areas have been less improved than those in other areas, and therefore their prices are artificially depressed. More likely, any depression in prices could be because of the erosion risk itself, and the properties are therefore of lower quality and will be lower priced anyway, irrespective of the direct risk factor.
- The approach adopted above (Section 7.4.3) avoids this problem by taking prices from properties with the same type without erosion risk but in the same general area, accepting the typical qualities found in these non-erosion-prone conditions. This assumes any difference in physical quality is due to erosion risk which depresses the overall value of properties.

The same argument is not appropriate to any price differential due to houses having a 'sea view'. Without coastal risk management, pieces of land and property will still have a sea view. They will just be different plots. There will be a redistribution of views: once the seaward properties have been lost, the landward properties will have the sea view that others once enjoyed.

Hence, no conclusive reason and no reliable method for making future predictions of long-term house price trends have been found, although it may be that the past relative price trend will continue. In the face of uncertainty, the standard approach of assuming constant relative prices is recommended.

An example of the effect of incorporating real house price changes into the calculation of the benefits of coastal risk management is provided in the 'Yellow Manual' (Penning-Rowsell *et al.*, 1992, 166–174) and this effectively amounts to a reduction in the discount rate.

7.5 INFRASTRUCTURE BENEFIT ASSESSMENT PROCEDURES

7.5.1 Types of infrastructure loss

The following types of infrastructure loss should be distinguished:

1 Infrastructure which is integral to the coast and existing coastal risk management works, in particular promenades and associated structures.
2 Infrastructure which simply serves erosion-prone property (access roads, electricity, water, etc.) and which is not itself erosion prone, other than adjacent to the erosion-prone property it serves.
3 Infrastructure which serves areas either not at risk from erosion or only at risk some considerable time after the service might be cut, typically through-routes and public utility main systems which, if cut, would affect larger areas.

The recommended methods for benefit assessment for each of these categories are given below.

7.5.2 Promenades and associated structures

Promenades and associated structures are continually threatening to fail or falling into a state of poor repair. The policy choice here is either between the decaying coast or a new project, and therefore the 'do nothing' scenario involves both these changes and the impacts of them in terms of reduced amenity and recreational opportunities.

- Therefore, it is recommended that a more appropriate treatment of the loss of such features (rather than treating them as assets with a declining value) is to assess the loss of the recreation facilities that they provide.
- Thus, the benefits of a coastal risk management scheme in relation to this infrastructure is to maintain the use that these facilities provide, in terms of access and transport (a roadway) or recreation and amenity (a promenade).
- Contingent valuation methods used for recreation lost or benefit estimation can include an assessment of the value to society of the maintenance of these infrastructure features.
- This means that any illustrations or scenarios used in conjunction with the contingent valuation method should incorporate the future condition of the infrastructure (e.g. the promenade) in the 'do nothing' or 'without project' scenario, and also should show the proposed (better) condition of the seafront and promenade in the 'with project' scenario.

In a less detailed or generalised appraisal, an estimate of the remaining value of the promenade (but not the existing coastal risk management works) can be used as an approximation (shadow value) for the loss (Penning-Rowsell *et al.*, 1992, 51). That value should be in current prices for a promenade of the current specification in 'reasonable' condition (rather than any new design). The benefit of delaying this loss by providing coastal risk management can then be calculated using the two equations in Section 7.4.2.

It is important to note that the value of existing coastal risk management works 'protected' by new works *cannot* be counted as a benefit of new work. If the old works were fully effective, there would be no need for a new project, whilst there is no direct benefit from saving defunct works from eroding since they have no value.

However, existing coastal risk management works may be incorporated physically into a proposed new project and this may have the advantage of reducing the costs of new defences. Hence, any residual value of old defences will be implicit in the cost calculation for a proposed new project, and would be double counted if it were also included in the benefit calculation.

7.5.3 Infrastructure integral to properties at risk from erosion

Where the loss through erosion of power cables, telephone lines, gas pipes and access roads would not affect properties other than those lost to erosion, and the impact would be approximately at the same time, there is no additional economic loss (because the use of those properties assumes power and other utilities are supplied). Extending the lives of these services will not bring an additional benefit to society, since they benefit only the properties which would, by that time, have been lost to erosion.

However, the timing of erosion is important here. It is crucial to identify the earliest time when the property would be deemed uninhabitable, and this may be when infrastructure services become unavailable owing to erosion. Hence, the alignment of these services should be mapped along with the erosion-prone property and the erosion contours. When deciding at what point in time in the future the property will become unusable in the 'do nothing' scenario, allowance should be made for the time when this infrastructure becomes lost.

7.5.4 Infrastructure serving areas not at risk from erosion at the same time

These other areas, which are not likely to erode when the infrastructure serving them will be lost, may be erosion prone themselves, or the demand served by the infrastructure at risk may be unaffected by erosion.

Infrastructure serving distant erosion-prone properties

If the infrastructure is erosion prone and this renders the property it serves uninhabitable then the point in time in the future when that property is deemed to be lost from use should be when the infrastructure is lost. This then changes the timing element within the equations in Section 7.4.2.

- However, where the infrastructure is vulnerable to erosion a substantial time before the properties are vulnerable, it is unlikely that those properties would be abandoned. In this case the benefit of protecting that infrastructure (and the use of the property it facilitates) is gauged as the costs of providing alternative services or infrastructure for the properties potentially disconnected.
- If the present value of these costs (using normal discount rates) is less than the difference between the present value of the property loss assessed for when the properties' services would be lost and when the properties themselves would be lost, then the least social cost response to erosion without a coastal risk management project would be to invest in the new infrastructure. Therefore, the cost of additional infrastructure investment should be included in the project benefits as a social cost averted by the project, and the property life continued until it is directly uninhabitable due to erosion, rather than uninhabitable owing to infrastructure loss.

Infrastructure serving areas free of erosion risk

Where infrastructure is providing a service to an area which is free from erosion risk, but is itself threatened by erosion, the provision of least-cost alternatives may be a suitable measure of the benefits of coastal risk management.

The rationale for this approach is that simply including an economic value for the annual use of the infrastructure (e.g. the value of the electricity supplied by a cabling system) may overestimate what the 'without project' economic loss would be, because that electricity could be supplied to those who need it by cabling capacity elsewhere in the grid system.

- If no spare capacity or existing alternatives exist, then the economic costs of providing new infrastructure, if feasible, are the appropriate benefits (cost avoided) of coastal risk management.
- Hence, the loss avoided by coastal risk management is the least-cost option by society in a 'do nothing' scenario. This may be the cost of building new infrastructure, the cost of increasing the capacity of existing facilities or congestion costs

from increased use of existing facilities (e.g. roads and through-routes).

In general, these infrastructure impacts are not likely to be substantial unless major sewers, or major road or rail communication links, are involved. In other cases it may be sufficient to discuss with local utility managers the implications of erosion and whether the existing system could cope with the altered usage in the 'without project' situation, or to estimate the likely costs of maintaining services by installing new works.

7.5.5 Infrastructure methods

The procedure given for infrastructure (Figure 7.5) presents, first, a question as to the nature of the infrastructure and its role at the coast. The answer to this question will influence whether recreation benefit methods (see Chapter 8) or the equations in Section 7.4.2 are employed. In all other cases (where the infrastructure loss could affect activities which would continue after the properties are lost by erosion) the least-cost methods need to be used and the avoidance of this cost is then the potential benefit from coastal risk management.

7.6 VALUING NON-BUILT-UP LAND

7.6.1 Scope of techniques

Erosion of the coast can result in the loss of agricultural land, recreational areas and amenity features. In general, these losses affect the economic enterprises using this land at the margin, rather than in totality. This makes the assessment of the benefits of protecting this land somewhat problematic, and care should be taken to avoid simply averaging the economic return from such areas in proportion to the economic return of the totality (e.g. the loss of a field or two on a farm does not always proportionately reduce the economic returns to the whole farm).

7.6.2 Agricultural land

Assessing the economic benefits to the nation of the outputs from agricultural land can be challenging, since some agricultural prices and financial returns to farmers may be distorted by a range of fiscal and land use planning factors (see Chapter 9).

- Hence, the current recommendation is that the value to society of protecting agricultural land from erosion is the annual economic margin arising from agricultural production that would be lost without coast protection. This, therefore, follows the methods outlined in Chapter 9.
- An estimate of the area of cultivable land lost in each erosion contour band by land type is the starting point for this type of appraisal. An average cropping pattern should then be compiled for that area, and predicted for future years for each category of land at risk. In general, the present value of the benefit is the discounted sum of the adjusted net margins between year p (when the land would erode in a 'do nothing' scenario) and year $p + s$ (erosion year plus project life). The calculation needs to be carried out for each erosion band, and in effect this is the same as applying the equations in Section 7.4.2.
- In some cases changes in fixed costs will be small and so agricultural benefits can be based on adjusted gross margins (i.e. adjusted revenue less variable costs). In other cases, where erosion is serious and covers large areas, there may be changes in overall farming productivity and farm scale fixed costs.
- Erosion control, soil conservation or restoration measures may be required for which costs will need to be identified.

An approach which may be simpler – but theoretically should give the same result – is to gauge the economic loss from erosion as the reduction in the market value of the land for the grade of land under consideration in the area (taking the price of the equivalent erosion-free land), adjusted according to the methods outlined in Chapter 9 . If large areas of agricultural land or land of particular strategic importance, such as Grade 1 and 2 land, are liable to erosion, a more detailed assessment will be required, again following the methods in Chapter 9.

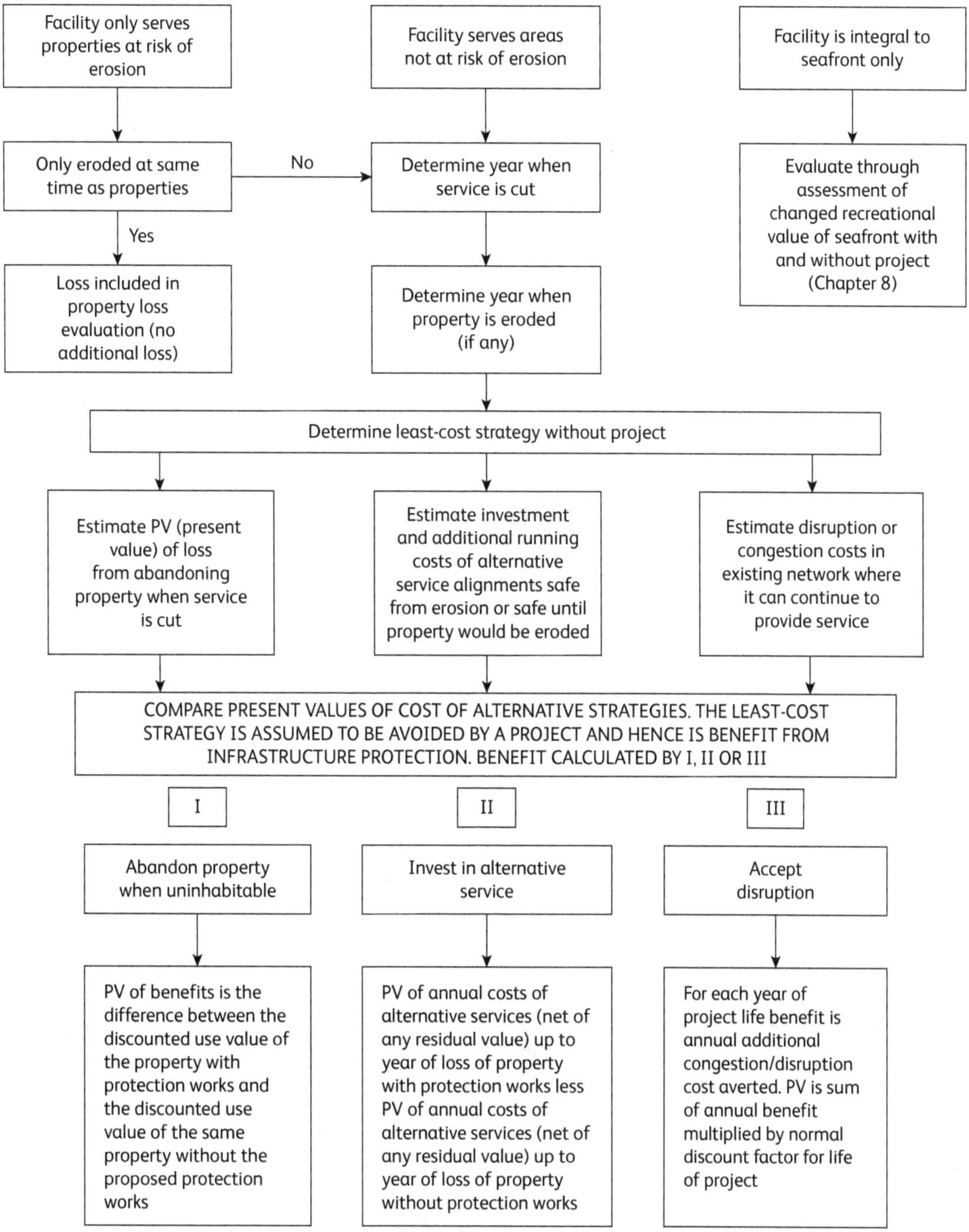

Figure 7.5 Schematic diagram illustrating the benefit assessment framework for infrastructure

7.6.3 Other open-space land

Open space at the coast often has high recreational and environmental values. Aspects of these are also discussed in Chapters 8 and 10.

- Changes in the recreational use as a result of erosion should be assessed using the methods recommended in Chapter 8. Some recreational uses are likely to be marketed, and therefore the costs of adjusting the relevant business to cope with change will be the loss, or the loss of net margin will be appropriate measures of the marginal losses.
- For example, an eight-hole golf course would have little value, but with some investment the lost ninth hole might be fitted into the reduced area following erosion. That investment cost is the benefit of erosion averted.
- Where losses are likely to become non-marginal, the alternative approach is to assess the cost of a shadow project as the economic loss being avoided. Where land of ecological value is at risk from erosion, the methods in Chapter 10 should be followed so that a qualitative and semi-quantitative assessment of the value of the site can be made, both with risk management and for several stages of erosion.
- In some circumstances costing the creation of an alternative site into any proposed project (the shadow project technique) is one possible approach. This approach does not estimate the loss to society, but does indicate the cost of maintaining the same number of ecological sites of that type.
- Where an ecological or geological feature relies for its value on continuing erosion (e.g. a site containing fossils that are uncovered as a cliff face erodes), it is the halting of this erosion that causes a loss, whilst allowing the erosion to continue creates a benefit.

Economic analysis in the cases of these open spaces' land uses will clearly be complex and challenging. In some cases it will be virtually impossible, through lack of data or lack of clarity concerning the impact of erosion on the areas concerned and their value to society. However, the appraiser should not do nothing in such circumstances. At the very least a clear statement of the likely losses involved with and without a project should be catalogued, and their timing should be presented.

7.6.4 The issue of contaminated land at the coast

One issue that needs consideration where protecting the coastline is concerned is the presence of contaminated land and landfill sites at risk from erosion and the associated environmental and societal costs associated with protecting – or not protecting – such sites (CIRIA, 2012).

The context here is that, over the years, processes of coastal erosion and flooding from the sea have resulted in waste from some sites being deposited on the foreshore or seeping into the coastal and marine environment. The result could be an adverse effect on public health and safety, or undesired physical, chemical and biological effects on the natural environment. If sea levels rise as a result of climate change, these adverse effects are likely to become more significant than they are today.

In order to prevent such adverse effects, investment may be needed to protect the sites from erosion by the sea, or from coastal flooding, and clearly this investment has to be justified in relation to the scale of these adverse effects and their likely propagation throughout the complexity of coastal ecosystems.

Guidance has been developed aimed at the interface between the well-established industry of waste management, pollution prevention and flood and coastal erosion risk management (CIRIA, 2012). Within this guidance, chapter 6 offers advice on appraising the options that are available to manage the risks presented by erosion and sea flooding of such sites, and the approaches for assessing the relative technical, economic and environmental merits of each within the context of a systematic options appraisal.

This is a new area for coastal risk management, and the available methods for evaluation and appraisal are almost completely untested. Nevertheless, the guidance described here gives appraisers a first sight of the coastal management issues involved and the likely methods they may use to decide on investment levels. One cannot but conclude that this is a difficult area for appraisers, as well as one in which the public is unlikely to support major investment without careful analysis of the costs and benefits involved. Clearly, it requires further investigation to establish the nature and severity of the risks, the options available for risk management and the costs and benefits of alternative strategies.

7.7 COMMON MISCONCEPTIONS AND LESSONS FROM EXPERIENCE

7.7.1 Some common misconceptions

Many common misconceptions arise when discussing coastal risk management, derived from an era when systematic project appraisals were not undertaken. For example:

1. Property must be protected at all costs.
2. Land and houses should be protected for ever.
3. Decisions in the future about coastal risk management should reinforce planning decisions made in the past.
4. A valuable promenade is a benefit if it is to be protected (even if it is falling down).
5. There is no merit in delay.
6. The sea will not win in the end.

In reality, systematic project appraisal will provide a more focused analysis of the options available and the best use of national resources.

7.7.2 Key lessons from experience

The assessment of the benefit of coastal risk management, and the appraisal of coastal risk management schemes, is not without considerable complications. Some lessons from experience are given below:

1. Flooding and erosion are often inextricably interlinked; probabilities can become very complex to calculate.
2. Unless they are very near the edges of cliffs, houses alone generally provide a poor base for the justification of major coastal risk management works.
3. Accurate and realistic erosion rates and probabilities are the key to accurate benefit estimation.
4. Prices of houses situated on the tops of cliffs do not accurately reflect their risk of falling into the sea.
5. House price inflation into the future must be treated with caution within project appraisals.
6. The loss of one person's view is another person's gain: the view itself is never lost.
7. The loss of infrastructure has different effects, depending on whether the property it serves erodes before or after it.
8. The environmental benefits of coastal risk management are mixed: some assets gain (e.g. eroding cliffs revealing important archaeological or geological sites), whilst others involve losses (e.g. the loss of habitats for bird species).
9. The recreational benefits of coastal risk management have been widely ignored and yet they are often key reasons for scheme implementation.
10. Delay is a real option that should be considered seriously.

In addition to the above, it needs to be borne in mind that a systematic comparison of investment versus no investment must start from a realistic evaluation of the 'do nothing' option.

8 Recreational gains and losses

This chapter looks at the changes in recreational enjoyment that can come about as a result of flood and coastal erosion risk management schemes. As such, we are examining 'use values' (Chapter 2) of natural and semi-natural resources (coasts and rivers).

This is not an easy area of project appraisal. First, the values are highly variable and site specific, depending on the resource and the potential for its change. Second, we have to estimate the numbers of people who will be affected and, in many cases, there is little firm data on which to base such estimates of these populations. Third, those who suffer 'losses' of recreational value will often transfer their time elsewhere, to other resources (a different beach or river, or a theme park). Such shifts are difficult to assess.

Nevertheless, over many years we have found that the recreation and amenity benefits and possible adverse effects from flood and coastal erosion risk management schemes can be substantial. This is particularly the case at the coast, but in certain fluvial environments they are also of considerable importance. They also can be highly significant in terms of local stakeholders' attitudes towards risk management schemes. Therefore, they should be pursued, notwithstanding difficulties mentioned here and detailed below.

This chapter reviews the literature in this area, and gives some details about current appraisal techniques. This volume of our 'Multi-Coloured Manual' does not contain any new empirical results, but we believe that the existing results as reported here (mainly from the period 1992 to 2005) continue to be valuable. A continuing programme of research is needed to ensure that the results remain as up to date as possible.

8.1 SUMMARY OF INFORMATION AND DEVELOPMENTS IN THE LAST DECADE OR SO

This chapter discusses the issue of recreation and amenity benefit assessment and the procedures and techniques recommended for use in assessing the potential recreation and amenity benefits of investment in coast protection or coastal or fluvial flood risk management. For brevity, the term 'recreation benefits' is used to cover benefits arising from the enjoyment of landscape, wildlife and natural amenities as well as from the enjoyment of recreational activities *per se*.

Research has indicated that, for coasts, recreation benefits, particularly the benefits of preventing the erosion of beaches – the first line of defence at the coast – can be among the most important benefits of coast protection in addition to the benefits of protecting property and other major infrastructure works, such as arterial roads and main sewers. In some locations, these benefits have been found to be equal to, or even greater than, the benefits of protecting property. One obvious reason is that beaches, promenades and other elements of coastal defence will be damaged or lost before the property which they protect is affected.

Much of the research and material in this chapter relates most explicitly to coastal erosion risk management. However, there are also likely to be recreation benefits from projects designed to alleviate coastal flooding. This will particularly be the case where beach recharge or other beach works, or sea walls and associated promenades, are proposed for coastal flood risk management reasons.

There is little research evidence on the recreation benefits of fluvial flood risk management projects, although schemes on rivers do appear to have considerable potential to yield recreation benefits. The Maidenhead flood relief channel, the Jubilee River, provides a good example of this potential. The research on the recreational and amenity benefits of river restoration, although now somewhat aged (Tunstall *et al.*, 1999; Eden *et al.*, 1999; Tapsell, 1995), also supports this view.

There are two key elements involved in assessing recreation benefits: estimating the number and type of users or beneficiaries; and estimating the value that they attach to recreation at a site under varying conditions. Section 8.2 provides a brief introduction to these two elements and to the methods that are used to derive estimates. Section 8.3 discusses the recommended approach to recreation benefit assessment in general terms, then Section 8.4 outlines the techniques recommended for use at initial study stages and for generalised investigations or strategy studies. Next, Section 8.5 describes site-specific data collection methods recommended for a study for a particular scheme. Figure 8.1 outlines the steps to be taken at both of these study stages. Additional guidance checklists, information and questionnaires are presented on MCM-Online (see Supporting Information for Chapter 8).

There is now a substantial body of research on coastal recreation benefit assessment. The results from this research are presented in Section 8.6. In addition, some results from relevant research on river restoration projects and other forms of river enhancement are included to indicate the ways in which methods have been applied in riverine settings. The Corton case study presented in Section 8.7 illustrates a reasonably recent application of the recommended techniques at a coastal site already seriously affected by coastal erosion. Finally, Section 8.8 sets out some of the lessons learned from experience of, and research into, recreation benefit assessment over the last ten to twenty years and indicates some of the continuing limitations of the available methods.

As with many other topics in this Manual, we are dealing with an evolving situation, regarding both risk management policy and appraisal technique. Within the last two decades there have been many new developments that have added new insights to those from the work that we first reported in our 'Yellow Manual' (Penning-Rowsell *et al.*, 1992). Increased awareness of the pressure on environmental resources worldwide (MEA, 2005) and in the UK (HM Government, 2011; Lawton *et al.*, 2010) has led to an even greater recognition of the need to value these resources properly, including recreation and amenity resources.

The developing ecosystems services approach (NEA, 2011; see also Chapter 10) potentially provides a framework within which the recreation and amenity changes associated with FCERM can be assessed using the methods presented in this chapter. Indeed, some researchers (e.g. Luisetti *et al.*, 2011b) have framed their studies of change at the coast fully in the ecosystem services context. In relation to other techniques, several publications have reviewed the now substantial fifty-year experience of contingent valuation in particular (Carson, 2011; Alberini and Kahn, 2006a) and of non-monetary valuation in general (Bennett, 2011). Over 7,500 contingent valuation papers have been published in over 130 countries – including developing countries – addressing a wide range of environmental issues but also other issues such as health; these are cited by Carson (2011).

This section on new developments since 2003 (when the first 'Multi-Coloured Manual' was drafted) focuses on stated preference valuation methods involving surveys: contingent valuation (CV) and choice modelling or choice experiments. Carson (2012) maintains that the big issues concerning the reliability of CV methods raised by critics in the 1990s have largely been resolved or have been shown to involve behavioural effects that routinely characterise markets. Others recognise that there is a continuing need to refine and test valuation methods (Day *et al.*, 2012; Luisetti *et al.*, 2011a; Bateman *et al.*, 2006; Alberini and Kahn, 2006b).

Among researchers, there has been increasing interest in choice experiments or choice modelling (formerly termed 'conjoint analysis') as compared with straightforward willingness to pay (WTP) methods as a means of eliciting values. Carson and other CV practitioners (e.g. Bateman, 2011; Whitehead,

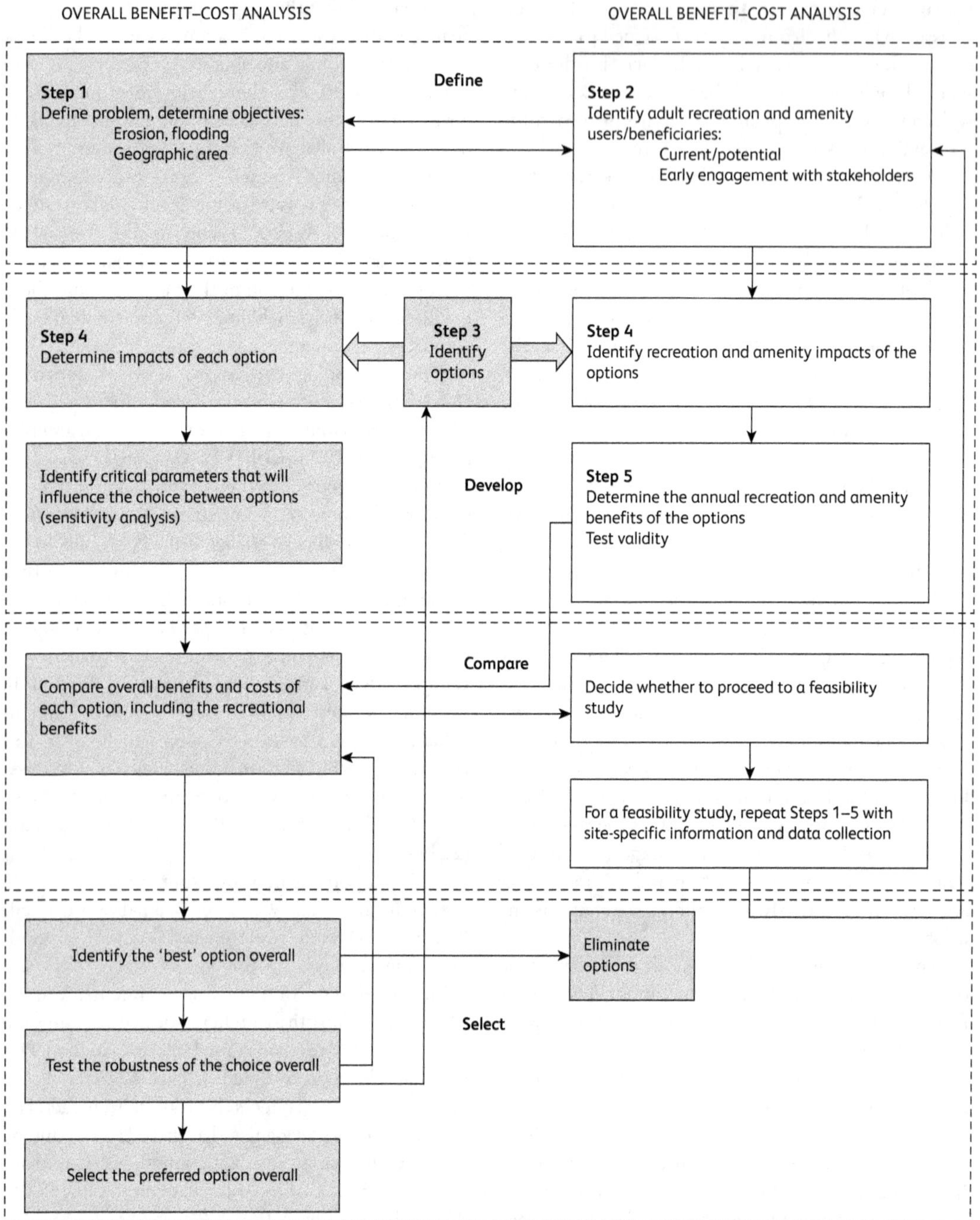

Figure 8.1 Stages in benefit–cost analysis

2006) argue that a key issue is the quality of a valuation study in terms of the nature and size of the sample, the design and testing of the survey instrument and the credibility of the representation of the environmental change, the survey method and quality of interviewing. These are all issues that are also emphasised in this chapter on the basis of our research and experience.

Developments in GIS and computer visualisations and animations may provide an effective alternative to drawings for representing environmental change in valuation scenarios and help also with visitor counts. Visualisations have been used by Appleton and Lovett (2003) for representing rural landscapes, and by Sheppard and Meitner (2005) for sustainable forestry. Several studies have considered these techniques in relation to flood risk management: Posthumus *et al.* (2008) for agricultural land use and flood risk; Pender and Néetz (2007) for flood modelling; Tunstall *et al.* (2009) for planning and flood risk. Bateman *et al.* (2009) found that visualisations and animations offered some advantages over numerical and verbal representations of land use change.

But since 2003 there have been few new UK valuation studies focused on recreation benefits on rivers and coasts, although databases indicate that coastal and river resources have been the subject of studies elsewhere. Outdoor recreation of all kinds has generated many studies and valuation databases in the USA on parks (Kaval, 2007) and in New Zealand (Kaval and Yao, 2007). Some specialist recreational uses have also been valued, including by: Paulrud and Laitilla (2004) on fishing in a Swedish river; Butler *et al.* (2009) on recreational fishing on the River Spey in Scotland; and Lawrence and Spurgeon (2007) on fishing in inland waters in England and Wales. The Water Framework Directive has stimulated interest in valuing water and ecological quality on inland waters and rivers in the UK (Lago and Glenk, 2008; NERA/Accent, 2007; Bateman *et al.*, 2006; Hanley *et al.*, 2006). River ecosystem restoration has also been the subject of valuation research by Holmes *et al.* (2004) in the USA.

The main shift in UK coastal recreation from holiday to day-trip destinations took place from the 1970s to the 1990s (Williams and Shaw, 1997; 2009). However, research has indicated an increase in specialist coastal activities, such as hiking, birdwatching and sports activities, requiring a separate valuation approach (NEA, 2011). On the coast, Barry *et al.* (2010) have examined willingness to pay for public access and improved trailways for coastal recreation on the west coast of Ireland. Trail walking – already a popular form of coastal recreation – is likely to become more popular as the Marine and Coastal Access Act 2009 (HM Government, 2009b) improves access to the English coast. A study on the East Anglian coast modelled beach use in detail (Coombes *et al.*, 2009a) and Tratalos *et al.* (2013) considered the impact of crowding on recreation and habitats at the coast.

Few UK studies have looked at coastal recreation in the context of FCERM. Luisetti *et al.* (2011b) considered the valuation of managed realignment sites on the East Anglian coast. A small study in Wales looked at beach users' evaluations of small changes to hard engineering coastal defences (Blakemore *et al.*, 2007). Since 2003, there appears to have been only one study published (Polyzos and Minetos, 2007) that uses the VOE approach to valuation presented in this Manual in relation to FCERM at the English coast, although there may be other studies buried in the 'grey literature'.

Given the limited number and high cost of new empirical studies, benefit transfer (now often termed 'value transfer') will continue to be important in recreation benefit assessment. There are now a number of valuation databases on which practitioners may draw. These include the Environmental Valuation Research Inventory (EVRI), a Canadian-run resource covering a large number of mainly US valuation studies, the CSERGE Water Database at the University of East Anglia, and the Environment Agency's Recreation Database (Eftec, 2010a, 31).

Estimating visitor numbers remains an important and under-studied element in river and coastal recreation benefit assessment. Visitor number modelling and estimation has received some attention in the context of woodlands (Jones *et al.*, 2010; O'Brien and Harris, 2010) and methods used there may offer some guidance on applications in the coastal and river contexts. The impact of climate change on coastal visit numbers has been considered at the regional UK level by Coombes *et al.* (2009b) and at the international level by Hamilton *et al.* (2005).

Practical guidance on survey and visit count methods can still be found in Scottish National Heritage's updated *Visitor Monitoring Manual*, which is supplemented by its detailed advice on monitoring access and recreation at sensitive Natural Heritage sites (Scottish National Heritage, 2007). GreenSpace (undated) has also produced a guide to automated methods and devices for counting visitors, including CCTV cameras for the Heritage Lottery Fund (HLF). The HLF increasingly requires authorities seeking funding for park or green space improvement to provide accurate visitor counts, so it may be able to provide information both now and in the future on organisations that have undertaken such counts.

8.2 DEVELOPING AND REFINING ESTIMATES OF RECREATION BENEFITS

Taking an iterative approach to project appraisal is particularly important when considering recreation benefits; it is advisable to determine early on in the planning or design process whether recreation benefits are likely to contribute significantly to scheme benefits (akin to a pre-appraisal 'sensitivity analysis', as suggested in Chapter 2). If this is the case then action will have to be taken early on in the appraisal process to refine the estimates of these benefits.

Recreation benefits are calculated by multiplying the value in pounds of a visit or for recreational use at a site by the number of visits or beneficiaries at or to that site. This usually involves multiplying what is commonly a comparatively small number that appears relatively more stable across sites (the money value per visit or for recreational use) by what can be a very large number (the number of visits or beneficiaries). Hence, the more important element in assessing the likely magnitude of the recreation benefits of a scheme or strategy is usually the estimation of the number of visits or beneficiaries (although both variables are obviously equally crucial in the actual calculation). Data on the number of visits to coasts and rivers are sparse, as are data on the pattern of visits over the year. For these reasons, particular attention should be paid to developing and refining this element in recreation benefit assessment. No single method for deriving estimates is recommended in this Manual; instead, a range of sources and methods are presented.

The recommended method for estimating the money value of a visit or of recreational use of a site and the basis for the standard data presented in this Manual is the contingent valuation (CV) method. This is essentially a questionnaire/survey in which respondents are asked directly, in carefully drafted questions, to say what value they place on the availability of a resource such as a beach or riverside recreation, or how much they would be willing to pay (WTP) for a change in the availability of that resource. This procedure was tested and endorsed as a method for coastal recreation benefit assessment in preference to other possible valuation methods such as the travel cost method (see Chapter 2) as part of the research for the 'Yellow Manual' (Penning-Rowsell *et al.*, 1992).

Since their first application in the USA in the 1960s, there has been a steady growth in the use of CV approaches, supported by institutional endorsement in the 1970s and a burgeoning of CV studies and reviews in the 1980s in the USA (Cummings *et al.*, 1986; Mitchell and Carson, 1989). More recent evaluations were discussed in Section 8.1. In the UK, the method has been extensively used to value a range of environmental changes by government departments and other major agencies, including the DETR (aggregates), the Environment Agency (low flows on rivers) and the Foundation for Water Research (FWR) (water quality), and it is now a much more widely accepted technique in the UK than it was a decade or so ago.

Some time ago we developed and tested a particular variant of the CV method – the value of enjoyment (VOE) per adult visit – as part of our research for the 'Yellow Manual' (1989–92). This specific approach, as opposed to the more commonly used WTP approach, was then recommended for use in coastal recreation benefit assessment in the 'Yellow Manual' and we continue to recommend it for that purpose now. In this approach, respondents are asked to say what value they put on their enjoyment of a day's visit under varying options in pounds and pence. (In the WTP approach, respondents are asked how much they would be willing to pay in entrance fees or in rates and taxes for a change, such as a river or coastal risk management scheme.) The VOE approach was developed as a method of meeting the requirements of three different groups:

1 The respondents who have to execute the valuation task by providing answers in the CV survey.
2 The decision-maker who wants accurate estimates of the economic value of a resource.
3 Economists who have theoretical expectations as to how people will approach the valuation task.

The respondent is faced with a difficult task because:

- people are not used to valuing in money terms those things that do not have a price; and
- a significant proportion of recreational visits to coasts and, particularly, to riversides are made on foot, so these visits do not incur a monetary cost that could serve as a guide to value.

So, the method adopted needs to help respondents decide what value they truly put on recreational use. In addition, if people are asked to pay for a visit that is currently free, the first consequence is that they should be expected to make fewer visits. Since the recreational value of a site is given by the value per visit multiplied by the number of visits made, the result of asking people to pay will be an underestimate of the current value of the site. This precludes phrasing the contingent valuation question in terms of an entrance or parking fee. The introduction and removal of charging from some national museums has dramatically illustrated the extent of the influence of charging on the numbers visiting. Furthermore, since the same income will now have to be spread over more things, the amounts people are prepared to pay for currently priced goods will fall, even if only fractionally.

For the decision-maker, asking people how much they would be willing to pay to preserve or enhance a site means that it is difficult to separate recreational value from all the other reasons people may have for wishing to protect a site. It is not always possible, as was the case in the Caister benefit assessment (Bateman *et al.*, 2001), to differentiate between an option offering additional recreation enhancements and those that simply protect property; most options will do both. The VOE approach is focused upon recreation and thus avoids the significant problem of double counting that may arise with the WTP approach.

Economists have been preoccupied with theoretical concerns, in particular the expectation that respondents will not tell the truth – that they will 'free-ride' – when asked how much they are willing to pay for the provision of a good. In practice, 'free-riding' has not been found to occur significantly (Marwell and Ames, 1979), although people have generally been found to answer questions as to how much they are willing to pay for a collective resource in ways that are different from economic theory (MacRae and Whittington, 1988).

The VOE variant of the CV method was developed at the Flood Hazard Research Centre in order to address the requirements of the respondent. It is open to the criticism that the values the respondents give are not constrained by income and that it does not remove the possibility that an individual respondent will seek to 'free-ride'. Conversely, unlike other approaches, it takes account of time as a fundamental constraint on consumption on the basis that the time spent making a visit cannot be used for other forms of consumption.

The VOE technique has now been used extensively and satisfies both external tests of validity (e.g. it should discriminate between different changes in the fluvial or coastal resource) and internal tests of validity (e.g. the money value of the enjoyment of a visit should be related to the degree to which the visitor reports having enjoyed his/her visit). Other CV formats, furthermore, do not generally measure recreational value alone.

In this Manual, the VOE approach is the recommended method for assessing the recreation benefits of coastal and flood risk management schemes. However, the use of the WTP approach is acceptable where it is judged to be more appropriate by those undertaking benefit assessments and where, particularly, double counting of direct benefits can be avoided. If the WTP approach is used, great care should be taken in designing the WTP scenarios to ensure that *only* the recreational value of the scheme is considered by respondents.

8.3 THE RECOMMENDED APPROACH AND TECHNIQUES: AN OVERVIEW

A two-stage framework for recreation benefit assessment is recommended. This involves:

- An initial study stage for initial examination of projects and for strategy studies. This will normally rely upon secondary source data and desktop methods, although it may be necessary to undertake some preliminary data collection where secondary sources are unavailable.
- A detailed study stage involving detailed, site-specific information and data collection methods: site-specific counts of visit/visitor or resident numbers and a site-specific CV survey to provide site-specific estimates of the value of recreation with each of the options. These methods will be used later, after it has been established in the initial

study stage that the proposed project has the potential to provide sufficient recreation benefits to justify the use of further resources in data collection. (CV surveys and count procedures are expensive and time-consuming activities to mount and manage.)

At both the initial and the detailed study stages it will be necessary to go through the same steps (see below), but at different levels of detail and site specificity.

The steps in recreational and amenity benefit assessment run parallel with – and interact with – those for benefit assessment in general, as outlined in Figure 8.1. Where, as is usually the case, there are other benefits to be assessed as well as recreation and amenity benefits, or environmental issues to be considered, there should be close interaction between these processes to ensure that they are well integrated. In particular, it is essential to ensure that the options presented in the CV scenarios and employed in the other parts of the benefit assessment are the same. Ideally, CV scenarios used in the recreation benefit assessment should also be the same as those used in any public consultation to avoid confusing stakeholders and members of the public.

8.3.1 Step 1: Define the problem and determine the objectives

Any benefit assessment starts with definition of the problem – the nature and rate of coastal erosion or degradation, or coastal or fluvial flooding – and with identification of the extent of the geographical area affected – its length and breadth, plus its characteristics and the types of changes to its physical characteristics that are likely to take place. Problems such as coastal erosion may be site specific or may affect a more extensive area, as in the case of the Holderness coast. Similarly, the flooding problems affecting a river may be present in much of the catchment or may be site specific. It is essential to consider problems and the options for dealing with them in the wider context in this way. Stakeholders and members of the public themselves may well be active and valuable in defining the problem and suggesting solutions.

8.3.2 Step 2: Identify adult recreation and amenity users or beneficiaries

The second step is to ascertain whether there is recreational use of the site (or potential for it) and to identify the range of recreational activities that are, or could be, undertaken there. A crucial determining factor is whether there is, or could be, public or some form of restricted access to the site. While there is a public right of access to the shore below high tide on British coasts, access is restricted along many riverbanks and riverside land is often privately owned.

Both Defra (2009) and the Environment Agency (2010a) recognise the need to ensure that public investment decisions are made with transparency and public involvement (see Chapter 2). Early identification of those who have an interest in recreation and amenity at the site is essential so that these stakeholders can be involved in the project: from the definition of the problem through to the identification of the options, the evaluation of their impacts on recreation, and the estimation of recreation gains and losses with the changes.

Although children may be important users of coasts and riversides, the benefit assessment methods recommended here apply to adult users or beneficiaries only because the valuation questions are judged to be too difficult for children to answer directly or for adults to answer on their behalf. Similarly, researchers should not make arbitrary assumptions about the values children might attach to their recreation.

The checklist of possible recreational uses of coastal and river sites on MCM-Online (Supporting Information for Chapter 8) can be used as a guide. The largest category of recreational user of beaches, cliff tops and river sites is generally the informal user who visits to walk or stroll, picnic, sunbathe or walk a dog. In cities, towns and even villages, this informal recreation may be only a part of, and contribute to, a visit for general tourism or other purposes. A second category is the specialist user whose activities require specific facilities, equipment or skills. Anglers, boaters and golfers are examples of this category of visitor.

Visitors may also be classified according to their origins. In the 'Yellow Manual' (Penning-Rowsell *et al.*, 1992) visitors were defined in terms of the following three categories, which remain valid:

- Local visitors: those living within a three-mile radius of a site, which is deemed to be a possible walking distance.
- Day visitors: anyone starting and finishing their trip from their permanent home, including some who may define themselves as local but who live more than three miles from the site.
- Staying visitors: anyone staying away from home for one or more nights. This category includes visitors not staying at or near the site but making a day trip there while staying away from home.

Definitions of these categories used in tourism or other recreation statistics may be slightly different. These different types of visitor may use coastal and river sites in different ways and have been found to attach different values to their recreation there (Penning-Rowsell *et al.*, 1992; Section 8.6, below). Therefore, it can add to the accuracy of the benefit estimates if these different types of user can be distinguished in the analysis.

A crucial issue in both the initial and the detailed studies is to establish the level of use of the site in terms of the number of visits it receives or the number of those who benefit from recreation there. The character of the site – large, commercial seaside resort, less developed resort or undeveloped coastal site – may provide some indication of the likely visit numbers. But some undeveloped 'wilderness' sites on the coast that are close to population centres – for example, Hengistbury Head – may attract large numbers of visitors specifically because of their undeveloped character. On rivers, some sites with visitor facilities (for example, those in country parks) may be 'honey pots' attracting large numbers of visitors. Town centre or town park riversides may also receive substantial numbers of visits.

8.3.3 Step 3: Identify options

This step involves identifying the options for dealing with the problem and their likely impacts on the physical characteristics of the site as well as the 'do nothing' option.

Coasts and rivers change naturally over time. The condition of coasts and rivers can be represented as a continuum reflecting the degree of change in terms of how 'good' or 'bad' the condition of the site is for recreational use and value. Any study of recreation and amenity benefits for a coastal or flood risk management project will normally be undertaken when some deterioration has taken place at the site. But a site will not usually be allowed to become totally degraded by erosion or flooding before the authorities intervene.

Thus recreation benefits (depending on the timing of the study, the nature of the problems and the options to deal with them) may have the following two components:

1. The prevention of further deterioration – *losses* with the 'do nothing' option. Almost all coastal risk management projects and some combined fluvial and coastal flood risk management projects will have this component. Options that only prevent further deterioration and simply reinstate the site to its current condition will involve only this loss component.
2. A reinstatement of the condition of the site from the current state to a better one – *gains*. For example, the replacement of hard river flood defence structures reaching the end of their lives with soft engineered defences may enhance the recreational value of a river site. Beach nourishment for coastal risk management purposes may result in a 'better' beach in recreational terms. Coastal protection and flood risk management schemes, for example off-shore reefs, new sea walls with promenades or river flood banks or retention lakes, may also change sites in ways that may provide new recreational opportunities and thus may enhance the recreational potential of the sites. Alternatively, changes associated with flood or erosion risk management may be viewed as having an adverse effect on landscape, wildlife and recreational opportunities and thus may reduce the enjoyment of a site for some users.

Central government grant support is provided essentially to reduce the risk of coastal erosion or flooding. All such schemes should seek to avoid environmental losses and provide gains where appropriate within the scope of the overall scheme. There are often opportunities to incorporate coastal management solutions within wider regeneration projects, and in these cases partnership funding should be sought.

8.3.4 Step 4: Identify the recreation and amenity impacts of the options

This step involves identifying the impacts on recreation and amenity of the changes to the physical environment resulting from the 'do nothing' and the 'do something' options. This process will benefit from the participation of the recreational stakeholders, particularly (but certainly not exclusively) at the initial stages. They may have particular insights into how changes will impact on their enjoyment.

These effects are likely to be broadly in four categories:

- on the perceived visual amenity or attractiveness of the site;
- on the type of recreational opportunities and the quality of the recreational experience;
- on safety at the site or perceptions of safety; and, crucially,
- on access.

Some of the possible effects are summarised for different environments on MCM-Online (Supporting Information for Chapter 8). There may also be indirect impacts of damage or protection of a recreational resource that should be considered.

8.3.5 Step 5: Determine the annual recreation and amenity benefits

Annual recreation benefits

This step involves first deriving estimates of the annual recreation benefits arising from the options, testing the validity of those estimates and comparing the benefits for the options.

Two components have to be estimated:

1. The value that individual adult users or beneficiaries place on the changes that would occur with the options in place, including the 'do nothing' option. These values will be derived from an application of the CV method using either the VOE per visit or the WTP approach.
2. The annual number of adult visits to the site (for the VOE approach) or beneficiaries who have an interest in the site (for the WTP approach).

The annual recreation benefits can then be determined as in Equation 8.1.

Equation 8.1

Annual benefits = £ value of the options (VOE gains and/or losses) or (WTP valuations) × number of visits per annum (VOE) or number of beneficiaries/visitors (WTP)

Where the options involve both VOE losses and gains, the annual benefits should be calculated separately for the losses and the gains because these may need to be treated differently for discounting (see 'Total recreation benefits').

National economic benefits, substitute sites and transfer payments

Recreation benefit assessment for use in benefit–cost analysis is almost always concerned with national economic benefits and not with benefits to a particular local economy. If changes to a particular coastal or riverine site simply result in a transfer of recreation from one site to another without any overall gains or losses in the value of recreational enjoyment, once travel costs have been taken into account, then no national gain or loss will be involved. The changes will simply be transfer payments – or different benefit share – because there will be no change in the national total of either all consumption or all resources required to generate that consumption (MAFF, 1999). The availability of substitute sites was taken into account in producing the data on recreational losses and gains in Tables 8.6 and 8.7 (see Section 8.6.1).

Recreation benefits, too, must be distinguished from the benefits of tourism to the local economy. The impacts of changes affecting recreation on local businesses and employment may be important local sustainability issues and might therefore constitute a significant element in a project appraisal. However, any benefits that tourism may bring to a local economy through multiplier effects should not be included in an assessment of project benefits in a benefit–cost analysis but may be evaluated separately, as in the Herne Bay case study (Tunstall *et al.*, 1990; Penning-Rowsell *et al.*, 1992). In exceptional circumstances, where it could be shown that tourism would be transferred to sites outside the UK, some allowance might be made for those losses to the national economy.

Total recreation benefits

The total recreation benefits of a project are estimated by discounting the annual benefits over the life of the project using the Treasury's recommended discount rates. A different approach and separate calculations are required where there are annual benefits from both VOE losses and gains with the options. This is because losses will begin to occur only when the site has deteriorated to the condition represented in the 'do nothing' scenario, and this may be after five, ten, fifteen or more years. For the VOE losses, the annual benefits will then accrue, discounted, over the remaining life of the scheme. In contrast, the VOE gains with the 'do something' options that involve a recreational enhancement over the current condition will arise as soon as a scheme is in place and will accrue, discounted, over the whole life of the scheme. The total recreation benefit for options involving both VOE gains and losses will then be the sum of the discounted annual benefits of the two components: losses and gains. This will represent the total recreation benefit with – compared to without – the scheme.

Where a scheme will only prevent further deterioration and maintain the current recreational value of the site, the annual VOE losses alone, appropriately discounted, will represent the total recreation benefit with – compared to without – the scheme. Thus, the total recreation benefit, like the other scheme benefits, compares benefits with and without the scheme, the only difference being that this benefit is derived as two components where a scheme option offers an enhancement in the recreation and amenity value over the current situation as well as preventing further deterioration.

8.4 THE INITIAL APPRAISAL STAGE: RECOMMENDED METHODS

8.4.1 Steps 1, 3 and 4: Define the problem, then identify the options and their impacts on recreation and amenity

At this stage, the geographic area of the project must be defined, with sufficient detail of the nature of the problem: the likely extent and rate of erosion or the frequency, depth and duration of flooding need to be known so that recreational stakeholders can be identified and the nature of the problem raised with them.

At this stage, it is essential only that the options should be sufficiently clearly defined so that their effects on recreation can be adequately represented to, and discussed with, recreational stakeholders.

8.4.2 Step 2: Identify adult recreation and amenity users or beneficiaries

The number of adult visits/beneficiaries at a site is a *primary indicator* of the likely magnitude of the recreation benefits. Therefore, any analysis should start by establishing:

1 Whether there is any public/restricted access to, and recreational use of, the site.
2 If there is no current access/use, whether there is potential for this with the project.
3 Estimates of the number of current adult visits/beneficiaries.
4 The availability of alternative sites in the area.
5 The type of adult user: general or specialist (see MCM-Online Supporting Information for Chapter 8) and their origins: local residents, day or staying visitors. This information is needed for the initial study stage to inform the decision as to what type of CV survey is appropriate: a survey of local residents at home, visitors on site, etc.
6 Estimates of seasonal variations in recreational use.

Sources and methods for obtaining some or all of this information – but particularly visit numbers – at initial and detailed appraisal stages are summarised in Table 8.1. Many of the organisations and individuals listed would normally be contacted as recreational stakeholders. Where no direct site-specific count or survey data are available (1–4 in Table 8.1) then the remaining indirect measures of site use (5–8 in Table 8.1) will have to be employed. All these data sources are likely to have severe limitations but will provide a first-order indication of the number of visits/beneficiaries, acceptable for CFMPs, SMPs, strategies and at an initial appraisal stage. Tables 8.2 and 8.3 provide some visit number data for coastal and fluvial sites and methods for estimating visit numbers at river locations.

Table 8.1 Sources and methods of information on recreation users/beneficiaries

Source/method	Comments
1 Long-period counts using people counters	Infra-red or other counters installed over a period (at least March to September). Counters are manually calibrated to relate passages to adult visits. Mainly applied in initial studies in conjunction with a CV survey (see Section 8.5.2).
2 Short-period manual counts/surveys	Manual counts/surveys over a period of days normally including the August bank holiday. At the initial stage, this method might be combined with site visits, and at detailed appraisal stage with a CV survey.
3 CV survey data	CV survey data on the frequency of visiting by local residents in conjunction with census data on the number of adult residents and staying visitors (in conjunction with managers' estimates of occupancy rates) can be used to generate visit number estimates. However, the tendency of survey respondents to overstate their visiting frequency has to be noted (see the Corton case study, Section 8.7).
4 Old survey/count data for the project	Planning, tourism or recreation departments of local authorities or local colleges or schools may have undertaken surveys or counts at the project site in the past, which can be updated to indicate current levels of use.
5 Inferred estimate	The number of visits to a coastal or river site is inferred from counts of visits to a related site nearby, such as: car and coach parks data multiplied by the average adult car or coach occupancy rate (Hengistbury Head), funfair, café, visitor centre, historic site or museum (Hurst Spit and Hurst Castle). This requires estimating the proportion of all visitors to the project site who also use the counted site and vice versa. At detailed appraisal level, this can be done in conjunction with a CV survey.
6 Visitor equations	A number of equations have been developed which predict distance–frequency functions so that from census data on the population in different zones a prediction can be made as to the number of visitors generated by the site.
7 Estimates from informed persons or sources	Written, telephone or personal contacts with: car park attendants, park rangers/wardens, visitor centre staff, staff at associated visitor attractions, local authority tourism, sport and recreation or planning staff, regional or local offices of organisations such as the English Tourist Board, National Trust and English Heritage and their Welsh equivalents, the Environment Agency's recreation and fisheries staff, managers of general recreation or staying visitor facilities or tourism business organisations that may have information on bed spaces and occupancy rates (see the Corton case study, Section 8.7), both commercial and club managers of specialist facilities (e.g. sailing, boating/sailboarding, fishing, birdwatching) and specialist organisations at national, regional and local level for information on the availability of alternative sites, e.g. for caravans or sailing.
8 Average number of visits to equivalent sites	This benefit transfer approach is suitable only for initial and strategic studies. The number of adult visits to the project site is estimated as being of the same order as the number of visits made to an equivalent site. However, there are few sites for which good data are available and little research to enable reliable identification of an equivalent site.

It is recommended that two or more methods/sources should be used where available to provide a range of visit numbers. Where no secondary data are available, a decision will have to be made as to whether it is worth collecting primary site-specific data on visit/beneficiary numbers and type. If the number of visits/beneficiaries is likely to be small – as may be the case for remote, undeveloped beach and cliff sites and many river sites – it may be worth collecting primary data at the initial appraisal stage only if the recreation benefits are likely to be significant in the overall benefit assessment.

Table 8.2 presents the visit numbers used in various coastal benefit assessment studies. References to the studies from which these estimates are derived are given in Table 8.5. The estimates for river sites are based on Green and Tunstall (1992).

These data should be used for the 'benefit transfer' approach (now termed 'transfer value'; Eftec, 2010a, 6fn) in strategic and initial appraisal studies only.

The *FWR Manual on Assessing the Benefits of Water Quality* (WRC/OXERA/FHRC, 1996) provides a desktop methodology for assessing visit numbers at riverside sites using the data in Table 8.3. Using this method involves establishing the category into which the site falls (e.g. local park or 'honey pot');

Table 8.2 Examples of visit numbers used for benefit assessment purposes

Site		Annual visit numbers	
Name	Characteristics	High estimate	Low estimate
Undeveloped coastal sites			
Hengistbury Head, Christchurch, Dorset	Natural headland, an SSSI, with nature, geology and archaeology sites	609,000	584,000
Hurst Spit, Hampshire	Undeveloped shingle spit with heritage site, Hurst Castle	107,000	880,000
Developed coastal sites			
St Mildred's Bay, Westgate, Kent	Small resort with promenade and sandy beach	212,000	–
Cliftonville, near Margate, Kent	Small resort with cliff tops and a mainly sandy beach	146,000	136,000
Corton, near Lowestoft, Suffolk	Small village resort with cliffs and partly sandy beach	97,000	75,000
River sites			
Local park	Park drawing visitors from 800m radius with no special attractions	30,000	60,000
'Honey pot' site, country park	Site drawing visitors from a 3km radius	60,000	250,000

Note: At all these sites, both coastal and riverine, almost all the visits involved informal use of the site for activities such as sitting, sunbathing and picnicking, strolling, dog walking, and, at coasts, playing informal games, playing in the sand and swimming or paddling. Very few visits involved specialist uses, such as angling, boating or sailboarding.

Table 8.3 Method for estimating the number of informal riverside visits

Site	Number of adults resident in catchment	Number of adult visits per year	
Local park	Number of adults within 500–800m radius of site	High:	27.6
		Medium:	21.3
		Low:	15.1
'Honey pot' site, country park	Number of adults within 3km radius of site	Medium:	17

establishing the number of adults living within the appropriate catchment; then multiplying the number of adults by the number of annual visits per adult (using the low, medium or high estimate as appropriate). Again, this approach is suitable only for strategic and initial appraisal studies.

8.4.3 Step 5: Determining the recreation benefits

For initial and strategic studies, it is acceptable to use standard values or data from existing CV studies. Several studies that have been carried out at coastal sites using the VOE and WTP CV techniques are discussed in Section 8.6. Data that can be used are presented in Tables 8.6 (on losses) and 8.7 (on gains), with various options at coastal sites and for rivers in Table 8.8. It will be necessary to consider carefully whether any of the projects presented in these tables approximate to the project case. The MCM-Online (see Supporting Information for Chapter 8) presents more detailed scenario/descriptions of the 'do nothing', 'change with erosion' and 'change with scheme option' scenarios summarised in Tables 8.6 and 8.7. It also includes a description of the current baseline scenario for the sites. These descriptions can be used to identify the most appropriate site data to use.

The valuation data should be updated to the project appraisal base date to reflect the rate of real growth in the economy using HM Treasury's GDP deflators (see HM Treasury, 2012a).

Using secondary source data on values is a very approximate approach. The research shows that recreational enjoyment and value depend upon a wide range of factors but there are, as yet, insufficient case studies to be able to identify the differences between sites that produce the differences in values. However, a benefit transfer approach on values is acceptable at the initial appraisal stage and for strategic studies only (see Eftec, 2010a). Therefore, where recreational value forms a significant fraction of the total benefit, it will probably be necessary to carry out a CV survey to derive site-specific WTP or VOE values. (In this regard, Eftec (2010a, ii) has stressed that its default (transfer) values 'will never replace original valuation work'.)

8.4.4 Deciding whether to proceed to a detailed appraisal study and on its emphasis

At this point, the question is whether the recreation benefits are likely to be large enough to make a difference to the decision on the scheme options. Key factors that will affect the possible size and significance of the recreation benefits are:

1 The number of visits/beneficiaries: if this is small (e.g. less than 10,000 per annum), it is unlikely that the recreation benefits will be substantial even if the individual places a high value on the recreation experience of any of the different scheme options.
2 The presence or absence of specialist users and facilities, such as sailing, golf courses and funfairs, that is likely to generate high values for recreation.
3 The rate and timing of change: the recreation benefits are unlikely to be sizeable if the changes to the recreation resources with the scheme occur only in the long term.
4 The availability of alternative sites: unique or special sites such as Hengistbury Head may be highly valued for that reason. Where there is very restricted access to the riverside generally, an accessible river site may be highly valued in the same way.

The following procedure is recommended as a way of assessing the range of possible recreation benefits for the scheme options under investigation. This will aid a decision as to whether it might be worth refining *either* the VOE/WTP *or* the visit number/beneficiary estimates (or both) at a detailed appraisal study stage, with site-specific data collection. This decision depends upon whether the benefit assessments under different assumptions would make any difference to the choice of option, that is whether the benefit–cost ratio overall would shift from robustly less than 1.0 to robustly greater than 1.0 under the different assumptions (A–D, opposite).

1 For coasts, consult Table 8.6 on losses and (where appropriate) Table 8.7 on gains, or Table 8.10 for WTP values.
2 For rivers, consult Table 8.9.
3 From these tables, identify the highest and the lowest VOE values for losses and/or gains or WTP values that are appropriate to the scheme options under investigation.

4 Using the methods presented in Table 8.1 and the visit number estimates presented in Tables 8.2 and 8.3, identify the highest and the lowest annual visit numbers/beneficiaries estimates appropriate to the scheme under investigation.
5 Go through the processes described in Step 5 (Section 8.3.5), which describes how to estimate the annual recreation and amenity benefits and total recreation benefits, to calculate the total recreation benefits under the following assumptions:

A – Lowest VOE/WTP with lowest visit number/beneficiary estimate;
B – Lowest VOE/WTP with highest visit number/beneficiary estimate;
C – Highest VOE/WTP with lowest visit number/beneficiary estimate;
D – Highest VOE/WTP with highest visit number/beneficiary estimate.

The range of total recreation benefits under the four assumptions (A–D) can then be considered in relation to the other benefits and costs of the scheme, to show the effects of the recreation benefit figures on the overall benefit–cost ratio. Table 8.4 illustrates what the effect might be for three hypothetical case studies.

In the first case (Table 8.4, Case 1), refining the estimates of recreation benefits would be justified because with high VOE and high visit numbers the benefit–cost ratio would be transformed (from <0.30 to >3.00). In the second case (Table 8.4, Case 2), any further development of this option should focus upon other areas of benefit and costs because the overall benefit–cost ratio is relatively stable across all assumptions and never robustly exceeds 1.00. In the final case (Table 8.4, Case 3), if assumption A is likely to be the case, then the scheme is not viable, but going further with benefit refinements would probably not be justified, particularly for small schemes.

The size and cost of scheme options are additional factors to be taken into account in deciding whether to proceed to a detailed CV or visit number study. If these are small, then the exact value of the recreation benefits would have to have a very marked effect on

Table 8.4 Evaluating the effect of assumptions upon the overall benefit–cost ratio

<table>
<tr><th></th><th colspan="4">Assumptions about the components of recreation values</th></tr>
<tr><th>Benefit–cost ratio (including recreation benefits)</th><th>A
Annual recreation benefit (low VOE/low visit numbers)</th><th>B
Annual recreation benefit (low VOE/high visit numbers)</th><th>C
Annual recreation benefit (high VOE/low visit numbers)</th><th>D
Annual recreation benefit (high VOE/high visit numbers)</th></tr>
<tr><td>Case 1</td><td></td><td></td><td></td><td></td></tr>
<tr><td>>3</td><td></td><td></td><td></td><td style="background:#ccc"></td></tr>
<tr><td>c 1</td><td></td><td></td><td style="background:#ccc"></td><td></td></tr>
<tr><td><0.3</td><td style="background:#ccc"></td><td style="background:#ccc"></td><td></td><td></td></tr>
<tr><td>Case 2</td><td></td><td></td><td></td><td></td></tr>
<tr><td>>3</td><td></td><td></td><td></td><td></td></tr>
<tr><td>c 1</td><td></td><td></td><td></td><td style="background:#ccc"></td></tr>
<tr><td><0.3</td><td style="background:#ccc"></td><td style="background:#ccc"></td><td></td><td></td></tr>
<tr><td>Case 3</td><td></td><td></td><td></td><td></td></tr>
<tr><td>>3</td><td></td><td style="background:#ccc"></td><td></td><td style="background:#ccc"></td></tr>
<tr><td>c 1</td><td></td><td></td><td style="background:#ccc"></td><td></td></tr>
<tr><td><0.3</td><td style="background:#ccc"></td><td></td><td></td><td></td></tr>
</table>

Note: Shaded cells indicate where the benefit–cost ratio would lie under each of the assumptions for each case.

the benefit–cost ratio before it would be worthwhile to undertake a detailed study because the cost of a CV survey is likely to be in excess of £30,000 (2004 prices).

The pattern of results in Table 8.4 also indicates whether it is necessary to concentrate attention in the detailed appraisal study on refining the estimate of visit numbers or upon better measurement of the value of enjoyment. In the first case (Table 8.4, Case 1), it is the value of enjoyment that makes a difference; in the third (Table 8.4, Case 3), it is the number of visits.

8.5 THE DETAILED APPRAISAL STUDY STAGE: RECOMMENDED METHODS

8.5.1 Steps 1, 3 and 4: Define the problem, then determine the options and their impacts on recreation

Definition and representation of the changes in the CV surveys

The project area must be defined so that it can be presented on a clear, simple map to be used in the CV survey. Detailed information on the changes affecting the site and the options proposed are needed so that these can be used to develop and present CV scenarios in the survey. It is very important that the options presented in the survey are sufficiently well developed to be likely to be those that will be used in the full benefit–cost analysis since the recreation benefits derived in the CV survey will apply only to the options as presented in the survey.

The CV scenarios need to describe in neutral, non-technical language the problems and the 'do nothing' and 'do something' options for the site. Carefully constructed drawings (and possibly photographs) of the same site under the various options should also be used to give a realistic view of the changes that may take place (Figure 8.2). Illustrations of flood or coastal risk management structures may be employed. A balance has to be struck between providing full information to the respondent and avoiding information overload. Great care must be taken with how and what information is presented, since it has been known for some time that this can affect the responses (Mitchell and Carson, 1989; Munro and Hanley, 1999; Bateman and Willis, 1999).

8.5.2 Step 2: Identify and estimate annual number of recreational users/beneficiaries

At the detailed appraisal study stage, accurate estimates of visit/beneficiary numbers are essential. Two methods – infra-red or other people counters and manual counts (Table 8.1, methods 1 and 2) – are recommended. It is also advisable to use at least two methods from Table 8.1, or variants of one method, so that a range of visit/beneficiary estimates can be compiled. Consideration should be given as to how visit/beneficiary numbers are to be estimated as soon as it is decided to proceed to a detailed appraisal study. This is because, ideally, people counters should be in place for twelve months and at least from March through to September to give an indication of seasonal variation.

Infra-red or other people counters

Where counters are to be installed, the following issues have to be addressed:

1. The number and location of counters to be installed at the site. The number needs to be sufficient to capture most of the visits but not so many as to lead to many multiple passages past counters. Counters should be sited at heavily used access points: on steps down to a beach or on stiles or gates on riverside paths.
2. Arrangements for checking, maintaining and collecting data from the counters. Daily or at least weekly checking and data collection are needed to ensure that too much data are not lost through counter malfunctioning.
3. Calibration counts. Manual counts are needed to translate the raw count data on the number of passages past a point into the number of adult visits/beneficiaries at a site in the form of a counter calibration factor. These can be undertaken in conjunction with the CV survey.
4. Correction factors. Estimates derived from site-specific questions included in the CV survey or a short site survey where no CV survey is being undertaken for correction purposes on: percentage of visitors missed by the counters; percentage of multiple passages past the counters by adult visitors.

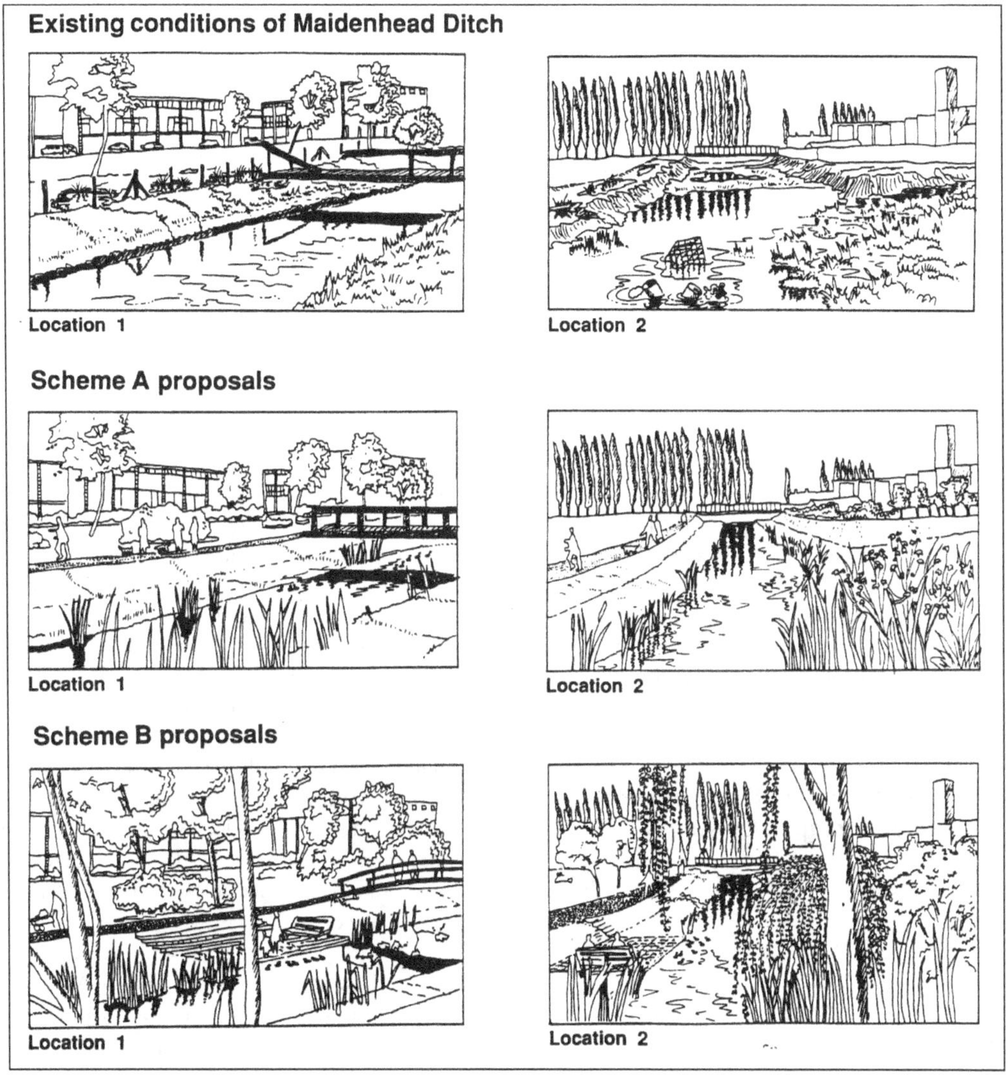

Figure 8.2 Example of diagrams used in CV study for river flood protection scheme

5 Correction for instances where data are missing due, for example, to counter malfunctioning.
6 Where less than a full twelve-month count is available a procedure to derive a full year's visit estimate from a short-term count is required.
7 Arrangements for analysing the data from these sources.

Further details and worked examples of how some of these issues can be dealt with are given on MCM-Online (see Supporting Information for Chapter 8).

Manual counts

1 Manual counts can be combined with CV survey interviewing on site over the survey period or can be mounted as a separate exercise. At Hurst Spit, interviewers counted those adults entering the site from the three available access points for ten minutes at each hour as part of their interviewing.
2 Counts should, where possible, cover all access points, recording adult visitors as they leave the site.

3 Where count data are missing for particular access points, similar procedures to those indicated for missing infra-red count data can be used (see MCM-Online Supporting Information for Chapter 8).
4 To estimate the total visits/visitors in the year, it is necessary to extrapolate from the short-period manual counts. The procedure for doing this is shown on MCM-Online (Supporting Information for Chapter 8) for infra-red counters and this can also be used for manual count data.

Scottish Natural Heritage (undated) has advised that, for each counter, a calibration survey should be undertaken over four or five periods of at least one hour covering a range of times of day and weekdays as well as weekends. Each separate survey period should cover at least one-hundred people and the length of survey time should be adjusted to ensure that that number is counted. The calibration for each counter should cover at least 300 to 450 counts. (Further details on the use of infra-red and other counters and manual counts are given in Scottish Natural Heritage (undated).)

8.5.3 Step 5: Determine the recreation benefits through a CV survey

The following issues need to be addressed in order to carry out the CV survey.

The conduct of the survey: pilot and main survey

1 A decision has to be taken whether to mount the survey in-house, to contract the work out to a market research company or consultancy, or to co-operate with local authority planning and recreation departments which may have the relevant survey expertise. Sufficient survey expertise needs to be available to design the sample and adapt the standard questionnaire(s) and materials (see MCM-Online Supporting Information for Chapter 8) for the specific site, supervise interviewers, and process and analyse the data.
2 Trained, experienced and briefed interviewers are required to conduct the interviews.
3 Decisions on how the data will be processed and analysed, and the software to be used, need to be taken at the outset.
4 Preliminary exploratory interviews or focus groups discussions, if not conducted as part of an initial appraisal study, should be conducted with visitors and/or local residents and a pilot survey of twenty to fifty interviews conducted to test the questionnaire and the response to the CV scenario materials.

The survey population and sampling

1 A decision has to be taken as to the relevant adult population to survey of those using or likely to benefit from the site. If the site provides an amenity or recreational resource mainly to local residents, as is likely to be the case for many small coastal sites or rivers, or if there is currently no access or use but the scheme might provide one or both of these, a survey of local residents at their home addresses using the Postcode Address file as a sampling frame may be appropriate. Instructions to interviewers should specify whether to select a random adult or to interview 'the householder or partner' at the address. Where a site attracts day and staying visitors, a visitor survey conducted on site is preferable. Methods for selecting randomly among visitors, such as interviewing the 'next person to pass', should be specified for site surveys. A local residents survey may have the additional benefit of being part of the process of involving and consulting local people in the decision-making. It is possible to mount both types of survey to ensure sufficient representation for all types of user.
2 The number of interviews to conduct. Valuations obtained in CV surveys have large variances, so large samples are required if good estimates are to be obtained. Samples of at least 500 or 600 people are required to provide valuations of a current visit to the coast which could be estimated with 95 per cent confidence to be within + or – £1.00 to £1.50 of the mean. There is insufficient river valuation data to enable a comparable estimate to be made. Samples of less than 200–300 are not recommended, although it is recognised that the sample size may be restricted where the number of visitors using the site is limited.
3 Timing of surveys. Ideally, since recreational use of sites varies over the year, surveys should be

carried out at different times of year to establish whether valuations vary with the seasons. Flood Hazard Research Centre research and other CV studies have not examined this possibility. If only one time can be selected for a survey on site then it should be carried out in the main summer holiday period when the largest number of people visit. Local residents surveys should preferably be undertaken outside the main summer and other holiday periods to enhance level of response. An absolute minimum of four months should be allowed to carry out the surveys from inception to completion of a report. The overall project appraisal programme needs to take these timing requirements into account.

Asking the valuation questions: the CV questionnaire

Two broad approaches to asking valuation questions have been developed and used to elicit recreation valuations (as discussed in Section 8.2):

1 willingness to pay (WTP); and
2 value of enjoyment (VOE) of a visit.

VOE remains the recommended approach, but WTP can be used where it is preferred.

For the VOE approach, the respondents in the survey should be asked the sequence of questions shown in the standard VOE residents and site visitors questionnaires, adapted and pilot tested as necessary for the project site, conditions and purposes (see MCM-Online Supporting Information for Chapter 8). Questions cover the following information, which is essential for eliciting estimates of economic gains and losses:

1 Frequency of visiting the site at present.
2 What visit or activity would give the same amount of enjoyment as today's visit or a visit in current conditions. A list of possibilities should be shown to the respondents. They are asked to think of what it costs them to make a visit to this alternative site and to use these costs as a guide to estimating the value of their enjoyment of today's/a current visit.
3 The value of enjoyment of today's visit (site visitor survey: current site may be based on visitor's experience or represented in a drawing and text) or the value of enjoyment of a visit in current conditions (presented in drawing and text, for residents interviewed at home).
4 The value of enjoyment of a visit under the 'do nothing' and each of the 'do something' options (presented in drawings and text).
5 Changes in enjoyment and in visit frequency and whether or not respondents would transfer visits to an alternative site with each of the options in place.
6 For those who would transfer their visits:

   i The value of enjoyment of a visit to the alternative site.
   ii The difference in costs associated with a visit to the alternative site compared with the current site.
   iii The convenience of a visit to the alternative site compared with the current site.

The first two questions are designed to help respondents to think over and arrive at a value of their enjoyment of their (unpriced) visit to a coastal or riverside site. The technique is based upon the economic concept of opportunity costs. The type of site chosen by respondents as offering equivalent enjoyment is itself an indicator of the nature of the recreational experience that the respondent seeks from a visit.

In all studies to date, the nearest equivalent to a river has been found to be a nature reserve. Conversely, in the case of coastal sites, there has been wide variation in the type of recreational facility seen as offering equivalent enjoyment. For example, for visitors to Spurn Head, 43 per cent stated that a visit to a nature reserve was the type of visit offering equivalent enjoyment. But for visitors to Scarborough, the most frequently cited alternative (22 per cent) was a leisure centre. Thus, the nature of possible equivalent sites varies markedly for coastal sites, and particular coastal sites serve very distinct markets. Equally it has been shown (Green, 1998) that there is a significant degree of market segmentation among visitors in the type of experience they seek.

If a site were to degrade or indeed simply change under the options then the existing visitors should be expected to do one of two things:

- visit less often because they gained less enjoyment from a visit; or
- visit less often but make more visits to a site offering a higher level of enjoyment.

Whether or not there would be economic losses associated with transferred visits will depend upon the value placed on a visit to the current site compared with the value of a visit to the substitute site and upon whether or not there are any additional costs in going to the alternative site over and above those associated with the current site. Where a perfect substitute site exists – that is, where visitors can transfer their visits to a nearby alternative site that gives them equal enjoyment at no additional cost – no loss will be incurred. The convenience factor in visiting an alternative site as compared with the current site which may not be fully covered by additional travel costs also needs to be taken into account. The questions outlined in points 5 and 6, above, are designed to elicit the information needed to describe and evaluate visit transfers under the different options.

It is to be expected that, with changes at the coast or riverside, some people who do not currently visit a site will start to do so instead of making visits elsewhere. Hence, a survey of existing visitors or of local residents will probably underestimate the benefits of a change or enhancement because new visitors or visits from people outside the local area will not be included. The CV survey procedure described in this Manual does not cover possible new users generated by changes.

The questionnaires we recommend for use are based on a theoretical model of recreational behaviour and determinants of valuations (Figure 8.3). Thus, they contain questions on the frequency and duration of visits, activities undertaken there, ratings of the site for various purposes, the subjective enjoyment of the site under the options compared to the site's current condition, and the characteristics of the respondents. All of these questions are needed so that the validity of the valuations obtained can be tested in the analysis.

For the WTP approach, a broadly similar questioning procedure is adopted (see WTP standard questionnaire on MCM-Online Supporting Information for Chapter 8).

Analysis of the survey data and assessment of the benefits

BASIC DATA ANALYSIS ISSUES

Initially, the frequency distributions and summary statistics for all the variables in the survey should be examined.

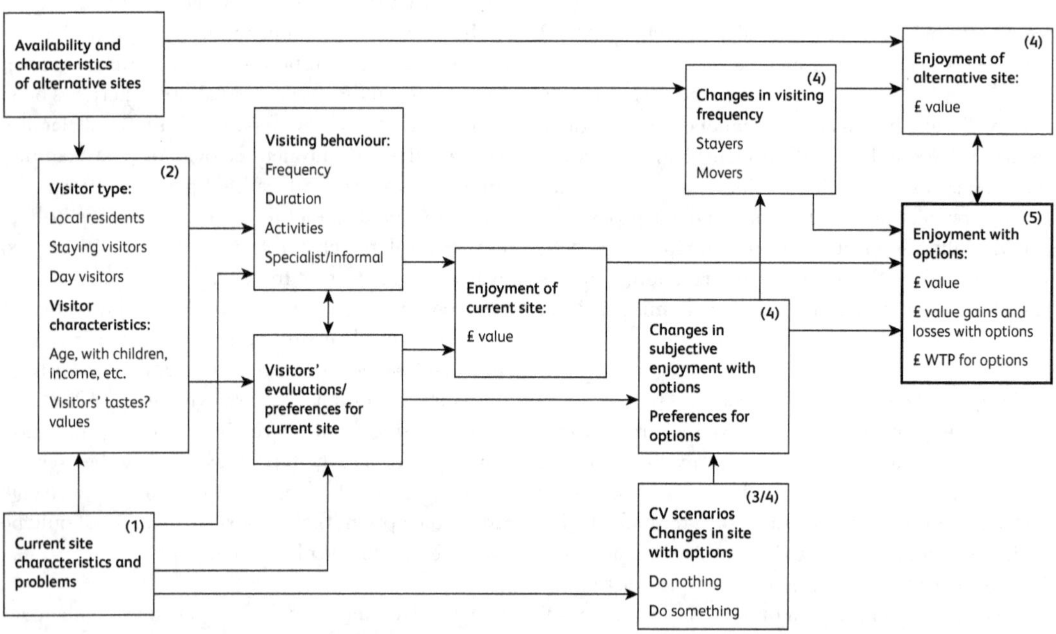

Figure 8.3 Theoretical model of recreational behaviour and determinants of valuation

Note: Figures in parentheses refer to steps in Figure 8.1

Two key issues in the analysis of the CV valuation variables, with both VOE and WTP valuations, are how outliers and non-response to the valuation questions should be treated. Where respondents have failed to answer the valuation question, a decision has to be taken as to whether to treat their responses as zero valuations and include them in the calculation of the mean valuations or exclude them from that calculation.

The survey should have elicited their reasons for not responding and these may suggest the appropriate course: where respondents have objected in principle to putting a value upon an unpriced resource, it would be wrong to assume that they placed no value on that resource. A further test is to compare the characteristics and ratings of the resource of those who did and did not offer a valuation. Where the non-respondents appear broadly similar in attitudes and characteristics to respondents, it is most appropriate to exclude them from the calculation of the means, thus assuming that they attach the average value to the resource rather than including them as zeros.

The mean values derived in CV surveys are sensitive to outliers: a few very high valuations. These high valuations may be legitimate, they may be a form of protest against the CV process, or they may be the result of recording or data entry errors. Responses on the questionnaires should be checked for such errors and the cases with extreme valuation responses should be checked against other survey responses, such as income and preferences to test the 'reasonableness' of the valuation responses.

Three possible approaches to the treatment of outliers are discussed in the CV literature:

1 Deletion of the outliers on the basis of the researcher's checks and judgement.
2 The use of robust statistical estimators, such as the alpha-trimmed mean (Mitchell and Carson, 1989), to exclude extreme values from the calculation of the means.
3 The use of regression diagnostics to identify whether or not the outliers should be regarded as part of the distribution or excluded.

Valuation estimates can be derived both including and excluding the outliers. Zero and negative valuations should be included in the calculation of the means. When using the VOE approach, the benefits of losses under the 'do nothing' option avoided and the gains under the 'do something' options should be estimated separately.

Losses under the 'do nothing' option: VOE approach

Two equations should be used for estimating possible losses (or gains) under the 'do nothing' option. Some respondents may enjoy the site under the 'do nothing' option more than the current site and therefore might gain.

Equation 8.2: Benefit for those who continue to visit

$$L_1 = Eo - E^1$$

Equation 8.3: Benefit for those who would visit an alternative site under the 'do nothing' option

$$L_2 = (Eo - Ea) + (Ca - Co)$$

Where:

L = the benefit per person (in cases 1 and 2)
Eo = the value of enjoyment of today's visit/a visit in current conditions
E^1 = the value of a visit under the 'do nothing' option
Ea = the value of a visit at the alternative site visited under the 'do nothing' option
Co = the cost incurred visiting the present site
Ca = the cost incurred visiting the alternative site under the 'do nothing' option

The difference between Co and Ca is derived from a question in the questionnaire.

Gains under the 'do something' option formulae: VOE approach

Two similar equations should be used for estimating possible gains (or losses) under the 'do something' options: some respondents may enjoy the site under the 'do something' option less than the current site, for example where there is a radical change in the appearance or recreational facilities with the option, and therefore might incur a loss. They might also wish to visit elsewhere instead.

Equation 8.4: Benefit for those who continue to visit

$$G_1 = Ex^n - Eo$$

Equation 8.5: Benefit for those who would visit an alternative site under 'do something' option n:

$$G_2 = (Eo - Ea^n) + (Ca^n - Co)$$

Where:

G = the benefit per person (in cases 1 and 2)
Eo = the value of enjoyment of today's visit/a visit in current conditions
Ex^n = the value of a visit under 'do something' option n
Ea^n = the value of a visit at the alternative site visited under 'do something' option n
Co = the cost incurred visiting the current site
Ca^n = the cost incurred visiting the alternative site under 'do something' option n

The difference between Co and Ca is given by a question in the questionnaire.

Using these equations, the losses and gains should be calculated for each individual in the survey and then the mean value should be calculated.

The axiom of economic rationality (Chapter 2) implies that the individual will choose to adopt whichever behaviour results in the least reduction or most gain in utility: that is, for example, to continue to visit the current site if the value of the benefit from Equation 8.2 is less than the value from Equation 8.3 and visit elsewhere if the reverse is the case and the value from Equation 8.2 is greater than the value from Equation 8.3. Under the axiom of rationality, the loss to those who choose to visit elsewhere under the 'do nothing' option cannot be less than zero: if an individual were to gain more enjoyment from the alternative site net of the increase in the cost of visiting there, then the economically rational individual should already be visiting there. The same applies to visiting alternative sites under 'do something' options.

Typically, a few individuals give 'irrational' responses: they should not be visiting the current site but should already be going elsewhere. These 'irrational' responses should be investigated. Possible explanations for them are measurement or data entry error, or imperfect information about the current site: respondents may have expected to gain more enjoyment from the current site when they decided to visit, as may happen with first-time visitors or where, as at Corton, a site known from the past had changed drastically.

Distortions may also arise owing to the fact that the model of visiting behaviour includes the cost but not the opportunity cost of travel time to another site – and convenience of the current site particularly for the local resident – and therefore may understate the losses of the movers. Where there are only several (fewer than ten) 'irrational' respondents, it is recommended that they be excluded from the analysis. Where, as in Corton, a substantial minority of the respondents are 'irrational', it is recommended that they be included in the calculation of means as zeros. The issue of the treatment of 'irrationals' is discussed further in the Corton case study (Section 8.7)

Checks on validity and reliability

The following are key ways in which the validity of the survey results and the valuations can be tested:

1 Theoretical or construct validity using regression analysis. The questionnaire and analysis should be based on a model of recreational behaviour and value, which defines the expected relationships between the variables in the survey (see Figure 8.3). A basic form of theoretical or construct validation is to test whether the results from the survey conform to this underlying theoretical model using correlation and regression analysis. In order to do this, variables may have to be transformed to normality using a logarithmic or other form of transformation. It is recommended that the gains or losses (or WTP valuations) of the options (the dependent variables or variables to be explained) should be regressed on a group of independent or explanatory variables, which are predicted by the model to be determinants. Then the sign and size of the beta-coefficients can be examined to test whether the results are consistent

with the theoretically expected relationships. The R^2 or proportion of the variance explained will indicate the extent to which the results conform to the underlying model but an analysis of residuals should also be undertaken to check this. An R^2 of the order of 0.20–0.50 can be taken as indicating a good level of fit or explanation (Mitchell and Carson, 1989).

2 Test of convergent validity. The convergent validity of the valuations can be tested by including in measures some alternative quantification of the same variable. However, WTP, VOE and travel cost method valuations measure somewhat different things and therefore are of only limited value as tests of convergent validity where questions to elicit all of these are available in the survey.
3 Validity testing by resource comparisons. A further test of validity is to compare the valuation results obtained using the same methods in different and comparable settings and for different and comparable resources. The data included in Tables 8.5–8.10 (for VOE and WTP) can be used for this purpose.

It is desirable, but rarely possible, to include in the design of CV surveys tests of reliability which require repeat measurements on the same population. There has, however, been some research to examine the stability of CV valuations over time, with mixed results (Tunstall *et al.*, 1999).

Aggregation to estimate the total annual recreation benefits (with checks on the representativeness of the sample)

Local residents samples can be checked for representativeness by comparing the characteristics of the respondents to the population from which the sample was drawn using census data.

For site visitors, data on the survey population are likely to be lacking. Manual count data or manual calibration count data may provide some information on the sample population which can be used to check the survey proportions, for example of men and women, or local, day or staying visitors. Data should be collected during the survey on non-response and on the characteristics of non-respondents; for example, gender, age, location and possibly visitor type. These data can then be examined to see whether there is any possibility of non-response bias. Non-response bias will be of concern only if it affects a sub-group that differs markedly in its valuations; for example, the elderly or those living at a significant distance from the site.

Once the survey sample has been checked for representativeness, the annual benefits of each of the options can be determined as in Equation 8.1:

£ value of each option (VOE gains and/or losses) or (WTP valuations) × number of visits per annum (VOE) or number of beneficiaries/visitors (WTP)

For the VOE approach, the losses with the 'do nothing' option and the gains with the 'do something' options should be calculated separately. For the WTP approach, the benefits of the 'do something' options only are more commonly evaluated and there may not be separate valuations of the 'do nothing' option. For a site survey, the benefits should be calculated separately for local, day and staying visitors to enhance the accuracy of the estimates. A range of estimates of visit numbers may be used to derive a range of annual recreation benefits.

Distributional impact analysis

The Treasury's 'Green Book' (HM Treasury, 2003) recommends, where it is necessary or practical, that in order to incorporate social equity considerations into flood and coastal erosion risk management appraisals, potential benefits should account for the distributional impacts (see Chapter 4, Section 4.9.4 for guidance on determining whether such analysis is necessary or practical). This section indicates that the use of total weighted factors by approximated social grade (shown in Table 4.36) is recommended only where AB or DE social grades are predominant because the total weighted factors for C1 and C2 social grades will have a negligible effect on the distributional impact assessment.

The CV survey questionnaires (see MCM-Online Supporting Information for Chapter 8) include a question designed to elicit the social grade of respondents. Experienced market research interviewers will be skilled in assigning respondents to a social grade on the basis of the information they elicit through such questions and will be able to code the social grade as they complete their interviews. Inexperienced interviewers will not be able to do this, so allocating respondents to the appropriate social grade should then be undertaken by someone trained in this procedure after the completion of the interviews.

The social grade distribution (of local, day and staying visitors, where appropriate) can be examined to see whether it is necessary to apply the weighting factors. Where AB or DE social grades are predominant, the weighting factors can be applied to the gains and losses of individual respondents according to their social grade. Then the mean valuations can be calculated in the normal way (for local, day and staying visitors, as appropriate). The assumption on which the distributional analysis is based (that higher social grades will have higher valuations) can be tested in the CV survey by examining the mean VOE gains and losses for the different visitor categories by social grade. However, it is important to bear in mind the size of the sub-groups on which mean values are based. Means based on very small numbers are usually unreliable.

The government guidance on distributional analysis indicates that both weighted and non-weighted results should be presented, with the former complemented by a sensitivity analysis. Where the distributional analysis will make very little difference to the scheme benefits (a difference of less than 20 per cent of scheme benefits, for example), it can be argued that it should not be presented.

8.6 RESULTS FROM THE SURVEYS

Since the early 1990s, a substantial number of studies have been carried out to evaluate changes in coastal environments but rather fewer relating to riverine changes and hardly any into changes due to fluvial flood risk management works.

As most of the coastal studies conducted since the 'Yellow Manual' (Penning-Rowsell *et al.* 1992) have been consultancy studies rather than part of a research programme, they do not properly represent all parts of the UK coast: there are no studies from Scotland, Wales and the south-west holiday areas, and few from the north of England. They do, however, cover a range of coastal conditions and scenarios. Beach erosion has remained the most common condition examined, but at three sites (Spurn Head, Hengistbury and Hurst Spit) we explored the public's response to a breach scenario. The Corton study provided a rare opportunity to examine the public's response to a managed retreat option on a developed coast.

The studies undertaken hitherto and the recreation benefit assessment methods presented are inevitably based on eliciting valuations from current residents and visitors. Where a significant change in the character of a coastal or river site is proposed, the constituency of visitors may be different from the current visitors. This may need to be considered in the analysis. A change to a more 'natural' river or coast, as at Corton, or to a coast with new recreational facilities, such as boating in Herne Bay, exemplifies these situations. However, there may be more scope at some sites than at others for attracting new and different visitors. The results from these studies are summarised in Tables 8.5–8.10.

8.6.1 Coastal sites: VOE methods

Using the procedure described in Section 8.5.3, generally quite high proportions of respondents have been able to provide a valuation of today's visit/a visit to the coast in its current condition (Table 8.5). The percentage ranged from 54 per cent at Peacehaven to 92 per cent at Corton and elsewhere. Furthermore, analyses included in five recent studies (Hurst Spit, Hengistbury, Corton, Cliftonville and St Mildred's Bay) have tended to show that the non-respondents to the valuation question on today's visit/a visit in current conditions were little different in their characteristics and attitudes to those who were able to respond. Most non-respondents simply felt that they could not value such things in money terms. These studies, therefore, have supported the exclusion of non-respondents from the calculation of mean valuations rather than their inclusion as zeros.

It is clear from Table 8.5 that the VOE method does produce different results in different coastal settings. Values of the enjoyment of today's visit/a visit in current conditions ranged from a low of £2.49 from local residents for Corton's severely damaged and restricted seafront to £17.20 for Clacton's beach. A meta-analysis is now required to explore whether the factors which explain the between-site variances as well as within-study variances in valuations can be identified.

The 'Yellow Manual' standard data on gains and losses covered only four sites affected by beach

Table 8.5 Money value of enjoyment of today's visit/a visit in current conditions for coastal sites

Site	Survey year	Sample size	Sample type	% able to value	£ mean value of enjoyment of today's visit – updated to 2013 [9]	£ mean value of enjoyment – at survey date
Undeveloped sites						
Hurst Spit [1]	1991	550	Site visitor	79	12.35	7.37
Hengistbury Head [2]	1996	625	Site visitor	64	13.91	9.48
Dunwich [3]	1988	101	Site visitor	61	14.19	6.87
Spurn Head [3]	1988	97	Site visitor	80	17.55	8.50
Resorts/ developed sites						
Corton (residents) [4]	2001	163	Resident	82	2.98	2.27
Corton (staying visitors) [4]	2001	304	Staying visitor	92	4.50	3.42
Herne Bay (residents) [5]	1990	189	Resident	83	6.35	3.59
Peacehaven (cliff tops) [3]	1988	214	Resident	54	7.23	3.50
Filey [3]	1988	88	Site visitor	88	7.52	3.64
Scarborough [3]	1988	101	Site visitor	83	10.19	4.93
Morecambe [3]	1989	150	Site visitor	92	11.05	5.76
Bridlington [3]	1989	151	Site visitor	86	£11.34	5.91
Cliftonville [6]	1993	528	Site visitor	81	11.77	7.47
Lee-on-Solent [7]	1995	–	Site visitor	–	13.07	8.63
St Mildred's Bay [8]	1992	462	Site visitor	71	14.23	8.77
Hastings [3]	1988	247	Site visitor	66	15.94	7.72
Hunstanton [3]	1989	152	Site visitor	90	16.78	8.74
Frinton [3]	1988	178	Site visitor	70	19.74	9.56
Herne Bay (visitors) [5]	1990	127	Site visitor	88	20.35	11.50
Clacton [3]	1989	146	Site visitor	67	20.18	10.52
Clacton [3]	1988	170	Site visitor	90	20.57	9.96

Sources (for this table and Tables 8.6 and 8.7): [1] Fouquet *et al.*, 1992; [2] Tapsell *et al.*, 1996; [3] Penning-Rowsell *et al.*, 1989, 1992; [4] Tunstall, 2001; [5] Tunstall *et al.*, 1990; [6] Garner *et al.*, 1994; [7] Whitmarsh *et al.*, 1999; [8] Costa *et al.*, 1993; [9] updated from 2005 prices to 2013 using the 27 March 2013 update of HM Treasury's GDP deflator series (HM Treasury, 2012a)

erosion and promenade deterioration where beach nourishment was proposed. Subsequent studies have extended the available data (Tables 8.6 and 8.7).

These studies show considerable variation in the economic losses experienced at different sites. This may reflect differences in the severity of the effects presented to respondents in the 'do nothing' scenario, particularly with regard to impacts on access and safety. For example, at St Mildred's Bay, the prospect of the esplanade becoming unsafe and closed to the public may account for the high value placed on the loss in recreational enjoyment there. At Corton, the current state of the seafront was so poor and access already restricted at the time of the survey that the scope for further deterioration, and thence recreational losses, was reduced. The differences may reflect differences in the availability of alternative sites to visit and in the behavioural response of respondents to changes. Visitors have a choice to continue visiting after changes or to go elsewhere. Those who choose to transfer their visits may not experience any loss.

Table 8.6 Money losses per visit with erosive changes at coastal sites

Site	Change with erosion	% expecting less enjoyment	£ loss mean per adult visit – updated to 2013 (£ loss mean per adult visit at survey date)				£ mean loss – all at survey date
			Local	Day	Staying	All	
Beach and promenade erosion							
'Yellow Manual' Standard data: 4 sites	Deterioration in beach and promenade	85	3.04	4.54	10.65	6.89	3.59
			(1.58)	(2.37)	(5.55)		
Lee-on-Solent	Shingle beach erosion	–	3.96	2.74	4.84	3.54	2.34
			(2.61)	(1.81)	(3.2)		
Herne Bay visitor survey	Deterioration in beach, sea wall and promenade collapsed in parts		3.43	3.22	13.38	6.61	3.74
			(1.94)	(1.82)	(7.59)		
Cliftonville	Cliff erosion, deterioration in beach, cliff-top promenade closed in parts		8.33	8.14	7.27	7.62	4.84
		83	(5.29)	(5.17)	(4.62)		
Corton (residents, staying visitors)	Cliff erosion, deterioration in beach and sea wall, very reduced access to and along beach and sea wall		2.73		1.34	2.49	1.89
		81	(2.08)	–	(1.02)		
St Mildred's Bay	Severe damage to esplanade wall, esplanade unsafe and closed in parts		8.90	10.08	10.59	9.91	6.11
		92	(5.48)	(6.21)	(6.53)		
Hastings	Beach deterioration	–	–	–	–	7.01	3.40
Breach scenarios							
Hengistbury Head	Breach, boat access only to head, reduced cliff-top area and paths		4.79	3.53	3.84	6.81	2.82
		62	(3.27)	(2.40)	(2.62)		
Hurst Spit	Breach to shingle spit, access by boat only		3.07	8.11	4.56	6.81	
		98	(1.83)	(4.84)	(2.72)		3.72

Table 8.7 Money gains per adult visit with coastal protection scheme options at coastal sites

Site	Change with scheme options	£ gain mean per adult visit – updated to 2013 (£ gain mean per adult visit at survey date)				£ gain mean – all at survey date
		Local	Day	Staying	All	
Beach and promenade erosion						
'Yellow Manual' Standard data: 4 sites	Nourished beach and promenade	2.00	3.46	2.51	2.86	1.49
		(1.04)	(1.80)	(1.31)		
Lee-on-Solent	(a) Shingle beach renourishment	1.66	1.59	1.64	1.64	1.08
		(1.10)	(1.05)	(1.08)		
	(b) Rock groynes with shingle beach renourishment	1.76	1.53	0.97	1.60	1.06
		(1.16)	(1.01)	(0.64)		
Herne Bay visitor survey	(a) Reef or jetty with no boat facilities	3.65	3.13	6.97	4.82	2.72
		(2.06)	(1.77)	(3.94)		
	(b) Reef or jetty with boat facilities	3.46	2.36	2.18	2.50	1.41
		(1.95)	(1.33)	(1.23)		
	(c) Higher sea wall, and promenade, rock groynes	–2.19	–3.13	–3.54	–3.10	–1.75
		(–1.24)	(–1.77)	(–2.00)		
Cliftonville	(a) Concrete lower promenade	2.25	2.04	5.40	4.29	2.73
		(1.43)	(1.30)	(3.43)		
	(b) Rock lower promenade	2.25	1.76	3.18	2.53	1.61
		(0.73)	(1.12)	(2.02)		
Corton	(a) Hold the line for a limited period: short-term protection to cliff, limited access to beach and along sea wall	2.62	–	2.42	2.46	1.88
		(1.99)		(1.84)		
	(b) Hold the line for a longer period >50 years: full access along renewed sea wall and onto all the beach from village	17.91	–	8.97	11.02	8.40
		(13.64)		(6.83)		
	(c) Managed retreat: sea defences and sea wall removed to leave a 'natural' seafront, direct access from village to beach	–0.26	–	2.38	1.71	1.30
		(–0.20)		(1.81)		
St Mildred's Bay	Improved beach and promenade	3.06	2.22	2.55	2.69	1.66
		(1.89)	(1.37)	(1.57)		

Table 8.7 continued

Site	Change with scheme options	£ gain mean per adult visit – updated to 2013 (£ gain mean per adult visit at survey date)				£ gain mean – all at survey date
		Local	Day	Staying	All	
Breach scenario						
Hengistbury Head	(a) 5 rock groynes: full cliff protection	–0.01	0.57	–0.24	0.05	0.03
		(–0.01)	(0.39)	(–0.16)		
	(b) 3 rock groynes: partial protection	–2.33	–1.16	–2.93	–2.31	–1.57
		(–159)	(–0.79)	(–2.00)		
	(c) Beach nourishment: annual disruption	–2.12	–3.43	–5.06	–3.48	–2.37
		(–1.44)	(–2.34)	(–3.45)		
Hurst Spit	Slightly enlarged shingle spit	1.05	0.42	0.75	0.66	0.39
		(0.63)	(0.25)	(0.45)		

There were differences in the losses experienced by local, day and staying visitors but they were not in the same direction in all the studies.

One finding that emerges from all the studies is that the great majority of the public see their enjoyment of the coast as adversely affected by the changes that follow from the 'do nothing' option (Table 8.6). Generally, the studies indicate that the public were reluctant to see natural processes taking their course at the coast and wanted the coast to continue to be maintained and defended as it had been in the past (Tunstall and Penning-Rowsell, 1998). However, at Hengistbury Head, 11 per cent of those who offered a valuation on the 'breach' scenario thought that they would enjoy a visit *more* if they had to reach the head by boat. At Hurst Spit, by contrast, hardly any visitors saw any benefits in the breach scenario, and all maintained that they would go elsewhere if the spit were breached rather than travel to the castle by boat.

The data indicate that, for most schemes, the losses are likely to be a more important component in the benefits than the gains with the options. Gains were greater than losses at only one of the studied sites, Corton, probably because the deterioration at the site was so well advanced at the time of the survey. Table 8.7 shows that economic gains varied between sites but most markedly between the various options within sites. These variations tended to reflect the preferences expressed for the options and level of subjective enjoyment that respondents anticipated they would get from the options.

For example, the scheme at Hurst Spit involved a slightly larger shingle spit which respondents considered would have little impact on their recreational enjoyment. This was reflected in a very small money gain at this site. At Hengistbury Head, respondents favoured the five-rock-groyne option because they understood it to be the only option that would offer the head and its current recreational assets full protection, although they did not see the scheme as offering great recreational benefits. They were against the three-rock-groyne option and beach renourishment because they thought both of these options would not offer sufficient protection and because of the regular disruption that renourishment would cause to recreation on the beach. Thus, these two options involved negative gains: losses.

At Cliftonville, respondents expressed a clear preference for the concrete promenade option because this was what they already had and some objected to the rock-faced promenade, which therefore yielded a lower gain.

What emerges from the studies is that respondents were supportive of major interventions to protect coasts from erosion, but that they generally had clear preferences among the options and among the different protection structures involved where more than one option was offered. At Corton, the managed retreat option was the least favoured of the three options presented because this scheme would not provide direct access from the village to the seafront for recreation and because of the property, business and local economic losses that were involved.

8.6.2 Rivers: VOE studies

The VOE approach has been used in a small number of studies of varied changes in river environments, including studies of low flows on rivers and river restoration. None of these studies focused on changes and recreation benefits resulting from a flood risk management scheme, although the river restoration schemes took flood risk concerns into account.

The proportions of respondents able to offer a money value for the enjoyment of a river visit have tended to be lower than for coastal sites. The money value of enjoyment of today's visit/a visit in current conditions to a river have also tended to be lower than for visits to coastal sites (Table 8.8). In contrast to the findings at the coast, especially those for Corton, in the two studies of river restoration schemes, respondents were supportive of the proposals to restore the rivers to a more natural condition and to take down hard defences while maintaining the level of flood protection. There were mean money gains per visit with both of these schemes, rather than losses (Table 8.9).

8.6.3 Willingness to pay studies

Many Flood Hazard Research Centre coastal studies have also included valuations using the WTP approach. In fact, in most of these studies, respondents have been asked whether they were prepared to pay, because our research has found that this reflects those who are willing and able to pay. In all the Flood Hazard Research Centre studies care was taken to legitimate a refusal to pay; respondents were also first asked whether they were prepared to pay, and only when they responded in the affirmative were they asked how much they were prepared to pay.

In some studies respondents were also reminded of other environmental improvements and policies for which they might wish to pay. The WTP formats used in the Flood Hazard Research Centre studies have evolved over time, so there is less consistency in the

Table 8.8 Money value of enjoyment of today's visit for river sites

Site	Survey year	Sample size	% able to value	£ mean value of enjoyment of today's visit – updated to 2013	£ mean value – at survey date
River Misbourne: Low flows [1]					
Visitors	1994	231	73	6.92	4.43
Residents	1994	150	74	5.79	3.75
River Wey: Low flows [1]					
Residents	1994	146	69	3.59	2.33
River Ravensbourne: Full river restoration [2]					
Visitors	1991	210	58	2.80	1.67
Residents	1991	183	52	2.16	1.29
Visitors and residents	1991	393	55	2.49	1.49
River Skerne: River restoration [3]					
Residents	1995	252	48	9.18	6.00

Sources (for this table and Table 8.9): [1] House *et al.*, 1994; [2] Tapsell *et al.*, 1994; Tapsell, 1995; [3] Tapsell *et al.*, 1997; Tunstall *et al.*, 1999

Table 8.9 Money value of losses and gains per visit for various changes at river sites

Site	£ mean value of loss – updated to 2013	£ mean value of loss – at survey date	£ mean value of gain – updated to 2013	£ mean value of gain – at survey date
River Misbourne: Low flows [1]				
Visitors	4.38	2.80	2.56	1.64
Residents	4.36	2.83	2.16	1.40
River Wey: Low flows [1]				
Residents	1.81	1.17	2.49	1.61
River Ravensbourne: Full river restoration [2]				
Visitors and residents	–	–	2.30	1.37
River Skerne: River restoration [3]				
Residents	–	–	2.91	1.9

WTP methods than in the VOE methods. In the earlier studies, low starting points were used in bidding game procedures and ladders, which may have constrained the WTP responses. (Responses were shown to be influenced by the starting points used.) There is a very real risk that WTP responses may reflect other factors as well as the recreation value of a site; for example, the benefits to local employment and businesses and the desire to avoid property losses through a scheme. There is a need to guard against this risk of double counting in WTP approaches through establishing very clearly in the CV scenarios that only the recreational value of the scheme is at issue. The overall WTP results that have been found in the various coastal studies are summarised in Table 8.10.

8.7 CORTON CASE STUDY: AN EXAMPLE OF A CV SURVEY

The Corton CV survey was carried out as part of the appraisal of the coastal management options for Corton village and its coastal frontage. An initial study was undertaken in 2000 of the potential recreation benefits involving interviews with key informants: residents and managers of holiday facilities. This established that there were likely to be sufficient recreation visit numbers and thence recreation benefits to justify a full CV survey (Tunstall, 2000). A coastal strategy study had also been undertaken by Halcrow which recommended such a study (Halcrow Group Limited, 1999). This study estimated that there were likely to be sufficient benefits to justify the likely costs, with a discounted benefit total of £4.7 million over fifty years arising mainly from cliff-top holiday businesses and residential properties threatened by soft cliff erosion along the Corton seafront (Figure 8.4).

8.7.1 The CV survey

At the time the survey was planned, it was expected that substantial parts of the seafront and the two most central of the three access points from the village to the lower promenade and beach below would be closed due to the collapse of sections of the coastal defences through undermining and erosion of revetments, the sea wall promenade at the base of the cliffs and an access ramp. Therefore, a site visitor survey and the installation of infra-red counters were not feasible. As the preliminary investigation had indicated that there were few day visitors, a survey of 163 local residents was carried out at addresses on the electoral register for Corton Parish and 304 staying visitors at the two main holiday villages and the

Table 8.10 Willingness to pay for coastal protection

Site	Survey date	Sample size and type	Payment vehicle	WTP format	% WTP	£ mean WTP – updated to 2013	£ mean WTP – at survey date
Peacehaven cliff top	1988	214 residents	Increased rates and taxes p.a.	WTP diagram	55 overall		
				50p starting point		3.78	1.83
				£1 starting point		5.85	2.83
Herne Bay	1990	189 residents	Extra national and local taxes p.a.	WTP diagram	73 overall		
				40p starting point		9.43	5.33
				80p starting point		11.28	6.37
Herne Bay	1990	143 visitors	Extra national and local taxes p.a.	WTP diagram	55 overall		
				40p starting point		6.06	3.43
				80p starting point		7.98	4.51
Hurst Spit	1991	550 visitors	Additional taxes p.a.	WTP payment ladder	74 overall		
				25p starting point		15.46	9.22
				£32 starting point		68.11	40.63
St Mildred's Bay	1992	462 visitors	Extra national and local taxes p.a.	WTP payment ladder and two starting points: 25p and £128	61 overall	51.02	31.44
Cliftonville	1993	528 visitors	Small increase in national and local taxes p.a.	WTP payment ladder with two starting points: 50p and £64	62 overall	42.62	19.47
Caister [1]	1997	452	Extra taxes every year	Open question	NA		
		Visitors				44.79	31.62
		Local residents				36.61	25.84

Source: [1] Bateman *et al.*, 2001; for other sources, see Table 8.5

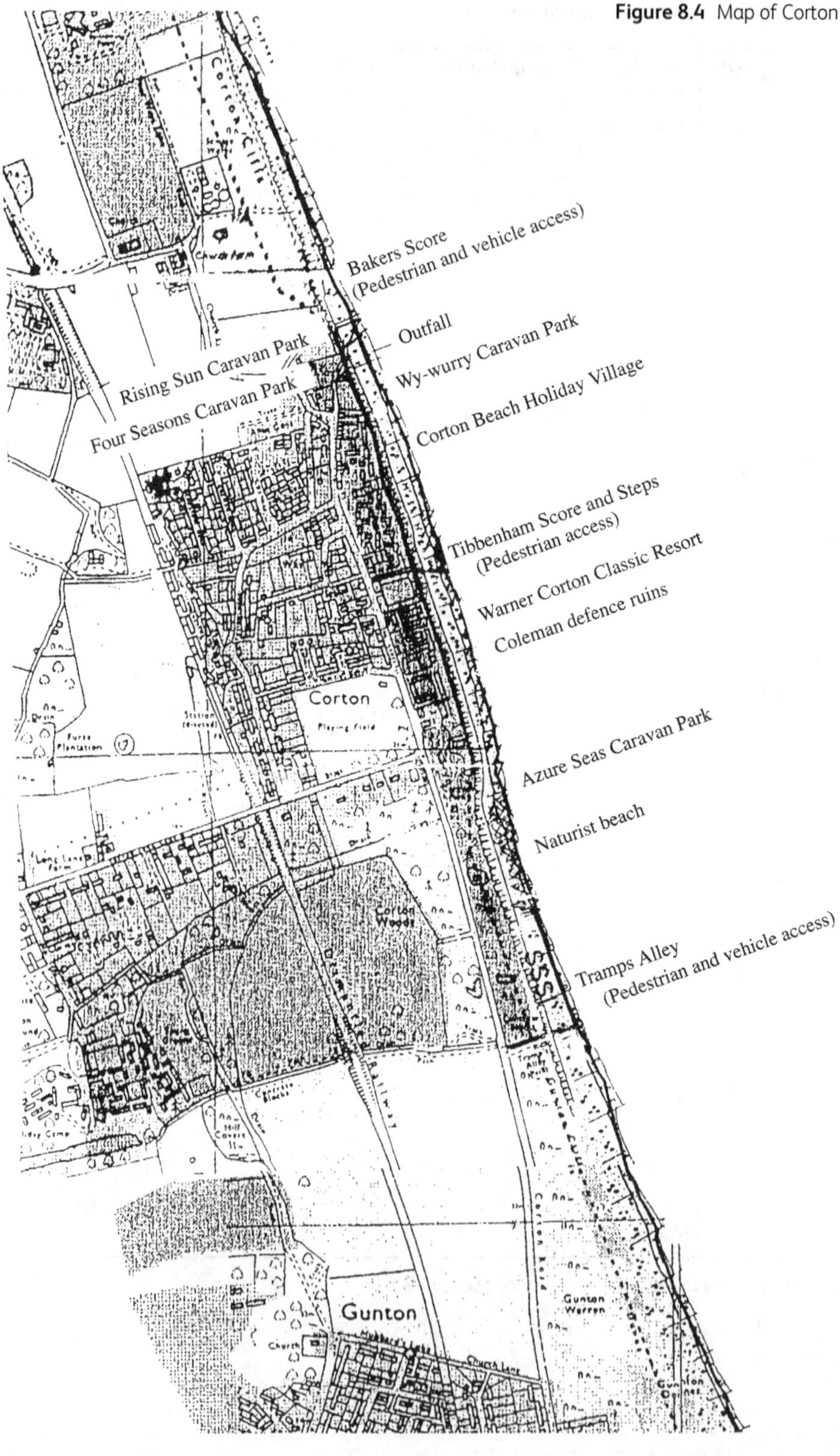

Figure 8.4 Map of Corton seafront

caravan sites in the village. This survey took place between 2 August and 28 September 2001 to cover the main holiday season.

8.7.2 Recreation benefits

The Corton seafront at the time of the survey was in a worse condition than had been the case in any other site where we had undertaken a CV survey, and this affected the responses. Respondents, particularly local residents, gave the Corton seafront much poorer ratings for walking, as a place to take children, for its beach and overall than is usually the case.

The respondents' valuations of a day's enjoyment of the seafront in its current state and with the 'do nothing' and three 'do something' options were obtained with the help of colour drawings produced by the project consultants, Halcrow, and descriptive texts agreed with Waveney District Council (these were necessary as the respondents were not able to view the current condition on site; see Figures 8.5 to 8.9) through a VOE questioning sequence (see the model questionnaires in MCM-Online Supporting Information for Chapter 8).

This involved initially suggesting that respondents should think of a visit or activity that gave them comparable enjoyment to a visit to the Corton seafront in its current condition, showing them a list of possible visits or activities (see Show Card D in MCM-Online Supporting Information for Chapter 8) and suggesting that they should use the costs associated with that visit or activity as a guide to the value of their enjoyment of the current seafront. The list shown had proved appropriate in all other surveys. However, at Corton, 47 per cent felt that there was nothing comparable and that the suggested places and activities were too positive and enjoyable to be compared to the current Corton seafront with its collapsed defences and restricted access. Moreover, the proportion unable to offer a valuation of the Corton seafront was low, partly because many readily gave the current seafront a zero value (Table 8.5). Many had simply given up visiting.

Respondents were then shown the drawings of the 'do nothing' and the three 'do something' options. With each option they were asked whether they

Figure 8.5 Drawing X: current Corton seafront

Figure 8.6 Drawing A: Corton 'do nothing' option

Figure 8.7 Drawing B: Corton 'hold the line for a limited period' option

Figure 8.8 Drawing C: Corton 'hold the line for a longer period' option

Figure 8.9 Drawing D: Corton 'managed retreat' option

Table 8.11 Money value of enjoyment of an adult visit with the Corton options (updated to 2013 prices)

Option	Residents mean £ value per visit	Visitors mean £ value per visit	All respondents mean £ value per visit
'Do nothing'	0.43	0.83	0.70
'Managed retreat'	1.66	3.24	2.82
Current seafront	2.99	4.50	4.01
'Hold the line for a limited period'	4.45	4.54	4.52
'Hold the line for a longer period'	24.41	13.35	15.93

would get more or less enjoyment from a visit to the seafront in the option drawing compared to the current seafront shown in its drawing, and to say how much enjoyment they would get from a visit in pounds and pence (Table 8.11). The mean money values attached to the enjoyment of a visit were more polarised for the local residents, who gave lower valuations than the staying visitors to the current seafront and the 'do nothing' and 'managed retreat' options and a higher valuation to the 'hold the line for a longer period' option. The mean values for the enjoyment of a day's visit were lower for all the options – apart from the 'hold the line for a longer period' – than has generally been the case for other coastal studies (see Table 8.11). It is likely that the low valuations reflect the deteriorated conditions and limited access associated with the other options.

After that, respondents were asked whether they would visit the seafront in the option drawing more or less than they would visit the current seafront. Those who would visit less often were asked whether they would go somewhere else instead, and if so, where. They were asked how much enjoyment in pounds and pence they would get from this alternative site and how much more or less it would cost them to go there. With these questions, four different groups were identified according to their behaviour. Their economic losses or gains were calculated separately, for each individual in the group, for each option. The overall average was then calculated for each option for residents and staying visitors separately.

With the 'do nothing' option, only a very small proportion of residents and staying visitors would continue to visit the Corton seafront as often as at present. The economic loss for this 'stayer' group was calculated as the difference between the money values of enjoyment of a visit to the current and to the 'do nothing' seafront. Most experienced no loss as they gave the seafront in both conditions a zero value and some preferred the 'do nothing' seafront to the present conditions, so there was a negative loss (or gain) with the 'do nothing' option for these respondents.

A more numerous category comprised those who would visit less often but would not transfer their visits to an alternative site. The loss for this 'homer' group was calculated in the same way as for the 'stayers' and was the most substantial. Some respondents did not visit the site currently and did not expect to do so under the 'do nothing' option. They were assumed to incur a zero loss as their enjoyment would not change.

A majority of the respondents said they would go somewhere else if the 'do nothing' option were to happen in Corton. There are plenty of local beaches close to Corton – at Lowestoft, Gorleston and further afield around Great Yarmouth. The losses for these 'movers' were measured by the difference between the enjoyment value of their alternative site (net of additional journey costs) and their enjoyment value for the visit to Corton in its current condition. Among the 'movers', an unusually substantial proportion (32 per cent of the 'movers', 13 per cent of all respondents) gave higher valuations for the alternative site over the current site, net of costs. These respondents have been characterised as 'irrationals' since the logic of economic theory would indicate that they should already be visiting the alternative site, rather than the current one.

However, the survey investigated another factor not evaluated in the CV procedure, which affects choices of sites to visit: the convenience of travelling to the alternative site. Travel costs do not necessarily capture this factor. Almost all of the 'irrational' local residents and a majority of the staying visitors reported that the alternative site would be less convenient for them to visit; factors such as summertime congestion along coastal roads would be involved. Furthermore, some of the staying visitors were visiting for the first time and others who knew the Corton seafront from past visits were disappointed to find it in such poor condition.

Normal practice is to exclude 'irrationals' from the calculations of the mean gains and losses. However, as these 'irrationals' were a significant minority for the 'do nothing' option in Corton, they were included as zeros in the calculation of the mean losses, which assumed that their gains in visiting the alternative site were exactly offset by the inconvenience of going there. This was a conservative assumption since excluding them would have resulted in a higher overall mean loss.

Under two other options – 'hold the line for a limited period' and 'managed retreat' – there were again significant proportions of respondents who would transfer their visits elsewhere and 'irrational' respondents. These were treated in the same way as indicated above.

The gains and losses with the options are shown in Table 8.12. The losses – usually the most important component in the recreation benefits – were much less, and gains with one option ('hold the line for a longer period') more, than has been the case in other studies, reflecting the fact that the Corton seafront had already deteriorated seriously by the time of the survey. It is notable that the 'managed retreat' option, which involved removing the coastal defences and returning the seafront to a more natural condition, did not yield substantial gains. Indeed, for local residents, it involved a loss, probably because there would be little direct access to the seafront from Corton village under such managed realignment. Local residents again had more polarised values, with larger losses under the 'do nothing' option and larger gains under the 'hold the line for a longer period' option than the staying visitors.

8.7.3 Visit numbers

In the absence of infra-red counter data, visit numbers had to be calculated from the survey data. Local residents were asked about their visiting frequency 'last winter and over the summer' and resident visit numbers were estimated from that data.

For staying visitors, information had been obtained in the preliminary investigation for the main tourist facilities (the two holiday villages and one large

Table 8.12 Losses, gains and annual benefit estimates with the options at Corton (updated to 2013 prices)

	Local residents			Staying visitors			All
	Adult visit numbers	£ value per visit	Annual loss £	Adult visit numbers	£ value per visit	Annual loss £	Annual loss £
'Do nothing'							
Halcrow	25,000	25,000	3	50,000	1.34	67,134	135,584
CV survey	38,447	38,447	3	58,702	1.34	78,818	184,086
Limited protection							
Halcrow	25,000	25,000	3	50,000	2.42	121,104	186,593
CV survey	38,447	38,447	3	58,702	2.42	142,182	242,894
Longer protection							
Halcrow	25,000	17.96	448,875	50,000	8.99	449,533	898,408
CV survey	38,447	17.96	690,316	58,702	8.99	527,771	1,218,086
'Managed retreat'							
Halcrow	25,000	–0.26	–6,582	50,000	2.38	119,130	112,548
CV survey	38,447	–0.26	–10,133	58,702	2.38	139,863	129,730

caravan site) on the number of units, the average number of adults per party, the estimated occupancy rate and opening period. From this, an estimated number of adult visitor days per year could be calculated. In the CV survey, staying visitors were asked how many days they were staying and on how many of those days they had visited or expected to visit the Corton seafront. From these data an estimated seafront visit rate could be calculated for each of the three facilities (0.254, 0.474 and 0.327, respectively). The estimated adult visitor days per annum multiplied by the visit rate gave the estimated adult seafront visits per annum at each holiday facility.

Both the resident and staying visitor estimates took some account of seasonality through estimated occupancy rates and opening periods in the case of staying visitors. Both estimates were based on current visiting and were therefore affected by the restrictions on access to the seafront. These estimates were used to assess annual benefits from the different options. Visit number estimates derived from tourist data for the area by Halcrow were included for comparison (Table 8.12).

8.7.4 Conclusions from the Corton study

Based on the interim results from the CV survey and from public involvement in the project, Waveney District Council and its consultants developed a preferred option: 'hold the line for an intermediate period'. This aimed to provide the recreation and amenity advantages, particularly in terms of access, that were a key factor in making 'hold the line for a longer period' the favoured option with the public. The Corton study thus illustrated how a CV study can contribute to understanding of public preferences and to the evolution of the options but also how this creates difficulties for the benefit assessment because any new or hybrid option will not have been directly evaluated in the CV study.

The study also shows the desire of the public to maintain the status quo at the coast and a rejection of the option of allowing natural erosive processes to take their course. In any case, the 'managed retreat' option at Corton – as on any developed and defended coast – would have been viable only if it had had substantial public support and recreation benefits since there were costs associated with removing the coastal defences and no direct benefits under that option.

The study highlights the limitations of the modelling of recreation behaviour and value underlying the CV questioning procedure in that significant minorities gave 'irrational' responses; it also indicates how these may be handled. Furthermore, it illustrates the problems that arise for a CV study when conditions at the site to be evaluated have already deteriorated and access has been affected, but shows some ways of dealing with these issues.

8.8 KEY LESSONS FROM EXPERIENCE

Some of the research reported here is quite old, but we nevertheless believe that the results and the lessons learned are still fully valid, not least because they concern data requirements and related issues that persist today and will continue to be problematic in the future.

A key lesson from the research experience is therefore that estimating the visit numbers or the number of beneficiaries deserves to be given as much attention as estimating the valuations, and this has not always been the case. The money values per adult visit for environmental changes are often small but they can yield quite high annual benefits where the number of visits or beneficiaries is large. Therefore, deriving accurate estimates of visit or beneficiary numbers is vital.

The research has also shown that the kinds of people who visit 'natural' undeveloped coasts are different in some respects from those who go to developed sites. If coastal sites were to be changed radically to a more 'natural' condition, they might draw on a different constituency of visitors (making the new visit numbers difficult to determine). However, sites may vary in the extent to which there is scope for them to attract 'new' and different visitors.

The research also shows that the public are generally supportive of measures to protect and defend the coast through major interventions, such as sea walls and off-shore reefs. Indeed, the coastal studies generally indicate that people who visit or live at the coast are reluctant to see natural erosive processes take their course, instead wanting the coast to continue to be maintained and defended as it has been in the past (Tunstall and Penning-Rowsell, 1998). This makes such options as 'managed realignment' or 'retreat' difficult to implement.

The few river restoration studies, in contrast, show that residents are supportive of, and attach value to, works to restore rivers to a more natural condition, including 'soft engineering' works used in place of hard defences where the level of flood risk is not increased.

Importantly, however, the research also shows that public responses to, and thus valuations of, options and structures at the coast, such as rock groynes – acceptable at Hengistbury Head but not at Herne Bay – vary from site to site in ways that are difficult to predict (making the 'value/benefit transfer' approach often difficult to apply here). Therefore, despite the substantial body of research on recreation benefits at the coast, there remains a vital need for site-specific CV surveys at detailed appraisal stages for both coastal and riverine sites. There is also a need for meta-analysis to explore the factors that affect responses across sites.

One virtually unavoidable problems that remains is that the recreation benefit assessment methodology recommended here does not take into account new visits (as opposed to transferred visits) that may be generated among local residents or more widely. Nor is additional visiting by current users easily allowed for (again not transferred visits). There may, therefore, be significant underestimates of the benefits of schemes which offer substantial recreational or amenity improvements or attractive new environments (such as the Jubilee River).

Nevertheless, CV surveys can make a valuable additional contribution to public consultation and participation on coastal and fluvial projects. However, as stressed in Section 8.3.2, those who have an interest in recreation and amenity at a site should be involved in the project from the start (as recommended by Eftec, 2010a) so that they can participate in the identification of the problems at the site, the identification of the options and the evaluation of the impact of the options on recreation and amenity. Unfortunately, though, it has to be recognised that this early engagement is an ideal that may be difficult to achieve: CV surveys will often be undertaken late in the process of option development – not least because new options emerge within the appraisal period – and therefore they are no substitute for other public consultation and participation processes. It is necessary to include in CV surveys rating questions on the different options and questions to establish the reasons for public preferences for those options.

These questions both help to establish the validity of the monetary valuations of the options and can add to the understanding of the public's views on these options. However, difficulties can arise when the survey information leads to a substantial rather than a minor modification of the scheme to meet the public's overall preferences, as happened at Corton. Recreation benefits are only valid for the options as presented in the CV scenarios in the survey, so valuations of a scheme that is modified in line with the public's preferences become problematic.

A final lesson is that we continue to learn that recreation and amenity changes are of vital public interest. There are few – if any – legal obligations in this respect on those promoting coastal and fluvial risk management, as opposed to the Water Framework Directive's strictures, but the issues still need very close attention. And, of course, further research is needed both to ensure that the results here remain as up to date as possible and to develop better techniques accurately to capture the public's aspirations regarding the benefits that arise when risk management agencies spend scheme funders' money on their behalf.

9 Appraisal of flood risk management for agriculture

The management of flood risk to farm land has been, and remains, an important element of support to the agricultural sector in Britain. Many floodplain and coastal areas have benefited from publicly funded flood control and land drainage schemes (Morris, 1992) that have reduced flood damage and provided opportunities for more productive farming. Indeed, approximately half of Britain's Grade 1 agricultural land depends on flood risk management schemes of one kind or another.

This chapter investigates the impacts of flooding on agriculture and reviews the functions and benefits of flood management for agriculture (Sections 9.2 and 9.3). Section 9.4 estimates the financial and economic value of flooding on agriculture before Section 9.5 provides a method for assessing the costs of a flood event. The chapter comments on Defra's guidance for the appraisal of agricultural flood risk management investment (Section 9.6) before reporting on the role and importance of both environmental protection (Section 9.7) and ecosystem services (Section 9.8) within flood risk management. The latter part of the chapter explores the data requirements and collection methods (Section 9.9) and agricultural flood risk management appraisal at the scheme level (Section 9.10) before a simplified worked example of an appraisal of a flood risk management project is provided (Section 9.11). Section 9.12 offers lessons from experience and the chapter concludes with a summary (Section 9.13).

9.1 INTRODUCTION

The trends and challenges affecting rural and agricultural flood risk management identified in MCM (2005) (Penning-Rowsell *et al.*, 2005a) have continued and further consolidated. Concern has grown about the vulnerability of food production systems to environmental degradation and climate change (Stern, 2006; IAASTD, 2009; Foresight, 2011; Winter and Lobley, 2009). International assessments of ecosystems (MEA, 2005; TEEB, 2010) and the UK's own National Ecosystem Assessment (NEA, 2011) have confirmed the diverse services provided by natural capital, including food production and the regulation of flooding, and the implications for ecological, economic and social well-being. The concept of ecosystem services, including the regulation of flooding, is now seen as a key element of policy and project appraisal and implementation (Defra, 2010; Everard, 2009; Natural England, 2011; HM Government, 2011).

With respect to agriculture, farm income support in England and Wales is no longer tied to commodity price support. Farmers receive annual Single (Farm) Payments that are conditional on compliance with good agricultural and environmental practice (Natural England, 2009). There is also a more comprehensive and targeted programme of Environmental Stewardship, with payments to land managers to protect the natural environment, including soil and water resources. Controls of diffuse pollution from agriculture, much of it associated with run-off from farmland, have been introduced under the EU Water Framework Directive, supported by initiatives such as the Catchment Sensitive Farming (Defra, 2012) and Demonstration Test Catchments (DCM, 2012) programmes.

With respect to flood risk management, following the strategic initiative embodied in *Making Space for Water* (Defra, 2005), there is much better recognition of the potential contribution of rural land to managing the conveyance and storage of flood waters in order to reduce flood risk to urban areas and critical infrastructure. Such 'non-structural' measures include the management of rural land by farmers to retain water in the general landscape and/or temporarily 'store it' in floodplains (Lane *et al.*, 2007). The use of natural processes to control the movement of water through the landscape is central to this approach (Environment Agency, 2010e; Morris *et al.*, 2010; *POSTnote*, 2011).

In this context, there is greater recognition of the scope, and indeed necessity, to integrate flood risk management in the rural space with other objectives, such as food production, nature conservation, water quality improvement and recreation (Posthumus *et al.*, 2010; SDF, 2012). The ecosystem framework referred to earlier can help to identify potential synergy as well as conflict between different services, and the range of 'stakeholder' interest that might be affected by any proposed flood risk management intervention (Reed *et al.*, 2009). Where land managers can offer flood risk management services for the benefit of others, it may be appropriate to reward them accordingly, applying the principle of 'payments for ecosystem services' (Wunder *et al.*, 2008; Beedell *et al.*, 2012).

There is now much interest in multiple partnership funding for flood risk management (Environment Agency, 2012). This demands an integrated approach at the appropriate catchment or sub-catchment scale, bringing together the various interested parties (Posthumus *et al.*, 2008). New and flexible arrangements are being sought to provide financial and other incentives to land managers to contribute to flood risk management (Beedell *et al.*, 2012). An understanding of the financial impact on farm businesses of the 'costs' of providing flood service is critical to any negotiation of compensation arrangements. The guidance provided here can inform this process.

Since the global economic crisis in 2007, agriculture in the UK has witnessed a revival of interest and fortune as a result of global and national concerns about food security (HM Government, 2010d) and associated strengthening of agricultural commodity prices (OECD/FAO, 2009). Most forecasts suggest that agricultural commodity prices up to 2020 will remain at least 25–30 per cent higher in real terms than pre-2007 levels. There is, however, uncertainty about the detail of outcomes of the current review of the Common Agricultural Policy (2013–20) (Posthumus *et al.*, 2009; EC, 2011). While the broad framework is likely to continue, there will be more targeted support for farms in 'marginal areas', for ecological restoration, and for strengthening agriculture's contribution to economic development. In the face of increased global demand for food and pressures on available land, there is a push towards 'sustainable intensive farming' (Royal Society, 2009; Foresight, 2011): improving farming productivity while protecting natural resources and the environment. Managing flood risk and land drainage in strategically important farming areas is central to this, especially under conditions of climate change. The promotion of ecological restoration, however, could further promote the retention and storage of flood waters in some extensively farmed wetland areas.

As explained later in this chapter, the basic principle of publicly funding flood risk management is that benefits should exceed costs at the Treasury's test discount rate (HM Treasury, 2011). In principle, investments for flood risk management in agricultural areas are treated in the same way as those in urban areas. There is now perceived to be little economic justification for public funding of agricultural enhancement schemes (just as Treasury guidance advises no justification for public investments that enhance the development value of urban land).

The current focus of agricultural flood appraisal is mainly to determine whether it is worthwhile from the viewpoint of the public purse to continue to fund flood risk management for agriculture, especially as previous capital works come up for rehabilitation. This involves comparing some hitherto existing standard with the 'do nothing' option. Alternatively, some intermediate option – neither the current situation nor 'do nothing' – may offer better value.

In some cases changes in catchment hydrology due to development and/or climate change expose agriculture to increased flooding, whether or not farm land has been protected by some previous scheme. Here, appraisal may explore the economic justification for flood risk management works to avoid damage costs to agriculture under changing conditions. Decisions about the balance of private and public contributions to funding such works can follow.

A further important case emerges that is likely to become more prominent in the future where agricultural land is a provider of flood risk management services, incurring economic penalties as a result. Here, the economic damages to agriculture are set against the benefits of avoided costs; that is, 'savings' in flood damage costs elsewhere, or savings in defensive expenditures, such as engineering works. Appraisal might identify rural flood storage as a preferred economic option. Decisions on funding, including beneficiary contributions, can follow.

The approach to appraisal, and the degree of detail required, depends on the value of the agricultural and rural assets and their related productivity that are put at risk, temporarily or permanently, as a consequence of changes in flood risk. The appraisal will at least require some comparison of the financial and economic performance of land use under relevant flood risk management options, and how these compare with the costs of delivering the options. The context of agricultural appraisals can be diverse. In some situations, options to deliver flood storage benefits or environmental enhancement should also be considered as candidates for appraisal, and an attempt made to include these non-agricultural benefits in the appraisal.

As explained later in the chapter, Defra provides guidance on the appraisal of agricultural flood risk management projects, distinguishing three scenarios, as follows:

- Where farming is impossible in the absence of flood control, the advice is to estimate economic loss (and therefore the benefits of flood risk management) in terms of the likely reduction in the asset value of agricultural land (see Section 9.6.1).
- Where the concern is with one-off damages to farm land due to temporary failure of flood controls, broad estimates of economic loss for major categories of land use can be used, although these should be locally relevant (see Section 9.6.2).
- Where the concern is to identify the impact of a permanent change in flood risk on farming, the appraisal process will be more complex (see Section 9.6.3). This will particularly be the case if there is a range of flood risk management options which differ in benefits not only for farming but for the natural environment and overall catchment flood risk management.

At a broad catchment scale, appraisals will at least require information on categories of land use, and the extent that these might be affected by a change in flood risk. Secondary data of the kind presented in this chapter may be adequate. At a detailed project appraisal level, however, there may be a need to collect primary data and undertake some detailed analysis to ensure that benefit estimates are robust and relevant to local conditions and the options being considered. The justification for greater detail must rest on a judgement whether further information or analysis is likely to influence the decision to undertake a scheme or option. A sensitivity analysis of the feasibility of scheme options to critical assumptions will help determine whether more work is needed on the appraisal.

Where agricultural schemes cover large areas and have strategic importance at the national or regional scale, the guidance referred to above may not be entirely appropriate and advice should be sought from Defra accordingly. The uncertainty facing agriculture associated with fluctuations in international commodity markets and the current review of the EU Common Agricultural Policy (2013–20) also requires that appraisal methods are kept under review.

9.2 FLOODING, DRAINAGE AND IMPACTS ON AGRICULTURE

Water is essential for agriculture, but too much of it limits crop growth and land use. In this respect the type and performance of farming activities and practices are critically dependent on the management of flood risk and land drainage (Smedema and Rycroft, 1983; Castle *et al.*, 1984). Flood risk management for agriculture, as further discussed below, refers to the provision of the degree of protection against flooding not only above the surface of the ground (that is, inundation) but also below the surface of the ground (that is, waterlogging). Thus, flood risk management for agriculture takes on a wider meaning than the alleviation of surface flooding and often includes measures to provide appropriate 'land drainage' conditions needed to support commercially viable farming. In river floodplains (Figure 9.1), flood management includes:

- Modification to channel capacity by engineering and/or vegetation management to reduce flood risk to adjacent farm land.

- Management of water levels in the river and arterial systems to provide outfalls for water draining from the soil profile, whether by natural movement of water through soils, through artificial subsurface drains or by pumping (or a combination of these).

In coastal areas and estuaries:

- Low-lying agricultural land is often protected from saltwater inundation by sea defences.
- Drainage to the sea may be assisted by pumping, or gravity through flapped outfalls which open at low tide.

Investments in flood risk management infrastructure for agriculture have traditionally been part funded through government grant aid and local authority precepts, by direct charges to beneficiaries in Internal Drainage Boards areas, and in the Anglian region by a general agricultural drainage rate. Responsibility for flood risk management works varies throughout the river and land drainage system. The Environment Agency, local government and Internal Drainage Boards have powers allowing them to manage river and arterial systems, but on-farm drainage systems are controlled by the individual land manager.

The benefits of flood risk management and improved drainage reflect the removal of constraints imposed by excess water and the processes of agricultural change that follow. Figure 9.2 provides a framework for assessing the impacts of a change, either positive or negative, in flooding and land drainage at the farm level.

Flood management benefits are associated with:

- 'reclaiming' land for commercial farming;
- improving performance of existing crop and livestock activities;
- reducing damage and saving costs of operations;
- switching to land uses with higher value added;
- addressing climate change impacts; and
- addressing the effects of land shrinkage/subsidence.

A reduction in flood protection, perhaps associated with 'do nothing' or 'set back' options, might involve the loss of these benefits previously obtained by a flood management scheme.

Furthermore, there may be benefits to the whole farm business due to improvements in part of its area, such as improved grass production providing winter feed for a large dairy herd or the release of land elsewhere on the farm for arable production.

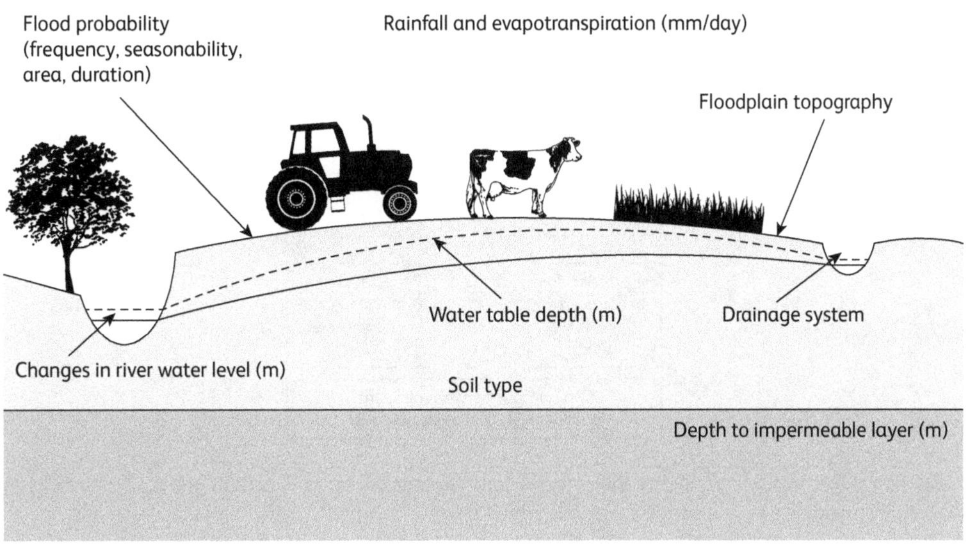

Figure 9.1 Flooding and drainage factors influencing agricultural productivity on floodplains

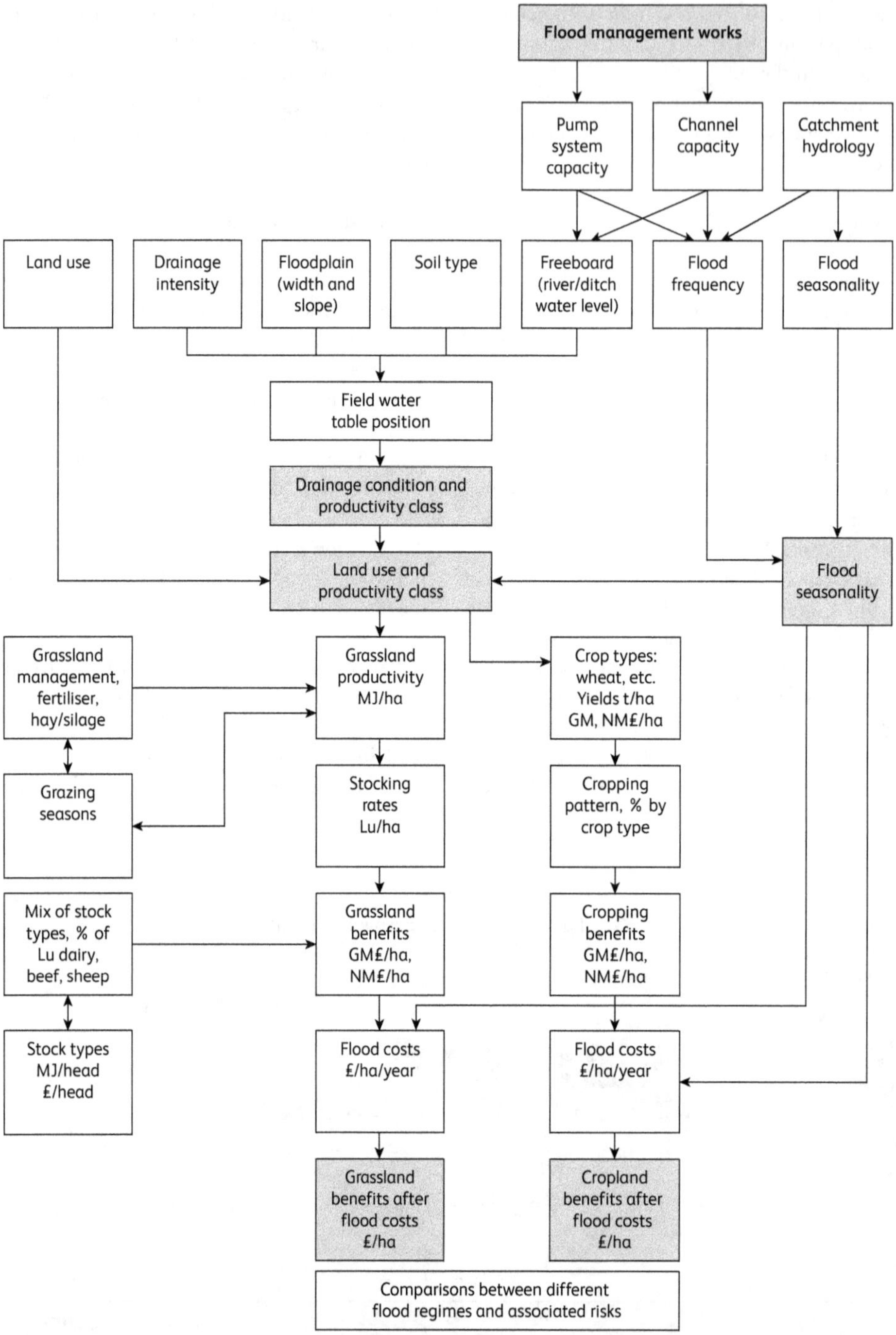

Figure 9.2 Flood management works and their influence on flood regimes, drainage conditions and the productivity and profitability of farming

Note: Stocking rates and yields as per Table 9.3; benefits expressed in gross margins (GM) and net margins (NM); MJ Mega Joules; Lu livestock units. See text for explanation.

9.3 FACTORS INFLUENCING FLOOD DAMAGE COSTS ON AGRICULTURAL LAND IN THE UK

The damage costs due to flooding on farm land depend mainly on the type and productivity of land use, the duration and seasonal timing of flooding and a number of local factors that influence damage costs at the farm scale (Morris and Hess, 1986; 1988; Pivot *et al.*, 2002; Posthumus *et al.*, 2009). The depth and turbulence of flood water can also affect the degree of damage to crop and livestock production. Flood waters contaminated with pollutants or carrying high sediment load can further increase damage to crops, livestock and soils. These factors are discussed in turn.

9.3.1 The flood risk management 'benefit' area

Total flood damage costs vary according to geographical extent or magnitude of flooding. A combination of hydrology and topographical factors determines the relative frequency of flood events of a given magnitude, usually represented as flood 'envelopes' or 'cells' associated with different return periods. It is important, therefore, to determine the farmed areas that are actually or potentially subject to flooding, and the likely frequencies of flooding within these areas. This constitutes the 'benefit' area of a flood risk management intervention, usually based on records of maximum known flood areas and/or predictive hydraulic modelling.

As discussed later, the proportion of the total farm that is subject to flooding can affect flood damage costs at the farm scale. If only small portions are affected, there may be little impact on the farm as a whole; but if the majority of the farm is inundated, the viability of the whole farm business may be placed at risk. In some cases parts of, rather than whole, fields may be flooded, although traditionally field boundaries tended to follow natural flood lines. As discussed below, changes in flood probability may also impact on arterial and field drainage systems, such that the 'benefit area' extends beyond the flooded area itself, probably incorporating areas that lie an additional 2m higher than the maximum known flood area.

9.3.2 Land use/crop cover and flood tolerance

The most critical factor determining flood damage costs is land use in the affected areas. Land use, in order of intensity of production and value added, is typically classified in the UK by crop cover, namely horticulture (soft fruit and salad crops), intensive arable (including root crops and vegetables), extensive arable (cereals and oil seeds), intensive grassland (mainly dairy) and extensive grassland (beef and sheep).

Generally, the more intensive is land use in terms of the value of agricultural output, the greater is the cost of damage for a given flood event. Thus, land use and the associated farming system determine the context for flood impact assessment and the 'counter-factual' – that is, the 'without flood event' – situation against which the incremental costs of the flood event can be assessed. There is usually a negative correlation between the intensity (and added value) of land use and the annual probability of flooding. High value-added land use, such as horticulture and intensive arable, does not usually occupy unprotected flood-prone areas (Table 9.1). High-value crops, such as

Table 9.1 Tolerance of flooding by agricultural land use in England and Wales

Agricultural land use type	Common tolerable flood frequencies: annual probability	
	Whole year	Summer April–October
Horticulture	1 in 20	1 in 100
Intensive arable, including sugar beet and potatoes	1 in 10	1 in 25
Extensive arable: cereals, beans, oil seeds	1 in 8	1 in 8
Intensive grass: improved grass, usually dairying	1 in 2	1 in 5
Extensive grass, usually cattle and sheep	At least once per year	1 in 3

potatoes and field vegetables, do, however, occupy low-lying areas of land with high agricultural potential that are served by flood control and land drainage schemes, including pumped areas. Thus, damage costs associated with extreme rainfall and flood events tend to be much higher in intensively farmed areas that are not usually subject to frequent flooding.

9.3.3 Land productivity and drainage

Observation and research show that the productivity of agriculture is critically dependent on water table levels and soil wetness as they determine crop growth conditions and field access (Hess *et al.*, 1989; Dunderdale and Morris, 1996a; 1996b). It is possible to prescribe drainage conditions (good, bad, very bad) to support given classes of agricultural productivity (normal, low, very low), as shown in Table 9.2. Waterlogging is associated with cold and anaerobic soil conditions that restrict crop growth and yields. Wet soils also have reduced strength and lower bearing capacity which, in turn, prevent or delay field access by machinery or grazing livestock.

The influence of water table levels on agricultural productivity is especially critical during the spring and early summer periods and, to a lesser extent, during the autumn. Water tables at or deeper than 0.5m from the surface are indicative of 'good' agricultural drainage conditions and will support 'normal' agricultural productivity. Conversely, persistently high water tables, within 0.3m of the surface, indicate 'very bad' agricultural drainage usually resulting in 'very low' levels of agricultural productivity associated with constraints on land use, limited field access, low yields and low financial performance. Such conditions, if they persist, would not support arable cropping and land use will mainly be confined to extensive grassland. Even so, turnout dates for grazing animals may be substantially delayed if damage to the soil surface and animal welfare problems are to be avoided. Water tables between these two extremes are associated with moderate impediment to drainage and productivity.

Just as flood probability is measured in terms of relative frequency, the wetness of soils can be defined in terms of the number of days that water tables exceed these critical levels during critical periods (SSEW, 1984). For example, a soil that has a water table within 0.3m of the surface for more than 80 per cent of the time in spring can be described as having 'very bad' drainage from an agricultural viewpoint. Farmers would describe such soils as 'commonly or permanently' wet.

Table 9.2 also shows the required freeboards in the drainage network required to provide given standards of drainage service (where freeboard here is the height difference between water in the river or ditch and the

Table 9.2 Drainage conditions for agriculture and water levels in fields and ditches in England and Wales

Agricultural drainage condition	Agricultural productivity class	Depth to water table from surface	Spring time freeboards in water courses (natural drainage)	Spring time freeboards in water courses (field drains)
Good: 'rarely wet'	Normal, no impediment imposed by drainage	0.5m or more	1m (sands) 1.3m (peats) 2.1m (clays)	1.2m (clays) to 1.6m (sands) (0.2m below pipe outfall)
Bad: 'occasionally wet'	Low, reduced yields, reduced field access and grazing season	0.3m to 0.49m	0.7m (sands) 1m (peats) 1.9m (clays)	Temporarily submerged pipe outfalls
Very bad: 'commonly or permanently wet'	Very low, severe constraints on land use, much reduced yields, field access and grazing season: mainly wet grassland	Less than 0.3m	0.4m (sands) 0.6m (peats) 1m (clays)	Permanently submerged pipe outfalls

adjacent field surface level). These vary according to soil type (as this defines hydraulic conductivity – the rate at which water moves through soils) and the presence of artificial under-drainage (as this defines clearance for pipe outfalls into the open drainage network). Freeboard provides gravitational fall for water to move through the soil profile to the water course.

Generally, sandy soils have high rates of hydraulic conductivity (rates of soil water movement at 2m/day) relative to clay soils (0.25m/day). It is noted that many soils in Britain's river floodplains are heavy clays where water levels in the rivers and ditches may have limited influence on water table levels in the field. Lowering water levels in water courses as part of capital or maintenance works will have little impact unless accompanied by field drainage; and even so, evacuation of excess water from soils may be slow. Furthermore, in storm conditions when heavy soils are already wet, rainwater will tend to run along the surface rather than into the profile of the soil, exacerbating the risk of flooding (O'Connell *et al.*, 2004).

The importance of land drainage conditions on land use and the productivity of farm land cannot be overstated. The ability to control field water levels is critical for arable cropping and intensive grassland.

Table 9.3 Indicative variations in agricultural productivity by field drainage conditions in England and Wales

	Field drainage conditions		
	Good	Bad	Very bad
Arable			
Yield as % of 'good' category			
Winter wheat and barley	100	80	50
Spring wheat and barley	100	90	80
Oil seed rape	100	90	80
Potatoes, peas, sugar beet	100	60	40*
Typical wheat financial gross margin £/ha	£800–900	£380–480	£200–300
Grassland			
Typical nitrogen use on grass kgN/ha	150–200	50–75	0–25
Grass conservation	2 cut silage	1 cut silage or graze	1 cut hay or graze
Typical stocking rates livestock units/ha	1.7–2.0	1.2–1.4	0.7–1.0
Typical livestock type	Dairy, intensive beef and sheep	Beef cows, 24-month beef, sheep	Fattening of 'store' cattle, and sheep
Typical financial gross margins £/ha (after forage costs)	£1,800–2000 (dairy) £500–900 (intensive beef/sheep)	£400–600	£200–400
Days' reduction in grazing season compared to 'good' category	none	Spring: 14 to 21 Autumn: 14 to 21	Spring: 28 to 42 Autumn: 28 No stock out in winter

Notes: * Not grown if persistently 'very bad'. Livestock units: dairy cow, 1 Lu; beef cow, 0.8 Lu; 24-month beef, 0.7 Lu; sheep plus lamb, 0.14 Lu. A grazing day is worth about £2.2/Lu in spring, £1.6 /Lu in autumn and £0.40/Lu in winter in terms of savings in housing costs and feed conservation costs.

Yields and land use, and likely financial returns, can vary with drainage conditions for arable crops and grassland. Table 9.3 provides examples of land use and productivity by field drainage conditions derived from farm surveys and experimentation (Morris *et al.*, 1984; Morris and Sutherland, 1993; Dunderdale and Morris, 1996b). Drainage critically affects not only the fundamental productivity of farm land but the cost of flooding when it occurs. The cost of a single flood on poorly drained land is likely to be low relative to costs on well-drained, more intensively farmed land.

It is noted that low water table conditions during the summer can cause water stress in crops (especially if they have poorly developed root systems because of waterlogging earlier in the year) and there may be benefit from raising water tables to provide sub-irrigation. Indeed, traditional water meadows require high groundwater levels in summer. But, for the most

Table 9.4 The impacts of flooding on farm land in the UK vary according to type of agricultural land use and the seasonality of the flood event

	Spring	Summer	Autumn	Winter
	March–May	June–August	September–November	December–February
Horticulture (soft fruits, salad crops)	Complete loss of soft fruits and winter/spring salads	Complete loss of annual production, possible loss of perennial stock	Loss of late season harvest, possible loss of perennial stock: replanting/reseeding	Damage to standing crops, and annuals/perennials
Intensive agriculture (including field vegetables and roots)	Delay in planting or loss of established crops	Likely complete loss of standing root crops, e.g. potatoes/onions/carrots	Loss of unharvested autumn crops, notably potatoes. Delayed planting or loss of winter crops, substituted by spring sown crops	Possible loss of winter harvest crops (Brussels sprouts and sugar beet). Yield loss on autumn sown crops
Extensive arable (cereals and oil seeds)	Loss or delay of spring sown cereals, yield loss on winter sown cereals, delayed spring treatments	Complete or partial loss of unharvested crops	Loss of unharvested autumn crops. Delayed planting or loss of winter crops, substituted by spring sown crops	Yield loss on autumn sown crops, reseeding with spring sown crops if severe damage
Grassland: intensive (mainly dairy)	Loss of grass yields, delayed stock turnout, delayed fertiliser applications. Grass reseeding if long-duration flooding	Loss of grass yields, partial or complete loss of hay/silage crop, loss of grazing, stock morbidity/mortality. Grass reseeding if long-duration flooding	Loss of autumn grazing, stock relocation/housing. Possible reseeding if long duration	Loss of winter 'accommodation' pasture
Extensive (mainly beef and sheep)	Loss of grass yields, delayed stock turnout, delayed fertiliser applications	Loss of grass yields, partial or complete loss of hay/silage crop, loss of grazing, stock morbidity/mortality	Loss of autumn grazing, stock relocation/housing	Limited impact on flood tolerant grass swards

part, spring/early summer water table levels are the defining water-related influence on the feasibility and profitability of agricultural land use in Britain.

In summary, the influence of soils and water level management, and the need for expenditure on arterial and field drainage, must be considered when assessing the potential benefits of flood risk management and land drainage schemes for agriculture. These fundamental aspects of soil-water regimes are also important for the design of water-level plans in environmentally sensitive areas, where the purpose is to maintain soil wetness conditions suited to natural species (Morris *et al.*, 2004).

9.3.4 Seasonality

The impact of flooding on agricultural land depends particularly on the timing of the flood with respect to crop growth stages and critical field operations. It is important to note the seasonal timing of the flood as this determines, for given land use, the type and cost of damage, and the scope for mitigation.

Based on experiments and field observations, Table 9.4 provides a high-level summary of the likely impacts of the seasonality of flooding according to land use. Spring and summer floods are much more damaging than autumn and winter floods. Autumn floods, however, can result in complete loss of pre-harvested late season vegetable and potato crops and delays to, and possible loss of, autumn sown crops. Once germinated, however, arable crops can endure short-duration flooding during the dormant winter period. Grassland can withstand some long-duration winter flooding, but it is sensitive to persistent flooding in spring and summer.

The seasonal distribution of flooding within a given year can, therefore, affect the expected annual cost of flooding. Table 9.5 contains estimates of flood probabilities by region in England and Wales based on available hydrological data (Roca *et al.*, 2010). Overall, approximately 76 per cent of floods are estimated to occur in winter (October–February). Smaller catchments (<25km^2), however, tend to have a more equal seasonal distribution of flooding, with approximately 60 per cent of floods occurring in winter and 40 per cent in summer. It is noted that the incidence of spring and summer flooding appears to have increased since the late 1990s due to extreme rainfall events, possibly associated with climate change.

Table 9.5 Estimates of the monthly distribution of flooding by region in England and Wales

Month	Anglian	Midlands, south and north-east	North-west, south-west and Wales	Thames	All
January	18%	17%	18%	13%	17%
February	13%	12%	12%	10%	12%
March	11%	9%	9%	8%	9%
April	8%	7%	5%	7%	6%
May	5%	4%	3%	7%	4%
June	4%	4%	2%	6%	3%
July	3%	3%	2%	6%	3%
August	3%	4%	3%	6%	4%
September	3%	5%	4%	6%	4%
October	7%	9%	10%	9%	9%
November	10%	11%	13%	10%	12%
December	14%	16%	19%	11%	16%

Source: Roca *et al.*, 2010
Note: Due to rounding, some percentages will not sum to 100.

9.3.5 Flood duration and season

The duration of flooding has a major influence on impacts, although this varies between crops and time of year in accordance with crop growth cycles. Research and empirical evidence suggest that most crops can recover from floods of less than one week during their growing periods, although yields are likely to be reduced. Flooding of one week or more results in much higher-yield penalties, possibly to the point where harvesting is not worthwhile (Table 9.6). Flooding between four and six weeks before harvesting can result in total crop loss, especially if crops are completely inundated. Flooding, no matter how short, of crops where quality rather than yield is crucial is likely to lead to complete loss of saleable yield. This applies to most fresh produce, such as salad crops and potatoes.

Grassland is tolerant of relatively long-duration flooding in winter of up to approximately one month, and longer for some flood-tolerant grass species. During spring and summer growth periods, when flood water temperatures rise and anoxic conditions quickly develop, 'improved' grassland can be severely damaged by flooding in excess of two weeks. The spring 2012 floods in Somerset bear witness to this, with complete loss of grass swards after flooding of two to four weeks in May (Morris and Brewin, 2013). Survey evidence of yield loss from the 2007 June and July floods in England confirmed the validity of the estimates in Table 9.6 (Posthumus *et al.*, 2009).

Long-duration flooding may have consequences for farming in subsequent years, especially if cultivations are delayed and crops require rescheduling. Long-duration and repeat flooding in Somerset in 2012, requiring reseeding in some areas, was estimated to reduce grass yields by at least 25 per cent in the following year (Morris and Brewin, forthcoming).

Table 9.6 Yield loss due to flooding of more than one week duration at various times of year as percentage of unflooded yield

Month	Yield loss as % of unflooded control						
	Crops*					Grass**	
	Winter cereals	Spring cereals	Roots	Oil seed rape	Spring peas	Grazed	Cut
	% loss	% loss	% loss	% loss	% loss	% loss	% loss
October	100 (30)ǂ	0	50	100	0	1 (0)	1 (0)
November	10 (0)	0	0	10 (0)	0	1 (0)	1 (0)
December	10 (0)	0	0	10 (0)	0	0	0
January	10 (0)	0	0	15 (0)	0	0	0
February	10 (0)	20 (13)	5	15(0)	0	0	0
March	10 (0)	60 (40)	10	15(0)	50 (5)	4 (2)	8 (2)
April	28 (19)	30	50	30	55 (15)	9	29 (10)
May	50	55	100	50	73 (42)	7	55
June	80	80	100	80	100	5	40
July	100	100	100	100	100	5	6
August	100	100	100	100	50	4	4
September	0	50	100	60	0	2 (0)	3 (0)

Notes: Figures in parentheses show yield losses for floods of less than one week where different; * Reduction in expected harvested yield in a year without floods; ** Reduction in expected annual yields of grass dry matter and energy from grass; ǂ Reseeding with a spring crop.

9.3.6 Water depth and quality

Water depth is positively correlated with crop damage, but not to the same degree as the flooding of built property. The impact of flood depth varies with seasonality: standing crops are more susceptible to deeper floods as they near the time of harvest. The turbulence of flood water can also exacerbate damage to crops, land surfaces and field infrastructure. Contamination by polluted water and deposited sediments from a variety of sources can render food crops unsuitable for harvest. In the 2007 floods, up to 50mm depth of sediment was deposited on Tewkesbury Ham, Gloucestershire, and poisoning of sheep by contaminated river silts from old industrial river depositions was reported in Oxfordshire. Flooding by nutrient-rich waters is associated with damage to species-rich natural grasslands (Posthumus *et al.*, 2010). Flooding in spring 2012 in Somerset of between two to four weeks resulted in anoxic conditions that severely damaged pastures, with loss of grass for the rest of the season and reseeding required in some areas (Morris and Brewin, forthcoming). Additional weights could be applied to the loss factors in Table 9.6 to allow for additional depth and contamination effects, if appropriate.

9.3.7 Multiple flooding

The effect of multiple, successive floods depends on the seasonal timing and duration of flooding, and the sensitivity of crops and their ability to recover. A long-duration flood in the main growing season would most likely destroy arable crops, such that subsequent floods within the farming year have limited effect. Successive floods could, however, limit the options for remedial action, including sowing of replacement crops, thereby increasing overall losses. Complete loss of yield is likely if successive floods occur closely together because crops have little chance to recover.

There is limited research evidence on the impacts of multiple floods (reviewed in Morris *et al.*, 1984). Work on the impacts of waterlogging on arable crops suggests that the cumulative impact of successive excess water events is probably at least additive. Experimental work in the 1980s, for example Cannell *et al.* (1980), showed that three differently timed waterlogging events on three separate plots of cereals each reduced yields by 15 per cent. When they occurred in succession on the same plot, overall yield loss was 46 per cent: more or less additive.

On grassland, much depends on the management system. For grazing where grass is continually cropped, short-duration multiple flooding is likely to be additive in terms of yield loss during the growing season. Where grass is 'closed' for hay/silage, a given flood is likely to have a greater impact on total yield loss than if grazed. The impacts of multiple floods on cut grass may, however, be mutually exclusive within a given cutting period. Once the hay/silage crop has been destroyed by flooding, a subsequent flood in the year may have limited additional impact.

9.3.8 Coastal saline flooding

Flooding by saline water usually causes more damage to crops and soils than flooding by fresh water (FAO,

Table 9.7 Crop yield penalties for coastal flooding

Crop group	Sensitivity to saline flooding	Yield penalty (extra penalty as a % of normal yield relative to fluvial)
Barley and oil seed rape	Tolerant	10%
Sugar beet	Tolerant	10% (including preplanting)
Winter cereals	Moderately tolerant	15%
Spring cereals	Moderately sensitive	10% before planting, 20% after
Spring peas	Moderately sensitive	15% before planting, 20% after
Horticulture and potatoes	Sensitive	30% before planting, 100% after

Source: After Roca *et al.*, 2010

1988). High salt concentrations cause crop stress, restricted growth and complete failure, especially during germination and establishment. Penalties for saline inundation are likely to be between 10 per cent and 20 per cent higher than for fluvial flooding, depending on salt tolerance (Table 9.7). Coastal flooding also tends to be longer duration than fluvial flooding.

Saline residues in soils may affect yields in subsequent years, depending on climatic and soil conditions, possibly ranging from one year on sandy soils to three to four years on clays. Planting salt-tolerant crops such as barley may help, with possibly a reduction in profitability. Remedial applications of gypsum ($CaSO_4 \cdot 2H_2O$) may be required to speed up reclamation. The scale and speed of coastal floods often result in higher livestock fatalities than is the case in fluvial floods.

9.4 ESTIMATING THE FINANCIAL AND ECONOMIC VALUE OF FLOODING ON AGRICULTURE

Flooding can affect the profitability of farm businesses, temporarily for a one-off, infrequent flood, or permanently if there is a persistent increase in flood frequency. Severe one-off events can, however, threaten the financial viability of already vulnerable, small-scale farming businesses. Simultaneously, depending on the scale of the event, flooding can impact on the local and national economy, with consequences for jobs, incomes and value added.

Two main perspectives can be taken to assess the monetary impacts of flooding on agriculture: financial and economic.

9.4.1 Financial appraisal

Financial analysis considers the impacts on the incomes and expenditures of individual farm businesses, expressed in market prices, including some imputed costs, such as extra unpaid family labour charged at market wage rates that may be required to mitigate flood impacts.

Monetary valuation of the performance of agriculture, and the impacts of flooding, can be assessed using accounting conventions and guidance. Tables 9.8 and 9.9 contain indicative examples of financial and economic returns for selected crop and livestock enterprises.

For crops (Table 9.8), gross margins per hectare measure the value of output, less variable costs such as seeds and fertiliser. Variable costs are directly related to each unit of activity, and can be avoided if that activity is not pursued. Gross margins show the monetary gain (or loss) associated with one more (or one less) unit of an activity, assuming other so-called 'fixed' resources available to the business, such as regular labour, machinery, buildings and land (and their associated costs), remain unchanged.

Gross margins are usually expressed per unit of limiting resource, such as land. They are useful for measuring the value of changes in the efficiency of individual enterprises as determined by revenues (yields and output prices) and variable costs (input use and input prices), and showing the advantage of substituting one enterprise for another, other things remaining fixed. For livestock enterprises (Table 9.9), it is usual to estimate gross margins per head of livestock and then per hectare according to the typical number of livestock per hectare.

UK farm management accounting conventions treat fixed cost items separately on the assumption that they are not easy to allocate to individual enterprises, are deemed not to vary directly with changes in individual enterprise size, and are unavoidable in the short term. In practice, however, some elements of 'fixed' cost behave more like the 'variable' costs used in the definition of the gross margin. These include inputs which can vary directly with scale of operation, such as labour for milking cows and potato harvesting, machinery fuels and repairs, and the use of contractors for farm operations.

It may be appropriate, therefore, to treat these as 'semi fixed' costs per head or hectare and charge them accordingly in livestock and crop budgets when considering the impact of marginal changes. Total fixed costs, by comparison, reflect overall average labour, machinery and buildings costs per unit of activity, including depreciation of capital assets, such as machinery and buildings. Thus gross output, gross margin, net margin after semi fixed costs and net margin after total fixed costs are, each in their own way, indicators of the monetary value of farm enterprises. They can be used to describe differences between alternative flood risk management options. Which one gives the best estimate depends on the situation and the type of change involved, as discussed later.

It is noted that while the gross margins for selected crop and livestock enterprises in Tables 9.8 and 9.9 are positive, some net margins appear to be negative once total fixed costs are deducted. Two points should be made here. First, it is more appropriate to charge fixed costs against the sum total of gross margins at the farm scale rather than against each enterprise: crops and livestock enterprises form part of a farming system involving crop rotations and internal transfers of livestock. Second, farmers receive an annual Single (Farm) Payment under the EU Common Agricultural Policy, currently around £200/ha for land previously eligible for subsidies. In return, they are required to comply with specified farming practices and maintain land in 'good agricultural and environmental condition'. This payment, 'decoupled' from support for production, is not included here and is assumed to be unaffected by any change in flooding regime.

Farmers may also receive payments under the Agri-Environment Schemes (AESs), supported by national rural development programmes with EU funding from the European Agricultural Fund for Rural Development, part of the Common Agricultural Policy (Natural England, 2009). AESs are voluntary agreements that pay farmers and other land managers to manage their land in an environmentally sensitive way. In 2012, approximately 60 per cent of agricultural land in England was registered under basic, entry-level agreements (paid at c. £30/ha/year), and a further 7 per cent was part of the Higher LEVEL Stewardship (HLS) scheme (Natural England, 2010) that targets particular environmental outcomes. HLS agreements operate in many extensively farmed floodplain areas in England and Wales, such as the Somerset Moors and Levels, prescribing fertiliser use, stocking rates, grazing and grass cutting dates, and in some cases ditch and water management levels. Payments range from approximately £50/ha to £400/ha, based on estimates of costs incurred and farming income forgone. The incentives and rewards to farmers, and the benefits to society, that environmental protection provides need to be considered in the financial and economic appraisal of flood risk management options in rural areas.

At the time of writing (February 2013), farming prospects are relatively buoyant. Following spikes in 2007/08, agricultural commodity prices have remained approximately 50 per cent higher than their 2005 levels, although farm profits have been squeezed by real increases in input costs, particularly for fossil-based fuels and fertilisers. Livestock farmers have been negatively affected by high feed prices. Fresh produce and vegetable prices have not increased at the same rate as bulk commodities such as grains and oil seeds such that some of the comparative advantage of intensive farming has been lost.

Since the last edition of the MCM (Penning-Rowsell *et al.*, 2005a), there has been further structural adjustment in the sector as a result of attempts to protect profit margins by reducing labour costs, greater use of contractors with specialist, large-capacity equipment, increased scale or specialisation of enterprises, and focus on niche markets (Angus *et al.*, 2009). There is continued polarisation into large-scale commercially oriented farms and small 'lifestyle' farms where farming is not the dominant purpose or source of income. International and national concerns about food security, combined with bullish markets in agricultural commodities, have, however, reinvigorated interest in farming as a business.

For these reasons, current estimates of farm revenues and costs may not be a reliable basis for future predictions, and they should be kept under continual review. Broadly speaking, in the absence of a major change in government policy, UK agriculture largely 'takes' its prices from international commodity markets. In this context, the financial and indeed economic prospects for agriculture are now firmly set in a global context, currently characterised by relatively strong market demand for food and fibre on the one hand and variable supplies due to climatic and political uncertainties on the other, and the volatility that this implies.

9.4.2 Economic appraisal

Economic analysis (usually building on a financial analysis) takes a broader view of impacts on the local and national economy as a whole, including indirect effects, such as impacts on businesses beyond the flood area, and on non-market goods, such as wildlife and the natural environment. Economic analysis uses economic rather than financial prices to reflect the real cost or benefit to society of resource use or loss of output. Market prices are adjusted to remove the distortion effects of taxes, subsidies and market interventions by government, such as price support under the Common Agricultural Policy. Economic

Table 9.8 Indicative financial and economic gross margins and net margins for selected arable crops

	£ 2013 values	Unit	Winter wheat	Oil seed rape	Spring wheat/barley	Beans	Sugar beet	Potatoes	Extensive arable [1]	Intensive arable [2]
a	Yield	t/ha	9	3.8	5.9	4.5	77	50		
b	Price	£/t	150	350	160	190	31	150		
c	Gross output (a x b)	£/ha	1,350	1,330	944	855	2,387	7,500	1,297	2,572
d	Variable cost	£/ha	480	430	330	260	985	3,000	448	994
e	Gross margin (c – d)	£/ha	870	900	614	595	1,402	4,500	849	1,578
	Fixed costs									
	Semi fixed									
	Labour	£/ha	83	81	90	68	234	270	81	140
	Machinery	£/ha	125	115	125	115	290	290	122	181
	Buildings and utilities	£/ha	25	20	20	20	40	40	24	30
	Other	£/ha	18	18	18	18	19	19	18	18
f	Sub total	£/ha	251	234	253	221	583	619	245	370
	Total fixed costs[3]									
	Labour	£/ha	230	230	230	230	350	350	230	350
	Machinery	£/ha	300	300	300	300	360	360	300	360
	Buildings and utilities	£/ha	65	65	65	65	90	90	65	90
	General expenses	£/ha	90	90	90	90	95	95	90	95
g	Total fixed costs	£/ha	685	685	685	685	895	895	685	895
	Financial returns									
	Gross margin (e)	£/ha	870	900	614	595	1,402	4,500	849	1,578
	Net margin [4]									
h	After semi fixed costs (e – f)	£/ha	619	666	361	374	819	3,881	604	1,208
i	After full fixed costs (e – g)	£/ha	185	215	–71	–90	507	3,605	164	683

	Economic returns (Defra PAG3, Scenario II: one-off loss)									
	Economic adjustment		none	none	none	remove sub [5]	treat as wheat [6]			
j	Adjusted gross margin (e or wheat)	£/ha	870	900	614	561	870	870	845	845
	Economic returns (Defra PAG3, Scenario III: permanent loss)									
	Adjusted gross margin (j)	£/ha	870	900	614	561	870	870	845	845
	Adjusted net margin									
k	After semi fixed costs (j – f)	£/ha	619	666	361	340	619	619	601	601
l	After full fixed costs (j – g)	£/ha	185	215	–71	–124	185	185	160	160

Notes:

[1] Extensive arable: wheat 70 %, oil seeds 20 %, beans and other 10 %.
[2] Intensive arable: wheat 66 %, sugar beet 17 %, potatoes and field vegetables 17 %, fixed costs at higher rate across all farmed area.
[3] Excluding land rents (or equivalent) that are omitted from economic analysis. Assumes average fixed costs and net margins per ha by farming system, either mainly cereals or general cropping, including sugar beet and potatoes.
[4] Arable farmers receive about £200/ha single payment income support for land previously eligible for production subsidies.
[5] Small protein subsidy at £34/ha, under review.
[6] Potatoes, sugar beet and vegetables treated as a wheat crop; see discussion.

Regional and local estimates vary according to circumstances and practices; winter barley gross margins are about 80 % of those of winter wheat.

Sources: Nix, 2012; ABC, 2012; Farm Business Survey, 2012; Defra farm management statistics, available at: https://www.gov.uk/government/organisations/department-for-environment-food-rural-affairs/about/statistics

Table 9.9 Indicative financial and economic gross margins and net margins for selected livestock enterprises

	£ 2013 values		Dairy cows	Beef cows [1]	Beef cattle [2]	Sheep	Beef and sheep [3]
a	Gross output	£/head	1,750	490	600	105	
b	Variable costs [4]	£/head	725	260	200	50	
c	Gross margin (a – b)	£/head	1,025	230	400	55	
d	Typical stocking rates: head per ha		2	2.1	3.2	11	
	Financial returns per hectare						
e	Gross output (a x d)	£/ha	3,500	1,029	1,920	1,155	1,368
f	Gross margin (c x d)	£/ha	2,050	483	1,280	605	789
	Fixed costs [5] semi fixed						
	Labour	£/ha	238	100	156	147	134
	Machinery	£/ha	207	104	104	104	104
	Buildings and utilities	£/ha	69	26	26	26	26
	General expenses	£/ha	18	11	11	11	11
g	Sub total	£/ha	532	241	297	288	275
	Total fixed costs						
	Labour	£/ha	640	380	380	380	380
	Machinery	£/ha	460	230	230	230	230
	Buildings and utilities	£/ha	210	80	80	80	80
	General expenses	£/ha	90	55	55	55	55
h	Total fixed costs	£/ha	1,400	745	745	745	745
	Financial net margin per hectare						
i	After semi fixed costs (f – g)	£/ha	1,518	242	983	317	514
j	After full fixed costs (f – h)	£/ha	650	–262	535	–140	44
	Economic returns (Defra PAG3, Scenario II: one-off loss)						
	Economic adjustment		wheat [6]	none	none	none	none
k	Adjusted gross margin	£/ha	870	483	1,280	605	789
	Economic returns (DEFRA PAG3, Scenario III: permanent loss)						
	Economic adjustment		wheat [6]	none	none	none	none
	Adjusted gross margin (k)	£/ha	870	483	1,280	605	789
	Adjusted net margin						
l	After semi fixed costs (wheat or i)	£/ha	619	242	983	317	514
m	After full fixed costs (wheat or j)	£/ha	185	–262	535	–140	44

Notes: [1] Beef cows: single sucklers, mix of autumn and spring born calves. [2] Beef cattle: finishing suckled calves on grass/silage, usually combined with suckler cows. It is noted that beef systems vary considerably: returns are lower for seasonal fattening of 'store' cattle. [3] Beef and sheep: beef cows, beef cattle and sheep in equal proportions by area, representative of extensive grassland. [4] Variable costs include average forage costs such as fertilisers on grass. [5] Land rent or land purchase costs are omitted from economic analysis. [6] Dairy grassland area is treated as equivalent to a wheat crop.

Some rounding errors; regional and local estimates vary according to circumstances and practices.

Sources: Based on Nix, 2012; ABC, 2012; Farm Business Survey, 2012; Defra farm management statistics, available at: https://www.gov.uk/government/organisations/department-for-environment-food-rural-affairs/about/statistics

damage to fixed assets, such as plant and machinery, is measured at loss of remaining value, rather than the financial costs of new replacements that may be required (or reimbursed by insurance). Furthermore, some loss of output and income suffered by businesses subjected to flooding may be substituted by others not affected by flooding, such that total economic loss is less than financial loss.

With respect to the valuation of agricultural commodities, since the 'decoupling' of farm income support and production subsidies, there is much better alignment of financial and economic agricultural commodity prices. Although some remaining import tariffs (on some livestock products) and quota restrictions (for sugar beet, for example) imply a degree of subsidy to UK farmers, most agricultural commodity prices are now based on international border prices.

Economic analysis requires valuation based on remaining value rather than replacement cost. With respect to impacts on assets, over 90 per cent of impacts of infrequent flood events (as shown by the 2007 floods) relate to output and cash flow items, rather than capital items that might have lower remaining economic value compared with their replacement cost (Posthumus *et al.*, 2009). Little adjustment between financial and economic costs is required for this purpose.

Economic analysis also requires that adjustments are made for 'additionality', 'displacement' and 'substitution' effects (HM Treasury, 2011). In the case of flooding, for example, damage cost could be attenuated by offsetting loss of output in flood areas by additional production in non-flood areas, either temporarily or permanently. For this reason, financial costs borne by private businesses (based on loss of sales) could exceed economic costs (based on loss of uncompensated production costs) on the assumption that production is made good from other sources.

With respect to additionality, much depends on the flood scenario, whether the assessment is considering infrequent events whose precise timing cannot be predicted, or some permanent and predictable change in exposure to flooding (as discussed below).

For infrequent, unpredictable events, there is limited opportunity to substitute lost agricultural production within the same production cycle/harvest period that is usually at least one year for most crops and many livestock systems. It is difficult for other UK farmers to step in to substitute for lost output. Producing more potatoes or wheat elsewhere to offset a spring or summer flood in the same year is not possible. The same is probably true for dairy production as dairy farmers are usually committed to target breeding, feeding and yield regimes, at least in the short term. There may be more scope for short-term production substitution in livestock fattening, assuming other farmers can quickly increase production to fill gaps by carrying more or heavier stock. Here it may be possible that a share of loss of sales could be made up by others, most likely at a marginally higher cost than the producers impacted in the flood. Evidence from the Somerset floods (Morris and Brewin, forthcoming), however, suggests that livestock farmers, where possible, sought to replace lost feed from grass in order to maintain sales of milk and fatstock, thus limiting flood costs to loss of production inputs rather than loss of outputs. So, again, limited adjustment is required to transpose financial into economic values.

If the flood impacts are significant at the regional or national scale, there would need to be some redirection of trade, or sales from store. Prices may rise to moderate demand, but in the short term any supply deficits would need to be met by imports valued at import/export parity prices (net of transport costs). Scope for selling from store is now much reduced since the decline of government intervention buying and buffer stocks. Here, again, loss of output valued at prevailing financial prices will provide a reasonable estimate of economic costs.

Where flooding at the national scale disrupts aggregate national supply, prices are likely to rise and this should be considered in assessing economic impacts of flooding. The 2007 floods affected approximately 1.6 per cent of the cereal area, with an average yield loss per hectare of approximately 40 per cent. This suggests a reduction of *c.*0.6 per cent in aggregate supply, probably insufficient in itself to affect aggregate supply and prices (Posthumus *et al.*, 2009). However, international and UK cereal prices doubled in 2007 due to shortages at the global scale and speculative commodity trading. The disruption to supplies caused by flooding exacerbated the volatility of already nervous markets.

It is noted that national cereal yields in 2012 were estimated to be approximately 15 per cent lower than expected due to climatic conditions during the year. Wet weather and waterlogging played a major part in this, rather than surface flooding. (In 2012, surface flooding on agricultural land was not as widespread as it had been in 2007, when 42,000ha were flooded.) The aggregate effect of saturated soils (causing a large share of the 15 per cent reduction) probably accounted for the majority of recorded yield loss, and probably greater in total than the £50 million attributed to surface flood damage costs in the summer 2007 event. This reinforces the importance of land drainage for agriculture, and the difficulty of separating flooding and waterlogging impacts in some areas.

Where permanent changes in flood frequency lead to changes in land use, there is an economic argument for adjusting for displacement effects. Here, increased flooding renders high-value cropping of horticultural, field vegetable and potato crops infeasible, as well as intensive grassland for dairy. It can be argued that potentially profitable crop and livestock production will move to flood-free areas, whether grown by the same or different farm businesses. Thus, there will not be a loss of output as such, although there may be some adjustment to value added if yields are lower and production costs are higher in the new location (which might be expected if enterprises are 'forced' to move). Commodities subject to government quota, such as sugar beet, and those tied to particular niche market outlets, such as organic produce, are also likely to move in the longer term. Defra (2008a) appraisal guidance assumes that high-value farming will relocate to other areas, displacing wheat as the most common agricultural land use in the process. For economic analysis, the displacement of high-value cropping due to a permanent increase in flooding, it is argued, should be valued in terms of the gross and net margins of wheat. These are shown in Tables 9.8 and 9.9 and further explained below.

Defra guidance, however, does recognise that the presumption of substitutability may not apply in all cases, especially for:

- large-scale schemes of more than 10,000ha, and smaller areas with particularly important agricultural assets, where changes in agricultural production and incomes could have economic implications at the regional or national scale, including national food security; and
- agriculturally less favoured areas where there could be significant impacts on vulnerable farming communities and local economies.

There is growing concern that the frequency of extreme weather events that cause flooding is likely to increase due to climate change. Indeed, more frequent and widespread flooding and waterlogging may call for a review of Defra's current guidance, which assumes that agricultural production permanently lost in one place can be readily offset by relocation elsewhere at little economic cost. This suggests the need for a strategic assessment of flood impacts on agriculture and food supply at regional and national scales. Furthermore, changes in the income support, compliance requirements and environmental incentives under the Common Agricultural Policy regime scheduled for 2013–20 may have implications for economic appraisal, but these new arrangements have yet to be worked out.

9.5 ESTIMATING THE AGRICULTURAL DAMAGE COST OF A SINGLE FLOOD EVENT

9.5.1 Estimating flood costs at the farm scale

Based on experimental research mainly undertaken in the 1960s and 1970s (reviewed in Morris *et al.*, 1984) and farm surveys conducted over the last thirty years (Morris and Hess, 1986; Morris, 1992; Dunderdale and Morris, 1997a; 1997b; Hess and Morris, 1988; Posthumus *et al.*, 2009; Morris and Brewin, forthcoming) of the impacts of flooding and associated agricultural drainage conditions, a general formula can be used to express the cost of a single flood event at the farm scale.

Equation 9.1

$$\text{FLOOD}_{\text{FARM}} = \text{ARABLE} + \text{GRASS} + \text{LIVESTOCK} + \text{OTHER}$$

Where:

- $\text{FLOOD}_{\text{FARM}}$ = cost of agricultural flood damage at the farm scale (£)
- ARABLE = cost of flood damage to arable (and horticultural) crop production (£)
- GRASS = cost of flood damage to grass production (£)

- LIVESTOCK = cost of flood damage to livestock production (£)
- OTHER = miscellaneous costs (£)

The ARABLE and GRASS components are best estimated per hectare flooded and grossed up according to respective areas at the farm scale. LIVESTOCK and OTHER costs are best estimated at the farm scale, and then averaged per hectare where they apply. $FLOOD_{FARM}$ costs can be aggregated for all agricultural areas affected by flooding. A summary of the components of $FLOOD_{FARM}$ is shown in Box 9.1.

BOX 9.1 FLOOD DAMAGE COSTS AT FARM SCALE

$$FLOOD_{FARM} = ARABLE + GRASS + LIVESTOCK + OTHER$$

Where:

- $FLOOD_{FARM}$ = cost of agricultural flood damage at the farm scale (£)

ARABLE and GRASS are best estimated per hectare flooded and grossed up according to respective areas at the farm scale. LIVESTOCK and OTHER are best estimated at the farm scale. $FLOOD_{FARM}$ costs can be aggregated for all agricultural areas affected by flooding.

ARABLE flood damage costs

$$ARABLE = CROPLOSS_{ARABLE} + INPUTS_{ARABLE} - HARVEST_{ARABLE} + LAND_{ARABLE} - REPLACE_{ARABLE} + RESIDUAL_{ARABLE}$$

Where:

- $CROPLOSS_{ARABLE}$ = reduced yield (t/ha) and quality at market prices (£/t) in the year of the flood (£/ha)
- $INPUTS_{ARABLE}$ = cost of additional agro-chemicals less savings in uncommitted costs (£/ha)
- $HARVEST_{ARABLE}$ = additional harvesting costs less savings in uncommitted harvest costs (£/ha)
- $LAND_{ARABLE}$ = cost of land restoration on arable land (cultivations, treatments) (£/ha)
- $REPLACE_{ARABLE}$ = value of output of replacement crop less costs of crop establishment and production (fertiliser, chemicals, labour and machinery), where relevant (£/ha)
- $RESIDUAL_{ARABLE}$ = carry-over impacts on yield and other costs to subsequent year(s) (£/ha)

GRASS flood damage costs

$$GRASS = FORAGE + INPUTS_{GRASS} - HARVEST_{GRASS} + LAND_{GRASS} + RESEED_{GRASS} + RESIDUAL_{GRASS}$$

Where:

- FORAGE = loss of energy based on estimates of forage dry matter (t/ha) and energy value (MJ/t) by grassland type and use, namely hay, silage and grazing – valued at barley feed equivalent (£/ha)
- $INPUTS_{GRASS}$ = cost of additional agro-chemicals on grass less savings in uncommitted costs (£/ha)
- $HARVEST_{GRASS}$ = net savings in forage harvesting costs (£/ha)
- $LAND_{GRASS}$ = cost of land restoration on grassland (£/ha)
- $RESEED_{GRASS}$ = cost of grass reseeding (seeds, fertiliser, labour and machinery) (£/ha)
- $RESIDUAL_{GRASS}$ = carry-over impacts on yield and other costs to subsequent year(s) (£/ha)

LIVESTOCK flood damage costs

LIVESTOCK = MOVE + HOUSING + FEED + SERVICES + MORTALITY + SALES

Where:

- MOVE = cost of labour and machinery to relocate livestock (£)
- HOUSING = cost of additional labour, machinery and buildings for housing of livestock (£)
- FEED = additional costs of conserving (and storing) feed for housed stock plus costs for purchased feeds over and above estimated forage losses (£)
- SERVICES = additional waste management, water and veterinary services associated with housed livestock (£)
- MORTALITY = cost of livestock morbidity and mortalities due to flooding (£)
- SALES = loss of value from forced sales (or reduced purchases) of fatstock and reduced milk sales due to flooding (£)

OTHER farm costs

OTHER = BUILDINGS + UTILITIES + MACHINERY + FIELD INFRASTRUCTURE + DEBRIS + FARM SERVICES + FINANCING + SUNDRY

Where:

- BUILDINGS = damage to farm structures, buildings and contents (£)
- UTILITIES = disruption and replacement of essential farmstead services, e.g. power and water (£)
- MACHINERY = damage and/or replacement of farm machinery and equipment, including irrigation and drainage equipment (£)
- FIELD INFRASTRUCTURE = damage to hedges, fences/gates, land drainage works, tracks (£)
- DEBRIS = clean-up and debris removal and disposal (£)
- FARM SERVICES = loss of net revenue from services, e.g. contracting (£)
- FINANCING = additional borrowing costs (£)
- SUNDRY = other farm-specific costs, emergency pumping, removals (£)

(Based on Morris *et al.*, 1984; Morris and Hess, 1988; Morris, 1992; Dunderdale and Morris, 1996a; 1996b; 1996c; Morris *et al.*, 2008; Posthumus *et al.*, 2009; 2010; Morris and Brewin, forthcoming)

9.5.2 Flooding on arable land (ARABLE)

Flooding can result in a loss of the expected crop yield and possibly the quality of the remaining crop, measured at market prices. It can also result in the 'loss' of the value of variable inputs already committed, such as fertilisers and seeds, but these already-committed costs can be ignored if yield reduction is being used to assess loss. There may, however, be some 'savings' in other scheduled but as yet uncommitted costs, such as harvesting. Flooding may also require additional expenditure associated with damage mitigation, restoration and clean-up, all of which should be accounted for.

The constituents of arable flood damage costs are given in Equation 9.2.

Equation 9.2

$$ARABLE = CROPLOSS_{ARABLE} + INPUTS_{ARABLE} - HARVEST_{ARABLE} + LAND_{ARABLE} - REPLACE_{ARABLE} + RESIDUAL_{ARABLE}$$

The components of Equation 9.2 are defined in Box 9.1 and discussed below.

Reduced arable yields ($CROPLOSS_{ARABLE}$)

A general statement for yield and quality reduction due to flooding on a standing arable crop in a given month is given by Equation 9.3.

Equation 9.3

$$CROPLOSS_{ARABLE} = YL_iP + Q$$

Where:

- Y = expected (unflooded) yield (t/ha)
- L_i = average yield loss from a flood i (fraction)
- P = market price of crop commodity (£/t)
- Q = losses due to reduced quality of (remaining) sales, if appropriate (£/ha)

Expected yields for non-flood conditions can be obtained from published sources, adjusted according to local soils and climate (Nix, 2012; Farm Business Survey, 2012). As previously explained, yield losses due to flooding vary according to crop type and the time of the flood relative to critical crop growth stages and critical field operations (Table 9.6).

Estimates of yield loss can be made for a single flood occurring in a particular month, if known, or for a single flood occurring in a given year by weighting the relative probability of flooding in each month, as shown in Equation 9.4.

Equation 9.4

$$\bar{L} = \sum_{i=12}^{i=1} L_i P_i$$

Where:

- $\bar{L}$ = average annual yield loss from a single flood occurring in a year
- L_i = yield loss from a single flood occurring in month i (fraction)
- P_i = relative probability of a flood in month i

Estimates of flood distribution by month (Table 9.4) can be combined with estimates of yield loss by month (Table 9.6) to derive estimates of expected damage for a single flood event occurring in a year (Table 9.10).

Using a distribution for monthly flooding for all England and Wales from Table 9.10, for example, gives an average annual yield loss for a crop of winter cereals of 29 per cent of expected yield. At 9t/ha and £150/t, for example, this equates to an average annual loss for a single flood occurring in a given year of approximately £390/ha for yield loss alone, before other adjustments.

Loss of crop quality (Q)

Flooding early in the season can affect the quality of the remaining harvested yield. Vegetables may be blemished or disfigured due to stress, and variations in grain quality may result in failure to meet quality standards for delivery contracts. As a result, affected produce may attract lower prices. In 2007, farmers reported that quality reductions in harvested crops added a further 10 per cent to overall yield losses at the farm scale. A cautious estimate of early season flood impacts on the quality of remaining yields are: 10 per cent reduction on price for remaining cereal yields and 5 per cent reduction for early season (non-destructive) flooding of horticultural and vegetable produce (see Table 9.11).

Extra production costs (excluding harvesting) ($INPUTS_{ARABLE}$)

Flooding may result in additional costs associated, for example, with reseeding of crops lost immediately after establishment, additional inputs such as replacement or additional fertiliser, and spray treatments, disposal of damaged crops, additional cultivations for land restoration, or application of gypsum to ameliorate saline flooding. There may be some savings in uncommitted input costs. Table 9.10 contains

Table 9.10 Estimated costs of flooding for a period of one week or more on arable crops by month of occurrence and expected weighted costs of a single flood event occurring within a year for monthly flood distributions in 'all England and Wales'

	Month of year												
	1	2	3	4	5	6	7	8	9	10	11	12	
Seasonality	17%	12%	9%	6%	4%	3%	3%	4%	4%	9%	12%	16%	
Yield loss													Weighted loss, %
Oil seed rape	15%	15%	15%	30%	50%	80%	100%	100%	60%	100%	10%	10%	**33%**
Roots and hort	10%	10%	10%	50%	100%	100%	100%	100%	100%	50%	25%	25%	**36%**
Spring cereals	0%	20%	60%	30%	55%	80%	100%	100%	50%	0%	0%	0%	**23%**
Spring peas	0%	0%	50%	55%	73%	100%	100%	50%	0%	0%	0%	0%	**19%**
Winter cereals	10%	10%	10%	28%	50%	80%	100%	100%	0%	100%	10%	10%	**29%**
Extra production costs													Weighted loss, £/ha
Oil seed rape	49	49	49	49	0	55	55	55	104	150	49	49	**58**
Roots and hort	0	0	99	234	270	270	270	270	270	270	135	0	**112**
Spring cereals	0	0	59	59	59	55	55	55	55	0	0	0	**19**
Spring peas	0	0	49	49	49	55	55	55	55	0	0	0	**17**
Winter cereals	59	59	59	59	0	55	55	55	55	175	59	59	**66**
Harvest costs*													Weighted loss, £/ha
Oil seed rape	0	0	0	−27	−45	−72	−90	−90	−54	−90	0	0	**−22**
Roots and hort	0	0	0	0	−300	−300	−300	−300	−300	−150	−75	−75	**−89**
Spring cereals	0	0	−48	−24	−44	−64	−80	−80	−40	0	0	0	**−17**
Spring peas	0	0	−43	−47	−62	−85	−85	−43	0	0	0	0	**−16**
Winter cereals	0	0	0	−25.2	−45	−72	−90	−90	0	−90	0	0	**−20**

Note: * Negative indicates saving in cost.

Table 9.11 Estimated indicative costs of a single flood of more than one week occurring in a year by type of crop, weighted by monthly probabilities of flooding

Crop type	Yield loss * £/ha	Quality loss ** £/ha	Extra production and restoration costs *** £/ha	Savings in harvest costs	Total ‡ £/ha
Horticulture *	3,500–4,500	700–900	150	−153	4,340–5,400
Roots: potatoes	2,450–2,890	430–506	106	−115	2,870–3,390
Roots: sugar beet	720–833	63–73	170	−63	890–1,020
Winter cereals	383–420	47–52	66	−20	480–520
Oil seed rape	376–492	38–50	58	−22	450–580
Field beans	217–272	27–34	25	−18	250–310
Spring cereals	163–192	27–31	19	−17	190–230
Spring peas	97–120	32–43	17	−16	130–170

Notes: * Yield losses based on Table 9.10 with range reflecting yield for ALC grade 1 land (highest in range) and ALC grade 2 land (lowest). Horticulture yield losses assumed at 1.5 potato losses; ** Quality losses assume 5 % reduction in price on remaining yield for all crops except horticulture and potatoes at 10 % of remaining yield; *** Restoration costs include replacement fertilisers, extra cultivations, reseeding and removal of damaged crop if required; ‡ Rounded to nearest 10 units.

estimates of changes in costs that vary according to crop type and the seasonal timing of the flood.

Changes in harvest costs (HARVEST$_{ARABLE}$)

Flooding may affect harvest and post-harvest costs. There may be 'savings' in harvesting and post-harvest costs, such as storage and drying costs, if a large proportion of the crop is lost, such as reduced combine-harvesting costs for cereals and oil seeds, and reduced picking, grading and marketing for potatoes, field vegetables and horticultural produce. In some cases costs may increase because damaged yet recoverable crops may be more expensive to harvest and to dry and store. The negative sign in Equation 9.2 indicates that net savings in harvest cost should be subtracted to derive an estimate of ARABLE flood costs.

Land restoration costs (LAND$_{ARABLE}$)

Land restoration may be needed, including removal of damaged crops, additional cultivations or treatments to neutralise contaminants, including application of gypsum to ameliorate saline flooding.

Replacement crop (REPLACE$_{ARABLE}$)

It may be possible to replace damaged crops if flooding occurs early in the crop production cycle. For example, a damaged winter-sown crop could be replaced by a spring-sown crop, thereby reducing the overall loss. The additional revenue from the replacement crop, net of additional costs, is subtracted to derive an estimate of ARABLE flood costs. Arable crops will be replaced only if a net gain is expected, hence the negative sign in Equation 9.2.

Residual costs (RESIDUAL$_{ARABLE}$)

Severe, long-duration flooding, particularly if accompanied by damage to soils due to wet conditions or saline inundation, may have a knock-on effect into subsequent years. Yield penalties, reduced cropping options or additional costs may result.

Total flood event costs on arable areas (ARABLE)

Using the preceding estimates, Table 9.11 summarises the estimated costs of a single annual flood, weighted by the monthly distribution of flooding for all England and Wales. The main costs are usually associated with loss of marketable yield as extra production costs and savings in harvest costs tend to cancel each other out. The range in cost estimates reflects yield differences on agricultural land classification grade (ALC 1 highest and ALC 2 lowest in the range).

9.5.3 Flooding on grassland (GRASS)

The impacts of flooding on grassland are diverse. Flooding can reduce the feed energy value of grassland by restricting or destroying grass production on grazing pastures and grass areas prepared for the conservation of winter feed. Lost feed must be replaced from other sources if livestock numbers and production are to be sustained. (The impacts of flooding on livestock production are considered separately below.)

The general formula for estimating the costs of flooding in grassland is given in Equation 9.5.

Equation 9.5

$$\text{GRASS} = \text{FORAGE} + \text{INPUTS}_{\text{GRASS}} - \text{HARVEST}_{\text{GRASS}} + \text{LAND}_{\text{GRASS}} + \text{RESEED}_{\text{GRASS}} + \text{RESIDUAL}_{\text{GRASS}}$$

The components of Equation 9.5 are defined in Box 9.1 and discussed below.

Loss of energy from grassland (FORAGE)

The loss of grass energy due to flooding is expressed in Equation 9.6.

Equation 9.6

$$\text{FORAGE} = YL_iP$$

Where:

- FORAGE = value of loss of grass energy (£/ha)
- Y = expected (unflooded) grass energy yield (MJ/ha)
- L_i = yield loss from a flood i (fraction)
- P = value of grass energy yield in terms of replacement feed or silage/hay sale value (£/MJ)

The estimation method assesses the 'normal' productivity of grassland and the way that this is affected either by a one-off flood event (considered here) or possibly by a permanent change in flood frequency.

The assessment of grassland productivity on a given site requires data not available on most farms (i.e. livestock numbers; ages and weights at grass; number of livestock grazing days; live weight gain or milk yields; weight and nutrient value of conserved grass). Using data from secondary sources and from farm surveys in the study area, it is possible to estimate the productivity of grass in terms of energy (MJ/ha) according to drainage condition (as previously defined), grass suitability class (as defined by climatic and site conditions (SSEW, 1984)), fertiliser use and grass conservation and grazing methods. Loss of grass yield can be valued either in terms of the cost of replacement by purchased feed or in terms of the loss of capacity to support grass-fed livestock, and the resultant loss of added value. The former cost-based approach is much simpler and is used here, but it may underestimate losses due to reduced output and sales.

Estimating expected energy from grassland

The first step is to estimate the feed energy value of dry biomass produced from grass. This depends on the suitability of climate, soils and general drainage conditions for grass production, as well as grassland management practices, notably the amount of fertiliser applied and whether grass is grazed by animals or conserved for winter feed.

Dry matter production

Total grass dry matter production can be estimated using information on nitrogen application, summer rainfall and soil available water capacity (Thomas and Young, 1982). The vast majority of floodplain soils in England and Wales fall into the high available water capacity class (>150mm). Therefore, the dry matter production can be estimated as in Equation 9.7.

Equation 9.7

$$Y = aN - bN^2 + c$$

Where:

- Y = dry matter yield, tonne dry matter (DM)/ha
- N = total nitrogen, kg nitrogen (N)/ha
- a, b and c are constants

The formula can be applied using the estimates in Table 9.12 to derive estimated tonne DM/ha. Typical applications, combining organic and chemical fertilisers, are approximately 50–75kg nitrogen (N)/ha for extensive grassland (beef and sheep) and 180–220kg N/ha for intensive grassland (mainly dairy).

Estimated dry matter production of grass can be converted into feed energy at standard rates according to whether it is grazed or conserved for winter feed (Table 9.13). There is considerable variation in practice, affecting quality of grass production and utilisation. Energy from grass can be valued at the cost of replacement feed (Doyle and Elliot, 1983), using feed barley as a straight substitute.

Grassland areas are typically used in equal proportions for both grazing and winter feed. Pastures initially 'closed' for conservation are usually opened up for grazing after grass has been cut. Overall, approximately 54 per cent of annual grass dry matter production is grazed and approximately 46 per cent conserved.

Table 9.12 Constants for grass production estimation based on summer (April–September) rainfall for soils with high (>150mm/m) available soil water

	April–September rainfall (mm) by region		
	<300	300–400	>400
	Anglian	Midlands, Southern, N East and Thames	N West, S West and Wales
Constant			
a	0.0316042	0.03419	0.0344199
b	0.0000252	0.0000296	0.000029
c	2.313	2.63	3.094
N applied	Estimated tDM/ha		
75kgN/ha	4.6	5	5.5
200kgN/ha	7.6	8.3	8.8

Table 9.13 Estimating the feed value of grass energy

	ME value MJ/tDM	Potential utilisation %	UME value MJ/tDM	Value £/tDM at £0.0123/MJ*
Grazed grass	11,500	70	8,885	110
Conserved hay	8,800	80	7,040	87
Conserved silage	10,800	80	8,640	106

Notes:

* Valued at equivalent cost of replacement barley feed: e.g. £162/tDM at 13,200MJ/tDM

ME = Metabolisable energy
UME = Utilisable metabolisable energy
MJ = Mega Joules
tDM = tonne dry matter

Table 9.14 Estimating the cost of a flood of one week or more on conserved and grazed grass weighted by monthly probabilities of occurrence; an example of intensive grassland (£150kgN/ha)

Grass for silage
Nitrogen applied 200 kg/ha
UME 71849 MJ/ha
Dry matter conserved 7.1 tDM/ha

Month of flood	Grass costs*					
	Loss of UME yield (%)	Reduced DM conserved (%)	Cost of replacement feed (£/ha)	Harvest costs (£/ha)	Prod costs (£/ha)	Total grass (£/ha)
January	0.0	0	0	0	0	0
February	0.6	0.6	5	–1	0	4
March	7.6	7.6	67	–19	26	75
April	29.0	27.8	256	–69	53	240
May	55.0	35.2	486	–87	53	452
June	39.3	17.6	347	–43	0	304
July	5.2	0	46	0	0	46
August	3.8	0	34	0	26	60
September	1.8	0	16	0	26	42
October	0.3	0	3	0	0	3
November	0.4	0	4	0	0	4
December	0.0	0	0	0	0	0

Grass for grazing
Nitrogen applied 200 kg/ha
Usable UME 70995 MJ/ha

Month of flood	Grass costs*				Livestock on grass costs**	
	Loss of UME yield (%)	Reduced DM conserved (%)	Prod costs (£/ha)	Total grass (£/ha)	Move (£/ha)	House (grazing days)† (£/ha)
January	0	0	0	0	0	0
February	0	0	0	0	0	0
March	3.5	31	26	57	15	12
April	8.6	75	53	128	15	47
May	6.4	56	53	108	15	47
June	4.8	42	53	94	15	47
July	4.6	40	53	93	15	24
August	3.3	29	26	55	15	24
September	1.7	15	26	41	15	24
October	0.4	3	0	3	15	12
November	0.4	3	0	3	0	0
December	0	0	0	0	0	0

Notes:
* Estimates here exclude LAND restoration, RESEED and RESIDUAL items that may apply, see text.
** Estimates here exclude other FEED, SERVICES, MORTALITY and SALES items that may apply, see text.

Livestock on grass costs**				Seasonally weighted totals			
Move (£/ha)	House (grazing days)† (£/ha)	Sales (dairy)‡ (£/ha)	Total livestock (£/ha)	Seasonal weights (all England and Wales) (%)	Grass (£/ha)	Livestock on grass (£/ha)	Total grass and livestock (£/ha)
0	0	0	0	17	0	0	**0**
0	0	0	0	12	0	0	**0**
0	0	0	0	9	7	0	**7**
0	0	0	0	7	17	0	**17**
0	0	0	0	4	18	0	**18**
0	0	0	0	3	9	0	**9**
15	24	63	101	3	1	3	**4**
15	24	63	101	4	2	4	**6**
15	24	31	70	4	2	3	**4**
15	24	0	39	9	0	3	**4**
0	0	0	0	12	0	0	**0**
0	0	0	0	16	0	0	**0**

		Seasonally weighted totals			
Sales (dairy)‡ (£/ha)	Total livestock livestock (£/ha)	Seasonal weights (all England and Wales) (%)	Grass (£/ha)	Livestock on grass (£/ha)	Total grass and livestock (£/ha)
0	0	17	0	0	**0**
0	0	12	0	0	**0**
0	27	9	5	2	**8**
31	93	7	9	6	**15**
62	124	4	4	5	**9**
62	124	3	3	4	**7**
62	101	3	3	3	**6**
62	101	4	2	4	**6**
31	70	4	2	3	**4**
0	27	9	0	2	**3**
0	0	12	0	0	**0**
0	0	16	0	0	**0**
			29	30	56

† House/grazing days assumes stocking at 2.2 Lu/ha x £2.2/Lu x 10 days lost grazing in peak season.
‡ Assumes 15% milk sales loss/month.
Rounding errors.

Extra grass production costs ($INPUTS_{GRASS}$)

Flooding may result in additional costs associated, for example, with replacement or additional fertiliser and spray treatments. There may be some savings in uncommitted input costs. Table 9.14 contains estimates of extra production costs on grass that vary according to the seasonal timing of the flood.

Changes in 'harvesting' costs for conserved grass ($HARVEST_{GRASS}$)

Reduction in the amount of dry matter conserved for winter feed results in 'savings' in hay or silage-making costs, charged here at £35/tDM. Hence, a negative sign is assumed for $HARVEST_{GRASS}$ in Equation 9.5. In some cases, however, there may be increased cost for harvesting part-damaged grass.

Land restoration costs ($LAND_{GRASS}$)

Land restoration may be needed following long-duration floods, including removal of spoiled biomass and additional cultivations or treatment to neutralise contaminants or ameliorate saline flooding.

Reseeding damaged grassland ($RESEED_{GRASS}$)

There may be a need to cultivate and reseed damaged grassland, involving costs of inputs and field operations, typically at around £200/ha.

Residual costs ($RESIDUAL_{GRASS}$)

Long-duration flooding, particularly during spring and summer, can damage grass swards to the extent that yields are affected in subsequent years. Reseeded grasslands may take time to re-establish. For example, repeat long-duration summer flooding in Somerset in 2012 required reseeding of large areas of improved grassland in the following spring. Estimated grass yields on reseeded grass reduced by about 25 per cent of normal in the year following the flood. Greater yield reductions were expected on some non-reseeded 'organic' grasslands.

Loss of grass sales

In some cases cut grass is sold for hay or silage, valued at market price less costs of sales. At current prices, this is very close to the value of grass as a replacement feed, so the estimation method used above for FORAGE can also apply. Loss of grass supply may also affect livestock sales (see below).

Total costs of flooding on grass (GRASS)

Table 9.14 contains an example costing of flood damage on intensive grassland for selected cost items, assuming the monthly distribution of flooding from Table 9.5 above. Average annual damage tends to be greater on grass cut for silage and haymaking than that used for grazing, by a factor of approximately two in the example shown, at £57/ha and £29/ha, respectively.

9.5.4 Flood damage costs on livestock (LIVESTOCK)

In addition to impacts on grassland, flooding may have diverse impacts on associated livestock systems. The general formula for estimation is given in Equation 9.8.

Equation 9.8

$$\text{LIVESTOCK} = \text{MOVE} + \text{HOUSING} + \text{FEED} + \text{SERVICES} + \text{MORTALITY} + \text{SALES}$$

The components of Equation 9.8 are defined in Box 9.1 and discussed below.

Relocation (MOVE), housing (HOUSING) and feeding (FEED) of livestock

Flooding may require emergency relocation of stock to flood-free areas (MOVE), involving extra costs of labour and machinery, estimated at *c.*£15/ha affected (shown in Table 9.14).

Flooding of grazing land can also delay the turnout of stock in spring or bring forward the date that stock must be housed for winter (HOUSE). Housed stock require feeding and tending. A spring 'grazing day' (excluding the value of feed itself) is worth approximately £2.20/grazing livestock unit (LU, a dairy cow equivalent) per day in terms of additional non-feed costs, such as labour and machinery and buildings expenses. Stocking rates range between 1LU/ha and 2.5LU/ha according to grassland productivity. For example, with stocking rates of 2.2 LU/ha, a ten-day loss of grazing in spring/early summer turnout can cost around £47/ha (Table 9.14).

Relocated livestock that otherwise would be feeding at grass may require additional feed costs in some cases (FEED), over and above the value of grass energy lost, especially if replacement feeds are generally short in supply as a result of flooding.

Additional livestock services (SERVICES)

Disruption to services, notably fresh water and waste treatment may result in significant additional costs, requiring, for example, transport of uncontaminated drinking and cleaning water for dairy animals. Additional veterinary services may be required.

Increased livestock morbidity and mortality (MORTALITY)

Flooding can cause health and disease problems in livestock, associated, for example, with mastitis in dairy cows, animal enteritis and liver fluke, negatively affecting yields or increasing veterinary costs. Rapid and large-scale floods, particularly in flat coastal areas with limited refuge, can result in high livestock mortality rates, especially amongst young stock. Mortality can be measured in terms of market prices of livestock, plus disposal costs.

Reduced sales (SALES)

Severe flooding and reduced availability of grazing land or winter feed can affect livestock sales revenues, due to premature sales of 'unfinished' stock or reduced purchases of livestock for fattening. Milk sales from dairy cows may be disrupted due to restrictions on grazing, changes in feed and stress. Here the loss is measured in term of loss of value added (sales minus cost of sales). Some dairy farmers reported losses of around 15 per cent in milk yields during floods on summer pastures (Morris and Hess, 1988; Posthumus *et al.*, 2009), equivalent at typical levels of output to *c.*£60/ha during the grazing season (Table 9.14). Loss of added value from livestock sales due to grass energy shortage is higher by a factor of approximately 1.5 than losses based on the value of replacing lost feed itself, making grass energy value added equal to *c.*£0.185/MJ (Morris and Brewin, forthcoming). A proportion of grass feed losses could be increased by 1.5 to reflect lost sales of livestock, if appropriate.

Following the Somerset floods in 2012, limited feed supply led some farmers to reduce numbers of beef cows (producing beef calves for fattening), dairy milk cows and young dairy cow replacements, thereby potentially affecting production, sales and costs in subsequent years.

9.5.5 Other costs (OTHER)

These items include farm-scale impacts, as shown in Equation 9.9.

Equation 9.9

OTHER = BUILDINGS + UTILITIES + MACHINERY + FIELD INFRASTRUCTURE + DEBRIS + FARM SERVICES + FINANCING + SUNDRY

The components of Equation 9.9 are shown in Box 9.1 and are discussed below.

Although the majority (typically over 80 per cent) of flood damage costs are directly associated with the impact on crop and livestock production, flooding can also generate OTHER costs at the farm scale. Flooding can cause damage to buildings and structures, including loss of contents (BUILDINGS). Estimates can be based on similar types of commercial buildings, whether used for storage or processing (e.g. dairy buildings, vegetable packing), allowing for seasonal differences in stored materials and activities. Disruption to utilities such as water and electricity may require alternative sourcing (UTILITIES), particularly to support livestock activities. Farm machinery may be damaged, including fixed equipment such as drainage or irrigation pumps, requiring repair or replacements (MACHINERY). Deep or turbulent

Table 9.15 Weighted average other agricultural damage costs due to summer 2007 floods (£/ha flooded)

	Horticulture	General cropping	Cereals	Mixed	Dairy	Grazing livestock	Pigs	All farms
	(n = 4)	(n = 20)	(n = 22)	(n = 11)	(n = 9)	(n = 9)	(n = 3)	(n = 78)
Buildings	97	8	129	11	10	8	10	**45**
Machinery	125	54	58	3	1	19	73	**40**
Field infrastructure	0	2	11	4	13	5	2	**6**
Drainage	0	33	8	11	3	1	8	**14**
Debris	9	8	5	2	4	7	2	**5**
Services/sundry items	56	62	4	1	22	2	0	**22**
Increased overdraft	0	11	2	13	27	22	61	**14**
Total cost (£2007)	**287**	**178**	**217**	**45**	**80**	**64**	**156**	**146**
Total cost (£2013)*	**318**	**197**	**240**	**50**	**89**	**71**	**173**	**162**

*Note:** Inflation adjusted using ONS GDP deflators 2007/8 to 2012/13 at 1.107.

Source: Posthumus *et al.*, 2009

flooding may require repairs to field infrastructure, such as fences and drainage networks (FIELD INFRASTRUCTURE), as well as clean-up of debris (DEBRIS). Flooding may also result in loss of net revenues from silage-making and harvesting services provided by contractors, some of whom are also farmers (FARM SERVICES). Lost revenues and extra costs can put pressure on farm cash flow, requiring increased borrowing (FINANCING). Other farm costs, such as hire of temporary pump equipment or removal of equipment or stores to avoid damage, may also be incurred (SUNDRY).

Average OTHER costs (weighted by area affected) were £146/ha (in 2007 prices) for the very severe 2007 floods in England (out of an average total cost of *c.*£1,200/ha) (Table 9.15). OTHER costs were *c.*£65/ha for the floods of four-week duration on grasslands in Somerset in early 2012. It is likely that OTHER costs for the annual flood will be lower than those observed for summer flooding. A default value at 70 per cent of the estimates given in Table 9.15 could be used.

9.6 DEFRA GUIDANCE ON THE APPRAISAL OF AGRICULTURAL FLOOD RISK MANAGEMENT INVESTMENTS

The principle of Defra (2008a) guidance is that, from an economic perspective, investments in agricultural flood risk management can be justified as long as extra benefits to the economy exceed extra costs at the test discount rate. This requires a comparison of the baseline 'counterfactual' – or 'do nothing' – option with one or more possible interventions and their outcomes, including environmental consequences.

Whether grant aid is provided for flood risk management is another issue. Much depends on whether control of flooding is regarded as a public good where society as a whole is better off, and it is difficult to discriminate between individual beneficiaries in terms of willingness or ability to pay. In the past, major public investments were made in agricultural flood defence as part of a strategic commitment to increasing food production in Britain (Morris, 1992). Since the 1980s, there have been few, if any, agricultural enhancement schemes funded by the public purse. Although farmers (along with other beneficiaries) may act collectively to fund a flood risk management project, the majority of appraisal for public investments are likely to concern:

- Major investments to replace/refit/rehabilitate previous agricultural flood defence (and land drainage) schemes that have reached the end of their design life.
- Investments to manage flood risk (and retain associated land drainage services) in the face of increased flood probability, possibly linked to changes in catchment hydrology due to non-agricultural land development and/or climate change.
- The purposeful use of farmland for the conveyance and storage of flood waters to mitigate flooding in urban areas, supported by compensation to land managers. This may include assessment of environmental benefits, such as those linked to wetland creation.

Reflecting the view that agricultural enhancement through publicly funded flood risk management investment is no longer a policy objective and hence unlikely on economic grounds,[1] Defra identifies three scenarios which reflect the nature of change, namely:

- permanent loss of agricultural land;
- one-off damages arising from infrequent flood events; and
- a permanent deterioration in flood management standards.

These scenarios justify different approaches to the assessment of flood risk management benefits, as shown in Table 9.16.

9.6.1 Scenario I: Abandonment of agricultural land

Here, agricultural land is lost for the foreseeable future, typically arising as a result of a permanent breach in a sea or river embankment.

Defra (2008a) advises that prevailing market prices for agricultural land of similar quality (for example, vacant possession, without buildings) can be used to assess the permanent loss of agricultural output. In a 'perfect' market, agricultural land prices reflect the income-earning potential of farm land over a reasonable time horizon; that is, the discounted present value of a series of future annual benefits (net of costs) from agricultural production.

Thus, in theory, the permanent loss of future agricultural output (£/ha) can be expressed in terms

Table 9.16 Defra advises that different assumptions are made for alternative agricultural flood risk scenarios

	Scenario I	Scenario II	Scenario III
	Permanent loss of agricultural land	Temporary, one-off loss of agricultural output	Permanent reduction in the value of agricultural output
All agricultural land use	Agricultural land lost is valued at market prices less 'transfer' payments		
Crops: cereals, oil seeds, beans/peas. Grassland: beef and sheep		Reduction in gross margins, adjusted for additional other costs and/or savings	Reductions in net margins associated with change in flood and associated land drainage conditions
Other: dairy, sugar beet, potatoes, high-value fruit/vegetables		As above, treated as though area occupied by wheat (if output loss is replaced)	As above, treated as though area occupied by wheat (assuming relocation of high-value cropping is feasible and possible no major regional and national effects)

Source: Based on Defra, 2008b; see text for explanation

of a loss of land value (£/ha). Land values may, however, be inflated by income support subsidies paid to farmers. Defra advises that an equivalent capital sum should be deducted from prevailing commercial land prices to reflect the present value of future annual Single (Farm) Payments received by farmers (net of costs associated with compliance with good agricultural practice).

Thus, the present value of land lost to agricultural production is as shown in Equation 9.10.

Equation 9.10

$$V^* = (V - p)$$

Where:

- V^* = adjusted present value of agricultural land (£/ha)
- V = market price for agricultural land according to location and type (£/ha)
- p = the present value of future annual income support payments made to farmers, taken here to be £600/ha after adjustments for compliance costs

Actual prices per hectare vary considerably according to the size and location of the land concerned, whether it is tenanted or vacant possession, whether it is 'equipped' with buildings and infrastructure, and other site-specific factors. Agricultural land prices have risen steeply since 2009, partly due to stronger agricultural commodity prices and the improved profitability of farming, and partly due to a decline in the attractiveness of other investment options.

Agricultural land prices often reflect a general shortage of land coming on to the market (less than 0.5 per cent of total stock per year) and strong demand for non-commercial ownership and use. In 2012, prices for bare, 'unequipped' prime arable land (typically ALC Grades 1 and 2) ranged between £14,500/ha and £17,500/ha; average arable and good-quality grassland (ALC Grade 3) between £12,500/ha and £15,000/ha; and poor-quality grassland (ALC Grades 4 and 5) between *c.*£7,500/ha and £8,500/ha. 'Equipped' land typically attracted an additional £1,000/ha to £1,500/ha.

Although the 'default' value for land lost might in most cases be that of bare land, where large-scale losses result in loss of buildings and other facilities it may be more appropriate to use prices for 'equipped' land. Published regional land prices are available from major land agents (VOA, 2011; Savills, 2012; Smiths Gore, 2013).

Where land receives EU Environmental Stewardship payments, Defra also recommends subtracting the discounted present value of these payments before applying the adjustment factor.

Care must be taken when using agricultural land prices to value agricultural losses, using values based on land sales that retain land in commercial agriculture, rather than acquisition for 'lifestyle' occupation that could force prices above agricultural values. Grade 4 land adjacent to urban areas may command higher prices than Grade 1 land in more remote farming areas.

In cases where land values are clearly dominated by non-agricultural factors, a reality check can be made by deriving estimates of the present value of net margins from farming. Table 9.17 shows that current market prices exceed the present value of net margins after deducting total fixed costs discounted at 3.5 per cent over fifty years for all land uses (based on estimates from Table 9.19). Land values are closest to productive values for intensive arable and dairy farming. Where loss of land reduces the size of existing farms that continue to operate at a smaller scale, the losses may be closer to the present value of net margins after semi fixed costs. Here, land prices could underestimate the value of agricultural losses for intensive farming.

These estimates of land value exclude non-agricultural benefits, such as recreational or nature conservation values, that require separate treatment. Defra advises, for example, that where land is in receipt of annual payments under the EU's Environmental Stewardship scheme, the present value of these should also be deducted from the market value of land to derive an adjusted agricultural asset value.

9.6.2 Scenario II: Occasional loss of output

This can occur as a result of a flooding incident, possibly associated with a temporary breach or the overtopping of an embankment.

Defra guidance advises that the impact of a one-off flood depends on whether the flood results in the

Table 9.17 Indicative estimates of land market prices and present value of agricultural benefits from farming by type of land use

£000/ha, 2013 values	Crops			
	Intensive arable	Extensive arable	Intensive grass (dairy)	Extensive grass
Land market price (2013)	14–18	13–15	13–15	7–10
PV* Gross margins	23–37	15–20	35–44	11–18
PV* Net margin after semi fixed costs	21–28	11–14	28–33	9–12
PV* Net margin after total fixed costs	12–16	2–4	6–12	0.6–1

Note: * Assumes fifty years at 3.5%, annuity 23.5. Thirty- and one-hundred-year annuities are 18.3 and 27.7, respecitvely, with −21% and +18% change in estimates, respectively.

complete or partial loss of a crop and whether land is rendered unusable for the remaining production year or longer. Where agricultural output is lost for a whole year, the gross margin can be used as a broad indicator of the loss of agricultural value added, if it can be assumed that fixed costs will not change. Much depends on when flooding occurs in the farming year, whether some variable costs have already been committed and therefore 'lost', and whether some have not yet been committed and therefore can be 'saved'.

For reconnaissance-level appraisal, where the purpose is to assess the possible impact of a serious annual flood, the gross margin (e.g. £870/ha for wheat in Table 9.8) can give a high-level approximation of loss.

For the appraisal of a specific flooding problem and proposed solution scheme, however, especially where the estimates of loss can draw on evidence of previous flooding, it is recommended that the $\text{FLOOD}_{\text{FARM}}$ loss equation (Equation 9.1) is used to assess the costs of a single flood event. This method can be used to evaluate the actual impact of a flood that has happened (*ex post*), or the likely impact if it does happen in the future (*ex ante*).

By way of an example for Scenario II, and using the methods described earlier, Table 9.18 contains broad-based estimates of the economic losses associated with a single flood event of more than one week's duration occurring on major land use types in England and Wales (such as horticulture and intensive arable). The estimates are weighted by the mix of land use types typically found in any one ALC grade, and the differences in expected yield for each ALC grade. Allowance has been included for 'other' costs as described earlier (assuming 70 per cent of the summer 2007 event costs apply for all-year flood events, adjusted to 2013 prices). For example, expected damage costs on ALC Grades 1 and 2 land for a single annual flood of more than one week's duration are £1,320/ha and £1,000/ha, respectively (inclusive of 'other' costs, as shown in Table 9.18). Alternative monthly distributions for flooding can be used if these are known.

9.6.3 Scenario III: Fall in agricultural output per hectare

This could occur because of a permanent increase in flooding or a rise in the soil water level, possibly associated with a decision not to rehabilitate flood embankments or pumping schemes to their original standard, or purposely to use farm land for flood storage.

In such cases there may be significant, long-term changes in land use and the type and mix of crop and livestock enterprises, leading to reductions in gross margins, possibly offset in the long term by some reduction in fixed costs. Although the frequency of flooding may increase, the cost of a single event may decline because land use becomes less intensive. This may not be the case, however, if the depth, duration or seasonality of flooding also changes adversely.

A key factor influencing the impact on a farm business of a permanent change in flood risk on any parcel of land is the proportion of the total area of the farm that is accounted for by this parcel. Generally,

Table 9.18 Indicative costs of a single annual flood of one week or more by Agricultural Land Classification Grade (ALC) and associated land use, assuming all England and Wales monthly distribution of flooding

ALC class		Land use					
		Horticulture	Intensive arable	Extensive arable	Intensive grass	Extensive grass	Flood costs £/ha
1	% of area	0	1	0			
	Flood cost (£/ha)	5,626	1,147	651			1,320
2	% of area	5	60	35			
	Flood cost (£/ha)	3,616	1020	601			1,000
3a	% of area		30	70			
	Flood cost (£/ha)		840	501			600
3b	% of area			50	50		
	Flood cost (£/ha)			501	180		340
4	% of area				100		
	Flood cost (£/ha)				180		180
5	% of area					100	
	Flood cost (£/ha)					100	100

Note: Average damage costs above include other costs at: Horticulture £226; Intensive arable £140; Extensive arable £171; Intensive grass £63; Extensive grass £50; in 2013 prices based on 2007 summer floods assuming 70% applies to whole-year floods.

the larger the proportion of the whole farm subject to change in flood risk, the bigger the impact on farm business as a whole. In some areas, there is a high degree of fragmentation of holdings, with small land parcels used or rented out by owners who have most of their land on higher ground. In other cases entire holdings may lie within the floodplain. Clearly, the impact of changes in flood risk on farm businesses will vary between these circumstances.

In the context of permanent changes in flood risk, it is important to determine whether the change will impact only on the gross margins, or whether there will also be changes in fixed costs at the farm level. For example, a change only in wheat yield can be represented by a change in gross margin, on the assumption that everything else remains the same. However, larger changes, such as a switch from intensive dairy to extensive cattle grazing, or a significant change in farm size due to land removed from production, can result in changes in the overall 'total fixed cost' structure of the farm. Here, the analysis must include changes in both gross margins and fixed costs.

Drawing on estimates of productivity given in Table 9.3 and gross and net margins in Tables 9.8 and 9.9, Table 9.19 gives financial (to the farmer) and economic (to the national economy) gross and net margins for major land use types for given drainage conditions assuming Scenario III. These estimates include allowance for changes in fixed costs associated with different levels of productivity. The economic estimates assume wheat as a substitute crop, as explained above, subject to the aforementioned caveats. It is noted that, excluding Single (Farm) Payments currently at *c.*£200/ha, net margins are negative for very bad conditions, rendering intensive grassland and arable farming commercially (and economically) non-viable. Extensive grassland also generates negative net margins under very bad drainage, but farmers adopting this 'residual' land use can potentially qualify for agri-environment payments for wet grassland.

Table 9.19 can be used to determine the financial and economic impacts of a change in land use and/or drainage condition associated with a change in flood

Table 9.19 Financial and economic returns by land use type and drainage condition

Land use type	£/ha: 2013 values						
	Drainage condition	Financial gross margin	Financial net margin (ASFC)	Financial net margin (ATFC)	Economic gross margin	Economic net margin (ASFC)	Economic net margin (ATFC)
1 Extensive grass	Good	789	514	44	789	514	44
Beef cows, cattle	Bad	631	398	–2	631	398	–2
and sheep	Very Bad	395	210	–104	395	210	–104
2 Intensive grass	Good	1,896	1,411	497	952	692	255
Dairy herd (80%) plus beef	Bad	1,517	1,062	202	685	444	38
cattle (20%)	**Very bad**	**948**	**525**	**–278**	**284**	**74**	**–282**
3 Grass/cereal rotation	Good	1,357	989	316	885	630	194
(grass as for 2 above)	Bad	1,050	704	67	634	394	–15
	Very bad	**595**	**275**	**–313**	**263**	**49**	**–315**
4 All cereal	Good	819	568	134	819	568	134
Winter (80%), spring (20%)	Bad	584	345	–67	584	345	–67
	Very bad	**241**	**25**	**–348**	**241**	**25**	**–348**
5 Extensive arable	Good	849	604	164	845	601	160
Wheat/oil seeds/beans or peas	Bad	624	392	–26	621	389	–30
	Very bad	**306**	**94**	**–287**	**302**	**91**	**–291**
6 Intensive arable	Good	1,578	1,208	683	845	601	160
Cereals, potatoes, sugar beet	**Bad**	**806**	**406**	**11**	**621**	**389**	**–30**
	Very bad	**173**	**–150**	**–458**	**302**	**91**	**–291**

Notes: These are indicative estimates within a range of about +/–£150/ha for gross margins and +/–£50 for net margins; farmers may receive single farm payments equivalent to about £200/ha on land previously eligible for production subsidies; intensive grassland and intensive arable assume fixed costs per ha for dairy farms and for mixed farms, respectively.

risk management. Thus, a switch from good to bad drainage on a grass/cereal rotation reduces financial net margin (after total fixed costs) by £249/ha and economic net margin by £209/ha. A switch from good drainage on cereal/grass rotation to very bad on extensive grassland, possibly associated with 'managed retreat', reduces financial net margin by £420/ha and economic net margin by £298/ha.

A change in land use due to increased flooding changes the cost of flooding when it occurs, broadly in proportion to the gross margin productivity of land use within a given land use class. Table 9.20 contains estimates of the average annual cost of flooding of one week or more by land use type and drainage condition, assuming all England and Wales monthly flood distribution. For example, a single flood of at least one week's duration results in an estimated damage of £650/ha on extensive arable under good drainage conditions (inclusive of approximately £170/ha 'other' costs), and £478/ha where drainage conditions are bad. The expected annual cost of a flood of a given frequency can be estimated by dividing by the return period in years. For example, Table 9.20 shows the expected annual damage costs of a single flood according to indicative 'tolerable' flood return periods by land use. The implications of changes in flood frequency for expected annual flood costs can be assessed accordingly.

Where, under Scenario III, a permanent change in flood probability is associated with a change in land use and a change in drainage condition, additional costs can be determined by comparing net margins less flood costs before and after the change. Hence, the incremental costs of Scenario III are given by Equation 9.11.

Equation 9.11

$$\text{Change in expected annual flood damage costs} = (\text{net margin}^{\text{before}} - (\text{FLOOD}_{\text{FARM}}^{\text{before}}/\text{flood return period}^{\text{before}})) - (\text{net margin}^{\text{after}} - (\text{FLOOD}_{\text{FARM}}^{\text{after}}/\text{flood return period}^{\text{after}}))$$

9.7 ENVIRONMENTAL PROTECTION AND FLOOD RISK MANAGEMENT

Agricultural and rural policy now seeks to achieve multiple benefits from rural land, including to food

Table 9.20 Estimated cost of a single annual flood and indicative average annual damage flood costs by land use and drainage condition, all England and Wales monthly distribution of flooding

	Drainage condition	Cost of a single annual flood, £/ha	Indicative flood return period by land use, years	Average annual cost of flood damage according to indicative return period, £/ha
1 Extensive grass	Good	100	1.0	100
	Bad	80	0.75	107
	Very bad	50	0.5	100
2 Intensive grass	Good	180	3	60
	Bad	50	2	25
3 Grass/cereal rotation	Good	405	8	51
	Bad	313	5	63
4 All cereal	Good	630	8	79
	Bad	449	5	90
5 Extensive arable	Good	650	8	81
	Bad	478	5	96
6 Intensive arable	Good	1,150	10	115

Note: Some rounding.

production, nature conservation, water resource management, recreation, tourism and public access to the countryside (Foresight, 2010). For example, the Higher Level Stewardship (HLS) scheme contains incentives to maintain or revert to wetland practices in floodplain areas. In Somerset, for example, the scheme offers annual payments for the adoption of management prescriptions to deliver wet grassland and raised water levels at approximately £225/ha and £430/ha, respectively.

Although agri-environment payments are not complete measures of the value of environmental benefits, they reveal a degree of willingness by society to pay for enhanced or protected landscapes and wildlife. Thus, HLS payments for wetland options might indicate the environmental benefit of a switch from arable to extensive wet grassland on the floodplain as a consequence of a change in flood management. Thus, economic losses from farming might be partly offset by gains for the environment. Entry-level payments are not specifically linked to flood and drainage conditions so they can be left out of the economic analysis for Scenarios II and III.

Farmland on floodplains can also offer scope for temporary storage of flood water, reducing the risk of flooding elsewhere while also providing a valuable washland environment for wildlife. Such options are being explored in low-lying areas such as the Fens of East Anglia and the Levels and Moors of Somerset. These initiatives, supported by incentives to land managers, could help to deliver sustainable flood risk management solutions which reconcile flood risk, environmental and farming interests (Morris *et al.*, 2004; 2008).

9.8 THE VALUATION OF ECOSYSTEM SERVICES AND FLOOD RISK MANAGEMENT

The UK's National Ecosystem Assessment (NEA, 2011) provides the first independent assessment of benefits provided by the natural environment for the well-being and prosperity of people. The ecosystems approach considers how stocks of natural capital (land, water, air and living systems) provide a diverse range of ecosystem services, many of which have hitherto been inadequately accounted for in decision-making. In particular, it classifies ecosystem services in terms of provisioning (e.g. food and fibre), regulating (e.g. flood control and carbon storage), cultural (appreciation of landscape) and supporting (biodiversity and soil formation). Defra (2007; 2010) and the Environment Agency (Everard, 2009) have adopted the ecosystem approach to development appraisal. HM Treasury (2012b) also recommends its use for valuing environmental impacts in benefit–cost analysis.

The ecosystems approach has a particular relevance for flooding as a key regulating service. It helps to support stakeholder participation and scenario analysis by revealing key trade-offs and synergies amongst ecosystem services and competing interests under different flood risk management options (Posthumus *et al.*, 2009; Rouquette *et al.*, 2011; Morris and Camino, 2011). Deriving comparative values of diverse ecosystem services is challenging, however, especially in the absence of complete scientific or economic information. Indeed, the approach calls for new quantitative and qualitative methods for valuation, some of which can be developed and tested during the assessment process itself.

9.9 DATA NEEDS, SOURCES AND COLLECTION METHODS

Data requirements for project appraisals vary according to whether the proposal involves a new scheme or a rehabilitation of an existing one, possibly involving alternative exposure to flood risk. In all cases options are compared with a 'do nothing' or 'do minimum' option (the latter possibly defined by legal obligations), which needs to be defined in terms of agricultural consequences. Furthermore, although the data requirements for the appraisal of schemes which increase protection to agriculture are much the same as for those which propose to reduce it (as for wetland creation or flood storage), the interpretation of the data is very different. Rather than looking for farmers to take up potential agricultural benefits, the purpose in these latter schemes is to confirm that farmers are willing to give them up. Insights into farmer motivation and factors influencing farmer behaviour are therefore important topics of enquiry.

Table 9.21 lists the type of information needed for the assessment of agricultural flood risk management schemes, and the sources and collection methods. The degree of detail required varies according to the stage of the planning process, whether reconnaissance level

Table 9.21 The assessment of the impacts of flood risk management on agriculture involves a range of information derived from site observation and discussions with land managers

1 Benefit area definition
Area over which flood risk management and drainage works exert influence in terms of flood risk and ground water levels, including tributary arterial systems affected directly or indirectly by the flood management works.
Relevant organisational aspects: internal drainage board areas, main rivers, local authority ditches, farmers' ditches.
History of capital and maintenance works, purpose and justification.
Evidence of changes in land use and farming practice over time, especially flood risk, records of events, aerial photos.
A set of 1:2500 series maps and/or Google Earth photo mosaics covering the benefit area are essential for farmer interview, and for determining field areas, water courses, scheme boundaries and the location of other salient features.
2 Farm level data (for the farm in the benefit area)
Percentage of the farm area in the benefit area, farm size (ha).
Topography, major soil types and agro-climate.
Cropping pattern (crops by area).
Livestock types, numbers, ages, systems.
Ownership type: owner, tenant, partnership.
Number of family and hired workers.
Major changes in farm circumstances and practice over the past five years.
Current development proposals and factors influencing/limiting.
Relevant environmental features and qualities: participation/eligibility in agri-environment schemes, especially Higher Level Stewardship.
Relevant fixed cost scenarios for expected changes in crop and livestock activities.
Typical land values: purchase, rents.
3 Farmer characteristics (as they affect motivation)
Age, length of occupancy, qualifications, experience.
Business orientation: commercial, lifestyle issues, environmental/conservation interest and actions.
Extent of non-farming commercial activities: dependence on farming.
Experience of farming elsewhere.
Inheritance: farming with sons or daughters.
Membership of local groups: farmer, drainage, conservation groups.
Importance attached to drainage for agriculture.
4 Field level (or blocks of fields with similar features)
Soil type, field size (ha).
Tenure status and length of occupancy.
Distance from, and contiguity with, farm.
Whether under-drained or not (pipe/mole drains, drain conditions, ditch system, outfall facility in ditch, standards of maintenance, pumping).

Table 9.21 continued

Predicted with and without scheme flood risk and drainage conditions by fields – groups: • Drainage condition; wetness class (four types: permanently wet, commonly wet, occasionally wet, rarely wet). • Flood frequency and areas by return period. • Seasonality (particular months of year). • Flood duration (days, weeks, months). • Flooding depth (mm). • Other flood water issues: quality, turbulences, damage to structures, livestock at risk. • Percentage of field liable to flooding. Field drainage requirements: type, design, costs, percentage of field in need of drainage. Evidence of flooding: crop damage, high water levels (flood tolerant species, stressed crops, tractor wheelings, muddy cattle, blocked drains/channels/debris, reports of 'muddy floods').
5 Land use and enterprise level data
Grassland Grassland site class: poor, average, good, very good (as defined by soils, climate and available soil water). Grass type: permanent, temporary, improved ryegrass, rough grazing, presence of sedge/rushes. Drainage condition: good, bad, very bad (reflecting ground water levels, high ditch levels, heavy soils, damaged artificial drains). Poaching risk: whether soils are liable to surface damage when wet by animals (and machines). Nitrogen use (kgN/ha) as it defines grass yields. Grass conservation type: graze, hay, silage, and timing and number of cuts. Stock types and performance levels: dairy, dairy replacements, beef cows and sucklers, 24–30 month beef, store cattle at grass, sheep. Typical yields for milk (l/cow/year, usual live-weights bought and sold, prices for milk, beef, sheep). Beginning and end of grazing season, spring and autumn, outside over-wintering of cattle and sheep. Fixed cost assumptions: as a result of the changes (either increase or decrease) in output due to flood risk will there be an increase or decrease in any of the following for grass and livestock production: breeding herd, labour, machinery, buildings? **Arable** Crop type: cereals, oil seeds, legumes, root crops (potatoes, carrots, sugar beet), vegetables (brassicas, onions), salad crops (lettuces, celery), soft fruit (strawberries), top fruit (apples, pears). Crop yield (t/ha), prices (£/t) and quality aspects as they affect prices. Crop rotation (crop type, including grass, and number of years for each). Fixed costs assumptions: labour, machinery, buildings: will any changes in flood risk affect 'fixed costs' such as labour, machinery, buildings, use of contractors?

Note: On large schemes, where there is homogeneity of farm type and practice, the formulation of whole-farm budgets which show the implications of major changes in farming activities and resource use is likely to be needed.

at a catchment scale or detailed appraisal of selected benefit areas or reaches.

It is recommended to start with an exploratory survey of the study area in order to define the geographical boundary of influence – that is, the benefit area – and to determine current approaches to flood risk management and issues arising. Broad categories of land use should be identified together with dominant farm types and systems, possible flood risk management options and outcomes, and the likely attitudes of key stakeholders, especially farmers, towards them. Six questions – Why? What? Where? When? How? and Who? – provide an alarmingly simple but effective checklist for open-ended questions to help understand local flood management issues and interests.

The exploratory survey will draw on a mix of secondary data and sources, such as:

- Ordnance Survey and remote-sensing sources, notably Google aerial photo mosaics and Bing bird's-eye and aerial maps (http://www.bing.com/maps/);
- Environment Agency floodplain maps and records, including catchment flood management plans and shoreline management plans;
- Internal Drainage Boards maps and records;
- local government environmental services (where relevant);
- wildlife and rivers trusts and conservation organisations;
- farmers' organisations and groups;
- rural business research data (Farm Business Survey, 2012);
- farm management pocketbooks (e.g. Nix, 2012; ABC, 2012);
- relevant research literature;
- consultants' reports; and
- site observations and discussions with key informants.

The key informants will include personnel with flood and conservation management interests in regional offices of the Environment Agency, lead local flood authority and Defra, and local Internal Drainage Boards, if relevant. Other informants include representatives of farmers' organisations (such as the National Farmers' Union), local advisors and land agents, and environmental organisations, such as local wildlife trusts, and farming and wildlife advisory Groups (FWAGs).

In most cases some kind of farm survey will be needed to confirm the link between flood risk, drainage conditions and farming practice and performance. Usually, a quota of representative farmers will be selected for interview to cover the major variations in farm circumstance (e.g. size, tenure, land type, flood risk), farm practices (e.g. enterprise mix, drainage improvements) and farmer characteristics (e.g. age, family circumstances, motivation). It may be appropriate to engage farmers as a group as part of a stakeholder-driven participatory process.

Although the farm survey should use a structured questionnaire, there should also be scope for some open-ended discussion to reflect the experience, insights and perceptions of land managers, many of whom will have been long-term occupants of the floodplain. Personal photographs and newspaper cuttings can provide records of memorable flood events. It is also useful to derive an insight into the interconnectedness or cohesion of the local farming community. This can vary considerably. The river can be more than a physical boundary between farmers. Community of place and interest can be critical where changes in flood management require collective action or agreement.

It is useful to carry out interviews in the field, where direct observation can inform the interview process. A good set of maps, showing water courses and field boundaries, is essential to support the survey. Field-level information on soil type, flooding, drainage condition, presence of field drains, pipe outfalls, observed ditch levels and crop and livestock practices can be scribbled directly on to these maps.

For agricultural enhancement schemes, the extent to which flooding and drainage currently constrain farming will be a focus of enquiry, together with the factors that are likely to encourage farmer take-up of potential benefits, if appropriate. For rehabilitation of existing infrastructure, the focus will be on the likely consequences for land use and farm incomes of increased flood risk and lower standards of agricultural drainage. Where proposals include options for extensive farming, flood storage, washlands and/or wetlands, the practicalities of reconciling these diverse objectives will be a focus of enquiry. Table 9.21 also contains a checklist of information commonly derived by farm interviews. Experience shows that it is best to focus on physical performance rather than detailed financial data, other than local price and market information. It is also important to determine whether farmers receive Single (Farm) Payments and participate in agri-environment or other schemes.

Appraisal of flood risk may be set within the broader framework of ecosystem services. This requires that data are collected to enable the synergies and trade-offs between flood control and other ecosystem services, such as storage of soil carbon, water purification and amenity (Posthumus *et al.*, 2010; Rouquette *et al.*, 2011; Morris and Camino, 2011; Maltby *et al.*, 2013). It may also require the 'transfer' of estimates of benefits, such as those associated with informal recreation, from studies outwith the appraisal site (Eftec, 2010b). Integrated approaches to appraisal require a multidisciplinary approach, combining natural, physical and social sciences.

This generates estimates of changes in net agricultural benefits attributable to the scheme, including allowances for benefits that might have arisen in any case without the scheme, and costs related to field drainage. These estimates of net benefit are then combined with data on scheme costs and other non-agricultural impacts to produce predicted flows of extra benefits and costs attributable to the scheme over its life.

Uptake of benefits, whether for agriculture or conservation management, critically determine the outcome of flood management schemes. Experience of uptake elsewhere can provide a guide. Project appraisal should use risk and sensitivity analysis to assess the validity of the key assumptions, and the extent to which scheme viability is sensitive to them.

9.10 AGRICULTURAL FLOOD RISK MANAGEMENT APPRAISAL AT THE SCHEME LEVEL

Figure 9.3 provides an overview of the appraisal at scheme level. Farm- and site-survey data feed into a benefit assessment model as outlined earlier in Figure 9.2, with data on financial or economic prices as required.

9.11 WORKED EXAMPLE

For the purpose of illustration, Table 9.22 contains a simple example of a high-level appraisal of proposed rehabilitation of an agricultural flood risk management scheme. Estimates of net returns from farming and flood costs are drawn from earlier tables. The existing flood control and supporting drainage infrastructure has reached the end of its design life. Table

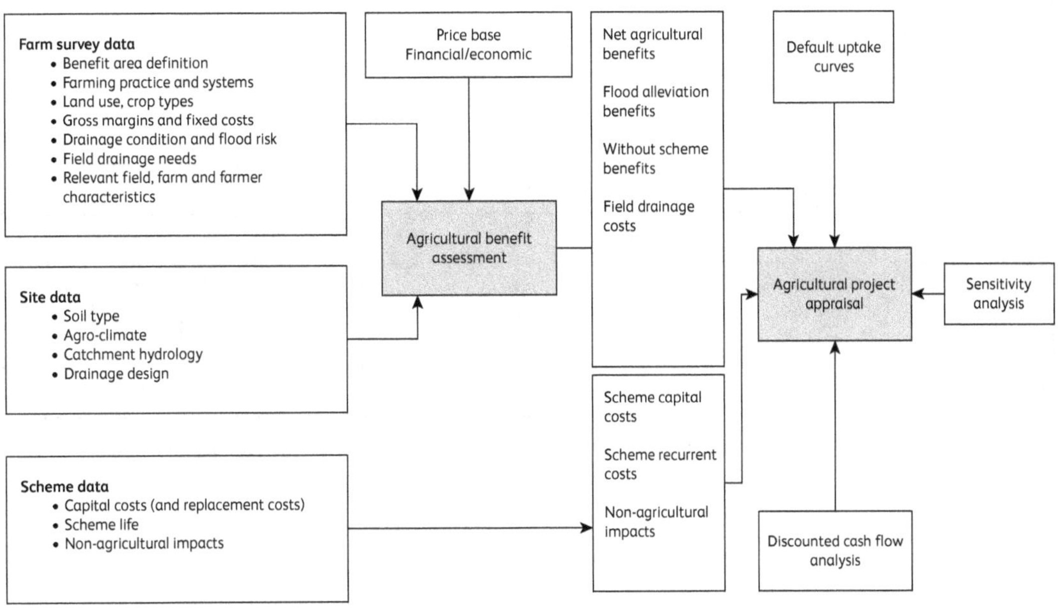

Figure 9.3 Field, farm and scheme level data are required to appraise the feasibility of agricultural flood management

Table 9.22 A simple worked example of the costs of a change in flooding and associated drainage conditions on agricultural land

	Sub-areas ha	Land use	Drainage conditions	Net margin before flood costs	
				Financial	Economic
				£/ha	£/ha
With Flood Protection Scheme (A)					
1	200	Intensive grass	Good	497	255
2	400	Extensive arable	Good	164	160
3	400	Intensive arable	Good	683	160
Total	**1,000**				
Without Flood Protection Scheme (B)					
1	200	Abandoned	–	0	0
2	400	Extensive grass	Bad	–2	–2
3	400	Extensive arable	Good	164	160
Total	**1,000**				
Difference (A – B)					
1	200				
2	400				
3	400				
Total	**1,000**				

Present value of benefits of continued protection for agriculture at 3.5 % DR				
	Economic		Financial	
period years	50	100	50	100
PV £000	5,895	6,507	10,485	12,362
PV £/ha	5,895	6,507	10,485	12,362

Note: * ALC Grade 3 at £13,000/ha less £600 adjustment, amortised at 3.5 % over fifty years (£529/years as shown) or one-hundred years (£448/year).

	Flood event cost	Flood risk return period	Annual flood damage	Net margin after flood costs			
				Financial	Economic	Financial	Economic
	£/ha	years	£/ha	£/ha	£/ha	£/sub-area	£/sub-area
	180	5	36	461	219	92,200	43,800
	650	10	65	99	95	39,600	38,000
	1,150	50	23	660	137	264,000	54,800
				396	**137**	**395,800**	**136,600**
	0	0.25	0	0	0	0	0
	80	0.5	160	–162	–162	–64,800	–64,800
	650	5	130	34	30	13,600	12,000
				–51	**–53**	**–51,200**	**–52,800**
				461	**529***	92,200	105,732
				261	257	104,400	102,800
				626	107	250,400	42,800
				447	**251**	**447,000**	**251,332**

9.23 shows that with rehabilitation existing land use will continue, involving three land use types under good drainage conditions, subject to different flood frequencies.

In the absence of rehabilitation, flooding increases, drainage conditions deteriorate and land use changes. In sub-area 1, land is abandoned due to frequent long-duration flooding. ALC Grade 3 land market values for good grassland are approximately £13,000/ha, less £600 economic adjustment. This capital sum is amortised to give an equivalent annual amount of loss due to abandonment (an annuity). Land in sub-area 2 switches from extensive arable to extensive grass with a decline in drainage condition and an increase in flood frequency. In sub-area 3, land use on Grade 1 land switches from intensive arable, including root crops, to extensive arable cropping due to a big increase in flood frequency. It is assumed the drainage conditions remain good (if drainage becomes bad, there could be a further £190/ha loss on this land).

Table 9.23 shows that the differences between annual financial and economic returns for 'with' and 'without' flood protection are approximately £450/ha and £250/ha, respectively, across the benefit area. Discounting these annual benefits of flood protection at 3.5 per cent over fifty years, the expected life of the rehabilitation scheme, gives a present value of £10,500/ha and £5,900/ha in financial and economic prices accordingly, equivalent to £10.5 million and £5.9 million, respectively, for the whole benefit area. The economic estimate is used to assess benefits from the point of view of the national economy. This can be set against the capital and maintenance costs of the project to determine feasibility. It is noted here that financial benefits to farmers exceed economic benefits by a factor of 1.8.

Sensitivity analysis can be carried out for all the assumptions used, notably the extent of changes in land use, flooding and drainage conditions, discount rates and the project's lifespan. Alternative assumptions can also be made about likely rates of deterioration from the assumed 'with' to 'without' situations, as well as the possible effects of catchment and/or climate change on the frequency and seasonality of flooding. The validity of assumptions regarding the capital valuation of land lost to agriculture can be assessed, and the assumption that intensive arable production (with root crops) should be treated as extensive arable in the economic analysis in this case. The analysis may be extended to include possible changes and differences in agri-environment benefits between the two situations.

This simple case illustrates the use of the preceding methods and indicative values to undertake a high-level appraisal of flood risk for agriculture. It indicates the type of information required, particularly regarding land use and the impact of changes in flood risk. It also shows the importance of understanding and accounting for variations in agricultural circumstances and practices and their interaction with flood risk management. It is critically important, although not done here for simplicity, to scrutinise key assumptions and, reflecting degrees of confidence, to produce a range of estimates rather than single values.

9.12 LESSONS FROM EXPERIENCE

During the period 1950–75, a large part of the public expenditure on flood risk management involved capital projects to provide flood protection for agricultural land. These were justified against prevailing agricultural policy objectives, but with little attempt to assess the efficacy of these investments of public funds in flood risk management for agriculture. In this context, some post-project appraisal research was reported in Penning-Rowsell *et al.* (2005a, Section 9.10), and the methods for the appraisal of flood risk presented in this chapter draw on this work.

9.12.1 Capital schemes

Site-based reviews, supported by field and farmer surveys, of twenty-two agricultural land drainage projects (Morris, 1992) confirmed that the main benefits to farmers were twofold:

- 'automatic' benefits: e.g. the relief of flood damage on a particular land use whether grass or arable; and
- 'potential' benefits which need to be 'taken up' by farmers and involve some action on their part, such as the switch to high-value cropping as a result of flood controls which otherwise would not occur.

By far the biggest benefits from agricultural drainage projects have been those associated with the take-up of potential benefits, especially the intensification of existing land use and the switch from grassland to arable (Morris and Hess, 1986; Morris, 1992). This take-up has often been associated with the installation of artificial field drainage, encouraged at the time by government subsidy and generally high levels of support to the sector. Thus, the performance of flood management investments was dependent on the perceived need for, and the resultant response by farmers to, improved drainage opportunities.

For the schemes reviewed, there was considerable variation in capital cost (from £770/ha of benefit area to £4,600/ha, in 2013 prices) and average annual benefits (from *c.*£35/ha to £590/ha in 2013 prices), but these two indicators were not necessarily correlated. Fifteen of the twenty-two schemes were considered to have met the 5 per cent discount rate criterion which applied at the time. The best-performing schemes were large pumped projects which mainly intensified existing arable production to include vegetables and root crops, or schemes which facilitated land use change from grass to arable. Schemes which involved grassland intensification gave modest returns but, with relatively low average costs per hectare, they provided benefits that were greater than costs. The poor performers were high-cost, low-impact schemes where drainage was not a major constraint and farmer interest and take-up were limited.

Formal (IDB) or informal drainage organisations were a critical determinant of scheme performance. Eleven of the fifteen 'profitable' schemes involved such organisations. It was here that farmers pushed for and were quick to exploit the potential benefits of improved drainage, and where drainage organisations ensured appropriate operations and maintenance. At individual farm level a complex array of farm and farmer characteristics were associated with uptake of benefits, but younger, better-educated, commercially oriented farmers tended to exploit benefits first.

The review showed that the better-performing, most cost-beneficial schemes were those that were done first, which made sense. Many of these early schemes were not subject to formal benefit–cost analysis (BCA). Of seven schemes subject to BCA, only two appeared to meet the test criterion applied at the time. BCA was applied to later, more marginal schemes, and it did not appear to give reliable predictions of performance. There was a tendency both to underestimate costs and to overestimate benefits, especially in those schemes requiring take-up by farmers.

9.12.2 Maintenance programmes

Considerable funds are committed to annual maintenance of the drainage system in rural areas, such as vegetation cutting and clearance, and de-silting. The methods here can be adapted to the appraisal of river maintenance, where benefits are the avoidance of losses associated with an increase in flood risk. In some cases maintenance is tied to the design standard of a capital scheme, but in many cases standards of maintenance have evolved over time as part of custom and practice such that it is difficult to determine the reason for the service that is provided and the benefit so derived. In some cases maintenance has been used to provide a higher standard of service than otherwise might prevail; that is, improvement rather than maintenance. In some cases this has encouraged farmers to switch to higher-value cropping.

A review of fifteen river maintenance programmes in six Environment Agency regions (Dunderdale and Morris, 1996b) revealed benefits to maintenance programmes (in 2013 prices) of between £6 and £86/ha/year (average weighted by area, £56/ha), highest in support of intensive arable. Maintenance costs varied from £4 to £93/ha (average, £41/ha). Twelve of the fifteen schemes appeared to be cost-effective using financial prices for farm commodities, but this was reduced to nine using economic prices.

This evaluation – almost alone in its aim of assessing past maintenance expenditure – confirmed that the more intensive the land use, the better the standard of drainage (either due to piped under-drainage or freely draining soils), the greater the sensitivity to standards of maintenance, and the greater the justification and scope for expenditure on river maintenance for agriculture.

9.12.3 Integrated washland schemes

The changing policy priorities alluded to earlier are encouraging an integrated, multi-purpose and potentially more 'sustainable' approach to flood risk management in rural floodplains. This seeks, for example, to combine storage of flood water with biodiversity management and extensive farming. Evidence suggests that there is considerable scope to exploit the potential synergy amongst flood risk management, conservation and farming livelihood objectives, but it is important that integration is 'designed-in' from the beginning rather than introduced as an afterthought (Morris *et al.*, 2004: 2005)

9.12.4 Evidence from assessment of major flood events in England

Summer flooding in June and July 2007 caused serious damage across 42,000ha of farmland in the West Midlands, Oxfordshire and Yorkshire at a time when agriculture is most vulnerable to flooding. Total agricultural damage costs were approximately £50 million. Although this amounted to just *c.*2 per cent of the estimated total economic cost of the flood events (£3.2 billion), the agricultural damage costs were very serious at the local scale (Chatterton *et al.*, 2010). It is also clear that flooding of farm land helped to reduce flood risk in urban areas.

A detailed visit survey of over eighty affected farmers using the survey methods outlined above provided estimates of damage costs (Posthumus *et al.*, 2009). Analysis confirmed that agricultural flood damage costs varied between crop types and regions, mainly associated with statistically significant ($p<0.05$) differences in the duration of flooding, yield loss and, to lesser extent, expected yields by region. Yield loss on cereal and oil seeds, which typically account for *c.*80 per cent of arable crop areas, were approximately 55–60 per cent of normal yields for fields in which flooding occurred. Adjusted for the proportion of fields actually flooded, typically 70 per cent, this increased yield loss to 77–86 per cent loss of normal yield on crops actually flooded. Farmers reported a further loss (equivalent to an extra 10 per cent loss) due to reduced quality on any recovered yield.

Flooding of potato fields, representative of high-value roots and fresh produce, led to the loss of the whole crop. Grass yield losses were approximately 40 per cent of annual production (46 per cent adjusted). Average damage costs were positively associated with the relative profitability of crops, for example for grass (£380/ha mean estimate for all region flood areas in 2007 prices), oil seed rape (£485/ha), cereals (£797/ha) and potatoes (£4,054/ha). These estimates of yield loss are consistent with those derived using the methods outlined above.

The summer 2007 flood study also showed significant differences in flood damage costs at farm level amongst farm types ($p<0.05$). Damage costs (in 2007 prices; Table 9.23) were highest for horticultural farms, at *c.*£7,000/ha flooded, followed by *c.*£2,000/ha for general cropping, including potatoes, *c.*£1,000/ha for cereals and dairy farms and *c.*£600/ha for grazing livestock farms. These damage costs broadly reflect the intensity and relative profitability of the main farming systems and are consistent with the guidance given earlier.

Flooding in spring (April–May) 2012 severely damaged grasslands on the Somerset Levels and Moors (Morris and Brewin, forthcoming). The seasonal timing of the flood and its relatively long duration (in excess of four weeks in some areas) led to major loss of grazing and winter feed for the 2012 farming year. Eleven farmers affected by flooding in the Currymoor drainage district were interviewed using the approach outlined above. The main financial and business impacts on farms (in 2012 prices) were associated with:

- complete loss of grass production for grazing or hay/silage that will need to be replaced by purchased feeds if output is to be maintained (typically £480/ha);
- loss of revenue from sales of hay (£400/ha) or finished livestock (typically £780/ha);
- loss of grazing days (£100/ha);
- need to reseed damaged pastures (typically £200/ha); and
- other miscellaneous costs (*c.*£60/ha).

Taking care to avoid double counting because of multiple and overlapping impacts, estimated average costs were approximately £900/ha (+/– £100) in the worst-affected areas, rising to over £1,100 (+/– £100) if the likely knock-on effects into 2013 are allowed for. In the worst-affected areas *c.*512ha of HLS were also seriously damaged. Average estimated environmental damage costs were between £213/ha and £317/ha based on the 'revealed' value of payments to farmers and scheme administration costs

Table 9.23 Weighted averages of agricultural damage costs due to summer 2007 floods (£/ha flooded)

	Horticulture	General cropping	Cereals	Mixed	Dairy	Grazing livestock	Pigs	All farms
	(n = 4)	(n = 20)	(n = 22)	(n = 11)	(n = 9)	(n = 9)	(n = 3)	(n = 78)
Arable production	6,592	1,760	530	223	163	7	715	828
Grass production	0	64	73	143	464	371	68	156
Livestock production	0	5	12	–3	328	152	4	58
Other	287	199	236	49	104	81	162	163
Total cost	**6,879**	**2,028**	**850**	**411**	**1,058**	**612**	**948**	**1,207**

Notes: Mean values are significantly different between farm types at $P<0.01$; General cropping included potatoes and field vegetables; Mixed farming combines arable and grassland. Rounding errors.

Source: Posthumus *et al.*, 2009

(and assuming that the intended outcomes of the scheme were not achieved following serious flood damage).

9.13 SUMMARY

This chapter has explained the importance of the management of flood risk and associated land drainage for agriculture. Although flood damage costs per unit area are much lower in agricultural areas than urban areas, they are substantial when spring and summer floods occur on intensively farmed land. There is a close connection between the management of flood risk for agriculture and the management of land drainage, as this affects the productivity of farm land. Managing flooding on farm land cannot be seen in isolation of managing waterlogging. The selection of methods for appraisal depends on the extent and permanency of change in flood risk on farm land, in the extreme case leading to a potential loss of agricultural land. The guidance referred to here is applicable to changes that are not perceived to have a major regional or national impact. Where they do have such an impact, a more strategic assessment may be required, including issues of food security. Furthermore, the appraisal of flood risk management in rural, predominantly agricultural areas is increasingly set in a broader framework that attempts to exploit the potential synergies and trade-offs amongst flood risk management and other beneficial ecosystem services, such as food production, nature conservation and access to the countryside.

NOTES

1 This is not to say that agricultural enhancement is not, of itself, a valid area for public investment if circumstances change.

10 Environmental gains and losses in flood and coastal erosion risk management

This chapter discusses the difficult and often contentious question of when it is useful to evaluate environmental impacts in economic and monetary terms.

Section 10.1 introduces the topic, lays out some guiding principles to be considered and emphasises the importance of taking a systems approach. The likelihood of encountering environment-to-environment conflicts is stressed, as is the changing scene in this area of appraisal methodologies. Section 10.2 summarises the relationship between the natural endowment and the economy, together with two problems in estimating the economic impact of some environmental change. Two arguments for evaluating all such changes in economic terms are judged to be weak.

Section 10.3 introduces stakeholder engagement and negotiation as a key component of decision-making, particularly where environmental issues are concerned. Since conflict was defined in Chapter 2 as a necessary condition for the existence of choice, two mechanisms which may nevertheless make it possible for the stakeholders to reach agreement are outlined. Section 10.4 sets out aspects of the wider policy context in which environmental impacts need to be considered.

A step-by-step approach to the assessment of environmental values is presented in Section 10.5, which conforms with Defra guidance and advice. A proposed but still experimental approach for taking account of environmental impacts is outlined in Section 10.6. This can avoid the many difficulties of formally evaluating environmental benefits and costs in economic terms unless it is essential to do so.

Section 10.7 identifies environmental costs, and Section 10.8 suggests a number of situations where it will not be sensible to avoid evaluating environmental benefits and costs in money terms and outlines how to extend the experimental approach set out in Section 10.6 in those circumstances.

Section 10.9 outlines the Environment Agency's recommended 'scoring and weighting' approach to be used when monetising non-marketed goods, and Section 10.10 summarises some of the lessons we have learned from our experience in this field.

10.1 INTRODUCTION

In Chapter 2 it was stressed that economics is the application of reason to choice and that the primary purpose of applying economic analysis is to gain a better understanding of what the choice involves. This comes very sharply into focus when considering environmental issues. The core message of this chapter is, first, to focus upon improving understanding of environmental systems and how they would be affected if a particular course of action were adopted, where a key part of that process of improving understanding is engagement with the stakeholders (Acland, 1990). An important second point is that there are considerable legislative and other constraints to actions concerning flood risk management works that can – and indeed must – override any consideration of economics (Figure 10.1). A third key message is that monetary evaluation of environmental resources poses problems, even (and sometimes especially) when the recently developed 'ecosystem services' approach is taken (Eftec, 2010a), and that this monetary evaluation may therefore often have to be relegated to a secondary, supporting role.

Having said this, this appraisal of environmental resources should be:

- approached positively to explore the case for flood and coastal erosion risk management schemes contributing to environmental improvement; and
- part of mainstream appraisal, both from the outset and throughout, not an afterthought.

Indeed, any appraisal should aim to assess all the costs and benefits, including those environmental costs and benefits which are not straightforward to value in monetary terms (Eftec, 2010a). The costs and benefits of goods and services that are not traded in markets must not be ignored just because they are more difficult to assess (e.g. nutrient capture or a breeding site for birds).

This chapter covers both the purely natural endowment (including sites of ecological, geomorphological and geological importance) and those which are of significance to us now partly because of – and as a record of – past human modification of that endowment (Figure 10.1). We suggest how to take account of the impact on the environment of such schemes in the appraisal of FCERM schemes (both positive/benefits and negative/costs), as an essential component of project appraisal (HM Treasury, 2003). We also discuss the many situations where legislation (particularly under the EU Water Framework Directive (WFD)) both impacts upon what can (or cannot) be achieved in the way of human intervention to alter flood risk and requires us to undertake 'WFD assessments' (now embedded in the Environment Agency's Strategic Environmental Assessments (SEAs)).

As with much of this Manual, we are dealing here with a changing situation. Nevertheless, the available methods for valuing environmental impacts of flood and coastal erosion risk management schemes have

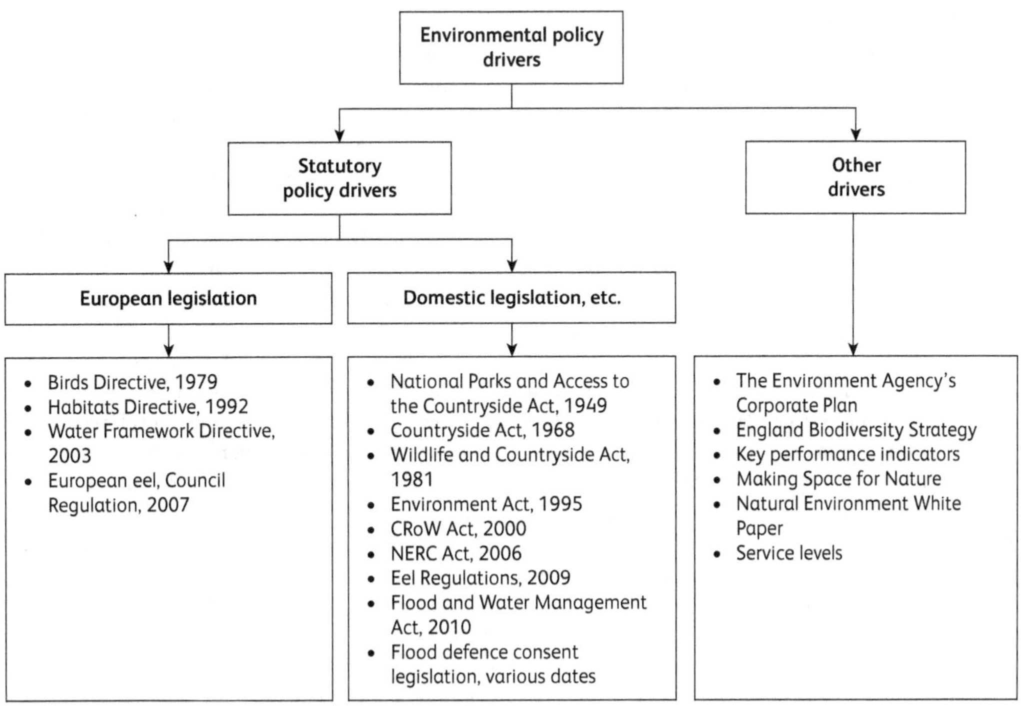

Figure 10.1 The relevant legislation and other drivers that must be followed in England and Wales when undertaking FCERM schemes

Source: Environment Agency, undated e, 2

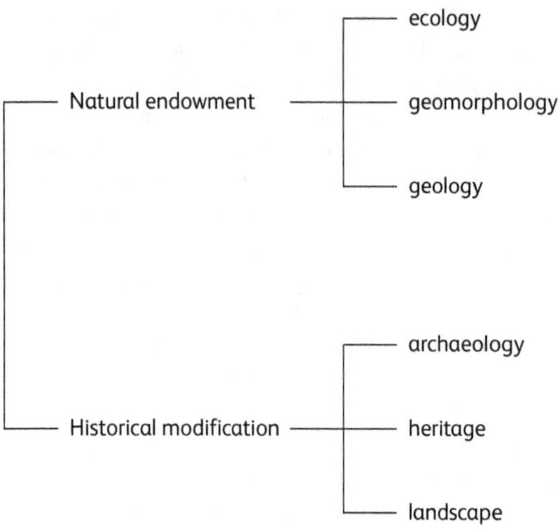

Figure 10.2 Environmental assets

not changed greatly since 2005, when the first 'Multi-Coloured Manual' (Penning-Rowsell *et al.*, 2005a) was published (i.e. the use of benefit transfer, willingness-to-pay evaluations, etc.), although the interest in and emphasis on environmental issues have continued to grow, with perhaps the more significant and certainly popular development relating to the ecosystem services approach (Eftec, 2010a). The UK Government, amongst many others, seems supportive of this direction of travel (Defra, 2007).

In this context, the Environment Agency has investigated – including with pilot 'trials' – whether and how the ecosystem services approach could be used to improve decisions across all its functions. The approach is seen as a potentially useful way to categorise and describe the impacts/services of ecosystems to society, although to date the main valuation techniques remain largely unchanged. The Agency published guidance in its *Economic Valuation of Environmental Effects* on how to use an ecosystem system service approach in FCERM (Eftec, 2010a), but this pre-dates the National Ecosystem Assessment (NEA, 2011) and is becoming somewhat outdated until alternative new guidance is published.

The 2010 guidance presents two levels of analysis:

- The 'first cut – a quick look at the economic value evidence' provides a series of default values for use in option development. The intention is to provide an indication of the magnitude of economic value evidence related to typical environmental effects associated with FCERM scheme options. This is particularly appropriate for preliminary assessments of an initial 'long list' of FCERM options, thus ensuring that an explicit account is made of their environmental costs and benefits. Depending on the requirements of the decision-making context, the evidence generated by this 'first cut' may be sufficient in a preliminary assessment. It is less likely to be sufficient in a main options assessment, when more detailed analysis is suggested.
- The 'second cut – value transfer' is a full-scale value transfer analysis in the specific context of FCERM schemes with the express intention of inputting to a benefit–cost analysis. This level of analysis requires more information and user effort than the first cut. It is reaffirmed here that the level of effort should, of course, be appropriate to the needs of the overall decision-making context, as highlighted above.

Taking a step back from this concern for 'how to do it', we need always to recognise that the natural endowment is a system of systems, changing and adapting both over time and space; this is part of the problem for appraisal, in that we are not dealing with a static situation. The first key element is thus that catchments and coasts are systems and must be managed as such: local action at one particular point usually has consequences elsewhere in this system, particularly downstream or downdrift. Catchments are systems of stocks and flows of water and sediment out of which emerges the morphology of the floodplain; coupled to those systems is that of the pollutants picked up, transported and distributed by the water and sediment. On the coasts, the geomorphological system, driven by wave, tidal and current energy, is the key system. Ecosystems then develop around the local water, nutrient or energy regime.

The consequences of intervention with schemes have to be evaluated in terms of the performance or functioning of the system as a whole, not just the local impact (although that is important, too). Hence, we support the shift to integrated water resource management (IWRM) and integrated coastal zone management (ICZM): the management of the catchment and coastal zones as wholes. The same philosophy also underlies the WFD, with its catchment focus, and its assessment and evaluation requirements, and it underlies the Environment Agency's position statement on biodiversity (Box 10.1).

BOX 10.1 THE ENVIRONMENT AGENCY'S POSITION STATEMENT ON CONSERVING, ENHANCING AND RESTORING BIODIVERSITY (ENGLAND AND WALES)

The Environment Agency's Flood and Coastal Erosion Risk Management (FCERM) work is required to be designed to conserve, restore and enhance biodiversity. The Agency aims to work with natural processes, maximising opportunities to conserve, enhance and restore biodiversity:

- on FCERM projects, schemes, plans and strategies from the start;
- ensuring FCERM projects and schemes provide effective mitigation or compensation for any significant adverse impact;
- screening (our strategies, plans, capital projects, schemes, maintenance works, flood defence consents and the applications for flood defence consent made by others) to make sure we avoid, mitigate or compensate for adverse impacts;
- designing projects which achieve multiple environmental benefits;
- stating in our approvals submission if/why we cannot identify restoration/enhancement opportunities;
- leading by example on our own land, acting upon opportunities to achieve biodiversity benefits;
- creating new partnerships and utilising other sources of complementary funding to achieve biodiversity benefits;
- consulting regularly with external partners on our capital schemes, strategies and routine maintenance programmes;
- monitoring our projects once constructed to ensure the planned benefits are achieved;
- undertaking adaptive management on sites where the anticipated benefits have not been achieved; and
- developing good-practice techniques.

In the case of IWRM, the intention is to integrate spatially across the catchment, between land and water, and across the different uses to which water is put (GWP, 2000). In Europe, the principles of IWRM are expressed in the WFD. The principles of ICZM are similar (CEC, 2000).

But this is not straightforward; the problem is often to discover those independencies within and between systems in the first place. Integration across the catchment is expressed in the catchment flood management plans (CFMPs; Environment Agency, 2004) which are now subsumed into flood risk management plans (themselves a Floods Directive requirement). The integration between functions will be found in the river basin management plans (RBMPs) as these are developed under the WFD (although integration across the two sets of plans – CFMPs and RBMPs – is not easy). Similarly, on the coast, the spatial aspects of erosion are covered by the shoreline management plans (SMPs). One implication of the drive towards functional integration should be expected to be a search for multi-functional or multi-scale solutions (Technical Support Unit, 2003).

The fact that we have to evaluate changes in the system, rather than a local change in isolation, has two important consequences:

1 The value of any element of the system takes its value from its relation to the whole of which it is part. Both functional and spatial relations are important, so, for example, the gains from improving a stretch of river to a standard that could support salmon depend upon both the condition of other parts of that river and whether it is inherently suitable for salmon. Similarly, the effects of saltmarshes on wave and energy attenuation depend upon relational factors (Moller and Spencer, 2002). Again, in some circumstances, erosion of part of the coast releases sediment which creates beaches downdrift so that the coastline has to be seen in terms of 'sacrificial anodes' which protect other areas. Where erosional material feeds sites downdrift, it is necessary to

count the loss of that provision in assessing any coastal protection works as one of the costs of the proposed works.

2 Non-linearities or thresholds in system response are critical so that any apparently small change may have major consequences.

Therefore, it should not be supposed that there is such a thing as a standard value applicable to every wetland or sand dune system. Again, the value of the gain of 100 hectares of wetland will depend upon where that gain takes place within a catchment, or coastal behavioural unit, as well as in which catchment or coastal behavioural unit.

Rivers (Newson, 1997) and coasts (Burgess *et al.*, 2002; Bowers, 2001) are naturally dynamic, both spatially and temporally: any local change can thus play out over both time and space. But, as locations to live and work, we would prefer them to be less dynamic, or to confine their change to certain limits. We have, in the past, sought to fix them temporally and spatially as an unchanging 'static' state. The emphasis now is on seeking to work more closely with those dynamic 'natural' processes (Defra, 2002; 2005) and there is an emphasis upon seeking systems that are resilient (de Bruin and Klijn, 2001; Nicholls and Branson, 1998).

Because they are dynamic systems, we have to consider the impact of an intervention on the dynamic response as a whole, where, for example, a flood is only one extreme of the variation of river flows over time. Any intervention must consider the impact on the overall pattern of variation in flows, and not simply on flood flows (Green, 2004).

For example, afforestation may reduce flood flows but only at the cost of creating a low-flow problem as well (Calder, 2003). Similarly, the consequences that reservoirs can have on downstream ecosystems as a result of changes in the flow regime are well known (Acreman *et al.*, 1999; Dyson *et al.*, 2003). But a wetland can have the same effect on the flow regime and hence the same impact on downstream ecosystems. In hydrological and hydraulic terms, inducing any form of storage has similar advantages and disadvantages. For example, as compared to the very peaked annual hydrograph of the Blue Nile, the annual hydrograph of the White Nile shows only a gentle rise as a consequence of the Sudd wetlands and Lake Victoria (Baecher, 2001).

The scientific interest in geomorphological and geological systems is in the processes of change; in the case of geology, this interest in change is limited to sufficient erosion occurring so that the surface exposure is kept clean of plants. Whilst ecology is centred upon change – ecosystems being dynamic (e.g. Holling, 1973) – ecologists have often in the past been reduced to seeking to maintain the state of those remaining sites of ecological importance; so much has been lost that they seek to hold on to whatever is left. But this situation is changing: the *Making Space for Nature* proposal (Lawton *et al.*, 2010) suggests the need for a 'whole-landscape' approach to biodiversity, by creating more and bigger interconnected reserves, such as 'green corridors'. Of those historical modifications of the environment the concern from archaeologists and those with similar interests is always to maintain what is there, to maintain the current state. The landscape and ecosystems of this country have been created by millennia of anthropogenic intervention (Hoskins, 1970); it is this artificial landscape which we value and which generally we want to preserve from change.

For example, some of our most valued wetlands (e.g. the Somerset Levels) are the result of drainage undertaken between the ninth and twelfth centuries (Williams and Williams, 2003). It takes active intervention to maintain those created landscapes and ecosystems in their current state. Hence, there are inherent conflicts of interest between change and seeking to maintain the current state. Other environment-to-environment trade-offs are commonly found particularly in coastal zone management. For example, without managed realignment, an area of mudflats might be lost through coastal squeeze, but managed realignment could mean the loss of brackish lagoons behind the existing coastal defences.

Thus, the conditions under which catchment and coastal zone management must be undertaken (where it is within those broader concepts of catchment and coastal risk management in which flood and coastal defence must be considered) are:

- dynamic systems;
- closely coupled systems; and
- conflict.

So, to summarise in this context, in principle, all environmental costs and benefits of flood and coastal risk management that can sensibly be valued in monetary terms should be included in an appraisal's benefit–cost analysis. The only exceptions that we would recognise are:

1 When environmental valuation is likely to be very difficult (or disproportionately expensive), and when a sensitivity test has clearly shown that it would make no difference to the decision about which scheme/option to develop.
2 Where no meaningful monetary valuation is possible. In this case the environmental costs and benefits should still be fully described and taken into account outside the benefit–cost analysis, so as still to have a bearing on the overall appraisal. Multi-criteria analysis can provide a framework for this description.

Even if it is not feasible or practical to value all costs and benefits of a proposed scheme, it is important to consider how the scheme options differ in environmental terms, and how just these differences might be best described and possibly valued in money terms.

10.2 THE ENVIRONMENT, THE ECONOMY AND ECONOMICS

It follows from the above that the first problem in assessing environmental change is to be able to take a systems approach to the environment. A further issue is the relationship between the environment and the economy.

There would be no economy without the environment, so the relationship between the two is best seen as 'multiplicative' (i.e. the one factors up the other): we use our ingenuity and human resources to maximise what can be achieved, given the natural endowment. There can be no fishing without a sustainable fishery, no matter how advanced our fisheries technology. Reduce that natural endowment and we have to increase the human resources we employ, or the ingenuity with which we use them, in order to 'stay in the same place'.

A pre-condition is therefore that there must be some technical means of increasing the efficiency of utilisation of those human resources to compensate for the reduction in the natural endowment. This demands, for example, that we invent a means of farming cod in order to replace declines in the stock of wild cod. If no such means exist then the environmental loss is unrecoverable. The fact that the relationship between the economy and the environment is multiplicative makes the problem of evaluating changes in the environment much more difficult.

The second problem is that there are marked differences between people in how they frame the problem: some individuals and organised groups will reject outright the utilitarian approach that underpins conventional economics and instead frame the problem in essentially moral, rather than technical, terms. Such people or groups see the whole process of economic evaluation of environmental resources as both involving a fundamental misunderstanding of what the choice is about and essentially 'wrong'.

Thus, when Burgess *et al.* (1997) interviewed members of the public who had previously been involved in a CV study of the Pevensey Levels, some of the respondents were outraged both that the decision would be based upon what people said they would be prepared to pay and that their views had been elicited for that purpose. The implication is that economic values are not useful unless they are widely accepted as meaningful and address the nature of the choice to be made.

We see two arguments that are sometimes made for economic evaluation as unsound. First, whilst many choices will imply some valuation of environmental consequences, an implied valuation derived from these circumstances does not necessarily point to the real value of that environmental asset or one that should be the basis for a new choice. As argued in Chapter 2, choice is a learning process, so our past choices cannot unambiguously prescribe our future choices; past choices only provide guidance.

Second, a particularly pernicious argument is that anything that does not have an economic value or cost attached to it will be omitted from the decision. We need to ensure, guided by Treasury advice (HM Treasury, 2003), that this should not happen.

The moral of this argument is that any decision process that excludes elements which some people consider to be important should be changed so as to try, as far as possible, to include those elements. In such cases deliberative multi-criteria analysis (DETR, 2000b; Green, 2003) is likely to be preferable to reliance solely upon benefit–cost analysis.

But we are dealing here with an evolving situation. Over the last ten years, a new, broader approach to valuing the natural environment has been – and continues to be – developed: the ecosystem services approach. This has its origins in the UN's Millennium Ecosystem Assessment (MEA, 2005), which sought to assess the consequences of ecosystem change on human well-being on a global scale. In response, the UK Government mounted a major study to assess the state of the UK's natural environment and the services it provides to people, considering the drivers for change in the past and into the future. Its report is published as the UK National Ecosystem Assessment (NEA, 2011).

The ecosystem services approach identifies a wider range of interrelated benefits to human well-being, the economy and society, or ecosystem services, that nature provides, directly or indirectly, than had previously been considered and stresses the importance of valuing these services properly. It divides ecosystem services into four broad categories:

1 Supporting services – the basic resources, such as soil formation and retention and water cycling, that are needed for other ecosystem services.
2 Provisioning services – products such as food and timber obtained from ecosystems.
3 Regulating services – the benefits arising from the regulation of ecosystem processes, such as erosion control by vegetation or flood and other hazard mitigation.
4 Cultural services – non-material benefits – spiritual, aesthetic, cultural and social – that people obtain from ecosystems; for example, from the landscape, biodiversity and recreation.

Eight UK broad habitats are noted as offering differing combinations of ecosystem services and goods. These include fluvial habitats, coastal margins and freshwaters, and it is obvious that these are particularly relevant to this Manual.

The NEA informed the government's White Paper on the natural environment, *The Natural Choice: Securing the Value of Nature* (HM Government, 2011). This document recognises nature as a complex, interconnected system and stresses the need to value fully the economic and social benefits of a properly functioning environment. In 2007, Defra produced an introductory guide to valuing ecosystem services, including an FCERM case study on Wareham, Dorset (Defra, 2007; see also Eftec, 2010a, Part 4). The Environment Agency has also published provisional guidance on how to use an ecosystem services approach in FCERM. This is very much an evolving area and further work is needed to develop the ecosystem services approach in policy and project appraisal (including building on the pilot 'trials' mentioned above). However, it does offer a potentially useful framework within which the specific benefit assessment techniques outlined in this chapter and in Chapter 8 can be applied.

10.3 NEGOTIATING CHOICES

These problems imply that the key is the process through which the choices will be made; that it is critical to engage with all the stakeholders (Audit Commission, 1999; Cabinet Office, undated) and the central role of project appraisal is to support this process of engagement by enhancing understanding of what the choice involves, and enabling the different stakeholders to share their understandings (see Chapter 2). In turn, this means that the scope of economic analysis should be defined and accepted by the stakeholders; it should be undertaken to the extent that they feel it to be helpful and meaningful. In particular, it is unlikely to be helpful to couch environment-to-environment trade-offs in terms of relative economic values.

Merely introducing stakeholder engagement does not mean that a consensus should be expected. Indeed, since conflict was argued in Chapter 2 to be one of the two conditions for the existence of a choice, we should logically not expect consensus to emerge in all circumstances and should be somewhat suspicious of apparent consensus if it does appear to emerge.

Arrow's 'Impossibility Theorem' (Arrow, 1963) proves that, under some reasonable assumptions, when different individuals have different rank orders of preferences between options, there is no option which they should be expected to agree is the best. Like the famous claim in the nineteenth century that science proves that bumble-bees cannot fly, everyday life demonstrates that there are ways around Arrow's theorem. The two hidden assumptions behind the theorem are: that the participants cannot talk to each other; and that each choice has to be taken in isolation from other present and future choices. Thus, that the participants cannot negotiate and that they have nothing about which to negotiate.

The two ways around the theorem involve introducing negotiation and something to negotiate about. One way to do this is to invent a new option; in general, there is nothing so useful in resolving choices as the invention of a new option. Including environmental enhancements or adopting a 'no net loss' approach (Heimlich, 1991) are both ways of introducing new options. The alternative is to couple one choice to others, but this can create more scope for trade-offs than can satisfy all parties.

For example, faced with a conflict as to which television programme to watch, buying another TV set or suggesting going out to dinner instead are ways to invent new options. Negotiating the choice of programmes over the night or the week is an example of the second approach. In our field, the adoption of a catchment or coastal zone approach should create a greater opportunity for negotiated outcomes than simply looking at each local problem in isolation.

But we believe there are contexts when some economic evaluation of an environmental change cannot be avoided; these contexts are discussed in Section 10.8. The necessary first step before stakeholder engagement or economic evaluation is to understand the relative importance of different types of site within the classes of assets described in Figure 10.2, the legislative requirements associated with each, and how the management of each fits within the wider policies of sustainable development.

10.4 THE POLICY CONTEXT OF THE ENVIRONMENT IN FLOOD AND COASTAL RISK MANAGEMENT

Since publication of the Brundtland Report (WCED, 1987), sustainable development has progressively become a central policy objective for the UK Government. *A Better Quality of Life: A Strategy for Sustainable Development for the UK* (DETR, 1999) sets out government thinking and identifies three objectives to achieve sustainable development:

- prudent use of natural resources (environmental);
- maintenance of high and stable levels of economic growth and employment (economic); and
- social progress that recognises the needs of everyone (social).

In 2011, the UK Government reaffirmed its commitment to sustainable development (Defra, 2011c, 2), indicating that this meant 'making the necessary decisions now to realise our vision of stimulating economic growth and tackling the [fiscal] deficit, maximising wellbeing and protecting our environment, without negatively impacting on the ability of future generations to do the same', building on the principles that underpinned the UK's 2005 sustainable development strategy. The 'refreshed vision' in 2011 recognised that the three 'pillars' of the economy, society and environment are interconnected and that 'Our long term economic growth relies on protecting and enhancing the environmental resources that underpin it, and paying due regard to social needs' (Defra, 2011c, 2).

These points were reaffirmed later in 2011 in *The Natural Choice: Securing the Value of Nature* (Defra, 2011d, 4), with the affirmation that:

> Too many of the benefits we derive from nature are not properly valued. The value of natural capital is not fully captured in the prices customers pay, in the operations of our markets or in the accounts of government or business. When nature is undervalued, bad choices can be made. We will put natural capital at the centre of economic thinking and at the heart of the way we measure economic progress nationally.

In this context, it is clearly important that schemes at the local level should contribute to the achievement of these sustainable development objectives by finding a viable balance between these environmental, economic and social objectives.

Principles of long-term sustainability depend on the maintenance of underlying environmental processes. This is important because, as noted earlier, the protection of current wildlife or geological assets may conflict with sustaining geological and geomorphological processes by working with geomorphic principles.

Along with other countries worldwide, the UK has commitments under the Rio Earth Summit agreements (UNCED, 1992) to foster practices for achieving sustainable development. Environmental impact assessment is one of the key tools identified for this purpose. It operates by ensuring that decision-makers are provided with appropriate information on the likely environmental, social, economic and health effects of development proposals before development consent is given. This allows decision-makers to weigh carefully the benefits and costs of developments, in both the short and long terms, before arriving at their decisions, and ensures that environmental costs can be taken into account even if they are not quantified in economic terms.

Government policy for sustainable flood and coastal risk management measures therefore emphasises a holistic approach which takes into account interrelationships within a catchment or coastal cell and provides flexibility for future generations to determine their own options. In addition, there are three other evolving areas of policy which also have to be taken into account. These are discussed below.

10.4.1 The concept of 'quality of life capital'

As part of the UK Government's sustainability strategy, the concept of 'quality of life capital' was developed as a tool jointly by the Countryside Agency, the Environment Agency, English Nature and English Heritage (CAG Consultants and Land Use Consultants, 2001). It is designed to help in determining the features and characteristics of the environment that are important, in resolving problems and in identifying opportunities.

Its key objectives are to:

- integrate environmental, social and economic issues;
- look for improvements in the quality of life, not just accept the status quo;
- value the commonplace as well as the unusual or rare; and
- link community planning, professional judgement and the concerns of local people.

In pilot studies the approach has been found to dovetail well with environmental assessment scoping studies and its principles should be incorporated into environmental assessments of coastal protection and flood risk management schemes at that stage. It provides a set of broad criteria for assessing the sustainability of projects which impact upon the quality of people's lives as opposed to the rather narrower environmental components defined in the introduction to this chapter.

10.4.2 The Water Framework Directive (WFD): a vital and overriding consideration

Introduction

The WFD is a European Directive which introduces a new strategic planning process that is designed to manage, protect and improve the water environment (Environment Agency, 2011b). The purpose of the WFD is to establish a framework for the protection of water bodies (including terrestrial ecosystems and wetlands directly dependent on them) which aims to:

- prevent further deterioration;
- enhance their status;
- promote sustainable water use;
- reduce pollution; and
- mitigate the effects of floods and droughts.

In this context river basin management plans (RBMPs) are statutory plans for protecting and improving the water environment. They describe the main issues for the water environment within each river basin district. They tell us, at a local level, which actions and measures the competent authority (the Environment Agency in England and Wales) and others need to implement to achieve the objectives of the WFD: RBMPs are the tool used to deliver WFD objectives.

Measures in RBMPs need to be implemented to prevent the deterioration in status of water bodies and to achieve good status or potential for all water bodies. In the WFD, a water body is defined as a discrete and significant element of surface water (such as a lake, reservoir, stream, river, canal, a transitional water – i.e. an estuary – or a stretch of coastal water).

The WFD requires organisations such as the Environment Agency to aim to achieve good status or potential in all water bodies. For surface waters, this means:

- good ecological status (GES) in water bodies; or
- good ecological potential (GEP) in water bodies designated as 'artificial' or 'heavily modified' water bodies (AWB/HMWB).

And:

- good chemical status.

The RBMPs define deadlines for achieving GES/P; typically actions need to be implemented by 2015, 2021 or 2027. Where water bodies are currently at less than GES/P, a series of improvement measures need to be planned (as, in England and Wales, described in Annexes B, C and D of RBMPs).

Good ecological status

GES is the WFD default objective for all water bodies and is defined as a slight variation from undisturbed natural conditions. The elements that make up ecological status include:

- biological elements (including fish, macro-invertebrates, macrophytes and diatoms); and
- supporting elements (made up of hydromorphology, ammonia, pH, phosphates, dissolved oxygen and eighteen pollutants, including some heavy metals and pesticides).

The term 'hydromorphology' is used in the WFD to describe the physical form and physical processes which occur in a water body, such as:

- flow patterns;
- the width and depth of river channels;
- features such as pools and riffles;
- sediment availability/transport; and
- interaction between the river and its floodplain.

This term includes both the hydrological and geomorphological characteristics that can support a healthy functioning aquatic ecosystem. Each of these elements contributes to the overall ecological status. A lowest common denominator rule is applied to the elements, so the lowest-scoring element denotes the overall status of the water body.

Good ecological potential

GEP is the WFD objective for AWB/HMWBs that are designated for a specific use, such as recreation, flood risk management or urbanisation.

Water bodies are designated as AWB/HMWBs when:

- the level of modification in these water bodies means the biological status is not able to achieve GES; and
- the use(s) for which the water body has been modified are still needed and cannot be achieved through 'other means'.

The AWB/HMWB designation accepts that the biological status of the water body has been impacted by its modification and so the alternative objective of GEP is set. This is the best ecological status an AWB/HMWB can achieve without compromising the use for which it was designated. No WFD action can be taken on these water bodies which will have a significant adverse impact on its use. So a water body which has been designated as having a flood risk management use should maintain that use. Only when all the relevant mitigation measures have been put in place can an AWB/HMWB be said to have reached GEP.

'No deterioration'

The WFD includes an obligation to prevent deterioration in the overall status of water bodies, referred to as 'no deterioration'.

New activities, such as flood alleviation schemes, maintenance activities and in-channel engineering works, may affect the water environment, impacting on its biological status, hydromorphology or chemistry. These impacts could lead to deterioration. New activities and schemes may also prevent proposed actions and mitigation measures included in RBMPs from being delivered. This may lead to a water body failing to meet its ecological objectives.

For new FCERM schemes, any hydromorphological impacts need to be fully assessed to establish if they will:

- cause deterioration; or
- prevent the achievement of ecological objectives.

To do this, a WFD assessment needs to be made, for which an eight-step process has been developed by the Environment Agency to help assess the compliance of new modifications with the WFD (Table 10.1). In many cases, however, it will not be necessary to follow all eight steps because at an early stage in the assessment process a project may be screened out of further consideration.

Article 4.7 of the WFD

Exceptionally, there may be situations where it is not possible for a scheme to be designed to prevent deterioration in ecological status/potential. Under these circumstances, the project needs to satisfy the

Table 10.1 An eight-step process to inform the compliance of new modification with the Water Framework Directive

Step	Action
1	Collect up-to-date water body baseline data
2	Collect proposed scheme baseline data
3	Preliminary assessment
4	Design and options appraisal
5	Detailed impact assessment
6	Apply Article 4.7 tests
7	Reporting
8	Follow-up post-project appraisal work

Source: Environment Agency, 2011b

exemptions criteria set out in Article 4.7 of the WFD. These criteria are summarised below:

- All practicable steps or measures are taken to minimise the impact.
- The reasons for the modification are explained in the RBMP.
- The reasons for the modification are of overriding public interest and/or the benefits to human health, safety or sustainable development outweigh the benefits of achieving WFD objectives.
- The benefits of the modifications cannot be achieved by another means (because these are not technically feasible or are disproportionately costly).

10.4.3 Defra's high-level targets

The high-level targets for operating authorities in the flood and coastal risk management area were first clearly set out in 1999 as elements in the environmental objectives for which they must aim (MAFF, 1999).

In the past, for biodiversity, these were:

- to avoid damage to environmental interests;
- to ensure no net loss of habitats covered by habitat action plans (under the framework of biodiversity action plans); and
- to seek opportunities for environmental enhancement.

These three targets provided the initial framework against which all schemes and scheme options for flood risk management and coastal defence were then assessed.

The metric changed after some ten years and six 'outcome measures' were subsequently established for the four-year period from April 2011 to March 2015 to monitor the outcomes from Defra's capital investment in flood and coastal erosion risk management. These measures are broadly similar to those used up to March 2011, continuing Defra's focus on the number of households receiving an improved standard

Table 10.2 Outcome measures set as targets for FCERM, together with the contribution these make towards Grant in Aid scheme allocations

OM 4	Statutory environmental obligations fully met through flood and coastal erosion risk management	Outcomes specifically funded under OM4:	
OM 4a	Hectares of water-dependent habitat created or improved to help meet the objectives of the Water Framework Directive	Water-dependent habitat created or improved	£15,000 per hectare
OM 4b	Hectares of inter-tidal habitat created to help meet the objectives of the Water Framework Directive for areas protected under the EU Habitats or Birds Directive	Inter-tidal habitat created	£50,000 per hectare
OM 4c	Kilometres of river protected under the EU Habitats or Birds Directive improved to help meet the objectives of the Water Framework Directive	Protected rivers improved	£80,000 per km of river bed

of protection from flooding or coastal erosion, the overall economic value of the government's capital investment, and the achievement of important environmental outcomes which are dependent on flood or erosion risk management. The new measures are consistent with the flood and coastal resilience partnership funding arrangements for the allocation of capital grants announced in May 2011 (Defra, 2011a). The relevant environmental measures are shown in Table 10.2 and the full list appears in Chapter 1 (Table 1.4).

10.4.4 Environmental assessment procedures

Environmental assessment provides a framework for project development by drawing together an assessment of a scheme's likely significant effects on the environment in a systematic way.

This approach should be applied whether or not a statutory environmental impact assessment (EIA) leading to an environmental statement (ES) is required. It helps to ensure that the importance of predicted effects and the scope for reducing them are properly understood by all those concerned in the decision-making process before decisions are taken.

This process enables environmental factors to be given due weight alongside economic and social factors. It is key to delivering sustainable development through managing risks and opportunities and involving all interest groups through a transparent system that can deliver environmental improvements, not just minimise impacts.

It is also a requirement to go through such a process to comply with European and UK legislation for Natura 2000 sites and most sites of national importance that are designated as SSSIs. In order to be applied effectively, it requires environmental assessments to be produced in partnership, and so maximise opportunities for 'adding value' in project development, not just to focus on minimising impacts. In addition, community participation in the process of environmental appraisal is delivering environmental benefits by protecting and enhancing biodiversity and environmental assets, both ecological and archaeological, as well as delivering wider social benefits (in terms of recreation and access).

Increasingly, the environmental assessment for an individual scheme should be set within a strategic environmental assessment (SEA) framework. The SEA is the formalised and systematic process of evaluating the environmental impacts of policies, plans, strategies and programmes, all of which set the framework within which flood risk management and coastal protection projects are promoted and implemented. SEA became a statutory requirement in 2004. Guidance has been published by the UK Government (ODPM, 2005), and Figures 10.3 and 10.4 give guidance as to whether and how the SEA process can operate.

More generally, the key steps in the environmental assessment process are summarised in Figure 10.5. The process is iterative, which means that many of the stages may need to be revisited in the light of new information discovered at a later stage. Consultation with the relevant environmental agencies, in particular, should be ongoing throughout the process.

10.5 DEFRA'S RECOMMENDED STEP-BY-STEP APPROACH

Proper assessment of environmental impacts will depend on a structured and rigorous approach to appraisal, which we recommend should include the four steps described in Defra's project appraisal guidance (MAFF, 1999): define; develop; compare; select and confirm. These are generally useful today and are discussed below.

However, it should be noted that for assessment at strategy, initial and detailed study levels, and to avoid disproportionate time and resources being spent on environmental benefit assessments (such as inappropriate use of willingness-to-pay surveys), two questions need to be asked at the strategy and initial stages:

- Is there an environmental concern significant enough to warrant such time and resources in assessment?
- Is option choice likely to hinge on the environment issues to be tackled?

If appraisers have evidence that impacts are significant then more consideration of them should take place at the initial and detailed appraisal stages, exploring any concerns confirmed at strategy stages (in this respect WFD assessments can be embedded here). There will normally be a pressing need for assessment at initial stages, although at detailed study levels the need may

Application of the SEA Directive to plans and programmes

This diagram is intended as a guide to the criteria for application of the Directive to plans and programmes (PPs). It has no legal status.

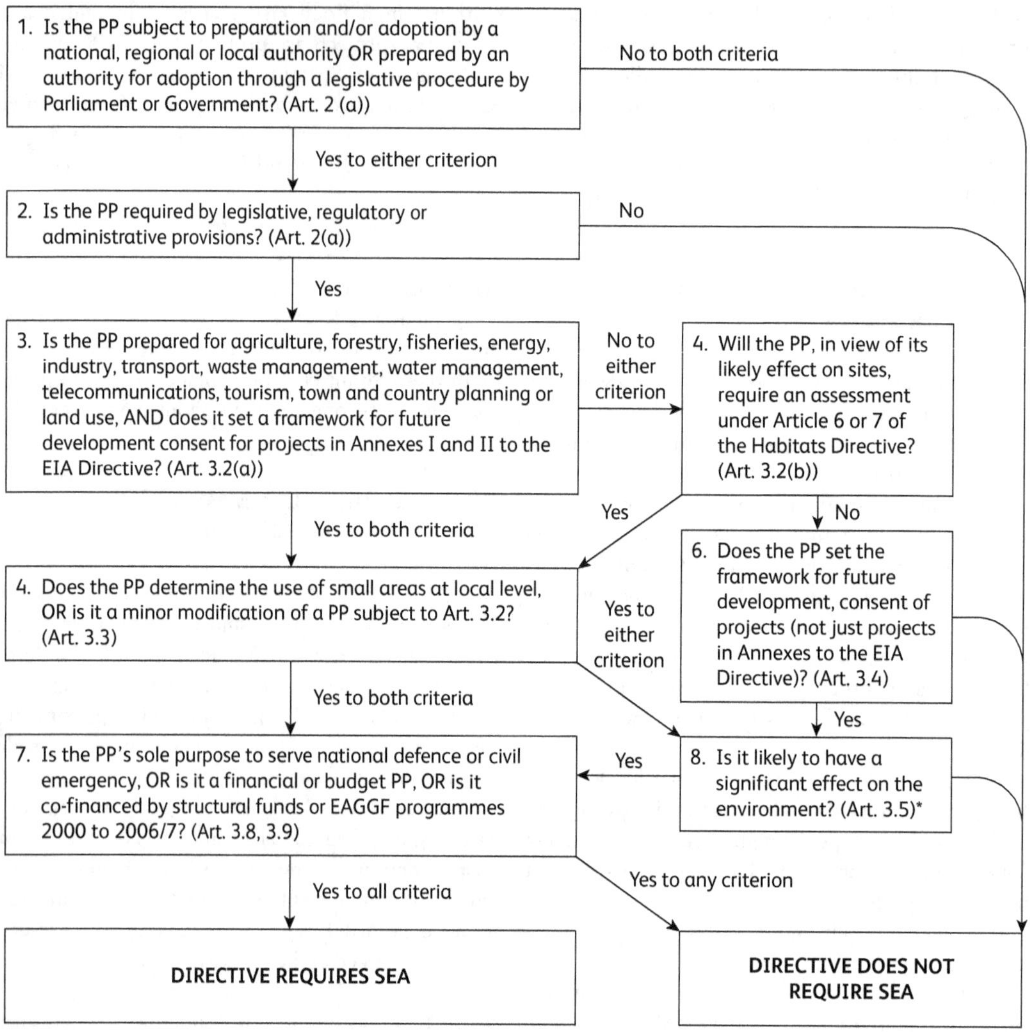

*The Directive requires Member States to determine whether plans or programmes in this category are likely to have significant environmental effects. These determinations may be made on a case by case basis and/or by specifying types of plan or programme.

Figure 10.3 Guidance as to whether the SEA Directive applies

Source: ODPM, 2005, Figure 2

Stages in the SEA process	
SEA stages and tasks	Purpose
Stage A: Setting the context and objectives, establishing the baseline and deciding on the scope	
Identifying other relevant plans, programmes and environmental protection objectives	To establish how the plan or programme is affected by outside factors, to suggest ideas for how any constraints can be addressed, and to help to identify SEA objectives.
Collecting baseline information	To provide an evidence base for environmental problems, prediction of effects, and monitoring; to help in the development of SEA objectives.
Identifying environmental problems	To help focus the SEA and streamline the subsequent stages including baseline information analysis, setting of the SEA objectives, prediction of effects and monitoring.
Developing SEA objectives	To provide a means by which the environmental performance of the plan of programme and alternatives can be assessed.
Consulting on the scope of SEA	To ensure that the SEA covers the likely significant environmental effects of the plan or programme.
Stage B: Developing and refining alternatives and assessing effects	
Testing the plan or programme objectives against the SEA objectives	To identify potential synergies or inconsistencies between the objectives of the plan or programme and the SEA objectives and help in developing alternatives.
Developing strategic alternatives	To develop and refine strategic alternatives.
Predicting the effects of the plan or programmes, including alternatives	To predict the significant environmental effects of the plan or programme and alternatives.
Evaluating the effects of the plan or programme, including alternatives	To evaluate the predicted effects of the plan or programme and its alternatives and assist in the refinement of the plan or programme.
Mitigating adverse effects	To ensure that adverse effects are identified and potential mitigation measures are considered.
Proposing measures to monitor the environmental effects of plan or programmes implementation	To detail the means by which the environmental performance of the plan or programme can be assessed.
Stage C: Preparing the Environment Report	
Preparing the Environmental Report	To present the predicted environmental effects of the plan or programme, including alternatives, in a form suitable for public consultation and use by decision-makers.
Stage D: Consulting on the draft plan or programme and the Environmental Report	
Consulting the public and Consultation Bodies on the draft plan or programme and the Environmental Report	To give the public and the Consultation Bodies an opportunity to express their opinions on the findings of the Environmental Report and to use it as a reference point in commenting on the plan or programme. To gather more information through the opinions and concerns of the public.
Assessing significant changes	To ensure that the environmental implications of any significant changes to the draft plan or programme at this stage are assessed and taken into account.
Making decisions and providing information	To provide information on how the Environmental Report and consultees' opinions were taken into account in deciding the final form of the plan or programme to be adopted.
Stage E: Monitoring the significant effects of implementing the plan or programme on the environment	
Developing aims and methods for monitoring	To track the environmental effects of the plan or programme to show whether they are as predicted; to help identify adverse effects.
Responding to adverse effects	To prepare for appropriate responses where adverse effects are identified.

Figure 10.4 Guidance as to stages in the SEA process

Source: ODPM, 2005, Figure 5

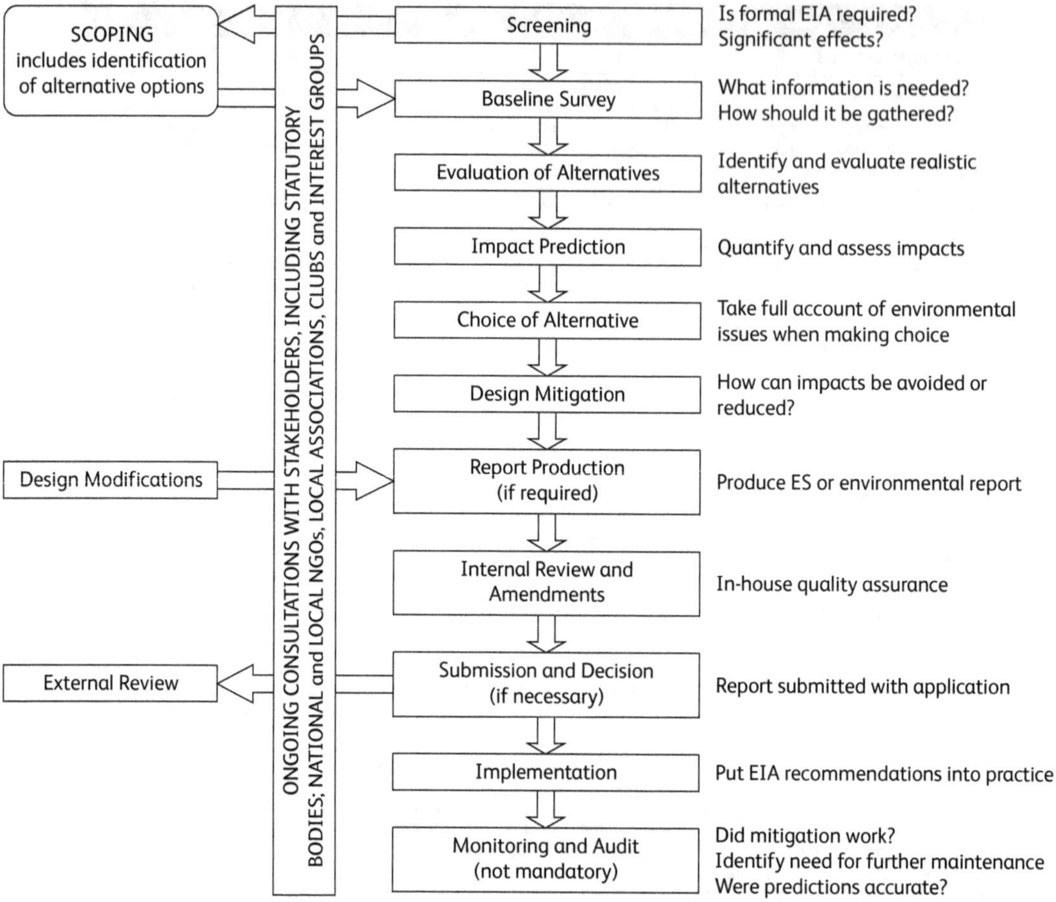

Figure 10.5 The key steps in the environmental business process

vary on a case-by-case basis, depending again on the size of the environmental impacts identified within the relevant area.

10.5.1 Step 1: Define: problem definition and objectives

This stage should define the full range of FCERM options.

In all cases the environmental consequences and objectives should be brought into the appraisal at the start. The most important aspect at this stage is an acknowledgement that avoiding environmental damage and achieving environmental gains are material considerations for scheme definition and objectives. They are just like any other category of benefit which may justify a flood or coastal erosion risk management scheme. In all cases the relevant stakeholders (e.g. Natural England and English Heritage, or their equivalents for Wales) should be included at this stage.

When considering environmental objectives, appraisers should identify:

- any critical environmental criteria, such as meeting legal requirements;
- any highly desirable objectives, such as meeting high-level targets (e.g. the WFD delivery on SSSIs); and
- any more general environmental outcomes that may be desired.

10.5.2 Step 2: Develop: preliminary appraisal

Having defined the FCERM options, a preliminary assessment should describe all the costs and benefits, including the positive and negative environmental impacts of all the alternatives.

When considering strategies and high-level plans, a scheme's SEA should help this task (Figure 10.4). The purpose here is not to attempt a monetary valuation or consider the balance of the costs and benefits (that comes later). But it is important here that descriptions of the effects are as clear and quantified as is practicable.

What needs describing (and later, if possible, valued) is the change (positive or negative) brought about by the options being considered, not an overall valuation of all aspects of the environment. Both the costs (damages) and the benefits of the 'do nothing' option should always be fully appraised.

One approach here is to use total economic valuation (Turner *et al.*, 1993; 2005). This comprises both 'use' and 'non-use' values (e.g. carbon sequestration and knowing that a wetland will be available for future generations) and takes a functional systems approach to establishing a total economic value for the effect that each option will have on the environment. In theory, this should capture most (but not all) values and avoid double counting. However, there are a number of practical difficulties and some of these – but by no means all – are rooted in quantifying environmental risks and uncertainties.

The next task is a preliminary appraisal and eliminating those options that are definitely not feasible, while ensuring that options with environmental benefits are not ruled out. Only options clearly not meeting the critical criteria, such as complying with legal requirements, should be eliminated here.

For example, a scheme having an adverse impact on a site designated under European Directives might be ruled out if there were an alternative solution not adversely affecting the site. Thus, schemes that are necessary to meet legal requirements may be assessed using cost-effectiveness analysis (Environment Agency, 2010a). The benefits of meeting the legal requirements are assumed to outweigh the costs and hence the focus can be shifted to achieving these objectives at least cost. However, often other types of benefit will differ between options which aim to meet the objective, in which case it may still be necessary to identify, describe, quantify and monetise the benefits, to the extent that they materially affect the choice.

Care should be taken not to let the appraisers' views or prejudices eliminate options that further analysis might justify. For example, until a realistic assessment is made of total benefits it might not be possible to say that the costs of a scheme with substantial environmental benefits, such as habitat creation, are disproportionate. Any grounds for ruling out options should be clearly reported. Appraisal summary tables may help structure this initial assessment to ensure that all environmental effects are captured.

10.5.3 Step 3: Compare: identifying the preferred option

A more detailed appraisal should be made of the options that have not been eliminated in Step 2.

This should include the statement describing the environmental costs and benefits of options together with a monetary valuation of those impacts where possible, subject to the principles described above. Care and rigour in the appraisal process will be needed to ensure that all relevant effects are captured and double counting is avoided.

Double counting is best avoided by recognising the impact pathway, the final impact on human welfare and the means of measuring this impact. For example, an environmental improvement that benefits anglers by improving fish nursery conditions and increasing fish stocks leading to higher catch rates should be evaluated via the change in the anglers' willingness to pay for these improvements. Other impacts, such as increased fish size, increased bait sales, consequential tourism impacts, etc., should already be reflected in this value, and separate estimations would lead to double counting.

A sequential approach should be used to decide on the method for:

- calculating a monetary value for an environmental cost or benefit; and
- ensuring that any impacts which cannot be included in the benefit–cost analysis are taken into account.

Following the principles outlined above, impacts on the environment should be valued (see Chapter 2) as described below.

Market prices

Market prices, where available, should be used to establish a value for environmental benefits/costs. Establishing monetary valuations should be relatively straightforward where there is a market price. For example, if a managed realignment increases fish stocks, this will have benefits to the local fishery, which can be valued.

However, many environmental goods and services do not have readily available market prices, in which case costings models and alternative means of establishing values will need to be considered (see below).

Benefit transfers

In some cases values from previous studies may be transferable. Care must be taken to allow for the fact that values may vary in differing circumstances, which may limit the validity of this approach. Where available, benefit functions should be used rather than unit benefits, as the former can take into account important variables, which may differ from site to site (Turner *et al.*, 2005).

As the number of valuation studies increases, the opportunity for drawing on their results should expand. If credible applicable values from previous studies are not available, plausible upper and lower bounds on values may be possible, helping consideration of whether it is worth commissioning further work to establish more robust values for the scheme in question.

Where no market price, acceptable proxy or robust transfer value is available, a scheme-specific study to establish values should be considered. Before undertaking this, though, an assessment should be made:

- to clarify whether the results are likely to affect the preferred option; and
- to clarify whether a meaningful monetary valuation is likely from that study.

Replacement costs

This method is to be used only where a prior decision has been made to maintain or replace a feature, either for policy reasons or to meet a statutory requirement. Then the cost of maintaining it *in situ*, relocating it or recreating it, whichever is the lower, can be used as a minimum value for the appraisal. However, this technique has limited applicability.

Therefore, the replacement cost method as an appraisal valuation technique is contingent on there being a prior decision to maintain, replace or relocate the feature being valued. What is then being assessed is the cost of complying with a policy/requirement, *not* the value of the feature (so these values cannot be used in benefit transfers). This is therefore not an acceptable measure of value where one is considering the merits of going beyond compliance with the policy/statute or assessing the acceptability of an option that would lead to the loss of a feature, whether or not it is protected by statute. Where the preferred option is to relocate or replace a feature, this method of valuation may not capture some potentially significant costs (dis-benefits), such as the loss of local amenity or historical significance. Such effects will need to be considered separately.

Willingness to pay

Where none of the above methods is applicable, a new study should be considered to establish values by calculating people's willingness to pay for the proposed environmental enhancement (Green and Tunstall, 1991).

At this stage, appraisers with experience and competence in environmental valuation need to:

- make a realistic assessment of the feasibility of such studies; and
- ensure that the values derived are credible.

The preferred method is to calculate the relevant population's willingness to pay as inferred by observing consumer behaviour (i.e. revealed preference using hedonic pricing). Where this is not feasible – and such a situation is likely – the alternatives are to ask people what they would be willing to pay for a particular benefit (stated preference) or to identify the compensation that they would require in order to accept a cost (willingness to accept).

Taking account of environmental costs and benefits that have not been valued in money terms

At this stage of appraisal all the environmental costs and benefits of all the options should be described and those that can be valued should have been valued.

If all the effects were included (through monetary valuation), the preferred option should be revealed by the scheme meeting Defra guidance on decision rules (see Chapter 3). Any environmental costs and benefits that it has not been feasible to include will need to be clearly identified, because they may still influence the decision about which option to choose. Again, appraisal summary tables can help here.

Where there are significant non-monetised costs and benefits, judgement will be needed as to whether they are sufficient to influence the preferred option. The most common framework for comparing unvalued costs and benefits is weighting and scoring (i.e. multi-criteria analysis). This technique can help rank options by taking account of both monetised and non-monetised costs and benefits (see Section 10.9).

Even if all the costs and benefits of an option cannot be valued, it is important to consider how the options differ and whether the difference between them can be valued. Switching analysis is one way of valuing the difference between options.

Thus, consider two alternative schemes, A and B. The whole-life cost of A is £10 million, compared to £8 million for B. However, A has significant additional environmental benefits. These environmental benefits would need to be at least £2 million for B to be preferable to A. Some 5,000 people live in the affected area, who might benefit from these environmental improvements. Each beneficiary would therefore need to be willing to pay £400 for these benefits to be sufficient to alter the choice based on the whole-life costs.

10.5.4 Step 4: Select and confirm: an appraisal of the preferred option

The final step is an appraisal to determine whether the preferred option is justified in terms of the Defra/Environment Agency funding criteria (see Chapter 1). Much of the work for this should already have been done in Steps 2 and 3.

However, if a scheme has been chosen on the basis of benefits that have not been valued in money terms, and the benefit–cost ratio is not sufficient to justify it, extra consideration may need to be given to ensure that the non-monetised benefits justify the required expenditure.

10.6 A PROPOSED EXPERIMENTAL METHODOLOGY

In parallel with and to complement the step-by-step approach outlined above, we have developed an experimental method that might guide decisions about the assessment in monetary terms of key environmental resources. This compares the benefit–cost ratios of schemes with and without consideration of these resources, to determine whether their quantitative assessment might be crucial to any decision that might be made.

In summary, the steps here are as follows:

1 Engage with the stakeholders to gain their views about all aspects of the possible scheme options, ensuring that these stakeholders include those with specialist environmental knowledge and understanding.
2 Determine the scheme that represents the best practical environmental option (BPEO), and the scheme that represents the next-best option (NBO).
3 Compare the benefit–cost ratio of the BPEO to that of the NBO, exclusive of the net environmental gains.
4 If the benefit–cost ratio of the NBO is robustly greater than 1.0 and that of the BPEO is greater than 1.0, adopt the BPEO. Other cases have to be considered further, as discussed below.

Choices will often involve conflicts between different environmental assets (say, herons versus fish). The stakeholders have to decide which is the most satisfactory trade-off, where this includes the provision of environmental enhancements. However, the WFD technically does not allow this kind of trade-off: if a scheme causes deterioration or prevents mitigation measures from being implemented, in legal terms it will only be progressed if you meet the requirements of WFD Article 4.7 (see above).

Figure 10.6 sets out the procedure that is proposed. The BPEO must be identified as such by the stakeholders; and, for example, must satisfy the requirements of the WFD as set out in the RBMP and

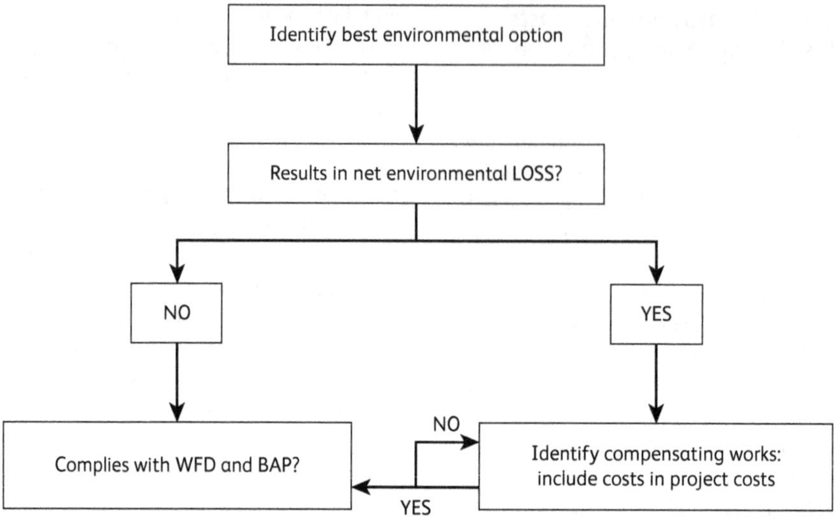

Figure 10.6 Identifying environmental best practical option

any relevant biodiversity action plan. The NBO is the cheapest option that provides adequate flood or coastal risk management; it provides no more than the legal minimum consideration of environmental issues.

Once the additional costs of the required environmental enhancements have been assessed, the benefit–cost ratio of the BPEO can be compared to that of the NBO. In the simplest situations, there are two possible outcomes.

Equation 10.1

$$\text{BPEO} > 1;\ \text{NBO} >>> 1$$

Equation 10.2

$$\text{BPEO} < 1;\ \text{NBO} > 1$$

In the first case where the benefit–cost ratio of the NBO is robustly greater than 1.0 and the ratio for the BPEO is greater than 1.0, adoption of the BPEO is justified. The benefits included in the assessment of the BPEO are identical to those of the NBO, and the only additional costs represent those of the environmental enhancements.

There are, however, other possible outcomes, as shown in Table 10.3. Those other cases (A and B) are considered further in Section 10.8.

Table 10.3 Possible outcomes

BPEO	NBO		
	b/c >> 1	b/c >1	b/c <1
b/c >> 1	Do BPEO	impossible	impossible
b/c > 1	Do BPEO	B	B
b/c < 1	A	B	B

10.7 ENHANCING AND MITIGATION COSTS

When considering the costs of environmental enhancement and mitigation, these could include:

- creation of new habitats;
- restoration of degraded habitats; or
- re-creation of previously lost habitats.

In the case of the protection of environmental assets, costs include:

- increased time for negotiation in the planning and design stages;
- increased land-take for the project;
- increased construction costs due to on-site mitigation measures during the operational stage;
- management after construction; and
- monitoring and management adjustments.

The creation of a substitute site as a replacement for what is being lost is not a straightforward process (and in the WFD the creation of substitute sites is not even an option: the legislation does not allow damage of a water body and compensation for it in another water body). The main costs relate to:

- land acquisition;
- set-up;
- ongoing management during the establishment stage;
- ongoing monitoring; and
- subsequent adjustment of management regimes over several years, depending on habitat type.

Estimates of habitat restoration and creation costs have been made for some, though not all, habitats relevant to flood risk management projects in the context of estimating costs of implementing biodiversity action plans (UK Biodiversity Group, 2000; see Table 10.4). However, monitoring of the first five years of implementation (Defra, 2002) indicates that there were considerable variations between predicted and actual costs, and more recent estimates (shown in Table 10.5) confirm the significant differences in obtainable valuations.

In most cases predicted indicative costs per hectare were found to be significantly underestimated. This was by a factor of five per hectare in the case of

Table 10.4 Predicted and actual habitat creation and management costs

Habitat type	Activity costed	Predicted cost (£/ha) [1]	Actual cost (£/ha) [2]	Actual cost (£/ha) [3]
Mudflats		Inadequate techniques to be costed		
Sand dunes	Scrub removal	1,000		
	Grazing reintroduction	800		
	Agri-environment scheme to promote vegetation restoration	245		
Cliffs and slopes	Scrub control and grazing	100		
	Agri-environment scheme to promote vegetation restoration	245		
Saltmarsh	Creation	400		
	Agri-environment scheme for management	50		
Coastal grazing marsh	Creation			800–1,200
Vegetated shingle	Restore shingle morphology	10,000		
	Re-establish vegetation	500		
Coastal lagoons	Creation		6,700*	4,200–57,000
Reed beds	Maintenance	100		
	Re-creation	620	3,200	2,800–7,700

Note: * Includes land purchase.

Sources: [1] UK Biodiversity Group, 2000; [2] Shepherd *et al.*, 2002; [3] Shepherd and Harley, 1999

Table 10.5 Unit value estimates for habitat gains and losses (£/ha)

Type	Low	Mid.	High
Woodward and Wui (2001) estimates			
Inter-tidal 'general habitat'	200	700	2,200
Inter-tidal 'bird watching habitat'	1,210	2,790	6,400
Brander *et al.* (2006) estimates			
Saltmarsh 50ha	–	990	–
Saltmarsh 250ha	–	830	–
Mudflat 50ha	–	1,680	–
Mudflat 250ha	–	1,410	–
Grazing marsh 50ha	–	390	–
Grazing marsh 250ha	–	260	–
Woodland 50ha	–	3,990	–
Woodland 250ha	–	2,670	–

Sources: Eftec, 2007, Wareham managed realignment case study taken from Defra, 2007; Woodward and Wui, 2001; Brander *et al.*, 2006

reed bed re-creation, and an even larger factor for restoration of the same habitat. For mudflats, inadequate techniques were available for even a 'bottom line' costing to be made. In the case of saltmarshes, figures used for predicted habitat creation could not be validated in practice but are likely to have been underestimated.

In the case of the replacement of environmental assets, costs should cover:

- land acquisition;
- initial site survey/feasibility study;
- background research, including species and population studies;
- removal and maintenance of plant species (ex-site conservation);
- seed bank creation from sources at site to be lost or damaged;
- reintroduction;
- habitat creation, including physical factors (e.g. hydrological and sediment regimes);
- habitat management/site wardening;
- control of competitors;
- monitoring over short, medium and long terms;
- site safeguards;
- ongoing advice to land managers; and
- publicity and public relations.

Estimates of habitat protection costs are difficult to identify because additional time costs in negotiating and implementing protection measures are often difficult to disaggregate from other tasks. Individual protection measures can be very site specific and there are insufficient examples for any figures to be given for general guidance.

Most available figures (Table 10.4) do not include any component of cost for land acquisition, which is highly variable and site specific. Yet land acquisition costs are normally the largest cost component and may be up to 85 per cent of all relevant costs. Land values may be considerably more than current agricultural land values would indicate, especially in coastal areas, where shortage of open land creates a scarcity and hence may inflate land values.

In assessing land values, care may be needed in determining the real economic costs, which may not equate to current market values. No standard figures are available for the area of land required in compensation for the loss of established habitat. These will be specific to each case and could be two, three or even more times the area of the site being lost.

The ratio will depend on:

1. The history of the existing site being lost or damaged, including cultural associations and length of time of recorded ecological history.
2. The extent to which the replacement site is an adequate substitute for what is being lost in terms of comparable habitat. Even if some transplanting of plant species can be undertaken, some allowance will need to be made for the uncertainty of their survival and the chances of adequate establishment.
3. The degree of uncertainty surrounding temporary loss of habitat where the replacement site is not being established before the existing site is changed.

10.8 CONTEXTS WHICH WE BELIEVE CAN REQUIRE THE ECONOMIC EVALUATION OF THE ENVIRONMENT

10.8.1 Trade-offs

The potential trade-offs that may be made between the environment and human systems are illustrated in Figure 10.7. Trade-offs between different forms of consumption and between consumption and resource use are generally readily embedded in benefit–cost analysis. We argued above that trade-offs between different aspects of the environment need to be resolved by the stakeholders seeking to define the best available environmental option.

This leaves the four remaining highlighted potential trade-offs (Figure 10.7):

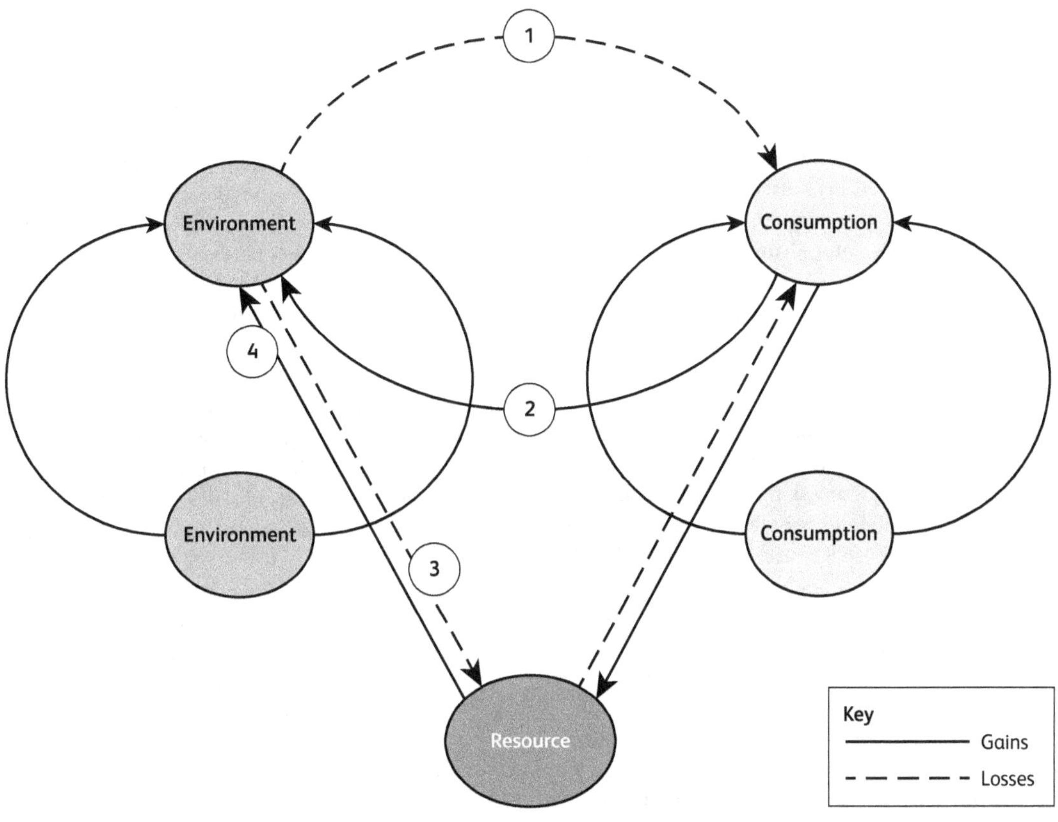

Figure 10.7 Potential trade-offs with environmental assets

1 An environmental sacrifice for a gain in consumption: the former practice of draining wetlands to provide improved farmland is an example of such a trade-off.
2 An environmental gain for a consumption sacrifice: an example is the banning of visitors to a site of environmental importance to avoid damage or disturbance.
3 An environmental sacrifice for a gain in resource: the extraction of sand from a sand dune system is an example of such a trade-off.
4 An environmental gain for a resource sacrifice: for instance, using pumps to maintain a desirable water regime in a wetland.

Conventionally, the distinction between consumption gains/losses and resource gains/losses is unnecessary but it is useful here in that some forms of consumption are unpriced.

Before discussing the suggested means of approaching each potential trade-off, an important role of economic analysis has been to show that first impressions as to the nature of a particular trade-off can be misleading. Much of the early work on the economic evaluation of wetlands (e.g. Bowers, 1983) was undertaken in order to show that the sacrifice of an area of wetland would not lead to a real gain in consumption or a real resource sacrifice. That is, the economic loss of the wetlands (Emerton and Bos, 2004) would be greater than the real gain in consumption. Functional or support values (Chapter 2) have been found to be critical here, where the difficulty of identifying all of the interdependencies was also discussed.

In some cases an apparent example of one of the four trade-offs above is not a real choice. Where a special area of conservation (SAC) or special protection area (SPA) site is at risk there is a legal obligation to protect that site and benefit–cost analysis is not applicable (MAFF, 1998); case 4 in the above list. The only instance of case 1 applying to an SAC or SPA is when there is both no alternative and overwhelming social reasons for accepting the sacrifice; or example, when there is no alternative way of reducing a substantial risk to human life.

Pure examples of cases 1 and 3 will now be rare; the recommended strategy is instead to include counterbalancing environmental enhancements so as to leave the trade-off on balance at least neutral and more usually converted to case 2 or case 4, respectively.

10.8.2 Unavoidable economic evaluation

There are a number of circumstances when we believe some implied or formal monetary valuation of the environmental consequences of an option cannot be avoided:

1 When the costs of the BPEO significantly exceed the costs of the NBO, particularly where the latter has a benefit–cost ratio that is robustly greater than 1.0 whereas the BPEO does not (see Section 10.6); for example, if constructing a multi-stage channel or an artificial wetland is being compared to replacing a sheet piled channel (case 4 above, and A in Table 10.3).
2 Where the benefits of the best 'do something' option are insufficient to justify the scheme but there are believed to be non-trivial environmental benefits associated with the best 'do something' option; for example, where a coastal risk management scheme would protect both an urban area and a locally important nature reserve (case 4 above, and B in Table 10.3).
3 Where the only technically feasible options have negative environmental consequences. In an urban area, the lack of space can restrict possible options, particularly when contaminated land is present (case 1). However, within the wider framework of a CFMP, it may then be possible to identify compensating environmental works elsewhere in the catchment, which would convert this case to fall into the first category above.

In those instances, any decision will explicitly or implicitly involve evaluating some of these environmental benefits. It is logical to start by evaluating those that can be most reliably estimated and are least controversial (see Section 10.5.3). In particular, economic evaluation is a two-edged sword. It is easy to adopt when it is expected that it will provide support for a course of action which seems intuitively desirable, but sooner or later it will imply the rejection of that course of action. Indeed, if benefit transfer (Section 10.5.3) is used then it will be the values elicited in the first instance that lead to the rejection of the intuitively preferred course of action in the second instance: not a very comfortable situation.

10.8.3 Functional values: the example of wetlands

Assuming that direct use values (e.g. recreational benefits) have already been evaluated, the next class of benefits which should be considered are functional (de Groot, 1987) or support values (Alcamo *et al.*, 2003). Table 10.6 lists the functions generally attributed to some wetlands, a particularly important environmental resource associated with both rivers and coasts. There is, however, a great variety of different types of wetland (Cowardin *et al.*, 1979; US National Research Council, 1995; Ramsar Convention Bureau, 1997) and the actual functions provided by a wetland depend upon its type and location (Mitsch and Gosselink, 2000).

In the UK, a highly simplified classification will generally be acceptable (Figure 10.8). So, for example, some wetlands are fed by groundwater and others provide groundwater recharge (Bullock and Acreman, 2003).

Table 10.6 Potential functions of a wetland

Potential functions
Flood attenuation and control
Prevention of saline water intrusion
Groundwater recharge and/or discharge
Flow regulation
Sediment retention
Storage and recycling of organic matter and nutrients
Storage and recycling of toxic material
Regulation of biological control mechanisms
Maintenance of migration and nursery habitats
Food web support
Maintenance of biological diversity
Storage of carbon dioxide (but freshwater wetlands may be net emitters of methane)
Provision of agricultural services (e.g. pasture, reeds)
Food resource provision: fishery and wildfowl
Medicinal resources
Shoreline stabilisation

Note: Not all wetlands of all types provide all functions and some functions are mutually exclusive of each other.

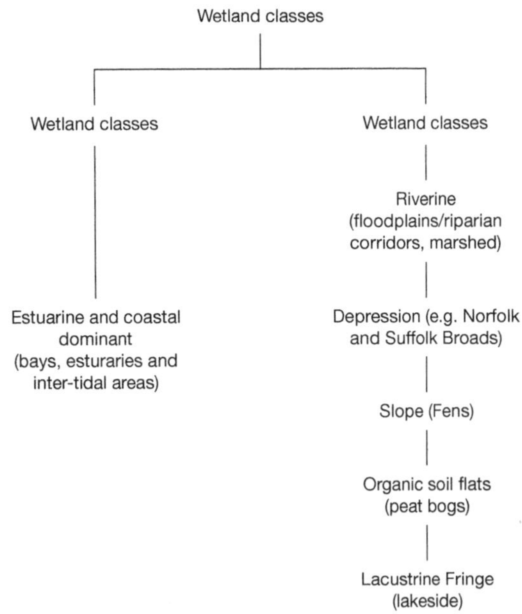

Figure 10.8 A simplified wetland classification for the UK

Source: Turner *et al.*, 2005

But it should not be thought that wetlands are a universal panacea; a new wetland will change the flow regime downstream where the current ecosystem is adjusted to the current flow regime. In some locations, wetlands may increase flood flows; in others, flows in the dry season are reduced (Bullock and Acreman, 2003). The Joint Statement on Wetlands (English Nature *et al.*, 2003) outlines how wetlands can contribute to flood risk management; those benefits can be evaluated in the normal way (MAFF, 1999).

In considering wetlands for river flood risk management, further considerations include, first, the location within the catchment of potential storage sites in the form of wetlands: the timing of releases from storage along tributaries influences the timing of the flood peak on the main stem. Second, on a number of rivers, since the design standard of protection for any embankments is only to the five–ten-year return period flood for agricultural land, that agricultural land is used to convey and store flood waters in the more extreme events; in consequence, the flood risk management benefits of converting that agricultural land to wetlands is small (Acreman *et al.*, 1999). Third, for a wetland to exist, it is the water condition in the growing season that is important (US National Research Council, 1995); where flooding is predominantly a winter phenomenon, it will be necessary to retain some of that water into the summer if a wetland is to develop.

When the channel downstream is a chalk or gravel river, the retention of sediment by a wetland is helpful, but sediment retained in a wetland is not available to maintain an active delta. Equally, whilst wetlands can act as carbon sinks (Mitra *et al.*, 2003), freshwater wetlands are emitters of methane (Houghton *et al.*, 1996; Mitsch and Gosselink, 2000), a particularly aggressive greenhouse gas. It is currently impossible to make any confident assessments as to which UK wetlands will act as net carbon sinks and which will act as sources (Turner *et al.*, 2005).

The economic evaluation of these different functions is reasonably straightforward using the techniques outlined by Turner *et al.* (2005), who provide more details on evaluating wetland functions and a number of examples. They emphasise the need to take an ecosystem functional approach and to avoid double counting. The dependence upon local conditions of the functional benefits provided by a wetland means that it is not currently wise to apply the benefit transfer approach (Chapter 2; Section 10.5.3) to evaluating a wetland.

10.8.4 Non-use values

We believe that seeking to evaluate 'non-use values' (Chapter 2) should be a last resort for several reasons:

1. Technically, it is difficult to do, with a wide range of biases being possible (e.g. see Diamond and Hausman, 1992).
2. Once a systems approach is adopted, it is quite difficult exactly to define the change that is being valued; it is not possible to ask respondents to put a value on a single SSSI in isolation. Or rather, whilst it is *technically* possible to ask them to put a value on a single SSSI, it is questionable whether the result has any meaning.
3. As with most benefits and costs, a value per person needs to be derived which is then multiplied by some estimate of the population who would be prepared to pay that amount. Thus, it is necessary to determine the size of the population either empirically or practically (i.e. the population who will pay the costs or the population sampled, where this is different). In the past, this task has been rather neglected, with somewhat arbitrary assumptions being made about the size of the relevant population.
4. It takes an exclusively utilitarian view when some of the stakeholders will argue that the choice is essentially a moral rather than a technical one.
5. The economic theory upon which it is based is highly speculative; there is very limited experimental evidence that motivations hypothesised to account for people's preparedness to pay for environmental conservation are actually people's motivations, or that other motivations can be excluded (Green, 2003).

In one instance, that of the Kakadu National Park (Imber *et al.*, 1991), the consequence of such a study being undertaken was to cause the Australian Government to close down the government agency which commissioned the study! The final government decision was then based upon an entirely different rationale (aboriginal rights). Sometimes, the more controversial the choice, the less likely it is that an economic evaluation will be able to resolve that choice or clarify the nature of the choice.

10.9 THE SCORING AND WEIGHTING APPROACH

10.9.1 The method in brief

A scoring and weighting methodology can be used to estimate monetary values for those impacts that are difficult to measure in this way (Environment Agency, 2010a). It has been developed for use in flood and coastal erosion risk appraisal work through five pilot trials.

The method is based on three main steps:

- Step 1: scoring of impacts;
- Step 2: weighting of impacts; and
- Step 3: calculating implied money values for the intangible impacts.

Along with:

- Step 4: verifying the implied values.

The methodology forms part of the current approach to project appraisal, relying on options that have been identified elsewhere and impacts (benefits and damages) that have been described using an appraisal summary table (see Environment Agency, 2010a, 178).

The approach requires scores to be assigned to each option and weights to be assigned to each category of benefit. Once all the scores and weights have been assigned, implied values can be calculated as a method for estimating the value of the intangible benefits (where intangible benefits are defined as those that are difficult to value in monetary terms, such as environmental and social benefits) relative to the tangible benefits (where tangible benefits are defined as those that can be readily valued in monetary terms, such as property damages avoided).

Scoring and weighting can, therefore, provide a less costly alternative to stated preference and other willingness-to-pay techniques as a method for directly eliciting money values for those impacts that are otherwise difficult to estimate in money terms. It is most useful when the intangible impacts are likely to be significant and could have a significant effect on the choice of preferred option.

Implied values could also be used alongside benefit transfer values as a method of validation/verification that the estimated benefits are reasonable. Overall, the use of implied values could help to keep down the costs of appraisal while still allowing all of the benefits to be valued.

The outputs from Step 3 have been designed to fit in with the current method for identifying the preferred option since they enable the net present value (NPV), benefit–cost ratio (BCR) and incremental benefit–cost ratio (IBCR) to be calculated.

10.9.2 An application to the Thames Estuary

An options appraisal was undertaken in the UK within the major TE2100 project to investigate the future of protecting London from flooding from the sea (Penning-Rowsell *et al.*, 2012). The overall economic analysis, within a benefit–cost framework complemented by multi-criteria analysis (MCA), showed that improving the existing flood risk management schemes and, in 2070, constructing a new Thames Barrier downstream from the existing one are the 'front-runner' options selected from many that were appraised for tackling the increase in flood risk to London that is anticipated in the future as a result of climate change-induced sea level rise and socio-economic change in this intensely developed urban floodplain.

The value of impacts other than those to which a monetary valuation can be applied directly was derived through scoring and weighting. The discussion here is limited to reporting the calculation of the present value aggregate implied value estimates (Figure 10.9), along with associated key caveats.

The net total of MCA-derived (implied) benefits was equivalent to approximately 60 per cent of directly monetised benefits, and the implied 'multiplier' to convert directly monetised benefits to total benefits is therefore around 1.6. The implied economic benefit estimates for 'Key infrastructure' and 'Transport' (other than navigation) are very large (Figure 10.9), mainly reflecting the major threat to London's transport system. And, unlike for direct property damage, there is a reasonably clear distinction between the options in terms of protection for key infrastructure and transport. In particular, the new barrier options (3.1 to 4.3 inclusive in Figure 10.9) bestow around £5–6 billion more benefit than options 1 and 2, mostly in relation to protecting transport assets, such as the underground railway system in central London.

Of the implied benefit estimates for environmental categories, the most dominant differentiating issue between the options is water quality, with the 'do minimum' (DM) and barrier with locks (BWL) options having strong dis-benefits (up to around £20 billion in present value terms). Although the water quality assessment includes impacts through the flooding of contaminated land (e.g. landfill sites), the vast majority of the dis-benefit value for the DM and BWL options comes from physical and chemical factors in the channel in the absence of flooding and the free natural flow of the river that the barriers and their closures inevitably create.

The BWL options fare badly here because the barriers are forecast to be closing on every tide towards the end of the appraisal period, whereas the other barrier options (including options 1 and 2, which retain the existing Thames Barrier) are closing less – fifty times or fewer per annum in the period before the year 2135, albeit rising rapidly to *c.*400 times per annum by 2170.

Options 3.1 and 3.2 are slightly worse in water quality terms than options 1 and 2, because the new

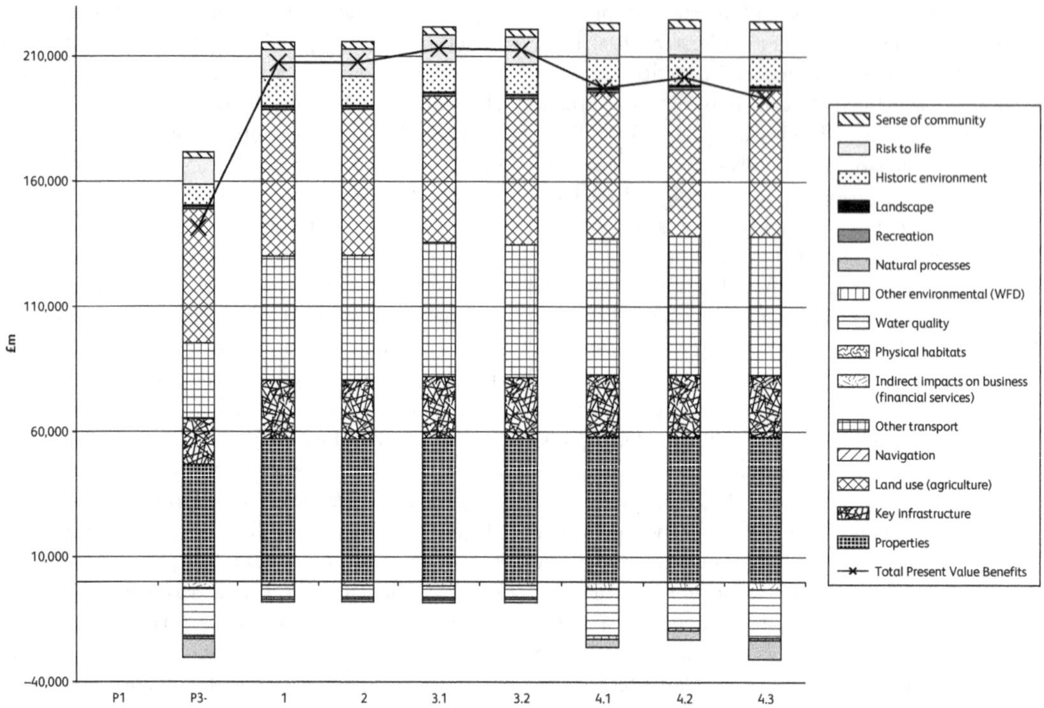

Figure 10.9 Thames Estuary implied environmental benefits derived using the scoring and weighting methodology to examine options 1 to 4.3 (£millions, PV to 2170). See Table 10.7 for the data displayed here.

Notes: For more detail, see Penning-Rowsell *et al.*, 2012

Tilbury or Long Reach barrier would be downstream of major sewage works and power stations, which would have a polluting effect in cases of frequent barrier closure and the impounding of water that this creates. This is partially offset, however, by the greater volume of water enclosed upstream of the Long Reach and Tilbury structures, compared with the existing Thames Barrier, which has a net benefit to the central London 'City Reach' in the estuary.

However, it is recognised that there is uncertainty surrounding the water quality assessment, as no formal modelling was carried out. Water quality issues would therefore need to be the subject of sensitivity analysis in any final appraisal of the long-term options (i.e. the new barriers).

The implied option benefit estimates for social categories (not including risk to life) show that differentiation between the 'do something' options on social grounds (other than safety and security, which is included in the directly monetised categories) is limited to around £200 million in total PVd terms. The new barrier options (3.1 to 4.3 inclusive) tend to perform marginally better than options 1 and 2, principally through the greater protection afforded to public buildings (which is the primary indicator of a 'sense of community' variable in this weighting and scoring application).

10.9.3 Thames Estuary: summary

In summary, this was a major application of the scoring and weighting approach in a project where risk is high into the future and the number of policy options is very substantial.

In this example the 'Properties', 'Key infrastructure', 'Other transport' and 'Indirect impacts on business' categories account for the vast majority of the flood risk management benefits that were calculated (*c*.87 per cent of the total positive benefits). The categories where the most differences between the options occur are 'Transport' and 'Water quality'.

Table 10.7 Present value benefits of the various Thames Estuary options

2007–2169 period	Option								
	P1	P3-	1	2	3.1	3.2	4.1	4.2	4.3
	Do nothing (walk away)	Do minimum (maintain)	Enhance existing system	Flood storage	Barrier at Tilbury	Barrier at Long Reach	Barrier with locks at Tilbury	Barrier with locks at Long Reach	Convert Thames Barrier to barrier with locks
Properties		46,892	57,173	57,084	57,473	57,173	57,936	57,937	57,924
Key infrastructure		18,318	23,325	23,341	24,526	24,432	24,745	24,921	24,810
Land use (agriculture)		1	1	1	1	1	1	1	1
Navigation		133	141	141	59	109	89	120	139
Other transport		29,972	49,556	49,826	53,636	53,121	54,452	55,405	55,202
Indirect impacts on business (financial services)		53,633	58,513	58,513	58,513	58,513	58,513	58,513	58,513
Physical habitats		−2,579	−1,431	−1,431	−1,760	−1,460	−2,764	−2,488	−3,023
Water quality		−19,390	−4,783	−4,783	−4,758	−4,656	−18,624	−15,824	−19,390
Other environmental (WFD)		−845	−845	−704	−845	−845	−1,406	−1,108	−845
Natural processes		−7,514	−1,097	−1,097	−1,113	−1,198	−3,217	−3,511	−7,519
Recreation		968	1,133	1,130	1,156	1,140	1,160	1,161	1,159
Landscape		453	460	421	424	424	676	676	672
Historic environment		8,366	11,530	11,537	11,963	11,964	12,001	11,993	11,740
Risk to life		10,578	10,856	10,856	10,843	10,844	10,842	10,844	10,843
Sense of community		2,518	2,996	2,996	3,115	3,150	3,153	3,191	3,190
Total present value benefits		**141,504**	**207,527**	**207,830**	**213,232**	**212,710**	**197,557**	**201,830**	**193,415**

The options with the highest overall net benefits are options 3.1 and 3.2, followed by options 1 and 2. Options 4.1–4.3 are significantly disadvantaged by their environmental dis-benefits as measured in this weighting and scoring application.

The method therefore appears to aid the differentiation between options and hence enhance understanding of the complex social, economic and environmental systems that make up the Thames Estuary and its current and future floodplain. Whether the absolute numbers are 'accurate' is debatable, of course, but clarifying this differentiation certainly helped to separate those options which appear to be unviable purely on environmental grounds (principally owing to water quality impacts) from those which warrant further investigation in the future.

10.10 LESSONS FROM EXPERIENCE

Assessing environmental costs and benefits is desirable (see Section 10.5) but in practice it is fraught with difficulties. We have learned some lessons and come to some conclusions about this process:

1 A more comprehensive approach to decisions affecting the environment over a longer time-scale, which fulfils criteria for sustainable management of the environment, needs to take a more holistic, as opposed to a piecemeal, approach, and should seek to manage dynamic systems as dynamic systems. Complying with the EU Water Framework Directive is the latest and most demanding requirement in this respect.
2 It is not always possible to predict the impacts of a specific course of action on the environment, even where there appears to be wide experience and monitoring data are available.
3 In addition to stakeholder engagement being both UK and European policy (Cabinet Office, undated; CEC, 2001), it can save time and resources spent on consideration of scheme designs which would have an unacceptable environmental impact. This is the first stage in an iterative process which should continue throughout the environmental assessment and on to the scheme implementation phase.
4 The 'acid test' question is now not one of how much of the environment to sacrifice in order to save money but how much we can afford to spend to enhance the environment.
5 Some wildlife sites are now designated at a European level. At these sites which are of international importance, environmental considerations can be given precedence over usual economic criteria in the initial appraisal process (MAFF, 1998).
6 The main changes during the past decade or so are that scheme-level decisions about specific sites are now set in a wider strategic context (e.g. shoreline management plans; coastal habitat management plans; catchment flood management plans; river basin management plans). This last is a legal document, with infraction fines for non-compliance. However, many of the other plans are not statutory documents, so although agencies will have been consulted during their production and will normally have agreed to them in principle, they may not have agreed fully with the details of their content when it comes down to practical application.
7 Only those criteria for which an internationally important site has been designated can be taken into account in the appropriate assessment. So, as long as a scheme does not have an impact on a designated species, it could affect other species or geological interests.
8 Environmental considerations may not be the same and there may be conflicts between the interests of different components of the environment or between maintaining current assets and processes. Decision-makers need to be aware of English Nature/CCW's responsibility for both ecology and geology/geomorphology, as possible conflicts may arise between these interests.
9 Existing environmental assets and current conditions may be highly significant but, in some circumstances, some loss may have to be accepted (for which a pass under the WFD Article 4.7 is required; see above). Mitigation measures could include some form of compensation for the loss and, although environmental gains are much harder to predict than environmental losses, it may be that the gains from some courses of action could be as valuable, in the longer term, as what may be lost.
10 European, national and local criteria for site importance should not necessarily be judged in the same way.

11 A scheme design which is preferred on technical and economic grounds can often be modified to minimise adverse impacts on the environment without compromising its engineering integrity.

12 If negotiations are inbuilt into the scheme design process then it may be possible to reach a compromise about the way a scheme is carried out, even on a designated site, particularly if only a relatively small part of the total site will be affected.

13 Consideration of wider issues of sustainability should incorporate factors such as the impacts on the environment at the source of beach nourishment material to ensure that other environmental resources, in this case fisheries, are not adversely impacted. Similar issues relate to the use of timber and the extent to which its sources are being sustainably managed.

14 Small schemes can be just as contentious as large ones and allowances for time in consultation exercises will not necessarily relate to the size of a scheme.

15 Difficult choices cannot be made into simple ones by some technical sleight of hand. Look for increased understanding through project appraisal rather than hoping that economic evaluation can, of itself, remove the difficulties.

References

AA (2011) 'Floods and heavy rain: advice from the AA's flood rescue team', http://www.theaa.com/motoring_advice/seasonal/floods-and-wet-weather.html, 24 October 2012, accessed 10 April 2013

AA (2012) 'Car depreciation', http://www.theaa.com/motoring_advice/car-buyers-guide/cbg_depreciation.html, 23 March 2012, accessed 10 April 2013

ABC (2012) *The Agricultural Budgeting & Costing Book*, 74th edn, Agro Business Consultants, Melton Mowbray

Abt, S.R., Wittler, R.J., Taylor, A. and Love, D.J. (1989) 'Human stability in a high flood hazard zone', *JAWRA Journal of the American Water Resources Association*, vol 25, no 4, pp881–890

ACC/ISGWR (1992) *The Dublin Statement and the Report of the Conference*, World Meteorological Organisation, Geneva

Acland, A.F. (1990) *A Sudden Outbreak of Common Sense: Managing Conflict through Mediation*, Hutchinson, London

Acreman, M.C. (2003) *The Impact of Floodplain Restoration on Flood Hydrology*, paper presented to the 38th Flood and Coastal Management Conference, University of Keele

Acreman, M.C., Barbier, E. and Birley, M. (1999) *Managed Food Releases from Reservoirs: A Review of Current Problems and Future Prospects, Project R7344, Report to DFID*, Institute of Hydrology, Wallingford

Adaptation Sub-Committee (ASC) (2012) *Climate Change: Is the UK Preparing for Flooding and Water Scarcity*, Adaptation Sub-Committee Progress Report, London, http://archive.theccc.org.uk/aws/ASC/CCC_ASC_2012_bookmarked_2.pdf, accessed 26 March 2013

ADB (Asian Development Bank) (2002) *Handbook for Integrating Risk Analysis in the Economic Analysis of Projects*, Asian Development Bank, Manila

Admiraal, J. (2011) 'Case study: vulnerability of the port of Rotterdam to climate change', in 'Flood damage to port industry', M.Sc. Thesis, Faculteit der Aard – en Levenswtenschappen, Vrije Universiteit, Amsterdam

Admiral (2012)Telephone conversation with Adrian Lloyd, Admiral Insurance, 18 July 2012

Agri-Food Trade Service (2011) 'The United Kingdom: a sophisticated retail sector', in *Market Analysis Report*, Agriculture and Agri-Food Canada, International Market Bureau, http://www.ats-sea.agr.gc.ca/eur/5735-eng.htm, February 2011, accessed 8 February 2013

Akerlof, G.A. (1970) 'The market for lemons: qualitative uncertainty and the market mechanism', *Journal of Economics*, vol 84, no 4, pp488–500

Alberini, A. and Kahn, J.R. (eds) (2006a) *Handbook on Contingent Valuation*, Edward Elgar, Cheltenham

Alberini, A. and Kahn, J.R. (2006b) 'Introduction', in Alberini, A. and Kahn, J.R. (eds) *Handbook on Contingent Valuation*, Edward Elgar, Cheltenham, pp1–6

Alcamo, J., Ash, N.J., Bennett, E.M., Biggs, R., Butler, D.C., Callicott, B.J., Capistrano, D., Carpenter, S.R. and Castilla, C.J. (eds) (2003) *Ecosystems and Human Well-being: A Framework for Assessment*, Island Press, Washington DC

Angus, A., Burgess, P.J., Morris, J. and Lingard, J. (2009) 'Agriculture and land use: demand for and supply of agricultural commodities, characteristics of the farming and food industries, and implications for land use in the UK', *Land Use Policy*, no 26S, ppS230–S242

AOL Autoblog (2009) 'Floods and cars: all you need to know', http://uk.autoblog.com/2009/11/30/floods-and-cars-all-you-need-to-know/, 30 November 2009, accessed 10 April 2013

Appleton, K. and Lovett, A. (2003) 'GIS-based visualisation of rural landscapes: defining "sufficient realism" for environmental decision-making', *Landscape and Urban Planning*, vol 65, no 3, pp117–131

Ardila, S., Quiroga, R. and Vaughan, W.J. (1998) *A Review of the Use of Contingent Valuation Methods in Project Analysis at the Inter-American Development Bank, ENV-126*, Inter-American Development Bank, Washington DC

Arrow, K.J. (1963) *Social Choice and Individual* Values, 2nd edn, Cowles Foundation for Research in Economics, Yale University (originally published by John Wiley and Sons, New York)

Arup (2012) *King's Health Partners' Cancer Centre at Guy's, Flood Risk Assessment*, Arup for Guy's and St Thomas' NHS Foundation Trust, London, http://planning online.southwark.gov.uk/DocsOnline/Documents/24 0659_1.pdf, 11 June 2012, accessed 26 March 2013

Association of British Insurers (ABI) (2012) Telephone conversation with Matt Cullen, Policy Adviser, Flooding, Association of British Insurers, 19 November 2012

ATOC (2012) 'National Rail travel bulletin', http://www.atoc.org/media-centre/previous-press-releases/national-rail-travel-bulletin---26-november-100772, 26 November 2012, accessed 26 March 2013

Audit Commission (1999) *Listen up! Effective Community Consultation*, Audit Commission, London

Audit Commission (2007) *Staying Afloat: Financing Emergencies*, Public Services National Report, December 2007, Audit Commission, London

Aysin, S. and Ozgen, A. (2009) 'Space efficiency in high-rise office buildings', *METU Journal of the Faculty of Architecture*, vol 26, no 2, pp69–89

Baecher, G.B. (2001) 'A pharaoh's plan for water management', *Water Resources Impact*, vol 3, no 5, pp3–7

Bangkok Post (2011) 'Flood-damaged machinery replaced', http://www.bangkokpost.com/learning/learning-from-news/269517/flood-damaged-machineryst, 6 December 2011, accessed 10 April 2013

Barry. L., van Rensburg, T. and Hynes, S. (2010) *Modelling Access to the Irish Coastline: A Contingent Behavioural Approach*, Working Paper No 162, Discipline of Economics, National University of Ireland, Galway

Bateman, I.J. (2011) 'Valid value estimates and value estimation validation: better methods and better testing for stated preference research', in Bennett, J. (ed.) *The International Handbook on Non-Market Environmental Valuation*, Edward Elgar, Cheltenham, pp332–352

Bateman, I.J. and Willis, K.G. (1999) *Valuing Environmental Preferences: Theory and Practice of the Contingent Valuation Method in US, EU and Developing Countries*, Oxford University Press, Oxford

Bateman, I.J., Turner , R.K., Klein, R. and Langford, I.H. (2001) 'The application of the cost benefit method to sea defence and coastal protection in England', in Turner, R.K., Bateman, I.J. and Adjer, W.N. (eds) *Economics of Coastal and Water Resources: Valuing Environmental Functions*, Kluwer, Dordrecht, pp113–142

Bateman, I.J., Cole, M.A., Georgiou, S. and Hadley D.J. (2006) 'Comparing contingent valuation and contingent ranking: a case study considering the benefits of urban river water quality improvements', *Journal of Environmental Management*, vol 79, no 3, pp.221–231

Bateman, I.J., Day, B.H., Jones, A.P. and Jude, S. (2009) 'Reducing gain–loss asymmetry: a virtual reality choice experiment valuing land use change', *Journal of Environmental Economics and Management*, vol 58, pp106–118

Beedell, J., Morris, J. and Hess, T.M. (2012) *Mobilising the Contribution of Rural Land Management to Flood Risk Management in Scotland*, report to Scottish Government, Ref CR/2010/14, Smithsgore, Peterborough, http://www.scotland.gov.uk/Resource/0039/00393714.pdf, accessed 7 June 2013

Bennett, J. (ed.) (2011) *The International Handbook on Non-Market Environmental Valuation*, Edward Elgar, Cheltenham

Black, A. and Evans, S. (1999) *Flood Damage in the UK: New Insights for the Insurance Industry*, University of Dundee, Dundee

Blakemore, F.B., Burrell, M. and Jones, S.D.R. (2007) 'Beach users' aesthetic and economic evaluation of a "minor change" to the hard engineering coastal defences at Wiseman's Bridge, Pembrokeshire, Wales', in Aravossis, K., Brebbia, C.A. and Gomez, N. (eds) *Environmental Economics and Investment Assessment*, WIT Press, Southampton, pp106–127

Bowers, J.K. (1983) 'Cost–benefit analysis of wetland drainage', *Environment and Planning A*, vol 15, pp227–235

Bowers, J.K. (2001) *An Economically Efficient Strategy for Coastal Defence and the Conservation of the Inter-tidal Zone*, report to the Joint Marine Programme of the Wildlife Trusts and WWF UK, WWF, London

Brander, L.M., Florax, R.J.G.M. and Vermaat, J.E. (2006) 'The empirics of wetland valuation: a comprehensive summary and a meta-analysis of the literature', *Environmental and Resource Economics*, vol 33, pp223–250

Brent, R.J. (1990) *Project Appraisal for Developing Countries*, Harvester Wheatsheaf, New York

Briene, M., Koppert, S., Koopman, A. and Verkennis A. (2010) *Financiële onderbouwing kengetallen hoogwater-schade*, Eindrapport, Rotterdam (in Dutch)

Bristol City Corporation, City Engineer and Planning Officer's Department (1971) *Land Use Survey 1971*, Research Section study paper, Bristol City Corporation, Bristol

Bristol Water (2012) 'Charges for hire of a water bowser, static tank or tanker', http://www.bristolwater.co.uk/pdf/customerService/generalInfo/Bowser_2012.pdf, accessed 26 March 2013

British Household Panel Survey (1991–2000) http://iser www.essex.ac.uk/ulsc/bhps/, accessed 10 April 2013

BRMB (1999) *Flood Warning Dissemination: Post Event Surveys*, report prepared for the Environment Agency, July, BRMB, London

BRMB (2001) *Flood Warning Dissemination: Post Event Surveys*, report prepared for the Environment Agency, August, BRMB, London

Brouwer, R., Langford, I.H., Bateman, I.J. and Turner, R.K. (1999) 'A meta-analysis of wetland contingent valuation studies', *Regional Environmental Change*, vol 1, no 1, pp47–57

Brown, C. and Graham, W. (1998) 'Assessing the threat to life from dam failure', *Water Resources Bulletin*, vol 24, no 6, pp1303–1309

Buijs, F.A., Segura, D., Sayers, P.B., Simm. J.D. and Hall J.W. (2005) 'Tiered reliability-based methods for assessing the structural performance of coastal defences', *Proceedings of the 8th International Conference on Coastlines, Structures and Breakwaters*, Thomas Telford, London, pp496–507

Building Maintenance Information (2001) *Review of Rehabilitation Costs*, Serial 307 BMI Special Report, December, Building Maintenance Information, London

Bullock, A. and Acreman, M. (2003) 'The role of wetlands in the hydrological cycle', *Hydrology and Earth System Sciences*, vol 7, no 3, pp358–389

Burgess, J., Clark, J. and Harrison, C.M. (1997) *I Struggled with this Money Business: Respondents to the Wildlife Enhancement Scheme CV Survey Discuss the Validity of their WTP Figures*, University College London

Burgess, K., Jay, H. and Hosking, A. (2002) 'FUTURECOAST: predicting the future coastal evolution of England and Wales', paper presented to EUROCOAST Conference, Littoral 2002: The Changing Coast, Faculty of Engineering, Porto

Burr, T. (2008) *Reducing Passenger Rail Delays by Better Management of Incidents*, report by the Comptroller and Auditor General, HC 308, Session 2007–2008, National Audit Office, http://www.nao.org.uk/wp-content/uploads/2008/03/0708308.pdf, 14 March 2008, accessed 26 March 2013

Butler, J.R.A., Radford, A., Riddington, G. and Laughton, R. (2009) 'Evaluating an ecosystem service provided by Atlantic salmon, sea trout and other fish species in the River Spey, Scotland: the economic impact of recreational rod fisheries', *Fisheries Research*, vol 96, pp259–266

Cabinet Office (undated) *Viewfinder: A Policy Maker's Guide to Public Involvement*, Cabinet Office, London

Cabinet Office (2002) *Risk: Improving Government's Capability to Handle Risk and Uncertainty*, Cabinet Office, London

Cabinet Office (2008) *Floods: Damaged School Buildings*, UK Resilience Reports, Cabinet Office, London

Cabinet Office (2010) *Strategic Framework and Policy Statement on Improving the Resilience of Critical Infrastructure to Disruption from Natural Hazard*, Cabinet Office, London, https://www.gov.uk/government/uploads/system/uploads/attachment_data/file/62504/strategic-framework.pdf, March 2010, accessed 26 March 2013

Cabinet Office (2011) *Keeping the Country Running: Natural Hazards and Infrastructure: A Guide to Improving the Resilience of Critical Infrastructure and Essential Services*, Cabinet Office, London, https://www.gov.uk/government/uploads/system/uploads/attachment_data/file/78901/natural-hazards-infrastructure.pdf, accessed 26 March 2013

CAG Consultants and Land Use Consultants (2001) *Quality of Life Capital, Managing Environmental, Social and Economic Benefits*, overview report, CAGConsultants, London

Calder, I.R. (2003) 'Forests and water: closing the gap between public and science perceptions', keynote address to the Water Symposium, Stockholm

Cannell, R.O., Belford, R.K., Gales, K., Dennis, C.W. and Prew, R.D. (1980) 'Effects of waterlogging at different stages of the development of the growth and yield of winter wheat', *Journal of the Science of Food and Agriculture*, vol 31, pp117–132

Carson, R.T. (2011) *Contingent Valuation: A Comprehensive Bibliography and History*, Edward Elgar, Cheltenham

Carson, R.T. (2012) 'Contingent valuation: a practical alternative when prices aren't available', *Journal of Economic Perspectives*, vol 26, no 4, pp27–42

Castle, D.A., McCunnel, J. and Tring, I.M. (1984) *Field Drainage Principles and Practice*, Batsford, London

Central Networks (2008) *Central Networks Stakeholder Engagement: High Impact Low Probability (HILP) Event, Record of the Day and Discussions Held at the Event*, Sutton Coldfield, http://www.eonuk.com/Central_Networks_HILP_presentation_and_workshop_discussion_record.pdf, 7 July 2008, accessed 26 March 2013

Chatterton, J. (undated) *The Value of Water Level Management in the Somerset Drainage Boards Consortium*, unpublished report prepared by J. Chatterton Associates for the Somerset Drainage Boards Consortium

Chatterton, J. (2008) *TE2100: Assessment of Damages in the Thames Estuary of Likely High Consequence from Tidal Flooding*, unpublished report prepared for Halcrow as part of the TE2100 project

Chatterton, J., Viavattene, C., Morris, J., Penning-Rowsell, E. and Tapsell, S. (2010) *The Costs of the Summer 2007 Floods in England*, Project SC070039/R1, Environment Agency, Bristol

CIRIA (2012) *Guidance on the Management of Landfill Sites and Land Contamination on Eroding or Low-lying Coastlines*, C718, CIRIA, London

Clawson, M. (1959) *Methods of Measuring the Demand for Outdoor Recreation*, 10th reprint, Resources for the Future Inc, Washington DC

Coase, R.H. (1991) 'The institutional structure of production', Nobel Prize lecture, Stockholm

Coast Insurance (2012) Telephone conversation with Coast Insurance, 17 July 2012

Coker, A. and Richards, K. (1992) *Valuing the Environment: Economic Approaches to Environmental Evaluation*, Belhaven, London

Commission of the European Communities (CEC) (2000) *Communication from the Commission to the Council and the European Parliament on Integrated Coastal Zone Management: A Strategy for Europe*, COM/2000/547/ Brussels, Commission of the European Communities, Brussels

Commission of the European Communities (CEC) (2001) *European Governance: A White Paper*, COM/2001/428/ Brussels, Commission of the European Communities, Brussels

Committee on Climate Change (2012) *Assessing the Economic Case for Property Level Measures in England*, Final Report 9X1055, July, Committee on Climate Change, London

Conway County Borough Council (2013) 'Resurfacing', http://www.conwy.gov.uk/doc.asp?cat=9089&doc=28955, accessed 15 April 2013

Coombes, E.G., Jones, A.P., Bateman, I.J., Tratalos, J.A., Gill, R.A., Showler, D.A., Watkinson, A.R. and Sutherland, D.A. (2009a) 'Spatial and temporal modelling of beach use: a case study of East Anglia UK', *Coastal Management*, vol 37, no 1, pp94–115

Coombes, E.G., Jones, A.P. and Sutherland, W.J. (2009b) 'The implications of climate change on coastal visitor numbers: a regional analysis', *Journal of Coastal Research*, vol 25, no 4, pp981–990

Costa, P., Green, C.H. and Tunstall, S.M. (1993) *St Mildred's Bay: An Assessment of Coast Protection Benefits*, Final Report to Thanet District Council, Flood Hazard Research Centre, Middlesex University, London

Coulthard, T., Frostick, L., Hardcastle, H., Jones, K., Rogers, D., Scott, M. and Bankoff, G. (2007) *The June 2007 Floods in Hull*, Final Report by the Independent Review Body, http://www2.hull.ac.uk/science/pdf/geogfloodsinhull3.pdf, 21 November 2007, accessed 26 March 2013

Council of the European Union (2002) 'Council Regulation (EC) No 2012/2002 of 11 November 2002 establishing the European Union Solidarity Fund', *Official Journal of the European Communities*

Cowardin, L.M., Carter, V. and LaRoe, E.T. (1979) *Classification of Wetlands and Deepwater Habitats of the United States*, Fish and Wildlife Services, US Department of the Interior, Washington DC

CSIS (Center for Strategic and International Studies) (2011) *Economic Damage of Thailand's Flooding*, http://cogitasia.com/economic-damage-of-thailands-flooding, 25 October 2011, accessed 17 April 2013

Cummings, R.G., Brookshire, D.S. and Schulz, W.D. (eds) (1986) *Valuing Environmental Goods: A State of the Art Assessment of the Contingent Valuation Method*, Rowman and Allanheld, Totowa, NJ

Daily Telegraph (2013) 'High street shop closures hit 14 a day in 2011', *Daily Telegraph*, 24 February 2013

Dawson, R.J., Dickson, M.E., Nicholls, R.J., Hall, J.W., Walkden, M.J.A., Stansby, P., Mokrech, M., Richards, J., Zhou, J., Milligan, J., Jordan, A., Pearson, S., Rees, J., Bates, P., Koukoulas, S. and Watkinson, A. (2009) 'Integrated analysis of risks of coastal flooding and cliff erosion under scenarios of long term change', *Climatic Change*, vol 95, nos 1–2, pp249–288

Day, B., Bateman, I.J., Carson, R.T., Dupont, D., Louviere, J.J., Morimoto, S., Scarpa, R. and Wang, P. (2012) 'Ordering effects and choice set awareness in repeat-response stated preference studies', *Journal of Environmental Economics*, vol 63, no 1, pp73–91

DCM (2012) 'Demonstration test catchments', http://www.demonstratingcatchmentmanagement.net/?page_id=11, accessed 26 April 2013

de Bruin, K.M. and Klijn, F. (2001) 'Resilient flood risk management strategies', *Proceedings of the IAHR Congress, Beijing*, Tsinghua University Press, Beijing

de Groot, R.S. (1987) 'Environmental functions as a unifying concept for ecology and economics', *The Environmentalist*, vol 7, no 2, pp105–109

Department for Communities and Local Government (DCLG) (2008) *Model Standards 2008 for Caravan Sites in England*, http://www.communities.gov.uk/documents/housing/pdf/modelstandards2008.pdf, accessed 20 August 2012

Department for Communities and Local Government (DCLG) (2010) *Planning Policy Statement 25: Development and Flood Risk (Revised March 2010)*, http://www.communities.gov.uk/documents/planningandbuilding/pdf/planningpolicystatement25.pdf, accessed 20 August 2012

Department for Communities and Local Government (DCLG) (2012a) *Park Homes: Commons Select Committee Working Paper*, http://www.publications.parliament.uk/pa/cm201213/cmselect/cmcomloc/177/17704.htm#a1, 20 June 2012, accessed 20 August 2012

Department for Communities and Local Government (DCLG) (2012b) *A Better Deal for Mobile Home Owners: Changes to the Mobile-Homes Act 1983: Impact Assessment*, http://www.communities.gov.uk/documents/housing/pdf/2128121.pdf, accessed 20 August 2012

Department for Communities and Local Government (DCLG) (2012c) *Bellwin Scheme 2012 to 2013: Guidance*, http://www.communities.gov.uk/publications/localgovernment/bellwin201213guidance, accessed 22 April 2013

Department for Environment, Food and Rural Affairs (Defra) (2001a) *Flood and Coastal Defence Project Appraisal Guidance: FCDPAG1 Overview*, May 2001, MAFF, London (includes general guidance)

Department for Environment, Food and Rural Affairs (Defra) (2001b) *National Appraisal of Assets at Risk from Flooding and Coastal Erosion, Including the Potential Impact of Climate Change*, technical report by Halcrow, HR Wallingford and John Chatterton Associates for MAFF, London

Department for Environment, Food and Rural Affairs (Defra) (2002) *Investigating Biodiversity Costs*, Defra, London

Department for Environment, Food and Rural Affairs (Defra) (2003) *Flood and Coastal Defence Project Appraisal Guidance FCDPAG3 Revisions to Economic Appraisal Procedures Arising from the New HM Treasury 'Green Book'*, Supplementary Note to Operating Authorities, March 2003, Defra, London

Department for Environment, Food and Rural Affairs (Defra) (2004a) *Making Space for Water: Developing a New Government Strategy for Flood and Coastal Erosion Risk Management in England: A Consultation Exercise*, Defra, London

Department for Environment, Food and Rural Affairs (Defra) (2004b) *Flood and Coastal Defence Project Appraisal Guidance, FCDPAG3 Revisions to Economic Appraisal on Reflecting Socio-economic Equity in Appraisal and Appraisal of Human Related Intangible Impacts of Flooding*, Supplementary Note to Operating Authorities, July 2004, Defra, London

Department for Environment, Food and Rural Affairs (Defra) (2005) *Making Space for Water: Taking Forward a New Government Strategy for Flood and Coastal Erosion Risk Management in England (Delivery Plan)*, Defra, London

Department for Environment, Food and Rural Affairs (Defra) (2006) *Flood and Coastal Defence Appraisal Guidance, FCDPAG3 Economic Appraisal*, Supplementary Note to Operating Authorities: Climate Change Impacts, October 2006, Defra, London

Department for Environment, Food and Rural Affairs (Defra) (2007) *An Introductory Guide to Valuing Ecosystem Services*, Defra, London

Department for Environment, Food and Rural Affairs (Defra) (2008a) *Flood and Coastal Defence Appraisal Guidance, Economic Appraisal*, Supplementary Note to Operating Authorities: Valuation of Agricultural Land and Output for Appraisal Purposes, Defra, London

Department for Environment, Food and Rural Affairs (Defra) (2008b) *Flood and Coastal Defence Appraisal Guidance, Social Appraisal*, Supplementary Note to Operating Authorities: Assessing and Valuing the Risk to Life from Flooding for Use in Appraisal of Risk Management Measures, May 2008, Defra, London

Department for Environment, Food and Rural Affairs (Defra) (2009) *Appraisal of Flood and Coastal Erosion Risk Management: A Defra Policy Statement*, Defra, London

Department for Environment Food and Rural Affairs (Defra) (2010) *What Nature Can Do for You: A Practical Introduction to Making the Most of Natural Services, Assets and Resources in Policy and Decision Making*, Defra, London

Department for Environment, Food and Rural Affairs (Defra) (2011a) *Flood and Coastal Resilience Partnership Funding: Defra Policy Statement on an Outcome-Focused, Partnership Approach to Funding Flood and Coastal Erosion Risk Management*, May 2011, Defra, London

Department for Environment, Food and Rural Affairs (Defra) (2011b) 'More flood and coastal defence schemes to go ahead under new funding system', Environment Minister, partnership funding announcement, http://www.defra.gov.uk/news/2011/05/23/more-flood-defence-schemes/, 23 May 2011, accessed 9 April 2013

Department for Environment, Food and Rural Affairs (Defra) (2011c) *Mainstreaming Sustainable Development: The Government's Vision and What This Means in Practice*, Defra, London

Department for Environment, Food and Rural Affairs (Defra) (2011d) *The Natural Choice: Securing the Value of Nature*, Defra, London

Department for Environment, Food and Rural Affairs (Defra) (2012) *Catchment Sensitive Farming*, http://webarchive.nationalarchives.gov.uk/20130123162956/http:/www.defra.gov.uk/food-farm/land-manage/nitrates-watercourses/csf/, accessed 10 April 2013

Department for Environment, Food and Rural Affairs (Defra)/Environment Agency (2004) *The Appraisal of Human Related Intangible Impacts of Flooding*, R&D Technical Report FD2005/TR, Defra, London

Department for Environment, Food and Rural Affairs (Defra)/Welsh Assembly Government (2008) *Security and Emergency Measures (Licensed Water Suppliers) Directions for Licensed Water Suppliers: Notification and Guidance*, http://dwi.defra.gov.uk/stakeholders/guidance-and-codes-of-practice/SEMD%20(LICENSED%20WATER%20SUPPLIERS)%20DIRECTION.pdf, January 2008, accessed 11 June 2013

Department for Environment, Food and Rural Affairs (Defra) (and Welsh Office) (2001) *Shoreline Management Plans: A Guide for Coastal Defence Authorities*, Defra, London

Department for Transport (DfT) (1981) *Traffic Appraisal Manual*, Department for Transport, London

Department for Transport (DfT) (2009) 'Unit 3.5.7: The reliability sub-objective', in *Transport Analysis Guidance (TAG)*, Department for Transport, London, http://www.dft.gov.uk/webtag/documents/expert/pdf/unit3.5.7.pdf, April 2009, accessed 26 March 2013

Department for Transport (DfT) (2011a) 'Table NTS9902', in *National Travel Survey 2011*, http://www.dft.gov. uk/statistics/tables/nts9902/, accessed July 2012

Department for Transport (DfT) (2011b) *New Approach to Transport Appraisal (NATA)*, http://www.dft.gov.uk/webtag/, April 2011, accessed 26 March 2013

Department for Transport (DfT) (2012a) *Vehicles Registered for the First Time by Generic Model, Table VEH0161*, http://www.dft.gov.uk/statistics/tables/veh0161/, accessed July 2012

Department for Transport (DfT) (2012b) 'UNIT 3.5.6: Values of time and vehicle operating costs', in *Transport Analysis Guidance (TAG)*, Department for Transport, London, http://www.dft.gov.uk/webtag/, October 2012, accessed 26 March 2013

Department for Transport, Local Government and the Regions (DTLR) (2001a) *Local Housing Statistics 2000–2001*, DTLR, London

Department for Transport, Local Government and the Regions (DTLR) (2001b) *Bellwin Scheme of Emergency Financial Assistance to Local Authorities: Report of the Review Group*, October 2001, DTLR, London

Department of Business, Enterprise and Regulatory Reform (2007) *Electricity Priority Users' Arrangements*, http://www.berr.gov.uk/files/file40466.pdf, accessed 26 March 2013

Department of Education (2012a) *2012 School Performance Tables*, http://www.education.gov.uk/schools/performance/, accessed 26 March 2013

Department of Education (2012b) 'Schools, pupils and their characteristics, Statistical First Release (SFR) reports January 2012', Table 2a, in *Datasets*, http://www.education.gov.uk/researchandstatistics/datasets, June 2012, accessed 18 February 2013

Department of Energy and Climate Change (DECC) (2006) *Revised Electricity Supply Emergency Code (Revised January 2005)*, Version 2.0, https://www.gov.uk/government/uploads/system/uploads/attachment_data/file/79195/RevElectSupply.pdf , November 2006, accessed 26 March 2013

Department of Energy and Climate Change (DECC) (2012) *Energy Trends: December 2012*, https://www.gov.uk/government/uploads/system/uploads/attachment_data/file/65751/3939-energy-trends-section-1-total-energy.pdf, accessed 26 March 2013

Department of Health (2012) 'NHS reference costs: financial year 2011 to 2012', https://www.gov.uk/government/publications/nhs-reference-costs-financial-year-2011-to-2012, accessed 27 March 2013

Department of the Environment, Transport and the Regions (DETR) (1998) *Security and Emergency Measures Direction 1998*, linked to the Water Industry Act 1991, http://archive.defra.gov.uk/corporate/about/how/contingency/documents/water-semd98.pdf, June 1998, accessed 3 April 2013

Department of the Environment, Transport and Regions (DETR) (1999) *A Better Quality of Life: A Strategy for Sustainable Development in the UK*, HMSO, London

Department of the Environment, Transport and Regions (DETR) (2000a) *Public Participation in Making Local Environmental Decisions: The Aarhus Convention Newcastle Workshop*, DETR, London

Department of the Environment, Transport and Regions (DETR) (2000b) *Multi-criteria Analysis: A Manual*, DETR, London

Department of the Environment, Transport and Regions (DETR)/Environment Agency (EA)/Institute for Environment and Health (IEH) (2000) *Guidelines for Environmental Risk Assessment and Management: Revised Departmental Guidance*, HMSO, London

Diamond, P.A. and Hausman, J.A. (1992) 'On contingent valuation measurement of nonuse values', in Cambridge Economics (ed.) *Contingent Valuation: A Critical Assessment*, Cambridge Economics Inc., Cambridge, MA, pp3–38

Doncaster Council (2008) 'Summer 2007 flood review consultation: key findings', March 2008, Overview and Scrutiny Management Committee, Doncaster Council

Donnelly, W.A. (1989) 'Hedonic price analysis of the effect of a floodplain on property values', *Water Resources Bulletin*, vol 25, pp581–586

Dora, J. (2007) 'Summer 2007: a Network Rail perspective', *Proceedings of the 42nd Annual Flood and Coastal Management Conference, 3–5 July 2007*, York

Doyle, C. and Elliot, J.G. (1983) 'Putting an economic value on increases in grass production', *Grass and Forage Science*, vol 38, pp169–177

Dunderdale, J.A.L. and Morris, J. (1996a) 'R&D Note 456', in *Guidelines for the Justification of River Maintenance*, National Rivers Authority, Bristol

Dunderdale, J.A.L. and Morris, J. (1996b) 'R&D Note 511', in *River Maintenance Evaluation*, National Rivers Authority, Bristol

Dunderdale, J.A.L. and Morris, J. (1996c) 'The economics of acquatic vegetation removal in rivers and land drainage systems', *Hydrobiologica*, vol 340, pp157–161

Dunderdale, J.A.L. and Morris, J. (1997a) 'Agricultural impacts of river maintenance: a method of assessment', *Journal of Agricultural Engineering Research*, vol 68, pp317–327

Dunderdale, J.A.L. and Morris, J. (1997b) 'The benefits and cost analysis of river maintenance', *Journal of Institution of Water and Environmental Management*, vol 11, no 6, pp423–430

Dyson, M., Bergkamp, G. and Scanlon, J. (2003) *Flow: The Essentials of Environmental Flows*, IUCN, Gland

ECLAC (Economic Commission for Latin America and the Caribbean) (2000) *A Matter of Development: How to Reduce Vulnerability in the Face of Natural Disasters*, Inter-American Development Bank, Washington DC

Eden, S.E., Tunstall, S.M. and Tapsell, S.M. (1999) 'Environmental restoration: environmental management or environmental threat?', *Area*, vol 3, no 2, pp151–159

Eftec (2007) *Policy Appraisal and the Environment: An Introduction to the Valuation of Ecosystem Services – Wareham Managed Realignment Case Study*, report prepared for the Environment Agency, Eftec, London

Eftec (2010a) *Flood and Coastal Erosion Risk Management: Economic Valuation of Environmental Effects*, Eftec, London

Eftec (2010b) *Valuing Environmental Impacts: Practical Guidelines for the Use of Value Transfer in Policy and Project Appraisal Case Study 3 – Valuing Environmental Benefits of a Flood Risk Management Scheme*, report for Defra, Eftec, London

Emerton, L. and Bos, E. (2004) *Value: Counting Ecosystems as an Economic Part of Water*, IUCN, Gland

EMRFRA (2009) *East Midlands Regional Flood Risk Appraisal (Consolation Report)*, http://southeastlincslocalplan.org/files/2012/02/EM-Regional-Flood-Risk-Appraisal.pdf, accessed August 2013

Emschergenossenschaft and Hydrotec (2004) *Hochwasser-Aktionplan Emscher*, Kapitel 1: *Methodik der Schadensermittlung*, report, Emschergenossenschaft and Hydrotec, Essen (in German)

Energy Networks Association (ENA) (2009) 'Resilience to flooding of grid and primary substations', *Engineering Technical Report (ETR 138)*, issue 1, http://www.spenergynetworks.co.uk/connecting_to_our_network/specifications.asp, accessed 26 March 2013

Energy Networks Association (ENA) (2011) 'Electricity networks climate change adaptation report' *Engineering Report 1*, issue 1, http://archive.defra.gov.uk/environment/climate/documents/adapt-reports/04distribute-trans/ena-networks.pdf, accessed 26 March 2013

Energy Networks Association (ENA) (2013) 'Electricity distribution map', http://www.energynetworks.org/info/faqs/electricity-distribution-map.html, accessed 27 March 2013

English Heritage (undated) *Floods in Historic Buildings*, English Heritage, London

English Nature/Environment Agency/Department for Environment, Food and Rural Affairs/Forestry Commission (2003) 'Wetland, land use change and flood management', joint statement

Environment Agency (undated a) *Supporting Spreadsheet to the Economic Appraisal for a Flood or Coastal Erosion Risk Management Project*, http://publications.environment-agency.gov.uk/pdf/GEHO0310BSDE-E-X.xls, accessed 8 April 2013

Environment Agency (undated b) *Adapting to Climate Change: Advice for Flood and Coastal Erosion Risk Management Authorities*, Environment Agency, Bristol

Environment Agency (undated c) *National Receptor Dataset*, Environment Agency, Bristol

Environment Agency (undated d) *Understanding Flood Risk: Our National Flood Risk Assessment (NaFRA)*, Environment Agency, Bristol

Environment Agency (undated e) 'Flood and coastal risk management – conserving, enhancing and restoring biodiversity (England and Wales)', Operational Instruction 635_08, Environment Agency, Bristol

Environment Agency (2001) *Lessons Learned: Autumn 2000 Floods*, Environment Agency, Bristol

Environment Agency (2004) 'Catchment flood management plans, Volume 1', in *Policy Guidance*, Environment Agency, Bristol

Environment Agency (2005) *Response to Flooding*, Environment Agency, Bristol

Environment Agency (2006) *Response to Flooding*, Environment Agency, Bristol

Environment Agency (2007) *Response to Flooding*, Environment Agency, Bristol

Environment Agency (2008) *Economic Evaluation of Damages for Flood Risk Management Projects*, Environment Agency, Bristol

Environment Agency (2009a) *Flooding in England: A National Assessment of Flood Risk*, Environment Agency, Bristol, http://www.tritonsystems.co.uk/pdf/EA_Flooding_in_England.pdf, accessed 27 March 2013

Environment Agency (2009b) *Synthesis Report (Revised), Phase 3, ii, Product 13, Thames Estuary 2100*, Environment Agency, London

Environment Agency (2010a) *Flood and Coastal Erosion Risk Management Appraisal Guidance (FCERM-AG)*, March 2010, Environment Agency, Bristol

Environment Agency (2010b) 'Supporting document for the appraisal summary table', in *Flood and Coastal Erosion Risk Management Appraisal Guidance*, Environment Agency, Bristol

Environment Agency (2010c) 'Guidance on applying the scoring and weighting methodology', in *Flood and Coastal Erosion Risk Management Appraisal Guidance*, Environment Agency, Bristol

Environment Agency (2010d) *Assessment of Coastal Erosion and Landsliding for the Funding of Coastal Risk Management Projects: Guidance Notes*, December 2010, Environment Agency, Bristol

Environment Agency (2010e) *Working with Natural Processes to Manage Flood and Coastal Erosion Risk: A Guidance Document*, Environment Agency, Bristol

Environment Agency (2011a) *Understanding the Risks, Empowering Communities, Building Resilience: The National Flood and Coastal Erosion Risk Management Strategy for England*, HMSO, London

Environment Agency (2011b) *Flood and Coastal Risk Management: An Introduction to Delivering the Water Framework Directive*, Operational Instruction 871_11, Environment Agency, Bristol

Environment Agency (2012) *Principles for Implementing Flood and Coastal Resilience Funding Partnerships*, Environment Agency, Bristol, http://www.environment-agency.gov.uk/research/planning/134732.aspx, accessed 9 April 2012

Environment Agency (2013a) *What Do the Flooding from Surface Water Maps Show and How Were They Created?*, http://www.environment-agency.gov.uk/research/planning/129324.aspx, accessed 23 April 2013

Environment Agency (2013b) *Flood Incident Management Investment Review: An Evidence Base – Main Report*, report prepared for the Environment Agency by Halcrow, Flood Hazard Research Centre and RAB Consultants, April 2013, Halcrow Group, Swindon

Environment Agency (2013c) Personal communication with G. Watson, Coastal Engineer, Norfolk and Suffolk, Environment Agency, April 2013

Environment Agency Wales (2009) *Flooding in Wales: A National Assessment of Flood Risk*, Environment Agency, Cardiff

Esure (2012) *Esure Insurance Flood Advice*, http://www.esure.com/media_centre/car_flood_victims.html, accessed August 2013

European Commission (EC) (2011) *The Common Agricultural Policy after 2013: Legal Proposals*, European Commission, Brussels

European Commission (EC) (2013) *EU Solidarity Fund*, http://ec.europa.eu/regional_policy/thefunds/solidarity/index_en.cfm, accessed 22 April 2013

European Parliament (1998) 'Report on proposal and amended proposals for a Council Directive establishing a framework for Community action in the field of water policy (COM(97)0049-C4-0192/97, COM(97)0614-C4-0120/98 and COM(98)0076-97/0067(SYN)', Committee on the Environment, Public Health and Consumer Protection, European Parliament, http://www.europarl.europa.eu/sides/getDoc.do?type=REPORT&mode=XML&reference=A4-1998-261&language=EN, accessed 11 April 2013

European Parliament and the Council (2007) 'Directive 2007/60/EC of the European Parliament and of the Council of 23 October 2007 on the assessment and management of flood risks', *Official Journal of the European Union*, vol L288

Evans, E., Ashley, R., Hall, J., Penning-Rowsell., E.C., Sayers, P., Thorne, C. and Watkinson, A. (2004) *Foresight Future Flooding, Scientific Summaries*, Volume 1: *Future Risks and Their Drivers* and Volume 2: *Managing Future Risks*, Office of Science and Technology, London

Evans, E.P., Johnson, P.J., Green, C.H. and Varsa, E. (2000) 'Risk assessment and programme prioritisation: the Hungary flood study', *Proceedings of the 35th Annual MAFF Conference of River and Coastal Engineers*, MAFF, London

Everard, M. (2009) *Ecosystem Service Case Studies*, Environment Agency, Bristol

Experian (2013) *3.9% Footfall Growth in Retail Parks Despite the Torrential Rain*, http://www.experianfootfall.com/3-9-footfall-growth-in-retail-parks-despite-the-torrential-rain/, accessed 8 February 2013

Farm Business Survey (2012) *Region Reports*, Rural Business Research University of Nottingham, Nottingham, http://www.farmbusinesssurvey.co.uk/, accessed 8 June 2012

Federal Emergency Management Agency (FEMA) (2010) *Multi-hazard Loss Estimation Methodology, Flood Model, HAZUS MH MR4, Technical Manual*, Department of Homeland Security, FEMA, Washington DC

Flood Hazard Research Centre (FHRC) (1992) *Datchet, Wraysbury, Staines and Chertsey Flood Study: Draft Floodplain Management Plan*, benefit refinement report, Thames Region National Rivers Authority, Reading

Flood Hazard Research Centre and RPA (2001) *The Role of Multi-criteria Analysis in the Appraisal of Flood and Coastal Defence Schemes*, Flood Hazard Research Centre, Middlesex University, London

Food and Agriculture Organisation of the UN (FAO) (1988) 'Salt-affected soils and their management', *FAO Soils Bulletin 39*, Food and Agriculture Organisation of the United Nations, Rome

Food and Grocery Experts (2012a) *Convenience Retail Marketing Overview*, http://www.igd.com/our-expertise/Retail/Convenience/3369/Convenience-Retailing-Market-Overview/, accessed 8 August 2012

Food and Grocery Experts (2012b) *UK Food and Grocery Retail Logistics Overview*, http://www.igd.com/our-expertise/Supply-chain/Logistics/3457/UK-Food--Grocery-Retail-Logistics-Overview/, accessed 8 August 2012

Fordham, M.H. (1992) 'Choice and constraints in flood hazard mitigation: the environmental attitudes of floodplain residents and engineers', unpublished Ph.D. thesis, Middlesex University, London

Fordham, M. (2000) 'Participatory planning for flood mitigation', in Parker, D.J. (ed.) *Floods*, Vol 2, Routledge, London, pp66–79

Foresight (2010) *Foresight Future Land Use Project*, Government Office for Science, London

Foresight (2011) *The Future of Food & Farming: Challenges and Choices for Global Sustainability*, Government Office for Science, London

Fouquet, P.-M., Green, C.H. and Tunstall, S.M. (1992) *Hurst Spit: An Assessment of the Benefits of Coast Protection, Final Report*, report for New Forest District Council, Flood Hazard Research Centre, Middlesex University, London

Friedman, M. (1957) *A Theory of the Consumption Function*, Princeton University Press, Princeton, NJ

Friedman, M. (1980) *Free to Choose*, Penguin, Harmondsworth

Gardiner, J.L. (ed.) (1991) *River Projects and Conservational Manual for Holistic Appraisal*, John Wiley and Sons, Chichester

Garner, J., Tunstall, S.M and Green, C.H. (1994) *Cliftonville: An Assessment of Coast Protection Benefits*, report for Thanet District Council, Flood Hazard Research Centre, Middlesex University, London

Garrido, A. (1999) *Agricultural Water Pricing in OECD Countries*, ENV/EPOC/GRRI(98)11/FINAL, OECD, Paris

Gende, S.M., Edwards, R.T., Willson, M.F. and Wipfli, M.S. (2002) 'Pacific salmon in aquatic and terrestrial ecosystems', *Biosciences*, vol 52, no 10, pp917–928

GFK (2008) 'The insurance gap', *On the Horizon*, issue 5, http://www.gfk.com/imperia/md/content/gfk_nop/financial/onthehorrizon/issue5theinsurancegap.pdf, March 2008, accessed 15 April 2013

Gibson, V. (2003) 'Flexible working needs flexible space? Towards an alternative workplace strategy', *Journal of Property Investment and Finance*, vol 21, no 1, pp12–22

Global Water Partnership Technical Advisory Committee (GWP) (2000) *Integrated Water Resources Management*, TAC Background Paper 4, Global Water Partnership, Stockholm

Gloucestershire Hospitals NHS Trust (2008) *Annual Report and Accounts 2007–2008*, presented to Parliament pursuant to Schedule 7, paragraph 25(4) of the National Health Service Act 2006

Gouldby, B., Sayers, P., Mulet-Marti, J., Hassan, M. and Benwell, D. (2008) 'A methodology for regional scale flood risk assessment', *Proceedings of the Institution of Civil Engineers: Water Engineering*, vol 161, no 3, pp169–182

Graham, W.J. (1999) *A Procedure for Estimating Loss of Life Caused by Dam Failure*, DSO-99-6, Dam Safety Office, Denver, CO

Green, C.H. (1997) 'Water, then environment and economics: what does experience teach us so far?', *Canadian Water Resources*, vol 22, no 1, pp57–70

Green, C.H. (1998) 'The environment: how can it be valued? Ought it to be valued? How does valuation assist coastal sustainability?', paper presented to the Annual Conference of the Standing Conference on Problems Associated with the Coastline (SCOPAC), Chichester

Green, C.H. (2001) 'Flood management of the Yangtze River: the lessons of economic appraisal', paper presented to the 36th Defra Conference of River and Coastal Engineers, Keele

Green, C.H. (2002) *The Economics of Integrated Water Resource Management: Flood Risk Management*, World Meteorological Organisation, Geneva

Green, C.H. (2003) *A Handbook of Water Economics*, John Wiley, Chichester

Green, C.H. (2004) 'Flood risk management in the context of integrated water resource management', keynote address to the Workshop on Flood Prevention and Control on the Yangtze River: State-of-the-art and Future Developments, Wuhan

Green, C.H. and Penning-Rowsell, E.C. (1989) 'Flooding and the quantification of intangibles', *Journal of the Institution of Water and Environmental Management*, vol 3, no 1, pp27–30

Green, C.H. and Tunstall, S.M. (1991) 'The evaluation of river water quality improvements by the contingent valuation method', *Applied Economics*, vol 23, pp1135–1146

Green, C.H. and Tunstall, S.M. (1992) 'The amenity and environmental value of river corridors in Britain', in Boon, P.J., Calow, P. and Petts, P.E. (eds) *River Conservation and Management*, Wiley, Chichester, pp425–441

Green, C.H. and Penning-Rowsell, E. C. (1999) 'Inherent conflicts at the coast', *Journal of Coastal Conservation*, vol 5, pp153–162

Green, C.H. and Penning-Rowsell, E.C. (2011) 'Stakeholder engagement in flood risk management', in Pender, G. and Faulkner, H. (eds) *Flood Risk Science and Management*, Wiley-Blackwell, Oxford, pp372–385

Green, C.H., Emery, P.J., Penning-Rowsell, E.C. and Parker, D.J. (1985) *The Health Effects of Flooding: A Survey at Uphill, Avon*, Flood Hazard Research Centre, Middlesex Polytechnic, London

Green, C.H., Tunstall, S.M., N'Jai, A. and Rogers, A. (1990) 'The economic evaluation of environmental goods', *Project Appraisal*, vol 5, no 2, pp70–82

Green, C.H., Nicholls, R.J. and Johnson, C.L. (2000a) *Climate Change Adaptation: An Analysis of Decision-making in the Face of Risk and Uncertainty*, Report No 28, National Centre for Risk Assessment and Options Appraisal, Environment Agency, London

Green, C.H., Parker, D.J. and Tunstall, S.M. (2000b) *Assessment of Flood Control and Management Options*, World Commission on Dams, Cape Town

Green, P.E. and Srinivasan, V. (1978) 'Conjoint analysis in consumer research: issues and outlook', *Journal of Consumer Research*, vol 5, pp103–123

Greenley-Polhemus Group (1992) *Guidelines for Risk and Uncertainty Analysis in Water Resources Planning*, IWR Report 92-R-1, US Army Corps of Engineers, Institute for Water Resources, Fort Belvoir, VA

GreenSpace (undated) *A Guide to Automated Methods for Counting Visitors to Parks and Green Spaces*, GreenSpace, Reading

GVA (2011/12) *What Will Influence UK Office Occupier Requirements in the Next Decade?*, GVA, London

Halcrow Group Limited (1999) 'Appendices E to G', in *Gorleston to Lowestoft Coastal Strategy Study Part Three*, consultation draft, Halcrow, Swindon

Halcrow Group Limited (2007) *Environment Agency National Flood Risk Assessment 2006*, project report, Halcrow Group Limited, Swindon

Hamilton, J.M., Maddison, D.J. and Tol, R.S.J. (2005) 'Climate change and international tourism: a simulation study', *Global Environmental Change*, vol 15, no 3, pp253–266

Hanley, N., Wright, R.E. and Alvarez-Farizo, B. (2006)

'Estimating the economic value of improvements in river ecology using choice experiments', *Journal of Environmental Management*, vol 78, no 2, pp183–193

HEC-FDA (2008) *Flood Damage Reduction Analysis: User Guide V1.2.4*, http://www.hec.usace.army.mil/software/hec-fda/documentation/CPD-72_V1.2.4.pdf, accessed 8 June 2013

Heimlich, R.E. (1991) *National Policy of No Net Loss of Wetlands: What Do Agricultural Economists Have to Contribute?*, Resources and Technology Division, US Department of Agriculture, Washington DC

Henderson, J.M. and Quandt, R.E. (1971) *Microeconomic Theory: A Mathematical Approach*, McGraw-Hill, Kogakusha, Tokyo

Hensher, D.A. (1994) 'Stated preference analysis of travel choices: the state of practice', *Transportation*, vol 21, no 2, pp107–133

Hertfordshire County Council (undated) *An Introduction to Highway Maintenance*, http://www.hertsdirect.org/docs/pdf/i/inthighmaint.pdf, accessed 15 April 2013

Hess, T.M. and Morris, J. (1988) 'Estimating the value of flood alleviation on agricultural grassland', *Agriculture Water Management*, vol 15, pp141–153

Hess, T.M., Leeds-Harrison, P.B. and Morris, J. (1989) 'Evaluation of river maintenance in agricultural areas', *Proceedings of the 11th International Congress on Agricultural Engineering, Dublin, 4–8 September*, pp501–509

Higgins, R.J. (1981) *Appraisal and Evaluation in Central Government: A Technical Guide for Government Departments*, HMSO, London

HM Government (1947) *Fire Services Act 1947, 10 & 11 Geo 6, Chapter 41*, HMSO, London, http://www.legislation.gov.uk/ukpga/Geo6/10-11/41, accessed 26 April 2013

HM Government (1949) *Coast Protection Act 1949, Chapter 74, 12, 13 and 14, Geo 6*, HMSO, London, http://www.legislation.gov.uk/ukpga/Geo6/12-13-14/74, accessed 26 April 2013

HM Government (1960) *Caravan Sites and Control of Development Act 1960: Model Standards*, HMSO, London

HM Government (1989) *Local Government and Housing Act 1998, Chapter 42*, HMSO, London

HM Government (1993) *Railway Act 1993*, HMSO, London, http://www.legislation.gov.uk/ukpga/1993/43/enacted, accessed 3 April 2013

HM Government (1996) *Gas Safety (Management) Regulations (1996), Statutory Instruments, Health and Safety, 1996, No. 551*, HMSO, London, http://www.legislation.gov.uk/uksi/1996/551/contents/made, accessed 27 March 2013

HM Government (2002) *Statutory Instruments, Electricity, 2002 No. 2665, The Electricity Safety, Quality and Continuity Regulations 2002*, HMSO, London, http://www.legislation.gov.uk/uksi/2002/2665/introduction/made, accessed 27 March 2013

HM Government (2004) *Civil Contingencies Act 2004*, HMSO, London, http://www.legislation.gov.uk/ukpga/2004/36/introduction, accessed 26 March 2013

HM Government (2008) *Climate Change Act 2008*, HMSO, London, http://www.legislation.gov.uk/ukpga/2004/36/introduction, accessed 3 April 2013

HM Government (2009a) *Flood Risk Regulations 2009, Statutory Instruments, Statutory Instrument 2009, no 3042, Environmental Protection*, HMSO, London, http://www.legislation.gov.uk/uksi/2009/3042/pdfs/uksi_20093042_en.pdf, accessed 21 January 2013

HM Government (2009b) *Marine and Coastal Access Act, Chapter 23*, HMSO, London

HM Government (2010a) *Flood and Water Management Act 2010, Chapter 29*, HMSO, London, http://www.legislation.gov.uk/ukpga/2010/29/contents, accessed 26 April 2013

HM Government (2010b) *Equality Act 2010*, HMSO, London, http://www.legislation.gov.uk/ukpga/2010/15/contents, accessed 26 April 2013

HM Government (2010c) *The Electricity (Standards of Performance) Regulations 2010, Statutory Instruments, Electricity 2010, no 698*, HMSO, London, http://www.legislation.gov.uk/uksi/2010/698/contents/made, accessed 27 March 2013

HM Government (2010d) *Food 2030*, Defra, London

HM Government (2011) *The Natural Choice: Securing the Value of Nature*, HMSO, London

HM Government (2013) *GB Road Traffic Counts*, http://data.gov.uk/dataset/gb-road-traffic-counts, accessed 3 April 2013

HM Government/Defra (2008) *Future Water: The Government's Water Strategy for England*, February 2008, TSO, London

HM Treasury (1991) *Appraisal and Evaluation in Central Government: A Technical Guide for Government Departments*, HMSO, London

HM Treasury (2003) *The Green Book: Appraisal and Evaluation in Central Government*, HM Treasury, London

HM Treasury (2011) *The Green Book: Appraisal and Evaluation in Central Government*, HM Treasury, London

HM Treasury (2012a) *Latest Figures, GDP Deflators at Market Prices, and Money GDP*, http://www.hm-treasury.gov.uk/data_gdp_fig.htm, accessed 27 March 2013

HM Treasury (2012b) *Accounting for Environmental Impacts: Supplementary Green Book Guidance*, HM Treasury/Defra, London

HMRC (2013) *HMRC Tax Calculator*, http://tools.hmrc.gov.uk/hmrctaxcalculator/screen/Personal+Tax+Calculator/en-GB/summary?user=guest, accessed 17 April 2013

Holling, C.S. (1973) 'Resilience and stability of ecological systems', *Annual Review of Ecology and Systematics*, vol 4, pp1–23

Holmes, T.P., Bergstrom, J.C., Huszar, E., Kask, S.B. and Orr, F. (2004) 'Contingent valuation, net marginal benefits and the scale of riparian ecosystem restoration', *Ecological Economics*, vol 49, no 1, pp19–30

Homelet (2013) *March 2013 Rental Index UK Regional Map*, https://homelet.co.uk/rentalindex/regional-map, accessed 29 April 2013

Homeseeker Homes (2012) E-mail correspondence with Lee Pearce, Design Manager, Homeseeker Homes, 18 July 2012

Honda (2005) *Honda Price Guide, April 2005*, http://www.honda.co.uk/brochure/download/cars/HondaCarPriceList.pdf, accessed July 2012

Horsfield, G. (2012) *Survey Methodology: Appendix B*, Office for National Statistics, http://www.ons.gov.uk/ons/dcp171766_288107.pdf, 4 December 2012, accessed 25 April 2013

Hoskins, W.G. (1970) *The Making of the English Landscape*, Penguin, Harmondsworth

Houghton, J.I., Meira Filho, L.G., Callander, B.A., Harris, N., Kattenberg, A. and Maskell, K. (eds) (1996) *Climate Change 1995: The Science of Climate Change*, Cambridge University Press, Cambridge

House, M.A., Tunstall, S.M., Green, C.H., Portou, J. and Clarke, L. (1994) *The Evaluation of the Recreation and Other Use Values from Alleviating Low Flows*, R&D Note 258, National Rivers Authority, Bristol

Housing Care (2012) *Information Sheet: Park Homes*, http://www.housingcare.org/jargon-park-homes-5e4ec.aspx, accessed 13 August 2012

HR Wallingford (2003) *Flood Risks to People Phase 1*, final report prepared for Defra/Environment Agency Flood and Coastal Defence R&D Programme, HR Wallingford, Wallingford

HR Wallingford (2005a) *Performance and Reliability of Flood and Coastal Defences: Best Practice Guidance*, HR Wallingford, Wallingford

HR Wallingford (2005b) *Flood Risks to People, Phase 2, FD2321/TR1: The Flood Risk to People Methodology*, HR Wallingford, Wallingford

HR Wallingford (2012) *Development of Spatial Indicators to Monitor Changes in Exposure and Vulnerability to Flooding and the Uptake of Adaptation Actions to Manage Flood Risk in England: Results 2012*, Report EX6739 R4, HR Wallingford, Wallingford, http://hmccc.s3.amazonaws.com/ASC/2012%20report/HR%20Wallingford%20flooding%20indicators%20results%20reportAMND.pdf, August 2012, accessed 27 March 2013

HR Wallingford, Royal Haskoning and Peter Lawton Consulting Engineer (2009) 'Assessment and measurement of asset deterioration including whole life costing', in *Science Summary*, SC060078/SR2, Environment Agency, Bristol

IAASTD (2009) *Agriculture at a Crossroads: Global Report*, Island Press, Washington DC

Imber, D., Stevenson, G. and Wilks, L. (1991) *A Contingent Valuation Method Survey on the Kakadu Conservation Zone*, Resource Assessment Commission Research Paper No 3, Australian Government Publishing Service, Canberra

Independent (2013) 'How recession killed the high street', *Independent*, 8 February 2013

Instant Offices Ltd (2011) *Global Serviced Office Review 2011*, http://www.instantoffices.com/global/en/solutions/globalreview.pdf, accessed 16 April 2013

Institute of Inspection, Cleaning and Restoration Certification (IICRC) (2006) *S500: Standard and Reference Guide for Professional Water Damage Restoration*, 3rd edn, IICRC, Washington DC

Institute of Mechanical Engineers (2012) *UK Manufacturing in 2012*, IME, London

IPHAS (2012) Telephone conversation with Alan Savory, Independent Park Home Advisory Service, 17 July 2012

JBA (2012) *Establishing the Cost Effectiveness of Property Flood Protection*, FD2657, final report, 7 August 2012, Defra, London

Johns, D. (2011) 'Understanding the risks, building resilience, empowering communities: a benefit-led, partnership approach to funding', PowerPoint presentation, Defra, London

Johnson, C. (2008) 'Modelling the cost of evacuation for flood risk appraisal', unpublished report

Johnson, C.L., Penning-Rowsell, E.C. and Parker, D.J. (2007) 'Natural and imposed injustices: the challenges in implementing "fair" flood risk management policy in England', *Geographical Journal*, vol 173, no 4, pp374–390

Jones, A., Wright, J., Bateman, I. and Schaafsma, M. (2010) 'Estimating arrival numbers for informal recreation: a geographical approach and case study', *Woodland Sustainability*, vol 22, pp684–701

Jones Lang Lasalle (2012) 'The Western Corridor industrial and warehouse market report', *On.Point*, Autumn 2012, Jones Lang Lasalle, London

Jonkman, S.N. (2003) *Loss of Life Caused by Floods: An Overview of Mortality Statistics for Worldwide Floods*, http://www.citg.tudelft.nl/fileadmin/Faculteit/CiTG/Over_de_faculteit/Afdelingen/Afdeling_Waterbouwkunde/sectie_waterbouwkunde/people/personal/gelder/publications/citations/doc/citatie161.pdf, accessed 15 April 2013

Jonkman, S.N. (2007) 'Loss of life estimation in flood risk assessment: theory and applications', Ph.D. dissertation, Delft Cluster, Delft

Jonkman, S.N. and Kelman, I. (2005) 'An analysis of the causes and circumstances of flood disaster deaths', *Disasters*, vol 29, pp75–97

Jonkman, S.N. and Penning-Rowsell, E.C. (2008) 'Human instability in flood flows', *Journal of the American Water Resources Association*, vol 44, no 4, pp1–11

Jonkman, S.N., Van Gelder, P.H.A.J.M. and Vrijling, J.K. (2002) 'Loss of life models for sea and river floods', in Wu, B. *et al.* (eds) *Flood Defence 2002*, Science Press, New York, http://www.citg.tudelft.nl/fileadmin/Faculteit/CiTG/Over_de_faculteit/Afdelingen/Afdeling_Waterbouwkunde/sectie_waterbouwkunde/people/personal/gelder/publications/papers/doc/paper120b-V10210.pdf, accessed 16 April 2013

Karvonen, R.A., Hepojoki, A., Huhta, H.K. and Louhio, A. (2000) *The Use of Physical Models in Dam-Break Analysis*, RESCDAM Final Report, Helsinki University of Technology, Helskini, http://www.ymparisto.fi/download.asp?contentid=13494&lan=fi, accessed August 2013

Kaval, P. (2007) *Recreation Values of US Parks*, Working Paper Series No. 07/012, Department of Economics, University of Waikato, Hamilton, New Zealand, http://ideas.repec.org/s/wai/econwp.html, accessed 22 April 2013

Kaval, P. and Yao, R. (2007) *New Zealand Outdoor Recreation Benefits*, Working Paper in Economics No 07/14, Economics Department, University of Waikato, Hamilton, New Zealand

Kelman, I. (2002) 'Physical flood vulnerability of residential properties in coastal, eastern England', Unpublished Ph.D. dissertation, University of Cambridge

Ketteridge, A.-M. and Green, C.H. (1994) 'Technical Annex for the Full Flood Impacts Module', in *EUROflood*, report for the European Commission, Flood Hazard Research Centre, Middlesex University, London

Ketteridge, A.-M., Fordham, M. and Clarke, L. (1996) 'Technical Annex 14', in *EUROflood: Evacuation*, report for the EU EUROflood project, European Union Environment Programme, Flood Hazard Research Centre, Middlesex University, London

Klijn, F., Baan, .P, de Bruijn, K. and Kwadijk, J. (2007) *Overstromingsrisico's in Nederland in een veranderend klimaat*, Verwachtingen, schattingen en berekeningen voor het project Nederland Later, Milieu- en Natuurplanbureau, http://www.rivm.nl/bibliotheek/digitaaldepot/WL_rapport_Overstromingsrisicos_Nederland.pdf, accessed 16 April 2013 (in Dutch)

Kreibich, H., Thieken, A.H., Petrow, T.H., Muller, M. and Merz, B. (2005) 'Flood loss reduction of private households due to building precautionary measures – lessons learned from the Elbe flood in August 2002', *Natural Hazards and Earth Systems Sciences*, vol 5, pp117–126

Kreibich, H., Muller, M., Thieken, A. and Merz, B. (2007) 'Flood precaution of companies and their ability to cope with the flood in August 2002 in Saxony, Germany', *Water Resources Research*, vol 43, no 3, pp1–15

Kreibich, H., Seifert, I., Merz, B. and Thieken, A.H. (2010) 'Development of FLEMOcs – a new model for the estimation of flood losses in the commercial sector', *Hydrological Sciences Journal*, vol 55, no 8, pp1302–1314

Kreibich, H., Seifert, I., Thieken, A.H., Lindquist, E., Wagner, K. and Merz, B. (2011) 'Recent changes in flood preparedness of private households and businesses in Germany', *Regional Environmental Change*, vol 11, pp59–71

Kron, A. (2007) 'Flood damage estimation and flood risk mapping', in Vassipoulos, A., Ashley, R., Zevenbergen, C., Pashe, E. and Garvin, S. (eds) *Advances in Urban Flood Management*, Taylor and Francis, London, pp213–235

Krueger, R.A. (1988) *Focus Groups: A Practical Guide for Applied Research*, Sage, Beverly Hills, CA

Krutilla, J.A. (1967) 'Conservation reconsidered', *American Economic Review*, vol 57, no 4, pp77–86

Lago, M. and Glenk, K. (2008) 'Delivering good status in Scotland: using choice experiments for the estimation of non-market benefits of the EC Water Framework Directive', unpublished working paper, Macaulay Land Use Research Institute, Aberdeen

Land Registry (2013) *December Market Trend Data from Land Registry 2012*, http://www.landregistry.gov.uk/media/all-releases/press-releases/2013/market-trend-data-december-2012, 29 January 2013, accessed 16 April 2013

Lane, S., Morris, J., O'Connell, P.E. and Quinn, P.F. (2007) 'Managing the rural landscape', in Thorne, C.R., Evans, E.P. and Penning-Rowsell, E.C. (eds) *Future Flooding and Coastal Erosion Risks*, Thomas Telford, London, pp297–319

Lawrence, K.S. and Spurgeon, J. (2007) *Economic Evaluation of Inland Fisheries: Welfare Benefits of Inland Fisheries in England and Wales*, Science Report SC050026/SRI, Environment Agency, Bristol

Lawton, J.H., Brotherton, P.N.M., Brown, V.K., Elphick, C., Fitter, A.H., Forshaw, J., Haddow, R.W., Hilborne, S., Leafe, R.N., Mace, G.M., Southgate, M.P., Sutherland, W.J., Tew, T.E., Varley, J. and Wynne, G.R. (2010) *Making Space for Nature: A Review of England's Wildlife Sites and Ecological Network*, report for Defra, London

Luisetti, T., Bateman, I.J. and Turner, K. (2011a) 'Testing the fundamental assumptions of choice experiments: are values absolute or relative?', *Land Economics*, vol 87, no 2, pp284–296

Luisetti, T., Turner, R.K., Bateman, I.J., Morse-Jones, S., Adams, C. and Fonseca, L. (2011b) 'Coastal and marine ecosystem services: managed realignment case studies in England', *Ocean and Coastal Management*, vol 53, no 3, pp212–224

McCoubrey, H. and White, N.D. (1993) *Textbook on Jurisprudence*, Oxford University Press, Oxford

MacRae, D. and Whittington, D. (1988) 'Assessing preferences in cost–benefit analysis: reflections on rural water supply evaluation in Haiti', *Journal of Policy Analysis and Management*, vol 7, no 2, pp246–263

Maltby, E., Acreman, M., Blackwell, M.S.A., Everard, M. and Morris, J. (2013) 'The challenges and implications of linking wetland science to policy in agricultural landscapes: experience from the UK National Ecosystem Assessment', *Ecological Engineering*, Elsevier, Amsterdam, http://www.sciencedirect.com/science/article/pii/S092585741200448X, accessed 16 April 2013

Marwell, G. and Ames, R.E. (1979) 'Experiments on provision of public goods 1: resources, interest, group size and the free rider problem', *American Journal of Sociology*, vol 84, pp1335–1360

May, P. and Chatterton, J. (2012) *Establishing the Cost Effectiveness of Property Flood Protection*, Final Report FD2657, Defra, London

Merz, B., Kreibich, H., Thieken, A. and Scmidtke, R. (2004) 'Estimation of uncertainty of direct monetary flood damage to buildings', *Natural Hazards and Earth System Science*, vol 4, pp153–163

Millennium Ecosystem Assessment (MEA) (2005) *Ecosystems and Human Well-being: A Synthesis*, Island Press, Washington DC

Ministry of Agriculture, Fisheries and Food (MAFF) (1998) 'International conservation sites to be protected from flooding', News Release No 292/98, 9 July 1988, MAFF, London

Ministry of Agriculture, Fisheries and Food (MAFF) (1999) *Flood and Coastal Defence Project Appraisal Guidance (FCDPAG3) Economic Appraisal*, MAFF, London

Ministry of Agriculture, Fisheries and Food (MAFF)/Welsh Office (1993) *Strategy for Flood and Coastal Defence in England and Wales*, Publication PB1471, September 1993, MAFF, London

Ministry of Housing and Local Government (1968) *The Principles of a Land Use Classification in a Computer Data Bank*, MHLG, London

Mintel (2003–2006) *Mintel Market Intelligence*, monthly editions, Mintel Marketing, London

Mishan, E.J. (1971) *Cost Benefit Analysis*, Unwin University Books, London

Mitchell, R.C. and Carson, R.T. (1989) *Using Surveys to Value Public Goods: The Contingent Valuation Method*, Resources for the Future, Washington DC

Mitra, S., Wassmann, R. and Vlek, P.L.G. (2003) *Global Inventory of Wetlands and Their Role in the Carbon Cycle*, Discussion Papers on Development Theory No 64, ZEF, Bonn

Mitsch, W.J. and Gosselink, J.G. (2000) *Wetlands*, John Wiley, Chichester

Modigliani, F. (1966) 'The life cycle hypothesis of savings, the demand for wealth and the supply of capital', *Social Research*, vol 33, no 2, pp160–217

Moller, I. and Spencer, T. (2002) 'Wave dissipation over macro-tidal saltmarshes: effects of marsh edge typology and vegetation change', *Journal of Coastal Research*, special issue 36, pp506–521

Morris, J. (1992) 'Agricultural land drainage: land use change and economic performance: experience in the UK', *Land Use Policy*, vol 3, no 9, pp185–198

Morris, J., Bailey, A.P., Lawson, C.S., Leeds-Harrison, P.B., Alsop, D. and Vivash, R. (2008) 'The economic dimensions of integrating flood management and agri-environment through washland creation: a case from Somerset, England', *Journal of Environmental Management*, vol 88, pp372–381

Morris, J. and Brewin, P. (2013) 'Impact of seasonal flooding on agriculture: the spring 2012 floods in Somerset, England', *Journal of Flood Risk Management*, D0120.1111/jfr3.12041

Morris, J. and Hess, T.M. (1986) 'Farmer uptake of agricultural land drainage benefits', *Environment and Planning A*, vol 18, no 12, pp1649–1664

Morris, J. and Hess, T.M. (1988) 'Agricultural flood alleviation benefit assessment: a case study', *Journal of Agricultural Economics*, vol 38, no 2, pp402–412

Morris, J. and Sutherland, D.C. (1993) *The Evaluation of River Maintenance*, Report for the National Rivers Authority, Severn–Trent Region, Silsoe College, Cranfield

Morris, J. and Camino, M. (2011) *Economic Assessment of Freshwater, Wetland and Floodplain Ecosystem Services*, working paper, UK National Ecosystem Assessment, WCMC, Cambridge, http://uknea.unep-wcmc.org/LinkClick.aspx?fileticket=lVLEq%2bxAI%2bQ%3d&tabid=82, accessed 9 April 2013

Morris, J., Hess, T.M., Ryan, A.M. and Leeds- Harrison, P.B. (1984) *Drainage Benefits and Farmer Uptake*, 4 vols, report for the Severn–Trent Water Authority, Silsoe College, Cranfield

Morris, J., Hess, T.M., Gowing, D.J., Leeds-Harrison, P.B., Bannister, N., Wade, M. and Vivash, R.M. (2004) *Integrated Washland Management for Flood Defence and Biodiversity*, English Nature Research Report No 598 for Defra and English Nature, Cranfield University, Silsoe

Morris, J., Hess, T.M., Gowing, D.J., Leeds-Harrison, P.B., Bannister, N., Vivash, R.M.N. and Wade, M. (2005) 'A framework for integrating flood defence and biodiversity in washlands in England', *International Journal River Basin Management: IAHR & INBO*, vol 3, no 2, pp1–11

Morris, J., Hess, T.M. and Posthumus, H. (2010) *Agriculture's Role in Flood Adaptation and Mitigation: Sustainable Management of Water Resources in Agriculture: Policy Issues and Approaches*, OECD, Paris

Motor Insurers' Bureau (2013) E-mail correspondence with Kate Slater, MIB Group, 25 March 2013

Munro, A. and Hanley, N.D. (1999) 'Information, uncertainty and contingent valuation', in Bateman, I.J. and

Willis, K.G. (eds) *Valuing Environmental Preferences: Theory and Practice of the Contingent Valuation Method in US, EU and Developing Countries*, Oxford University Press, Oxford, pp258–279

N'Jai, A., Tapsell, S.M., Taylor, D., Thompson, P. and Witts, R.C. (1990) *Flood Loss Assessment Information Report (FLAIR)*, Flood Hazard Research Centre, Middlesex Polytechnic, London

Naess, A. (1993) 'The deep ecological movement: some philosophical aspects', in Armstrong, S.J. and Botzler, R.G. (eds) *Environmental Ethics: Divergence and Convergence*, McGraw-Hill, New York

National Audit Office (2000) *Hip Replacements: Getting It Right First Time*, HMSO, London

National Audit Office (2001) *Inland Flood Defence*, HC299, NAO, London

National Audit Office (2005) *Improving School Attendance in England*, report by the Comptroller and Auditor General, HC212, HMSO, London, http://news.bbc.co.uk/2/shared/bsp/hi/pdfs/04_02_05_truancy.pdf, 4 February 2005, accessed 27 March 2013

National Ecosystem Assessment (NEA) (2011) *The UK National Ecosystem Assessment Technical Report*, UNEP–WCMC, Cambridge

National Grid Gas (2010) *Climate Change Adaptation Report*, Gas Transmission and Distribution UK, National Grid Gas PLC, http://www.nationalgrid.com/NR/rdonlyres/C456F00F-1063-43CF-ACF6-85313D9D78F0/45161/nationalgridccagasreport100928.pdf, September 2010, accessed 27 March 2013

National Oceanic and Atmospheric Administration (NOAA) (1993) 'Report of the NOAA Panel on Contingent Valuation', *Federal Register*, vol 58, no 10, pp4601–4614

Natural England (2009) *Farming with Nature: Agri-environment Schemes in Action*, Natural England, Peterborough

Natural England (2010) *Higher Level Stewardship, Environmental Stewardship Handbook*, 3rd edn, Natural England, Peterborough

Natural England (2011) *No Charge? Valuing the Natural Environment*, Natural England, Peterborough

NERA/Accent (2007) *Report on the Benefits of Water Framework Directive Programmes of Measures in England and Wales*, final report for Defra on CRP Project 4b/c, Collaborative Research Programme on River Basin Management Planning Economics, NERA, London

Network Rail (2011) *Network Rail Climate Change Adaptation Report*, report prepared in response to the UK Government's adaptation reporting power, Network Rail, London, http://archive.defra.gov.uk/environment/climate/documents/adapt-reports/06road-rail/network-rail.pdf, accessed 3 April 2013

Newson, M. (1997) *Land, Water and Development: Sustainable Management of River Basin Systems*, Routledge, London

NHS (2012) *Western Sussex Hospitals NHS Trust, Trust Board Meeting, Agenda, Minutes and Public Papers*, http://www.westernsussexhospitals.nhs.uk/EasysiteWeb/getresource.axd?AssetID=434148&type=full&servicetype=Attachment, 26 July 2012, accessed 27 March 2013

Nicholls, R.J. and Branson, J. (1998) 'Coastal resilience and planning for an uncertain future: an introduction', *Geographical Journal*, vol 164, no 3, pp255–258

Nix, J. (2012) *Farm Management Pocketbook 2013*, 43rd edn, Farm Management Pocketbook, Melton Mowbray

O'Brien, L. and Harris, J. (2010) *Forest Research*, Research Agency of the Forestry Commission, Farnham

O'Connell, P.E., Beven , K.J., Carney, J.N., Clements, R.O., Ewen, J., Fowler, H., Harris, G.L., Hollis, J., Morris, J., O'Donnell, G.M., Packman, J.C., Parkin, A., Quinn P.F. and Rose, S.C. (2004) *Review of Impacts of Rural Land Use and Management on Flood Generation*, R&D Technical Report FD2114, Defra/Environment Agency, London

OECD/FAO (2009) *OECD/FAO Outlook for Agriculture 2009–2018*, Organisation for Economic Cooperation and Development and the Food and Agriculture Organisation of the United Nations, Paris and Rome

Office for National Statistics (ONS) (2001) *Living in Britain: Results from the 2000/01 General Household Survey*, ONS, London

Office for National Statistics (ONS) (2002a) *Social Trends No.32*, ONS, London

Office for National Statistics (ONS) (2002b) *Housing in England 2000/1: A Report of the 2000/2001 Survey of English Housing*, ONS, London

Office for National Statistics (ONS) (2008) *Family Spending: A Report on the 2007 Expenditure and Food Survey*, Palgrave Macmillan, Basingstoke

Office for National Statistics (ONS) (2011) *UK Relative Regional Consumer Price Levels for Goods and Services for 2010*, http://www.ons.gov.uk/ons/rel/cpi/regional-consumer-price-levels/2010/index.html, accessed 26 April 2013

Office for National Statistics (ONS) (2012a) 'Annual survey of hours and earnings, 2012: provisional results', *Statistical Bulletin*, http://www.ons.gov.uk/ons/dcp171778_286243.pdf, 22 November 2012, accessed 17 April 2013

Office for National Statistics (ONS) (2012b) *House Price Index*, http://www.ons.gov.uk/ons/rel/hpi/house-price-index/december-2012/stb-december-2012.html#tab-Data-tables, December 2012, accessed 20 April 2013

Office for National Statistics (ONS) (2013) *The National Statistics Socio-economic Classification (NS-SEC Rebased on the SOC2010) User Manual*, http://www.ons.gov.uk/ons/guide-method/classifications/current-standard-classifications/soc2010/soc2010-volume-3-ns

-sec—rebased-on-soc2010--user-manual/index.html, accessed 26 April 2013

Office of Rail Regulation (ORR) (2012) *Track Access and the Network Code*, http://www.rail-reg.gov.uk/server/show/nav.201, accessed 27 March 2013

Office of Rail Regulation (ORR) (2013) *The National Rail Trends (NRT) Portal*, http://dataportal.orr.gov.uk/, accessed 27 March 2013

Office of Science and Technology (OST) (2004) *Future Flooding in Scotland*, Foresight project, Office of Science and Technology, London

Office of the Deputy Prime Minister (ODPM) (2002) *Economics of the Park Homes Industry*, http://webarchive.nationalarchives.gov.uk/20120919132719/http://www.communities.gov.uk/documents/housing/pdf/138274.pdf, accessed 26 April 2013

Office of the Deputy Prime Minister (ODPM) (2005) *A Practical Guide to the Strategic Environmental Assessment Directive*, ODPM, London

Ofgem (2009) *Electricity Distribution Price Control Review: Final Proposals – Incentives and Obligations*, http://www.ofgem.gov.uk/Networks/ElecDist/PriceCntrls/DPCR5/Documents1/FP_2_Incentives%20and%20Obligations%20FINAL.pdf, 7 December 2009, accessed 27 March 2013

Ofgem (2011) *Typical Domestic Energy Consumption Figures*, Factsheet 96, http://www.ofgem.gov.uk/Media/FactSheets/Documents1/domestic%20energy%20consump%20fig%20FS.pdf, 18 January 2011, accessed 27 March 2013

Ofwat (2008) *The Guaranteed Standards Scheme, UK*, http://www.ofwat.gov.uk/consumerissues/rightsresponsibilities/standards/gud_pro_gss08.pdf, accessed 27 March 2013

Ofwat (2009) *Future Water and Sewerage Charges 2010–15: Final Determinations*, http://www.ofwat.gov.uk/pricereview/pr09phase3/det_pr09_finalfull.pdf, accessed 27 March 2013

Ofwat (2012) *Resilience: Outcomes Focused Regulation, Principles for Resilience Planning*, report by Mott MacDonald for Ofwat, http://www.ofwat.gov.uk/sustainability/climatechange/adapt/rpt_com120503mottmacresil.pdf, May 2012, accessed 27 March 2013

Ofwat (2013) *Water Companies' 'Strategic Direction Statement (SDS)'*, http://www.ofwat.gov.uk/pricereview/sds/, accessed 3 April 2013

Ordnance Survey (2011) 'Ordnance Survey data highlights the changing face of the high street', http://www.ordnancesurvey.co.uk/oswebsite/news/2011/01/data-highlights-changing-face-of-high-street.html, accessed 8 June 2013

Parker, D.J. (1991) *The Damage-reducing Effects of Flood Warnings*, Flood Hazard Research Centre, Middlesex Polytechnic, London

Parker, D.J. (1995) 'Floodplain development policy in England and Wales', *Applied Geography*, vol 15, no 4, pp341–363

Parker, D.J. and Tunstall, S.M. (1991) 'Managing flood warning systems: the United Kingdom experience', in Association of State Floodplain Managers *Inspiration: Come to the Headwaters, Proceedings of the 15th Annual Conference of the Association of Floodplain Managers, Denver,* June 10–14, 1991, (1991), MHI Publications, Paper 92, pp168–171

Parker, D.J. and Priest, S.J. (2012) 'Flood warning chain fallibility: can Europe's flood warnings be effective?', *Water Resources Management*, vol 26, pp2927–2950

Parker, D.J., Green, C.H. and Thompson, P.M. (1987) *Urban Flood Protection Benefits: A Project Appraisal Guide* (the 'Red Manual'), Gower Technical Press, Aldershot

Parker, D.J., Tunstall, S. and McCarthy, S. (2007) 'New insights into the benefits of flood warnings: results from a household survey in England and Wales', *Environmental Hazards*, vol 7, no 3, pp193–210

Parker, D.J., Priest, S.J., Schildt, A. and Handmer, J.W. (2008) *Modelling the Damage Reducing Effects of Flood Warnings*, Final Report T10-07-12, HR Wallingford, Wallingford

Parker, D.J., Penning-Rowsell, E.C. and McFadden, L. (2012) *Business Disruption and Recovery Planning in Relation to Coastal Flood and Erosion Risks: Conceptual Dimensions and Field Survey Evidence*, Research Report IDWT4.3, EU THESEUS Research Programme, Flood Hazard Research Centre, Middlesex University, London

Parkers (2012) *Car Valuations*, http://www.parkers.co.uk/cars/prices/, accessed August 2013

Passenger Focus (2012a) *National Passenger Survey TOC Report for Arriva Trains Wales Spring 2012*, http://www.passengerfocus.org.uk/research/publications/national-passenger-survey-spring-2012-train-operating-company-toc-reports, accessed 27 March 2013

Passenger Focus (2012b) *National Passenger Survey TOC Report for Arriva Trains Wales Autumn 2012*, http://www.passengerfocus.org.uk/research/publications/national-passenger-survey-autumn-2012-nps-train-operating-company-toc-reports, accessed 27 March 2013

Paulrud, A. and Latilla, T. (2004) 'Valuation of management policies for sport-fishing on Sweden's Kaitum River', *Journal of Environmental Planning and Management*, vol 47, no 6, pp863–879

Pearce, D.W. and Turner, R.K. (1990) *Economics of Natural Resources and the Environment*, Harvester Wheatsheaf, London

Pender, G. and Néetz, S. (2007) 'Use of computer models of flood inundation to facilitate communication in flood risk management', *Environmental Hazards*, vol 7, no 2, pp106–114

Penning-Rowsell, E.C. (1978) *The Effect of Salt Contamination on Flood Damages to Residential Properties*, Flood Hazard Research Centre, Middlesex Polytechnic, London

Penning-Rowsell, E.C. (1996) 'Incorporating above design standard (ADS) benefits into the appraisal of flood alleviation schemes', *Proceedings of the 31st MAFF Conference of River and Coastal Engineers, MAFF,* London, 5.1.1–5.1.9

Penning-Rowsell, E.C. and Chatterton, J.B. (1977) *The Benefits of Flood Alleviation: A Manual of Assessment Techniques* (the 'Blue Manual'), Gower Publishing, Aldershot

Penning-Rowsell, E.C. and Fordham, M. (1994) *Floods across Europe: Flood Hazard Assessment, Modelling and Management*, Middlesex University Press, London

Penning-Rowsell, E.C. and Green, C.H. (2000a) 'New insights into the appraisal of flood-alleviation benefits, 1: flood damage and flood loss information', *Water and Environment Journal*, vol 14, no 5, pp347–353

Penning-Rowsell, E.C. and Green, C.H. (2000b) 'New insights into the appraisal of flood-alleviation benefits, 2: the broader context', *Water and Environment Journal*, vol 14, no 5, pp354–362

Penning-Rowsell, E.C. and Green, C.H. (2000c) 'Enhanced appraisal of flood alleviation benefits: new approaches and lessons from experience', in Parker, D. (ed.) *Floods*, Volume 1, Routledge, London, pp214–237

Penning-Rowsell, E.C. and Wilson, T. (2006) 'Gauging the impact of natural hazards: the pattern and cost of emergency response during flood events', *Transactions, Institute of British Geographers*, vol 31, no 2, pp9–15

Penning-Rowsell, E.C. and Pardoe, J. (2012a) 'Who loses if flood risk is reduced: should we be concerned?', *Area*, vol 44, no 2, pp152–159

Penning-Rowsell, E.C. and Pardoe, J. (2012b) 'Who benefits and who loses from flood risk reduction?', *Environment and Planning C: Government and Policy*, vol 30, no 3, pp448–466

Penning-Rowsell, E.C., Winchester, P. and Bossman-Aggrey, P. (1987) *Flood Alleviation for Maidenhead: An Evaluation of Three Levels of Non-structural Approach*, Flood Hazard Research Centre, Middlesex Polytechnic, London

Penning-Rowsell, E.C., Coker, A., N'Jai, A.N., Parker, D.J. and Tunstall, S.M. (1989) 'Scheme worth-whileness', in Institute of Civil Engineers, *Coastal Management*, Thomas Telford, London, pp227–241

Penning-Rowsell, E.C., Green, C.H., Thompson, P.M., Coker, A.C., Tunstall, S.M., Richards, C. and Parker, D.J. (1992) *The Economics of Coastal Management: A Manual of Benefit Assessment Techniques* (the 'Yellow Manual'), Belhaven/Wileys, London

Penning-Rowsell, E.C., Chatterton, J.B. and Winchester, P. (1994) *River Irwell Flood Control Scheme: Benefit–Cost Analysis*, Flood Hazard Research Centre, Middlesex University, London

Penning-Rowsell, E.C, Chatterton, J., Wilson, T. and Potter, E. (2002) *Autumn 2000 Floods in England and Wales: Assessment of National Economic and Financial Losses*, Flood Hazard Research Centre, Middlesex University, London

Penning-Rowsell, E.C, Johnson, C., Tunstall, S., Tapsell, S., Morris, J., Chatterton, J., Coker, A. and Green, C. (2003) *The Benefits of Flood and Coastal Defence: Techniques and Data for 2003*, Middlesex University Press, London

Penning-Rowsell, E.C., Johnson, C., Tunstall, S., Tapsell, S., Morris, J., Chatterton, J. and Green, C. (2005a) *The Benefits of Flood and Coastal Risk Management: A Manual of Assessment Techniques* (the 'Multi-Coloured Manual'), Middlesex University Press, London

Penning-Rowsell, E.C., Johnson, C., Tunstall, S., Tapsell, S., Morris, J., Chatterton, J. and Green, C. (2005b) *The Benefits of Flood and Coastal Risk Management: A Handbook of Assessment Techniques*, Middlesex University Press, London

Penning-Rowsell, E.C., Viavattene, C., Pardoe, J., Chatterton, J., Parker, D. and Morris, J. (2010) *The Benefits of Flood and Coastal Risk Management: A Handbook of Assessment Techniques – 2010*, Flood Hazard Research Centre, Middlesex University, London

Penning-Rowsell, E.C., Haigh, N., Lavery, S. and McFadden, L. (2012) 'A threatened world city: the benefits of protecting London from the sea', *Natural Hazards,* vol 66, no 3, pp1383–1404

Pitt, M. (2008) *Learning Lessons from the 2007 Floods* (the 'Pitt Review'), Cabinet Office, London

Pivot, J.-M., Etienne, J. and Martin, P. (2002) 'Farms' adaptation to changes in flood risk: a management approach', *Journal of Hydrology*, vol 267, no 1, pp12–25

Planning Advisory Service (PAS) (2010) *Background Briefing: Park Homes*, http://www.pas.gov.uk/pas/aio/165 2186, accessed August 2012

Podar, M., Smith, V.K., Van Houtven, G.L., Pattanayak, S. and Bingham, T.H. (2000) *Improving the Practice of Benefit Transfer: A Preference Calibration Approach*, US Environmental Protection Agency, Washington DC

Polome, P. (2003) *DELOS: Extracting a Benefit Transfer Function from CV Studies*, Final Report D11, Delos Project, University of Twente, Twente

Polyzos, S. and Minetos, D. (2007) 'Valuing environmental resources in the context of flood and coastal defence project appraisal: a case study of Poole Borough Council seafront in the UK', *Management of Environmental Quality*, vol 18, no 6, pp684–710

Posford Duvivier/Met Office/J. Chatterton Associates/ADAS (2002) *Research to Determine Impact of Severe Weather (2000 to 2002)*, Met Office Project No Met1B/2767, The Met Office, Bracknell

Posthumus, H., Hewett, C.J.M., Morris, J. and Quinn, P.F. (2008) 'Agricultural land use and flood risk management: engaging with stakeholders in North Yorkshire', *International Journal of Agricultural Water Management*, vol 97, no 7, pp787–798

Posthumus, H., Morris, J., Hess, T.M., Neville, D., Phillips, E. and Baylis, A. (2009) 'Impacts of the summer 2007 floods on agriculture in England', *Journal of Flood Risk Management*, vol 2, no 3, pp182–189

Posthumus, H., Rouquette, J.R., Morris, J., Gowing, D.J.G. and Hess, T.M. (2010) 'A framework for the assessment of ecosystem goods and services: a case study on lowland floodplains in England', *Ecological Economics*, vol 65, pp1510–1523

POSTnote (2011) 'Natural flood management', *POSTnote*, no 396, December 2011, Parliamentary Office of Science and Technology, London

Priest, S.J., Wilson, T., Tapsell, S., Penning-Rowsell, E.C, Viavattene, C. and Fernandez-Bilbao, A. (2007) *Building a Model to Estimate Risk to Life for European Flood Events*, Milestone Report T10-07-10 for FLOODsite, Flood Hazard Research Centre, Middlesex University, London

Queensland Government, Department of Natural Resources and Mines (2002) *Guidance on the Assessment of Tangible Flood Damages*, DNRM, Brisbane

RAC (2012) E-mail correspondence with RAC Technical Team, July 2012

Rail and Safety Standards Board (RSSB) (2012) 'Managing incidents, floods and snow, Module 3, GE/RT8000/M3', in *Rule Book*, no 1, Rail Safety and Standards Board, London, http://www.rgsonline.co.uk/Rule_Book/Rule%20Book%20Modules/M%20-%20Mishaps,%20Incidents%20and%20Extreme%20Weather/GERT8000-M3%20Iss%201.pdf, March 2012, accessed 27 March 2013

Ramsar Convention Bureau (1997) *The Ramsar Convention Manual: A Guide to the Convention on Wetlands*, Ramsar Convention Bureau, Gland

Ramsbottom, D., Sayers, P. and Panzeri, M. (2012) *Climate Change Risk Assessment for the Floods and Coastal Erosion Sector*, report prepared as part of the UK Climate Change Risk Assessment, January 2012, HMSO/Defra, London

Reed, S., Graves, A., Dandy, N., Posthumus, H., Hubacek, K., Morris, J., Pell, C., Quinn, C.H. and Stringer, L.C. (2009) 'Who is in and why? A typology of stakeholder analysis methods for natural resource management', *Journal of Environmental Management*, vol 90, pp1933–1949

Ricability (2012) *Car Measurement Guide*, http://www.ricability.org.uk/consumer_reports/mobility_reports/car_measurement_guide/CarMeasurementList.asp, accessed August 2013

Rightmove (2012) *Consumer Rental Forecast*, http://www.rightmove.co.uk/news/files/2012/09/Rightmove-Rental-Market-Report-September-2012-Q3-CCS.pdf, accessed 23 April 2013

Roca, M., Bast, H., Panzeri, M., Hess, T. and Sayers, P. (2010) *Developing the Evidence Base to Describe the Flood and Coastal Erosion Risk to Agricultural Land in England and Wales*, Defra R&D Technical Report FD2634/TR, May 2010, HR Wallingford, Wallingford

Rosen, S. (1974) 'Hedonic prices and implicit markets: product differentiation in pure competition', *Journal of Political Economy*, vol 82, pp34–55

Rosenberger, R.S. and Loomis, J.B. (2001) *Benefit Transfer of Outdoor Recreation Use Values*, RMRS-GTR-72, Rocky Mountain Research Station, Fort Collins, CO

Rouquette, J.R., Posthumus, H., Morris, J., Hess, T.M., Dawson, Q.L. and Gowing, D.J.G. (2011) 'Synergies and trade-offs in the management of lowland rural floodplains: an ecosystem services approach', *Hydrological Sciences Journal*, vol 56, no 8, pp1566–1581

Royal Society (2009) *Reaping the Benefits: Science and the Sustainable Intensification of Global Agriculture*, Royal Society, London

RPA/FHRC (2004) *The Appraisal of Human Related Intangible Impacts of Flooding*, R&D Technical Report FD2005/TR, Defra/Environment Agency, London

Sadique, M.Z., Adams, E.J. and Edmunds, W.J. (2008) 'Estimating the costs of school closure for mitigating an influenza', *BMC Public Health*, vol 8, no 135, http://www.biomedcentral.com/1471-2458/8/135, accessed 27 March 2013

Samuelson, P.A. (1970) *Economics*, McGraw-Hill, New York

Savills (2012) *Market Survey of Agricultural Land*, Savills Research, UK Rural, Savills, London

Sayers, P., Galloway, G.E., Penning-Rowsell, E.C., Yuan, Y., Shen, F., Chen, Y., Le Quense, T., Wang, L. and Guan, Y. (2013) 'Strategic flood risk management: comparing international challenges and developing common approaches', in Klijn, F. and Schweckendiek, T. (eds) *Comprehensive Flood Risk Management: Research for Policy and Practice*, CRC Press, Taylor and Francis Group, London, pp937–944

Scawthorn, C., Flores, P., Blais, N., Seligson, H., Tate, E., Mifflin, E., Thomas, W., Murphy, J. and Jones, C. (2006a) 'HAZUS-MH flood loss estimation methodology, I: overview and flood hazard characterization', *Natural Hazards Review*, vol 7, no 2, pp60–71

Scawthorn, C., Flores, P., Blais, N., Seligson, H., Tate, E., Mifflin, E., Thomas, W., Murphy, J. and Jones, C. (2006b) 'HAZUS-MH flood loss estimation methodology, II: damage and loss assessment', *Natural Hazards Review*, vol 7, no 2, pp72–81

Scottish Natural Heritage (undated) *Visitor Monitoring Manual*, http://archive.snh.gov.uk/vmm/Glossary/VMM%20print%20version.pdf, accessed 20 March 2013

Scottish Natural Heritage (2007) *Guidance on Monitoring Access and Recreation at Sensitive Natural Heritage Sites*, http://www.snh.org.uk/pdfs/publications/heritagemanagement/GuidanceMonitoringAccessSensitivesitest.pdf, accessed 18 April 2013

SDF (2012) *Sustainable Development of Floodplains*, European Centre for River Restoration, http://www.ecrr.org/sdfproject/sdfproject.htm, accessed 18 April 2013

Sear, C. (2012) *The Bellwin Scheme*, Standard Note SN/PC/00643, House of Commons Library, London

Shepherd, P.A. and Harley, D. (1999) *Preparation and Presentation of Habitat Replacement Cost Estimates, Using Examples of the Restoration and Creation of Coastal and Floodplain Grazing Marsh, Reedbeds and Lagoons*, ENRR345, Natural England, Peterborough

Shepherd, P.A., Gillespie, J., Garrod, G. and Willis, K. (2002) *An Initial Investigation of the Actual Costs of Implementing UK Biodiversity Action Plans*, Defra, London

Sheppard, S.R.J. and Meitner, M. (2005) 'Using multi-criteria analysis and visualisation for sustainable forest management planning with stakeholder groups', *Forest Ecology and Management*, vol 207, nos 1–2, pp171–187

Smart Driving (2012) *Driving through Floods*, http://www.smartdriving.co.uk/Driving/Driving_emergencies/Floods.htm, accessed August 2013

Smedema, L.K. and Rycroft, D.W. (1983) *Land Drainage*, Batsford, London

Smith, D.I. and Greenaway, M.A. (1988) 'The computer assessment of urban flood damage: ANUFLOOD', in Newton, P.W., Sharp, R. and Taylor, M.A.P. (eds) *Desktop Planning: Advanced Microcomputer Applications for Physical and Social Infrastructure Planning*, Melbourne, Hargreen, pp239–250

Smith, V.K. (1992) 'Arbitrary values, good causes and premature verdicts', *Journal of Environmental Economics and Management*, vol 22, pp71–89

Smiths Gore (2013) *Review of the English Farmland Market, December 2012*, Smiths Gore, Peterborough

Society of Motor Manufacturers and Traders (SMMT) (2012) *Motor Industry Facts 2012*, http://www.smmt.co.uk/2012/03/motor-industry-facts-2012/, accessed 20 July 2012

Soil Survey of England and Wales (SSEW) (1984) *Soils and Their Use*, Regional Bulletins 10–15, SSEW, Harpenden (now National Soil Resources Institute, Silsoe)

Somerset County Council (SCC) (2013a) *Closure of the A361 Document: Transport Economics Review*, Somerset County Council, Taunton

Somerset County Council (SCC) (2013b) *Somerset: Extreme Weather Resilience Briefing Note*, Somerset County Council, Taunton

Squire, L. and van der Tak, H. (1975) *Economic Projects*, Johns Hopkins University Press, Baltimore, MD

Stately Albion (2012) Telephone conversation with Nigel Derrick, Sales and Production Manager, Stately Albion park home manufacturer, 10 August 2012

Stern, N. (2006) *The Economics of Climate Change*, Cambridge University Press, Cambridge

Stiglitz, J.E. (2008) 'Information', in *The Concise Encyclopaedia of Economics*, http://www.econlib.org/library/Enc/Information.html, accessed 18 April 2013

Suleman, M.S., N'Jai, A., Green, C.H. and Penning-Rowsell, E.C. (1988) *Potential Flood Damage Data: A Major Update*, Flood Hazard Research Centre, Middlesex Polytechnic, London

Tapsell, S.M. (1995) 'River restoration: what are we restoring to? A case study of the Ravensbourne River, London', *Landscape Research*, vol 20, no 3, pp98–111

Tapsell, S.M. (2000) *Follow-up Study of the Health Effects of the 1998 Easter Flooding in Banbury and Kidlington*, report for the Environment Agency, Flood Hazard Research Centre, Middlesex University, London

Tapsell, S.M. and Tunstall, S.M. (2001) *The Health and Social Effects of the June 2000 Flooding in the North East Region*, report for Environment Agency, Flood Hazard Research Centre, Middlesex University, London

Tapsell, S.M., Tunstall, S.M., Costa, P.L. and Fordham, M. (1994) *Ravensbourne River Queens Mead Recreation Ground Survey*, final revised report, Flood Hazard Research Centre, Middlesex University, London

Tapsell, S.M., Ketteridge, A.-M. and Tunstall, S.M. (1996) *Benefit Cost Appraisal: Hengistbury Head and Warren Hill Coast Protection Proposals*, final report for Bournemouth Development Services, Flood Hazard Research Centre, Middlesex University, London

Tapsell, S.M., Tunstall, S.M. and Eden, S. (1997) *Public Responses to River Restoration: The River Skerne and the River Cole*, report for the River Restoration Centre, Flood Hazard Research Centre, Middlesex University, London

Tapsell, S.M., Tunstall, S.M., Penning-Rowsell, E.C. and Handmer, J.W. (1999) *The Health Effects of the 1998 Easter Floods in Banbury and Kidlington, Oxon*, Flood Hazard Research Centre, Middlesex University, London

Tapsell, S.M., Penning-Rowsell, E.C., Tunstall, S.M. and Wilson, T.L. (2002) 'Vulnerability to flooding: health and social dimensions', *Philosophical Transactions of the Royal Society, London*, vol 360, pp1511–1525

Tapsell, S.M., Priest, S.J., Wilson, T., Viavattene, C. and Penning-Rowsell, E.C. (2008) 'A new model to estimate risk to life for European flood events', paper presented to FLOODrisk: The European Conference on Flood Risk Management Research into Practice, Keble College, Oxford

Tebodin (1998) *Schade bij inundatie van buitendijkse industrie*, Document No 334393, Tebodin Consultants and Engineers, Netherlands (in Dutch)

Tebodin (2000) *Schadecurves industrie ten gevolge van overstroming*, Document No 334666, Tebodin Consultants and Engineers, Netherlands (in Dutch)

Technical Support Unit (2003) *Integrated Flood Management: Concept Paper*, Associated Programme on Flood Management, World Meteorological Organisation/Global Water Partnership, Geneva

TEEB (2010) *The Economics of Ecosystems and Biodiversity: Mainstreaming the Economics of Nature: A Synthesis of the Approach, Conclusions and Recommendations of TEEB*, Progress Press, Malta

Thames Region/National Rivers Authority/Lewin, Fryer and Partners (1992) *Maidenhead, Windsor and Eton Flood Alleviation Scheme, Benefit Assessment*, National Rivers Authority, Maidenhead

Thibaut, J. and Walker, L. (1975) *Procedural Justice: A Psychological Analysis*, Erlbaum, Hillsdale, NJ

This is Money (2010) *Where to Buy a Holiday Home in Britain*, http://www.thisismoney.co.uk/money/mortgageshome/article-1691099/Where-to-buy-a-holiday-home-in-Britain.html, accessed August 2013

Thomas, C. and Young, J.W.O. (eds) (1982) *Milk from Grass*, ICI/GRI joint publication, Billingham Press, Cleveland

Thompson, P.J., Wigg, A.H. and Parker, D.J. (1991) 'Urban flood protection post-project appraisal in England and Wales', *Project Appraisal*, vol 6, no 2, pp84–92

Thurston, N., Finlinson, B., Breakspear, R., Williams, N., Shaw, J. and Chatterton, J. (2008) *Developing the Evidence Base for Flood Resistance and Resilience*, R&D Technical Report FD2607/TR, Defra, London

Towergate Insurance (2011) *Flood Loading Reduced for Residents in West Sussex and Warwickshire Parks*, http://www.parkhomeownersclub.co.uk/news/flood-loading-reduced-for-residents-in-west-sussex-and-warwickshire-parks, accessed 26 April 2013

Tratalos, J.A., Sugden, R., Bateman, I.J., Gill, J.A., Jones, A.P., Showler, D.A., Sutherland, W.J. and Watkinson, A.R. (2013) 'The conflict between conservation and recreation when visitors dislike crowding: a theoretical and empirical analysis of recreational beach use', *Journal of Environmental and Resource Economics*, http://link.springer.com/article/10.1007%2Fs10640-013-9634-2, February 2013, accessed 8 June 2013

Trend Tracker (2013) *Automotive News and Press Releases*, http://www.trendtraacker.co.uk/media, accessed 8 February 2013

Tunstall, S.M. (2000) *Corton Village Coastal Protection Recreational Works: Preliminary Investigation*, Flood Hazard Research Centre, Middlesex University, London

Tunstall, S.M. (2001) *Corton Village Coastal Protection Recreational Survey: Draft Final Report*, Flood Hazard Research Centre, Middlesex University, London

Tunstall, S.M. and Penning-Rowsell, E.C. (1998) 'The English beach: experiences and values', *Geographical Journal*, vol 164, no 3, pp319–332

Tunstall, S.M., Green, C.H., Lewin, P. and Coker, A. (1990) *Herne Bay Coastal Protection Benefit Assessment, Report III: The Recreational Benefit Assessment*, study for Canterbury City Council, Flood Hazard Research Centre, Middlesex Polytechnic, London

Tunstall, S.M., Tapsell, S.M. and Eden, S. (1999) 'How stable are public responses to changing local environments? A "before" and "after" case study of river restoration', *Journal of Environmental Planning and Management*, vol 42, no 4, pp527–547

Tunstall, S.M., Johnson, C.L. and Penning-Rowsell, E.C. (2004) 'Flood hazard management in England and Wales: from land drainage to flood risk management', paper presented to the World Congress on Natural Disaster Mitigation, New Delhi

Tunstall, S., Tapsell, S. and Fernandez-Bilbao, A. (2007) *Vulnerability and Flooding: A Re-analysis of FHRC Data, Country Report England and Wales*, FLOODsite Technical Report T11-07-11, Flood Hazard Research Centre, Middlesex University, London

Tunstall, S.M., McCarthy, S. and Faulkner, H. (2009) 'Flood risk management and planning policy in a time of transition: the case of Wapshott Road Planning Inquiry, Surrey, England', *Journal of Flood Risk Management*, vol 2, no 3, pp159–169

Turner, R.K., Pearce, D. and Bateman, I. (1993) *An Introduction to Environmental Economics*, Harvester Wheatsheaf, Hemel Hempstead

Turner, R.K., Georgiou, S., Burgess, D. and Jackson, N. (2005) *Development of Economic Appraisal Methods for Flood Management and Coastal Erosion Protection*, Defra, London

UK Biodiversity Group (2000) *Tranche 1 and 2 Action Plans, Cost Estimates: A Summary Report*, English Nature, London

UK Parks (2012) *Residential Park Homes*, http://www.ukparks.com/residentialhome.asp, accessed 15 August 2012

UK Water Industry Research (UKWIR) (2011) *Carrying out Willingness to Pay Surveys*, Report No 11/RG/07/22, UKWIR, London

United Nations (2000) *Guidelines on Sustainable Flood Prevention*, United Nations, Geneva

United Nations Conference on Environment and Development (UNCED) (1992) *Agenda 21: The United Nations Programme of Action from Rio*, United Nations, New York

Used Car Expert (2012) *Expert Price Guide*, http://www.usedcarexpert.co.uk/cars/ford/focus/price-guide/, accessed August 2013

US National Research Council (1995) *Wetlands: Characteristics and Boundaries*, National Academy Press, Washington D.C

van der Veen, A., Steenge, A.E., Bockarjova, M. and Logtmeijer, C.J.J. (2003) 'Structural economic effects of large-scale inundation: a simulation of the Krimpen dike breakage', in van der Veen, A., Arellano, A.L.V. and Nordvik, J.-P. (eds) *In Search of a Common Methodology on Damage Estimation*, EUR 20997, DG Joint Research Centre, Ispra, pp53–77

Vauxhall (2005) *Vauxhall Car Price Range 2005*, http://

www.vauxhallproductforum.co.uk/category.aspx?mid=2&sid=4&pid=89&cid=256 , accessed July 2012

VOA (2011) *Property Market Report 2011: The Annual Guide to the Property Market across England, Wales and Scotland*, Valuation Office Agency, London

VOA (2012) *Private Rental Market Statistics*, http://www.voa.gov.uk/corporate/statisticalReleases/121213_PrivateResidentialRentalMarketStatistics.html, accessed 4 February 2013

Vrouwenvelder, A.C.W.M. and Steenhuis, C.M. (1997) *Secondary Flood Defences in the Hoeksche Waard, Calculation of the Number of Fatalities for Various Flood Scenarios*, Report TNO 97-CON-R0332, TNO Bouw, Delft (in Dutch)

Waarts, P. (1992) *Methode voor de bepaling van het aantal doden als gevolg van inundatie*, Report TNO B-91-1099, TNO Bouw, Delft (in Dutch)

Walkden, M.J. and Hall, J.W. (2005) 'A predictive mesoscale model of the erosion and profile development of soft rock shores', *Coastal Engineering*, vol 52, pp535–563

Walkden, M.J. and Hall, J.W. (2011) 'A mesoscale predictive model of the evolution and management of a soft-rock coast', *Journal of Coastal Research*, vol 27, no 3, pp529–543

Water Research Centre/Oxford Economic Research Associates/Flood Hazard Research Centre (WRC/OXERA/FHRC) (1996) *FWR Manual on Assessing the Benefits of Water Quality*, Foundation for Water Research, Marlow

Water UK (2007) *UK and Water Sewerage Services Map*, http://www.water.org.uk/home/resources-and-links/links/water-operators/sewerage-operators/a4-sewerage-map.pdf, September 2007, accessed 27 March 2013

Water UK (2011) 'Technical Guidance Note No 11: Storage and use of bottled water', in *Regulations on Use of Bottled Water by Water Companies*, http://www.water.org.uk/home/policy/publications/archive/drinking-water/priciples-of-water-supply-hygiene/tgn-11-storage-and-use-of-bottled-water-final.pdf, accessed 27 March 2013

Western Sussex Hospitals NHS Trust (2012) 'Flooding at Worthing Hospital', press statement, http://www.westernsussexhospitals.nhs.uk/EasysiteWeb/getresource.axd?AssetID=422317&type=full&servicetype=Attachment, 11 June 2012, accessed 27 March 2013

What Car? (2012) *Depreciation Index*, http://www.whatcar.com/car-depreciation-calculator, accessed 18 April 2013

Whitehead, J.C. (2003) 'One million dollars per mile? The opportunity costs of hurricane evacuation', *Ocean and Coastal Management*, vol 46, pp1069–1083

Whitehead, J.C. (2006) 'A practitioner's primer on the contingent valuation method', in Alberini, A. and Kahn, J.R. (eds) *Handbook on Contingent Valuation*, Edward Elgar, Cheltenham

Whiteley, D. (2008) *Report on the Lessons Learned from the Summer 2007 Flooding Experiences from an Estates and Facilities Perspective*, DH Gateway Reviews and Estates and Facilities Division, Department of Health, HMSO, London

Whitmarsh, D., Northen, J. and Jaffry, S. (1999) 'Recreational benefits of coastal protection: a case study', *Marine Policy*, vol 23, nos 4–5, pp453–463

Wicks, J., Clarke, J., Lovell, L., Priest, S. and Parker, D. (2012) *Update to the Framework, Tool and Data*, Benefits of Flood Incident Management and Land Use Management research project, Halcrow, Swindon

Willatt, J. (2006) *Environment Agency: NaFRA 2006; Report on Changes to Method*, Environment Agency, Bristol

Williams, A.M. and Shaw, G. (eds) (1997) *The Rise and Fall of British Coastal Resorts: Cultural and Economic Perspectives*, Pinter, London

Williams, A.M. and Shaw, G. (2009) 'Future play: tourism, recreation and land use', *Land Use Policy*, vol 26, Suppl.1, S326–S335

Williams, R. and Williams, R. (2003) *The Somerset Levels*, Ex-Libris, Bradford on Avon

Williamson, O.E. (1981) 'The economics of organisation: the transaction cost approach', *American Journal of Sociology*, vol 87, no 3, pp548–577

Winter, M. and Lobley, M. (eds) (2009) *What is Land for? The Food, Fuel and Climate Change Debate*, Earthscan, London

Woodward, R.T. and Wui, Y. (2001) 'The economic value of wetland services: a meta-analysis', *Ecological Economics*, vol 37, pp257–270

World Commission on Environment and Development (WCED) (1987) *Our Common Future* (the 'Brundtland Report'), Oxford University Press, Oxford

Wunder, S., Engel, S. and Pagiola, S. (2008) 'Taking stock: a comparative analysis of payments for environmental services programs in developed and developing countries', *Journal of Ecological Economics*, vol 65, pp834–852

Zhai, G., Fukuzono, T. and Ikeda, S. (2006) 'An empirical model of fatalities and injuries due to floods in Japan', *Journal of the American Water Resources Association*, vol 42, no 4, pp863–875

Zurich Insurance (2012) *Flood Advice for Motorists*, http://www.zurich.co.uk/severeweather/floodinghelp/coping/floodcar.htm, accessed August 2013

Index

For Product Safety Concerns and Information please contact our EU
representative GPSR@taylorandfrancis.com
Taylor & Francis Verlag GmbH, Kaufingerstraße 24, 80331 München, Germany

www.ingramcontent.com/pod-product-compliance
Lightning Source LLC
LaVergne TN
LVHW080121011026
849160LV00013B/1150

9781032919768